Nuclear Engineering

Nuclear Engineering

A Conceptual Introduction to Nuclear Power

Malcolm Joyce

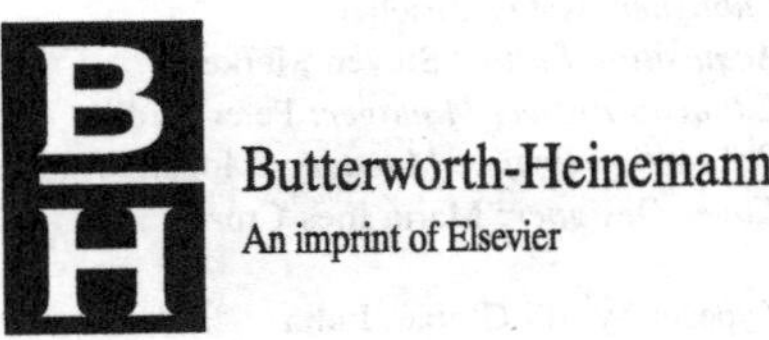

Butterworth-Heinemann is an imprint of Elsevier
The Boulevard, Langford Lane, Kidlington, Oxford OX5 1GB, United Kingdom
50 Hampshire Street, 5th Floor, Cambridge, MA 02139, United States

Portraits courtesy of Graham Lowe, artist and illustrator, www.grahamloweartist.com

Library of Congress Cataloging-in-Publication Data
A catalog record for this book is available from the Library of Congress

British Library Cataloguing-in-Publication Data
A catalogue record for this book is available from the British Library

ISBN: 978-0-08-100962-8

For information on all Butterworth-Heinemann publications visit our website at https://www.elsevier.com/books-and-journals

Publisher: Katey Birtcher
Acquisition Editor: Steven Merken
Editorial Project Manager: Peter Jardim
Production Project Manager: Mohana Natarajan
Cover Designer: Maria Ines Cruz

Typeset by SPi Global, India

Contents

CHAPTER

FUNDAMENTAL CONCEPTS

1

1.1 SUMMARY OF CHAPTER AND LEARNING OBJECTIVES

The aim of this chapter is to introduce the fundamental concepts associated with nuclear engineering that support the detail explored in the chapters that follow. Extensive detail is avoided in preference to a focus on the foundation principles associated with, for example, the order of magnitude of quantitative aspects of the field, the definition of terms used later in the book and the general aspects associated with nuclear reactor design. The concepts and principles selected at this stage are given further elaboration later in the text; they constitute features that in the author's experience pervade the nuclear engineering discipline irrespective of the specific aspect of the field in which they tend to arise. The subjects discussed in this chapter also represent concepts that are rather specific to nuclear engineering and do not always arise in the study of the other, more general branches of engineering. Many of these stem from the disciplines of nuclear physics and radiochemistry but nonetheless arise frequently in nuclear engineering.

The objectives of this chapter are to:

- introduce the main distinctions of *nuclear engineering* over other engineering disciplines
- review the structure of the *atom* and introduce the concept of the *atomic nucleus*
- discuss the interplay between the *Coulomb force* that exists between the protons in the nucleus and the *strong nuclear force* that holds the nucleus together
- describe the *extreme properties* of the nucleus in terms of its density and the miniscule space it occupies in nature
- introduce the *chart of the nuclides*, what this represents and the properties of the nucleus that it highlights
- provide a comprehensive description of the concept of the *generic nuclear reactor* and its components including *fuel*, *cladding*, *coolant* and *moderator*
- introduce the concept of the *reactor cycle* and the distinction of *direct* and *indirect* cycles

Nuclear Engineering. https://doi.org/10.1016/B978-0-08-100962-8.00001-9

1.2 HISTORICAL CONTEXT: ERNEST RUTHERFORD 1871–1937

Among Ernest Rutherford's first nuclear-related achievements was the proof of there being two distinct forms of radiation emitted by uranium (α and β particles), having separated them with thin layers of aluminium. Along with Robert McClung, he calculated that a significant amount of energy was radiated by radium in the form of α particles and, perhaps most significantly, he observed that the physical and chemical properties of the α particle were consistent with what would come to be known as the *nucleus* of a helium atom. For this, Rutherford (Fig. 1.1) was awarded the Nobel Prize in Chemistry in 1908.

Subsequently, Rutherford observed that the properties of radioactive disintegration were independent of the chemical and physical characters of the emitting substance and were thus atomic in origin, rather than molecular. He defined the correspondence of the α-decay process in terms of the related chemical changes of the parent substance, studied thorium leading to the discovery that it emitted radon, he developed the disintegration theory of radioactive decay (along with Frederick Soddy) and devised a means for detecting single α particles along with Hans Geiger. In 1910, he observed that α particles could be repelled by materials through large angles and hence postulated that the atomic mass was concentrated in the form of a miniscule but extremely dense 'nucleus'. This theory (albeit with notable improvements relating to concepts derived from the work of Neils Bohr and Werner Heisenberg) is that which is used to this day. This observation set the scene for the cascade of discoveries that followed, including those of the neutron and nuclear fission. Subsequently, along with Henry Moseley, he observed that atoms emitted characteristic X-rays in response to excitation from which a

FIG. 1.1

Ernest Rutherford.

number, their atomic number, could be assigned which was related to the chemical properties of the corresponding element. Rutherford also observed that some light elements could be disintegrated by the effect of α particles, providing the first evidence of deliberate transmutation of one element to another.

1.3 INTRODUCTION

A significant distinction of nuclear engineering from other branches of engineering is that nuclear systems deal with material that is or has the potential to become radioactive. In this context, many of the subsystems and general concepts of engineering science that an engineer needs to be aware of in a nuclear context correspond directly to those in general engineering disciplines. However, there are some very important issues that are unique to the nuclear engineering field; these arise because of the requirement to manage, process and be aware of nuclear materials and the radiation that can arise from them. Often the nuclear properties of only a relatively small number of the isotopes are exploited in most nuclear engineering systems. While the objective of nuclear operations might often be simple in terms engineering fundamentals, this can be complicated significantly by the imperative to manage the risk associated with the radioactivity of the substances involved to levels that are safe and considered acceptable to society.

1.4 THE NUCLEAR LANDSCAPE

1.4.1 ATOMIC RADIATION AND NUCLEAR RADIATION

Radiation can arise from processes associated with the electron shell structure of the atom; for instance, we are perhaps all familiar with the electromagnetic spectrum of radiation that encompasses everything from the frequencies received and transmitted by mobile phones, the infrared responsible for the images fire crews use to find people with heat-seeking cameras, through to the visible photon spectrum with which we see the world and on to the ultraviolet and X-ray components. These examples are not the *nuclear radiation* that we are concerned with in this book and in the wider nuclear engineering discipline because the origin of this radiation is *atomic* rather than *nuclear*. Nuclear radiation is, by definition, emitted by processes *in* the atomic nucleus rather than those in the electron shell structure of the atom that surrounds the nucleus. This distinction is important because where, for example, processes in the nucleus are associated with the emission of electromagnetic radiation, the potential exists for the emission of radiation at much higher energies (at shorter wavelengths) than that generally associated with processes at an atomic level. This can have implications as to how we might manage and protect ourselves from the risk posed by excessive radiation exposure. There is also the additional and important possibility that the nucleus might emit particles as forms of radiation. These can be highly ionising and tend not to occur in the case of the atom (electrons are an exception).

Nuclear radiation has the potential to *ionise* the matter that it interacts with and the extent to which this occurs is dependent on the energy, mass and charge of the radiation. Relatively significant amounts of energy can be imparted to matter by ionising radiation on a microscopic scale, and this has the potential to change the composition of the substances in which it is deposited. This can, in turn, change the properties of these materials; where said material is living tissue the effect of ionising radiation can cause it to behave differently or even to kill the cells of which it is comprised. However, it is important to appreciate that for the vast majority of nuclear processes and operations, the radiation environment experienced by people working in it bears no difference to natural background levels of radiation. These comprise those sources that we are all subject to, largely unavoidably, largely as a result of radiation emitted from naturally radioactive minerals present in the Earth's crust and from sources in outer space.

One further important distinction is that while atomic processes of radiation emission, such as fluorescence and phosphorescence rarely have lifetimes longer than a few minutes or hours, nuclear radiation can be associated with the decay of atomic nuclei that span enormous ranges in lifetime, from picoseconds through to many billions of years. Notwithstanding the possibility of transmuting long-lived radioactive isotopes into others with shorter lifetimes that we will discuss in Chapter 15, it is not possible to change the lifetime of a radioactive substance. For this reason, many requirements in nuclear engineering are associated with the management of radioactive materials to ensure that people are protected from the risk of harm, while we harness the potential of these materials for the benefit of civilisation.

The phenomena that astonished Rutherford little more than a hundred years ago occurred when a minority but nonetheless a significant number of α particles were detected being reflected from a thin metal foil. He is said to have remarked that it was as if he had 'fired a 15-in. shell at a piece of tissue paper and it came back'. This demonstrated that almost all of the mass of the atom, and thus the vast majority of all of the mass of visible matter, must be concentrated in a small and dense *nucleus*. This was in great contrast to the more dilute and dispersed arrangement that had been widely postulated at that time but which had not been proven outright. From the extensive research that followed Rutherford's observation based on nuclear scattering, it was possible to infer the dimensions of the nucleus. This led to the model of the atom that is now accepted universally and that has been the basis for much scientific discovery and related engineering that followed in the 20th century.

1.4.2 THE NUCLEUS

The nucleus is composed of an approximately equal number of *protons* and *neutrons* and these are known collectively as *nucleons*. The exception is the case of hydrogen, the lightest isotope of the lightest element, which has a nucleus composed of just one proton. Protons are positively charged, and neutrons (as their name suggests) are neutral; each chemical *element* is distinguished by the corresponding number of protons, and this corresponds to the atomic number. Each *isotope* of a given element has the same number of protons but differs in terms of the number of neutrons it possesses. An atom comprises a nucleus surrounded by a number of electrons equal in number to that of the protons in the nucleus. The charge and mass data for neutrons, protons, electrons and photons are given in Table 1.1.

Table 1.1 Fundamental Properties of the Major Subatomic Particles

	Proton	Neutron	Electron	Photon
Mass (kg)	1.673×10^{-27}	1.675×10^{-27}	9.11×10^{-31}	-
Charge (c)	$+1.602 \times 10^{-19}$	0	-1.602×10^{-19}	-

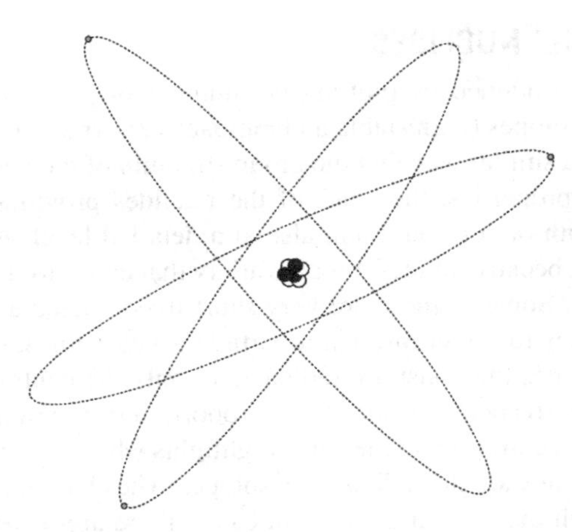

FIG. 1.2

A schematic diagram of the ^{7}Li atom, by way of example, to illustrate atomic composition. The scale of the nucleus as depicted is magnified by a factor of approximately 10,000 to render it visible. Note: electrons are shown as discrete entities whereas they are better approximated as diffuse charge clouds, with protons shown as black and neutrons white for the purposes of this schematic illustration. Diagram not to scale.

Since the protons have like charges, they are subject to the force of *Coulomb repulsion* acting to force them apart; this form of the *electromagnetic force* is generally weaker at the short ranges associated with the dimension of the nucleus than the force that binds the nucleus together. The latter, known as the *strong nuclear force,* is one of the four fundamental forces (along with gravitation and the weak interaction) but which is attractive at dimensions of the order of the size of the nucleus, typically of radius r where $r < 10^{-15}$ m. For the case of isotopes that are not susceptible to radioactive decay, these nuclei exist in a state of stable equilibrium. In this state, the protons are repulsed by one another but the nucleus is held together as a whole by the cohesive, strong force. A schematic illustration of a lithium atom is given in Fig. 1.2.

Rutherford's discovery that matter is concentrated into miniscule nuclei remains profound because it challenges our everyday experience of the density of matter that we are familiar with. The observation of nuclei of a diameter of, say, 10^{-15} m suggested something very different indeed. Given the size of the atom of the order of 10^{-10} m and given the mass of the electron of 9.11×10^{-31} kg, it is clear that the density of the nuclear material that constitutes 99.95% of the atomic mass but only ~1/100,000 of the atomic dimension must be extremely high, that is, of the order of ~10^{18} kg m^{-3}. Substances with the highest densities witnessed in our day-to-day experiences are of the order of 10^{4} kg m^{-3} (e.g. for the case of lead or tungsten), and hence, it is clear that the density of nuclear material is extremely high. From this observation, we can conclude that the strong nuclear force is indeed very strong because it acts to keep the protons and neutrons in such a tight bundle. Also, it is short ranged not acting much further beyond the dimensions of the nucleus itself.

In addition to the density, the nature of solid matter implied by the scale of the nucleus is also a little counter-intuitive. If the atomic nucleus were represented on this page to be the size of the head of a pin, the electrons surrounding it that constitute the size of the atom would be ~100 m from the position from which you are reading, if represented on the same scale. It is clear from this observation that the vast majority of matter, that is, 99.999999999999%, is actually free space.

1.4.3 THE CHART OF THE NUCLIDES

The chart of the nuclides is obtained by plotting the number of protons (Z) versus the number of neutrons (N) for all known isotopes (both stable and radioactive), where $A = N + Z$ is the atomic mass. This is shown in Fig. 1.3. In a similar way that the group structure of the periodic table of the elements can infer general chemical properties, the chart of the nuclides provides valuable insight into the properties of the nucleus, both on a general and also at a detailed level. A prominent feature of this summary of all matter is that, because of the isotopic variety that exists for many elements, it composes of several thousand nuclides. Some of these are very short-lived indeed and, in some extreme cases, only exist for the time it takes to travel through a particle accelerator at close to the speed of light. Second, the majority of isotopes that exist are radioactive; only the central spine of the chart, represented in black in Fig. 1.3, corresponds to the stable proportion of nonradioactive species. The chart resembles a cloud around this central spine; the spine highlights what is often referred to as the *valley of stability*. Either side of this valley are the radioactive isotopes. The chart narrows to a limit for very low masses and also for very high masses where, in both cases, there are fewer known isotopes for each element than in the central region. At the high-mass extreme, the valley of stability breaks up with the most massive isotopes not having a stable isotopic variant at all; this region is a key area of interest for nuclear engineers as it is where the *actinide* series of isotopes resides.

A subtle feature of the chart of the nuclides is that the line along the valley of stability is not straight. Rather, it bends because of the natural trend for heavier isotopes to have a higher proportion of neutrons than they do protons, and hence, the following proportionality holds between Z and N,

$$Z \propto N^p \tag{1.1}$$

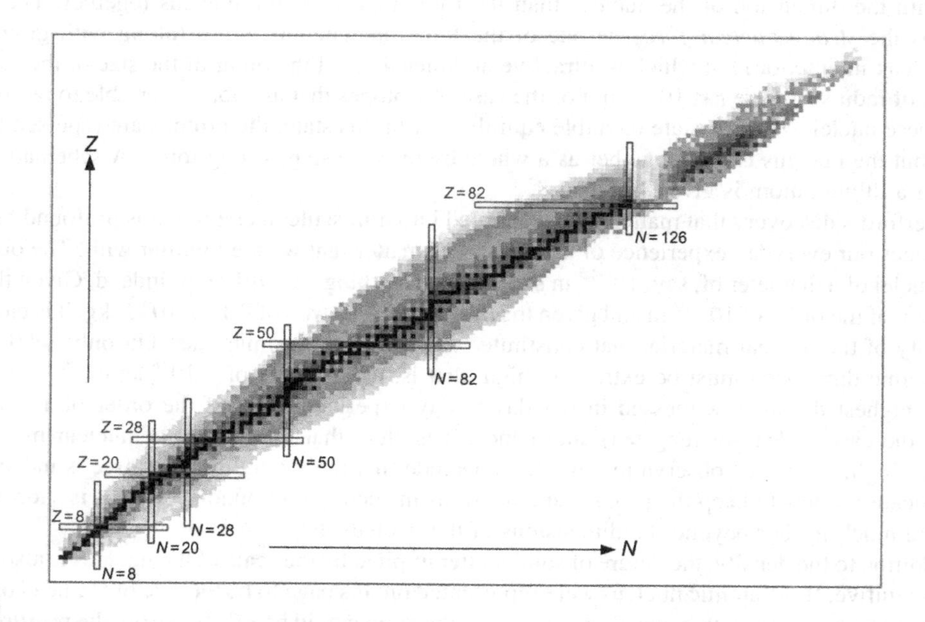

FIG. 1.3

The chart of the nuclides, stable isotopes depicted in black.

Taken from the National Nuclear Data Centre web site: http://www.nndc.bnl.gov/chart/.

where p is less than 1 since the trend of the valley of stability is sublinear. At low isotopic masses, the chart is linear with $Z \sim N$ and hence p tends to unity. The departure from linearity becomes particularly apparent for higher mass isotopes reflecting a particularly salient nuclear property that, as isotopic mass increases, stability is favoured by isotopes for which $N > Z$ rather than for $N = Z$ as might be expected on the basis of the trend for light isotopes.

Why is this? Our knowledge of the intricacies of the properties of nuclear matter is based on many years of complex research investigations by generations of nuclear scientists. The results of these studies lead us to conclude that the properties of the nucleus are dependent on many different structural phenomena, a full discussion of which is beyond the level of this text. However, one feature is of specific relevance to nuclear engineering: as proton number increases from light to heavy masses (thus moving up the vertical axis in Fig. 1.3), the force of electrostatic repulsion between the protons increases. The same is not true of the nuclear force (as discussed in Chapter 4 with reference to nuclear binding energy) because its influence per nucleon extends only to nucleons that are nearest neighbours and thus it does not generally increase with A *per se*. Consequently, as atomic mass increases, the repulsive action of the Coulomb force competes with the strong nuclear force more effectively. The significance of this in the context of nuclear energy is that the nuclei of very heavy nuclei are weakly bound relative to their lighter cousins, with the repulsion of the abundance of protons at these masses acting to push them apart and destabilise them.

While, as mentioned earlier, the structure and behaviour of nuclei at the upper mass extreme is generally very complicated, the instability due to the large number of protons is common to all isotopes to a greater or lesser degree, as $A \to \infty$. Hence, a natural propensity towards the complete breakup of the nucleus is observed at these extremes of mass. However, it is also true that a *universal* swing towards catastrophic breakup of all nuclei at this extreme is not observed; rather, the nuclei of some isotopes breakup readily, others can be encouraged to do so by excitation while others favour a variety of other modes of decay instead. Such is the complexity of the underlying nuclear structure at this extreme that only broad generalisations can be made with confidence, with neighbouring isotopes often exhibiting stark contrasts in behaviour to one another.

1.4.4 UNITS OF ENERGY ON A NUCLEAR SCALE

It was clear to the early discoverers of ionising radiation that the energies associated with the radiation they observed were vanishingly small relatively to their everyday experiences associated with, for example, bulk heat transfer and the motion of everyday objects. The traditional units of energy that had been adopted to describe the thermodynamic phenomena central to the industrial revolution were vast by comparison. Derived from the experiments of the day with electromagnetics (given this is how many new isotopes had been discovered and remains a popular way of detecting nuclear radiations), quantities associated with the ionisation of gases and the influence of evolved charge on thin metal filaments, as used in early electrometers, were used in many cases to describe what was observed in the context of ionising radiation. In particular, a convention based on the relationship between the energy E acquired by an electron with charge e accelerated by a potential difference V is used as follows:

$$E = \text{eV} \tag{1.2}$$

which defines the unit of the *electronvolt* or eV. This unit is adopted universally to quantify the very small subatomic energies associated with microscopic nuclear processes, particularly those that yield ionising radiation. On this basis, a single electron (with a charge of magnitude of 1.6×10^{-19} c) accelerated through 1 V would acquire an energy of 1 eV corresponding to 1.6×10^{-19} J. The normalisation provided by dividing through by the energy acquired by an electron accelerated by 1 V makes for a much easier comparison of energies at these scales without the need for constant reference to many orders of magnitude. It also reduces the potential for mistakes in juggling many very small numbers in calculations.

In the context of nuclear engineering, energies in the 'eV' domain are actually at the lower end of the general range we tend to encounter; energies in the keV range are often considered intermediate while in the MeV range they would be considered at the middle-to-higher end of the energy spectrum.

Let us consider the case of the radioactive isotope potassium-40 (^{40}K). This isotope decays so very slowly that much of it still present naturally in the Earth's crust from the formation of the universe some ~13.8 billion years ago. It is taken up by the food we eat in harmless quantities featuring in our diet particularly via its natural occurrence in bananas. The ^{40}K nucleus of this isotope decays via a number of transitions between quantised energy levels, which results in the emission of electromagnetic radiation in the high-frequency range in the form of γ rays. The photons associated with this exhibit a relatively small number of discrete energies in the range 50 keV through to a few MeV. We will focus specifically on the 1491 keV transition for the purposes of this example. The 1491 keV transition has an equivalent energy in SI units as per,

$$E = 1491 \times 1000 \times 1.602 \times 10^{-19} = 2.39 \times 10^{-13}\,\text{J} \tag{1.3}$$

which highlights the extremely small scale of the energies of nuclear radiation relative to what we are used to at a macroscopic level.

1.4.5 NUCLEAR BINDING ENERGY

Further to the earlier discussion of the strong nuclear force and its critical role in holding the nucleus together, energy is required to overcome this force to break up the nucleus. Taken to its logical extreme, the energy required to disassemble the nucleus into its constituent neutrons and protons is known as the *nuclear binding energy*. It might appear quite a sophisticated task to measure this energy if it were necessary to separate a given nucleus using, for example, a particle accelerator. However, the energy that binds the nucleus together is manifest as a difference in mass between the sum of its constituent parts (the mass of the neutrons and protons) and the mass of the bound nucleus. This difference is known as the *mass defect* and can be determined via Einstein's famous relationship between energy and mass given in Eq. (1.4).

$$E = mc^2 \tag{1.4}$$

We return to this important concept in more detail in Chapter 3.

1.5 THE GENERIC NUCLEAR REACTOR

There are many different types of nuclear reactor: just as for the heat engines that came before them, such as steam engines and combustion engines, there are a variety of processes and arrangements by which the energy from nuclear fission can be harnessed. To aid our understanding of this wide and often contrasting field of engineering, it is illustrative to first consider a *generic nuclear reactor design*, while not being specific about the materials, cycles and processes until later chapters.

A nuclear reactor might be defined thus: 'apparatus in which a nuclear fission chain reaction can be initiated, sustained and controlled for generating heat or the production of useful radiation'. Hence, in the definition of a nuclear reactor, the emphasis is clearly on *control*, since without this, the very high energy density afforded by a nuclear reaction cannot be dissipated in a form and at a rate that is useful. However, the reader will also note that we do not restrict ourselves to the production of electricity, *per se*, since there are many reactors in use that are not dedicated to power production, as were the very first reactor systems, but that are used for research and materials applications. Also, note that a reference to scale is not implied because nuclear reactors can vary widely in terms of size depending on the application for which they have been designed.

In the most simple of terms, a nuclear reactor is composed of *fuel*, in which heat is generated as a result of a self-sustaining nuclear chain reaction, and a *coolant* that is necessary to transport the heat away from the fuel so that this energy can be used to perform useful work. Often a material is also included to reduce the energy of the neutrons sustaining the reaction because this makes it easier to sustain the reaction in relatively dilute quantities of uranium; this substance is known as a *moderator*. In some reactor designs, the moderator and the coolant are the same substance.

The reader should note that in our description of the generic nuclear reactor, and in reference to it later in the chapters that follow, we do not refer to the control mechanisms, emergency instrumentation systems, coolant pumps, pressurisers, driers, condensers and so on. While these components are extremely important to the operation of specific reactor designs, the operation and arrangement of them is too specific to be included in a preliminary, generic overview at this point. For the purposes of this generic basis, it is assumed that:

(1) The moderator and coolant are *separate substances* and not one and the same, although the latter arrangement is a popular and very successful design variant that we shall consider later in this text.

(2) The reactor is a *heterogeneous* design such that the fuel is separated into relatively narrow elements and distributed uniformly throughout the moderator and coolant systems. This is a common feature of all of the low-enrichment power reactor systems that make up the world's fleet of nuclear power generating systems as illustrated schematically in Fig. 1.4.

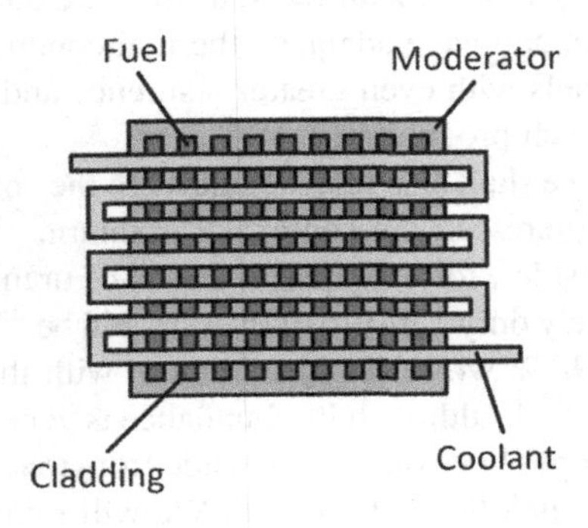

FIG. 1.4

A much simplified, schematic diagram of a generic nuclear reactor design.

1.5.1 FUEL

The first of the generic nuclear reactor components we shall consider is its fuel. Like all engines, energy cannot be generated without a source of fuel and nuclear reactors are no exception. All nuclear reactors in operation in the world today rely on uranium in some form or other (with variety in both isotopic composition and physical form) as the primary source of their fuel. This is because uranium is the only element that occurs in sufficient natural abundance with isotopes that are susceptible to the production of energy via splitting (*fission*) that is stimulated with slow neutrons. It is noteworthy that thorium, while frequently the focus of study for use in reactors and topical at the time of writing, is not usable directly in reactor systems until converted to sufficient quantities of ^{233}U as discussed in Chapter 11. Similarly, where plutonium is used it is either as a mixture with uranium, in the form of a mixed-oxide (MOX) fuel, or as the direct product of ^{238}U in uranium-fuelled reactors.

The naturally occurring uranium isotope of choice for nuclear energy production is ^{235}U because it is this isotope that has the dual benefit of being sufficiently abundant naturally (0.71% wt. of naturally occurring uranium is ^{235}U) and susceptible to *stimulated fission*. *Stimulated* or *induced* fission implies that neutrons are used to provoke the ^{235}U nucleus to split; this is essential in terms of sustaining a reaction based on this phenomenon. Other isotopes, such as californium-252 (^{252}Cf) exhibit *spontaneous fission* in which the isotope splits randomly as a form of radioactive decay. This is not very useful for reactor systems because, like other forms of radioactive decay, it cannot be controlled. In this context, we often refer to isotopes such as ^{235}U with this susceptibility to stimulation by slow neutrons as being *fissile*.

Uranium used in reactors and in combination with other fissile isotopes can take on a variety of geometries including: rods, plates and slabs with a variety of chemical forms and isotopic compositions. Early in the nuclear era, materials preparation techniques were more primitive than today. Often in this era, the only option was for the fuel to be used in metallic form. This was a reasonable solution at the time and many of the early reactors used fuel of this type. However, it was soon realised that the vulnerability of the fuel to the effects of heat cycling and exposure to high levels of radiation was a critical limiting factor in terms of reliability and longevity of fuel use in nuclear reactor systems. Consequently, alternatives were developed to appeal to the requirements of long life and tolerance to heat and radiation exposure. Many of the world's reactors are now reliant on ceramic forms of fuel usually comprising uranium dioxide (UO_2). These are more resilient to the rigors of reactor use and provide for longer periods of use than metal fuels, as they are more resistant to radiation damage and high temperature degradation, and have adequate thermal conductivity to access the heat. As reactor use continues to develop, fuels with even greater resilience and *accident tolerance* are a significant focus of many current research programmes.

In the context of nuclear fuel, we shall also refer regularly to the concept of *enrichment*. As referred to early, the specific isotope of interest in the context of uranium, ^{235}U, is only present naturally in relatively dilute proportions. Consider, for example, if atoms of uranium of natural enrichment were laid out in front of us, approximately only 1 in 140 of them would be ^{235}U. The vast majority of the rest would be ^{238}U that constitutes 99.3% wt. of natural uranium with the remainder being a very small amount of ^{234}U. Neither ^{238}U (nor ^{234}U although its abundance is very small) are fissile to a practically significant level. While these isotopes have other uses, aside from reactor use, they generally serve as a matrix that is relatively inert in which the ^{235}U is held. We will return to this debate to highlight an important exception in Chapter 11 associated with ^{238}U and breeding.

While sustained fission is feasible in uranium fuels of natural enrichment, that is, 0.71% wt. ^{235}U (this was the main approach for many of the first nuclear reactor designs), it can be difficult to achieve and can limit the choice of the materials used for the other components in a reactor. Fuel of natural enrichment tends to have been desirable in cases where access to enrichment facilities has not been readily available or desirable in terms of national energy policy.

Most nuclear fuel currently in use for power production is slightly enriched in ^{235}U as this eases the operation of the reactor and offsets the effect of neutrons absorbed in the coolant and moderator. Fuels of low enrichment in such cases are classified typically as <5% wt. ^{235}U. Considering the scenario described earlier, 5% wt. enrichment would correspond to 1 in 20 of the uranium atoms present being ^{235}U; clearly, a relatively small increase in enrichment can significantly increase the relative proportion of the fissile isotope. Some research reactors use fuel of higher enrichment, typically <20% wt. ^{235}U, while some exotic reactor designs have used much higher enrichments. The enrichment of uranium consumes energy; the higher the enrichment the greater the level of consumption and hence the cost of the fuel, other factors such as the natural enrichment level of the ore notwithstanding. Also, higher enrichments can be more prone to the risk of proliferation, that is, diversion to illicit use. Thus low enrichments provide a reasonable compromise between easing reactor operation, optimising burn-up, keeping costs competitive and the prevention of proliferation.

1.5.2 CLADDING

Nuclear fuel in a reactor is contained in a material known as the *cladding* or *clad*. The cladding does not contribute to power generation in the way that the fuel does, but it is an integral part of the fuel design and performs a number of very important functions. These include:

- Containing radioactivity in the fuel to prevent it from contaminating other materials in the reactor, most immediately the coolant.
- Being the interface between the fuel and the coolant in almost all reactor designs and thus constituting the medium by which heat generated in the fuel is transferred to the coolant.
- Enabling neutrons to pass unhindered from one element of fuel to another via the moderator.
- Resisting the effects of radiation, particularly changes to its physical properties such as the response to stress from long-term exposure to intense fields of neutrons and γ rays.
- Withstanding the effects of frequent cycling in terms of temperature, radiation damage and minimising the influence of these effects in the fuel such as expansion, contraction and deformation, without breaking or becoming deformed.

It is beneficial, particularly in respect of the eventual clean-up of a reactor, if the materials from which the cladding and the components associated with it are made are resistant to neutron activation. Otherwise, disposal of the cladding can be more complicated and can constitute the generation of unnecessary waste.

Many different types of cladding have been used since nuclear reactors were first developed. The most widespread include alloys of zirconium and to a lesser extent magnesium, aluminium and stainless steel. The choice of cladding material is influenced by several other factors such as the type of coolant and the thermohydraulic conditions of a given reactor's operation, for example: temperature, pressure, flow rate and so forth. Since reactors exploiting light water as both coolant and moderator have become the most widespread throughout the world, the corresponding cladding of choice has centred on zirconium alloy as this is used in these designs. This material is relatively resistant to corrosion in these environments where light water is used at elevated temperatures and pressures while providing the functions listed above. With the potential for advanced reactor designs operating at higher temperatures and thermodynamic efficiencies in the future, research continues into the development of cladding materials compatible with these operating conditions.

1.5.3 COOLANT

In our generic reactor system, we include the concept of a *coolant* as it is usually necessary to transport the heat from the reactor to a place where it can be used to perform work or at least so that it can be dissipated safely. This might include being used to generate electricity, propulsion or both. In some reactors that are operated for research purposes, such as for materials testing or the production of medical isotopes, the heat is not usually used for a specific purpose although it is always necessary to configure the system so that the heat is transferred from the fuel. This can be either passive or active, with the latter necessary in operating power reactors where the yield of heat is very significant.

Again, as for the case of fuel cladding, the requirements of a coolant in a nuclear reactor are readily defined. It should:

- not encourage corrosion of fuel cladding and other reactor components
- have good mass transport properties
- have a high specific heat capacity
- not be prohibitively expensive
- be readily available in significant quantities

A variety of media have been used as coolants throughout the years of reactor development. Among the most common include air, light water, heavy water and carbon dioxide. Further, a variety of rather more exotic materials have been used including liquid sodium, mixtures of sodium and potassium, liquid lead, lead and bismuth and helium gas, predominantly for fast reactor applications. It is noteworthy that many reactor system designs are based on the use of two coolants or even three, as we shall consider below when we introduce the concept of the reactor circuit; such an arrangement allows one coolant stream to be isolated from another. The coolants of choice do not have to be same substance.

1.5.4 MODERATOR

In most nuclear reactors, a substance known as the *moderator* is also usually required. This part of the reactor system is necessary because the fission nuclear reaction that is exploited in the production of nuclear energy is much more likely to occur for neutron energies that are reduced, especially for fuels where the quantity of ^{235}U is relatively low. The neutrons used to propagate and sustain the fission reaction usually arise from fission events with energies much higher than this level (typically a factor of 10^8 higher than is desirable in most cases). The moderator is necessary to slow these neutrons down so that the probability of them causing a fission reaction is increased. It performs this function by providing a matrix with which the neutrons interact and impart their excess energy to its atomic structure. For reasons discussed in Chapter 6, the ideal properties of a moderator include that it must comprise light elements, it must scatter neutrons effectively and it should not absorb them to too great an extent. It is advantageous if it is cheap, abundant and benign, that is, not flammable, chemically reactive or corrosive.

1.5.5 REACTOR CIRCUIT

The core of a nuclear reactor comprises the nuclear fuel and there are relatively few moving parts. However, a power-generating nuclear reactor is often a relatively complex thermohydraulic system in which a variety of liquids are used, as per the descriptions pertaining to cooling and moderation described above. Further, although it is not always desired, it can be feasible for the phase of these liquids to change with the operation of the reactor. This can result in mixed-phase flow, that is, a mixture of liquid and gas in the reactor system. There are frequently significant extremes in terms of temperature, pressure and radiation levels that need to be accommodated, especially for example during start-up and shutdown of the reactor.

Just as there are many different variants of combustion engine and electric motor, there are several alternative nuclear reactor circuit designs. Some were developed as prototypes and taken no further beyond initial feasibility or testing while others have become very successful and are in widespread use throughout the world today. A further important factor in nuclear engineering is that the Second World War played an indelible role in the development of nuclear technology. The historical context that followed this has influenced national trends in terms of favoured types of reactor circuit in contrast with another, for example, due to factors such as the availability of technologies, materials and production facilities derived in specific countries at a time of heightened secrecy. Since nuclear reactors are relatively long-lived facilities with some operating beyond 50 years, this historical background still exerts a legacy we see today, albeit often in decommissioning of these early plant. By contrast, the current era exhibits a greater degree of design diversification as reactor systems have been developed on commercial terms and licensed internationally.

In general, reactor circuits are referred to in terms of being either *direct* or *indirect* systems, with the emphasis here being on how the system is configured to allow the coolant to come into contact with the reactor core. Coolants necessarily have to be readily transportable and therefore are usually either liquids or gases. This is not an essential requirement for moderators and so our emphasis in the context of reactor circuits is usually on the transport mechanism for the coolant. The two alternative reactor circuit types are shown in Fig. 1.5.

Direct reactor circuits are those in which the coolant that is in contact with the core is also that which is used to derive useful work (most often to drive turbines to generate electricity). In contrast, indirect circuits usually employ two or more coolant circuits that are physically separated from one another and a *heat exchanger* is used to transfer heat from one to the other. The circuit that is in contact with the core is usually referred to as the *primary* circuit with the *secondary* circuit being that which is used to drive the power take-off system; the secondary is often referred to as the *feedwater system* or *feedwater loop* in indirect cycles.

A significant advantage of the direct system is that a great deal of the engineering infrastructure necessary for indirect circuits is not needed, such as the process pipework, pumps and heat exchangers. This can render direct-cycle plant cheaper and quicker to build than indirect systems, notwithstanding a range of other design factors that might influence cost. A significant advantage of indirect reactor circuits is that the coolant in contact with the nuclear fuel (albeit via the fuel cladding) is physically

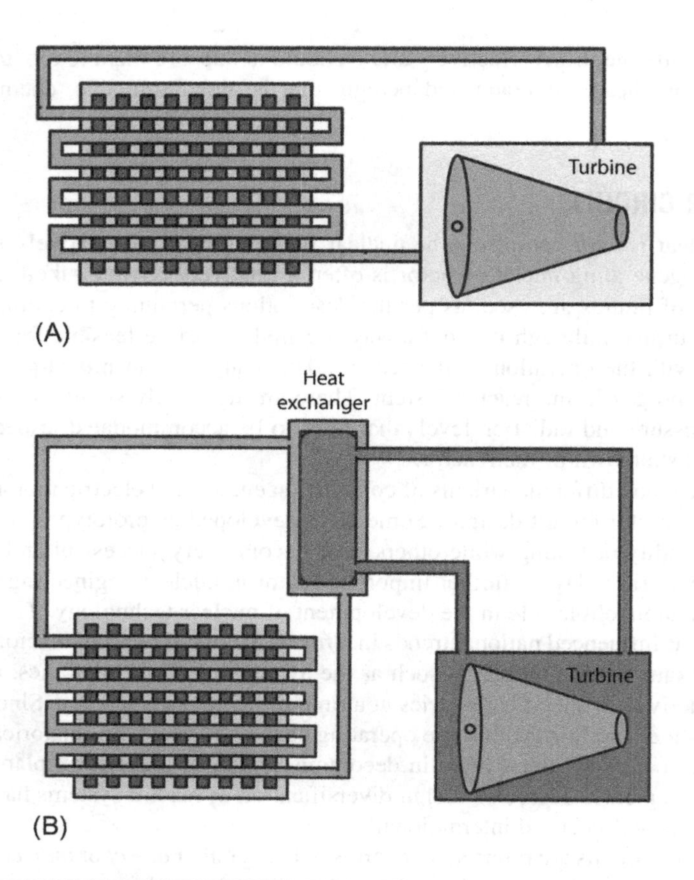

FIG. 1.5

Schematic illustrations of the two alternative reactor circuit designs: direct (A) and indirect (B) systems. The primary circuit in the indirect system is shown to the left and the secondary circuit is shown to the right.

separate from the coolant that drives the power generation system. Thus, in the event of cladding failing or an accident in which there is a risk that irradiated fuel is exposed to the coolant in the primary circuit, the extent of contamination by radioactive materials from the core is limited to just the primary circuit. The indirect cycle also has benefits from a decommissioning perspective since requirements relating to radioactive contamination are restricted to what is often referred to as the 'nuclear island', for example, the reactor, waste transport facilities and so on. The rest of the nonnuclear site can usually be dismantled relatively quickly and easily, with a much reduced risk of exposure, simplified regulatory administration and reduced quantities of radioactive infrastructure for disposal. Direct reactor circuits have the disadvantage of usually requiring highly purified reactor coolant to reduce the potential for contamination via neutron activation of minerals in the coolant that is in contact with the core. Both direct and indirect types of reactor system are in operation and construction across the world today.

1.6 ELEMENTARY NUCLEAR PHYSICS CONCEPTS

1.6.1 CONVENTIONS FOR THE EXPRESSION OF MASS

To make comparison across the range of isotopic properties as straightforward as possible, isotopic mass is expressed in terms of *atomic mass units* (u). This is based on the definition of a ^{12}C atom having a mass of 12 u and thus 1 u being one-twelfth the mass of a ^{12}C atom. Given that a mole is the number of atoms in 12 g of ^{12}C, where 1 mol is Avogadro's number (6.023×10^{23}), $1\ \text{u} = 1.66 \times 10^{-27}$ kg.

It is important to specify the units when dealing with isotopic masses, that is, ^{238}U has an *isotopic mass* of 238.0507 u, which is equivalent to a mass of 3.95×10^{-25} kg. The isotopic or atomic mass should not be confused with the *molar mass*; the latter is the mass of a mole of the isotope in question.

A further convention is used to express mass in the equivalent form in terms of MeV/c^2 via where 1 atomic mass unit follows as being equivalent via Eq. (1.4),

$$m_u = 931.5\ \text{MeV}/c^2$$

to one decimal place.

1.6.2 INTRODUCTORY CONCEPTS OF RADIOACTIVITY

In this short section, a summary of the accepted terminology is provided to set the scene for the chapters that follow. Where appropriate, specific terminology is introduced at the start of the section where it arises later in the text.

The distinction of nuclear engineering over other strands of the engineering discipline is that the substances and materials that are used and processed are often *radioactive*. This implies that they have the propensity towards *radioactive decay*; this is a natural process by which matter gains greater stability. In so doing, atoms of one type can change into atoms of another. For such a case, we refer to the isotopic species before decay as the *parent* and that which follows as the *daughter*. It is also possible for a given isotope to transition from an excited nuclear state to a lower excited state without changing its composition. Radioactive decay usually results in the emission of *ionising radiation*. This can comprise *electromagnetic radiation* (usually of a type known as γ rays) and subatomic particles such as *β particles* (electrons), neutrons but also *α particles* (the atomic nuclei of helium atoms). Often, the daughter product is also unstable towards radioactive decay. This constitutes the start of what is known as a *decay chain*. Stable materials can be made radioactive by the interaction of radiation upon them via a process called *activation*; the most relevant example of this phenomenon in the current context is that resulting from the interaction of neutrons with matter that is known as *neutron activation*.

With reference to the example of neutron activation, the likelihood that nuclear reactions of this type, that is, one caused by the interaction of a neutron resulting in a change in the isotopic composition of the matter that is exposed, is dependent on the energy of the neutron and the susceptibility of the specific isotope with which the neutron is interacting; some reactions are more likely than others. This is usually quantified in terms of a parameter known as the *cross section*, and this is a very important concept in nuclear engineering. The cross section is equal to the rate at which a given interaction takes place divided by the product of the rate at which neutrons are incident with the number of nuclei in the material in which the interaction is taking place. It is thus the microscopic probability that a given nucleus will undergo a specified interaction with a neutron. Hence it is called the *microscopic cross section* and is discussed in more detail in Chapter 4.

In contrast to our everyday experiences with much of the world around us, many of the properties associated with matter at the nuclear scale are *quantised*, such as energy, charge, angular momentum and so on. Consequently, we tend not to measure continuous trends of behaviour but rather we observe individual quanta, such as *photons* being the quanta of γ radiation. Similarly, *quantum mechanics* governs behaviour at this scale with changes often governed by a quantised structure of states that matter can occupy, particularly in terms of *excitation*.

It follows that most subatomic phenomena cannot be anticipated with certainty. Rather, they are understood and measured in terms of the *probability* of them occurring. The probability of radioactive decay is a property of the specific isotope in question. Therefore the *rate of decay* is directly proportional to the *amount of substance* that is present; the more entities with the same probability of decay that there are in a given sample the greater the probability that one of them *will* decay. The time taken for the amount of a substance to decay to an amount $1/e$ is known as the *lifetime*.

A more widespread measure of the propensity of a substance to decay is the *half-life*, which is the time taken for a quantity of radioactive material to decay by half. The rate of decay of a given quantity of radioactive substance is known as the *activity*, and this is measured in terms of the number of *disintegrations per second* or *Becquerels* (Bq). This quantity when normalised to the mass of substance is known as the *specific activity* and is measured in units of Bq g^{-1}. Historically, an alternative unit for activity was defined and is still used corresponding to the equivalent specific activity of natural radium, and this is called the *Curie* (Ci). One Curie is equal to 3.7×10^{10} Bq.

CASE STUDIES

CASE STUDY 1.1: CALCULATING THE DENSITY OF NUCLEAR MATERIAL

Given the atomic mass of carbon (^{12}C) is 12 u and that its nucleus has a radius of 10^{-15}m, it is possible to estimate the density of nuclear material for the case of carbon.

Assuming the ^{12}C nucleus is spherical, the volume V is:

$$V = \frac{4}{3}\pi r^3 = \frac{4}{3}\pi (10^{-15})^3 \sim 4 \times 10^{-45}\,m^3$$

The mass m_C of 1 atom of ^{12}C is

$$m_C = 12 \times 1.66 \times 10^{-27} \sim 2 \times 10^{-26}\,kg$$

Hence, the density ρ is as follows:

$$\rho = \frac{2 \times 10^{-26}}{4 \times 10^{-45}} = 5 \times 10^{18}\,kgm^3$$

To put this into familiar context, we could estimate the mass of 1 mm^3 of 'nuclear' carbon.This would be

$$m = 5 \times 10^{18} \times 10^{-9} = 5 \times 10^9\,kg$$

or 5 million tonnes.

CASE STUDY 1.2: RADIATION IS QUANTISED IN VERY SMALL AMOUNTS OF ENERGY

The photon energy for the 1491 keV line in ^{40}K of 2.39×10^{-13} J illustrated in Section 1.4.4 is a vanishingly small quantity, as the following calculation demonstrates. Here we estimate by how much the temperature of 1 kg of water would be raised if a 1491 keV γ ray from ^{40}K was absorbed totally by it. Water is a relevant example because the majority of living tissue is comprised of water and therefore it is a good first approximation to a tissue-equivalent substance when we come to consider radiation dose. However, it might take more than 1 kg to absorb a ^{40}K γ ray in practice because γ rays with energies in this range are very penetrating.

Given the specific heat capacity of water is $C = 4.2$ kJ/kg °C and the relationship between enthalpy H, specific heat capacity and the temperature change ΔT is

$$H = mc\Delta T$$

Then it follows that

$$\Delta T = \frac{H}{mC} = \frac{2.39 \times 10^{-13}}{1 \times 4.2 \times 1000} = 5.7 \times 10^{-17}\text{°C}$$

which is not very much certainly and not easily measurable!

REVISION GUIDE

On completion of this chapter you should:

- understand the distinction between *nuclear radiation* and atomic radiation
- understand Rutherford's discovery of the nucleus and its implications for the density and scale of the nucleus
- understand that a wide range of lifetimes is possible for radioactive substances
- understand the general composition of the nucleus in terms of its constituent nucleons, its size and density
- understand the interplay between the Coulomb force and the strong nuclear force that contributes to the trend in instability observed in very massive elements
- be able to draw a sketch of the chart of the nuclides and highlight the various features on it, including its nonlinearity, the valley of stability and the implications of this for the stability of the heavier isotopes
- be able to calculate the density to nuclear matter and relate this to our everyday existence
- be able to convert parameters of energy between the units of Joules and electronvolts
- appreciate the very small quantities of energy we tend to deal with associated with individual quanta of ionising radiation
- understand the four fundamental components of the generic nuclear reactor and the two key reactor cycle types that are used in the power reactors operating today
- have a good understanding of the elementary nuclear physics terminology that is used in the nuclear engineering field

PROBLEMS

(1) Given the atomic mass of uranium (^{238}U) is 238.0289 u and that it has a nuclear radius of 7.5×10^{-15}m, calculate the density of the nucleus of ^{238}U. Hence, estimate the mass of a fuel pellet of diameter 10 mm and height 8 mm if it were possible for it to be comprised entirely of uranium with the composition of its nucleus.

(2) Caesium-137 decays via β decay to an excited state of barium-137, which decays subsequently via the emission of a single γ ray having an energy of 662 keV. Convert this energy to joules. How many ^{137}Cs γ rays would need to be absorbed by 1 kg of water to raise its temperature by 1°C if its specific heat capacity is 4.2 kJ kg^{-1} °C^{-1}?

(3) Explain why the Chart of the Nuclides is shaped in the way it is. From a consideration of the chart, would you expect a nuclide with a mass $A \sim 200$ to have more neutrons than protons or vice versa? Justify your answer by explaining your choice.

(4) Describe the difference between a nuclear reactor design that exploits a *direct* cycle as opposed to one that exploits an *indirect* cycle. What are the advantages of the indirect cycle?

(5) From your appreciation of the performance requirements of the fuel cladding and coolant, highlight three properties of these components of a generic reactor system that are essential to the role that they perform.

CHAPTER

HISTORICAL CONTEXT

2

2.1 SUMMARY OF THE CHAPTER AND LEARNING OBJECTIVES

The adoption of technological advancements on a global scale rarely occurs based purely on the technical merits of a given engineering solution. Most importantly, critical matters associated with the potential impact on safety, health and the environment play a significant and important role, as does economics. In many cases, historical developments and particularly investments derived from military imperatives can influence technological developments significantly, and can accelerate research in areas that bring with them important benefits to allied fields of endeavour.

Commercial nuclear power is an important example of this development process, but there are others: Rudolf Diesel's vision of a simple and flexible engine running on biofuels was integrated into the design of the first submarines soon after his death while now, albeit running on fossil fuel, this invention constitutes the most common internal combustion engine on Earth. History plays an important role in the way in which engineering systems are adopted. In this chapter, the history associated with the current status of nuclear energy in the world is summarised. The aim of this chapter is to provide a context for the more technically focussed chapters that follow, and to provide the necessary background for the later chapters associated with nuclear safety, regulation, acceptability and economics.

The objectives of this chapter are to:

- summarise the origin of nuclear power as a source of energy, particularly fission-based nuclear power both in natural and man-made forms
- introduce uranium as an element in general terms with a history that predates its use in electricity production, and to describe its early uses
- provide a summary of the pioneering scientific discoveries that led to the realisation that a controlled and sustained nuclear chain reaction might be harnessed for the production of energy
- describe the development and operation of the first nuclear reactor
- summarise the impact of the military influence on the development of nuclear energy, particularly that of the Second World War, but also of the Cold War that followed
- describe the historical context behind the development of alternative nuclear fuels and fuel cycles
- review the current reactor classification system

Nuclear Engineering. https://doi.org/10.1016/B978-0-08-100962-8.00002-0

2.2 HISTORICAL CONTEXT: ENRICO FERMI 1901–54

Enrico Fermi's scientific achievements span some of the most prominent discoveries in the development of nuclear energy. For example, he discovered that radioactivity could be induced with neutrons and that slow neutrons were more effective at this than fast neutrons. For this he was awarded the Nobel Prize in 1938. He postulated the extent of the energy released by fission and subsequently led the development of the world's first self-sustaining nuclear reaction and reactor at the University of Chicago. He pioneered the *six-factor formula* as the basis for understanding neutron populations in self-sustaining reactions, which still forms the introductory basis for understanding this phenomenon today (Fig. 2.1).

However, Fermi's influence extends beyond the engineering of nuclear fission: He formulated the framework describing statistical behaviour central to quantum mechanics (*Fermi*-Dirac statistics) applicable to particles referred to as *fermions*. Beyond his accomplishments with neutrons, he made important contributions to our understanding of β decay, postulating the existence of the neutrino long before it was observed experimentally and hence the *weak interaction*. In this, he conceived the basis for one of the four known fundamental interaction mechanisms. Following his pioneering work in Italy, Fermi immigrated to the United States after receiving the Nobel Prize, as racial legislation at that time implicated the freedom of his family. His research in the United States was to continue to be very important in shaping the passage of world history, particularly in wartime.

FIG. 2.1

Enrico Fermi.

2.3 INTRODUCTION

On an evolutionary timescale, nuclear energy exploited by man is a recent development. To perfect its use on a commercial scale required a significant investment of resources and the profound ingenuity of a large number of exceptional people. The result was that the most important developments in its history were accomplished in less than a decade. This phenomenon was partly due to the startling insight of the early pioneers, but also because of the significant effort that flowed due to its association with nuclear weapons and this development in World War II. More so than perhaps any other technological development, commercial nuclear power has been shaped by historical events and past national priorities. This is evident in the power reactor designs in use today, the legacy issues that face the industry (particularly that associated with waste disposal), and also in terms of the engineering options for the future. However, there is geological evidence for the existence of nuclear energy as a natural phenomenon dating back several billion years which occurred entirely without human intervention. This provides a suitable place to start this chapter.

2.4 NATURAL REACTORS

The generation of energy from nuclear fission can be traced back to almost 2 billion years to the era when it is strongly suspected that natural reactors operated in Gabon, Africa [1,2]. The current isotopic abundance of ^{235}U, even in geological regions of where the relative abundance of elemental uranium is high by today's standards, is not sufficient to sustain a nuclear chain reaction; the proportion of ^{235}U is too low and thus the ^{235}U atoms are too widely dispersed.

The current isotopic proportion of uranium is the product of many billions of years of radioactive decay. By definition, those relevant isotopes with half-lives that are short relative to the age of the planet at ~4.5 billion years such as ^{234}U and ^{236}U have decayed away more rapidly than longer-lived examples such as ^{235}U and ^{238}U. It is these latter longer-lived isotopes that remain prominent today in natural isotopic distributions of uranium. Most importantly, in terms of fission, the significantly higher activity of ^{235}U (reflected by its shorter half-life) in comparison with ^{238}U has resulted in it having decayed away more rapidly. This has left behind the significant disparity in isotopic abundance present in the Earth's crust today, favouring the heavier isotope by the ratio ^{235}U to ^{238}U of approximately 1:140 (local variations not withstanding).

Almost 2 billion years ago the isotopic situation was very different. The natural abundance of ^{235}U relative to that of ^{238}U was much greater than it is today (this is explored quantitatively in Chapter 5). At that time, given that the requirements described below were met, it has been shown that a chain reaction could have been sustained for several thousand years in the ground. These requirements are as follows:

- The abundance of fissile uranium that is ^{235}U needs to be sufficient to constitute the necessary critical mass.
- There needs to be an abundance of light isotopes to provide neutron moderation (e.g. perhaps the presence of flowing groundwater through a porous deposit with a uranium-containing mineral).
- Sympathetically low levels or (ideally) the absence of species that absorb neutrons (particularly boron due to the very high neutron absorption probability of the relatively abundant isotope of boron, ^{10}B, and the abundance of boron in the Earth's crust).

FIG. 2.2

A photograph of the Oklo region in Gabon, Africa, reproduced with permission of the European Nuclear Society. The site of a natural reactor is depicted by a man-made construction built to contain resident radioactivity (https://www.euronuclear.org/e-news/e-news-9/issue-9-print.htm). Several individual natural reactors are suspected to have existed in this area.

To date, geological evidence suggests that these conditions were met in an isolated handful of locations in a specific part of Africa. These requirements provide useful insight into nuclear reactor systems because similar conditions are necessary to synthesise the production of nuclear energy. In Fig. 2.2, a photograph of one of the sites of a suspected natural reactor in Gabon, Africa, is shown.

That natural reactors could come about in this way is supported by the evidence of characteristic isotopic signatures associated with both fission products and actinide residues found in nearby geological deposits, and also because the isotopic composition of ^{235}U in the vicinity is depleted relative to the surroundings. It is suspected that natural reactors operated for sustained periods of time, perhaps in excess of 100,000 years generating power of the order of 100 kW, and potentially starting and stopping in sympathy with the rise and fall of the moderating hydration in the deposit and its temperature.

These regions in Africa are currently the only place in the world where the geological evidence has been preserved sufficiently well to be of use in research today. Not surprisingly, the uranium abundance in these regions in Gabon remains high relative to elsewhere and has long been associated with 20th-century uranium extraction for nuclear fuel used in man-made reactors. A major distinction of the man-made approach is that extensive refinement of the ore and often isotopic enrichment is necessary; it is anticipated that natural levels of ^{235}U enrichment might have been >3% wt. ^{235}U 2 billion years ago. Although predominantly of historical interest in the context of the natural environment, these reactors are being studied to benefit our understanding of the migration and containment of fission products and minor actinides in the context of nuclear waste management and disposal. This is discussed further in Chapter 15.

2.5 EARLY USES OF URANIUM

Relative to the era of the world's natural reactors, it took mankind quite a long time to harness uranium for power production. Discovered much earlier in 1789, uranium was put to other uses more than a century before its properties of radioactivity and nuclear fission were discovered. In the form of mined mineral deposits of autunite, coffinate, pitchblende, torbernite and uraninite, uranium was one of the several dense metals that came under what was relatively intensive mining at the end of the 19th century, along with lead and bismuth. Although the small mines associated with the recovery of these metals are not comparable in scale with the much larger operations that would serve the nuclear-power industry in the 20th century, significant quantities of these materials of the order of several hundred tonnes were recovered from them.

The primary purpose of uranium mined up until the 1930s was as an additive to glass and ceramic glazes; with its variety of valence possibilities uranium was an effective colourant as depicted by the example of associated glassware in Fig. 2.3. As the vacuum electronics revolution began to develop, uranium was also used as one of the several metallic additives that enabled effective glass-to-metal seals to be made for the vacuum electronics devices used in early radios and transmitters.

Latterly, after the discovery of radium by Marie Curie in 1898, it was the same uranium-containing mineral deposits that served as the feedstock for the material used in the pioneering discoveries that followed. A radium-based industry developed quickly to support its uses in luminous dials used in watches and aircraft instruments. Radium was also used in early, sometimes unfounded medical applications. Over time, as the significance of the radiotoxicity of radium and its related isotopes and decay products was better understood, its use fell almost entirely out of favour. In many cases, its

FIG. 2.3

A photograph of several examples of glass coloured with uranium, under ultraviolet light (reproduced under limited worldwide licence from Oak Ridge Associated Universities (ORAU), https://www.orau.org/ptp/collection/consumer%20products/vaseline.htm).

use has now been effectively banned by law. Some of the early medical uses were not only found to be without scientific basis but also to be a risk to human health. When, following the discovery of fission in 1938, the nuclear arms race began in the late 1930s, the unregulated recovery and processing of uranium and its ores was halted to prevent its proliferation while the world was at war. Many of the smaller mines that had constituted the primary sources of uranium for its early uses and pioneering scientific discoveries (such as South Terras in Cornwall in the United Kingdom) were closed before mainstream nuclear power production began; early workings of this type do not offer the logistical benefits in terms of scale and abundance to compete economically with the larger uranium mines established to provide the feedstock for nuclear fuel. It is noteworthy that the mine workings associated with these early activities, although long disused, remain significant sources of radiation exposure from, for example, radon. Prolonged exposure in these environments can result in doses above limits specified in the Ionising Radiation Regulations [3], for example.

2.6 THE SEARCH FOR TRANSURANIC ELEMENTS AND THE DISCOVERY OF FISSION

Scientific discovery in nuclear physics continued rapidly at the turn of the 20th century: Marie Curie, inspired by the discoveries of Becquerel and Roentgen and the birth of the new field of radioactivity, chose to focus her doctorate thesis in this area. In turn, her discovery of radium provided Rutherford with exactly what was needed next: a relatively pure, long-lived source of α particles with which to probe the physical properties of the atom and which enabled his pioneering discovery of the atomic nucleus. Chadwick, a close contemporary of Rutherford on the threshold of what was to become known as the *golden age of nuclear physics*, was able to make quantitative sense of Rutherford's conceptual description of the atom. In particular, he realised that data for the atomic mass and atomic number could not be reconciled without what he postulated must be a neutral particle with a mass similar to that of a proton. This was the *neutron*—which he discovered in 1932. The neutron, in turn, provided Hahn and Strassman with a means to stimulate what Meitner and Frisch would interpret successfully as being *fission*, although they could not have appreciated this at the time they began their experiments. Their discovery along with the contributions of many other contemporaries, for many, represents the birth of the nuclear age, which has come to symbolise the extent of mankind's intellectual grasp of the natural world in the 20th century.

Following Chadwick's discovery of the neutron, an important area of scientific focus at the time was on the discovery of new isotopes and, where possible, new elements. Curie's discoveries of polonium and radium filled two long-standing vacancies in the periodic table at the time, with protactinium following soon after by other researchers to fill the gap identified by Mendeleev between thorium and uranium. Thorium and uranium had long been known, having been discovered much earlier.

Subsequently, attention turned to whether isotopes beyond uranium could be formed; the so-called *transuranic* elements. Not only would the existence of new, heavier isotopes be of interest in itself, but also this would demonstrate an extension of the periodic table into a region of mass where it was not yet clear at the time whether atomic nuclei would be sufficiently stable to exist for long enough to be observed. Although the discovery of these isotopes did follow just a few years later made with the cyclotron at the Laurence Berkeley Laboratory by Glenn Seaborg and his team, at the time that nuclear fission was discovered an accepted view was that such isotopes might be formed via *accretion*, that is, by adding neutrons to existing heavy isotopes. Otto Hahn, having had success confirming the existence of protactinium with Lise Meitner, joined the search for the transuranic elements assisted by Fritz Strassman in Berlin. However, rather than focus on a new isotope, fate would intervene: it would be from their knowledge of the chemistry of the well-known rare-earth isotope barium that their most significant breakthrough would arise.

When Hahn and Strassman exposed uranium to neutrons and then refined the products of this experiment, they found a number of radioactive isotopes with chemical properties similar to radium. Clearly, this was not evidence of a new element as radium was a product of the discoveries of the previous century. However, neither did it make scientific sense for a neutron to be able to strip four protons from a uranium isotope as would be necessary to yield radium. They also observed that both thermal and fast neutrons gave similar results and thus incident neutron energy did not appear to have a strong influence on the isotopic outcome, which they had considered it might if the radium-stripping hypothesis were valid.

To resolve this conundrum, Hahn and Strassman precipitated the reaction products from their irradiated uranium media with various barium-based salt compounds. They then used fractional crystallisation to attempt to concentrate and separate any radium products that might be present from barium salt solutions. They expected to observe a correlation with what was known to occur for radium in such cases because of the difference on its solubility with its host solution; it was anticipated that some barium compounds would act to increase radium separation and others the opposite. However, they observed no such correlation and thus concluded that the compounds could not comprise radium and must instead be radioactive isotopes of barium [4]. Lise Meitner and Otto Frisch, drawing on their knowledge of Bohr's liquid drop model of the atomic nucleus, were able in exile to provide the physical basis for what Hahn and Strassman had observed: they concluded that the nucleus had separated into two just as a droplet of liquid might when under significant vibrational excitation, a mechanism they coined as 'fission'.

In the late 1930s, when Hahn, Strassman and Meitner were discussing these results, Adolf Hitler's fascist policies had already been established for several years. Many scientists, including Einstein, Fermi and Meitner among others, had chosen to flee fascist rule and racial oppression. Hence, Meitner corresponded by letter and occasional clandestine meetings with her colleagues to finalise the fission hypothesis. Subsequently, Siegfried Flügge was the first to report on the possibility of an applied use of the energy from fission.

2.7 THE INFLUENCE OF WORLD WAR II AND THE RACE FOR THE ATOMIC BOMB

It is difficult to dispute the significance of the impact of the Second World War on the development of nuclear power. When Hahn, Strassman and Meitner published their research in the open literature before the outbreak of war, such was the scientific interest in the illusive transuranic elements that many contemporaries were immediately, fully conversant with the implications of their discovery. The relatively straightforward calculation of the energy release derived from the mass defect before and after fission (as will be discussed in Chapter 4), indicated that here was a significant and highly concentrated form of energy: a million times greater than any observed from a chemical reaction and hence with the potential for much greater devastation than any known chemical explosive.

The connection of the new *nuclear* energy with a new and significantly more devastating weapon was appreciated by many in the immediate scientific community. However, it was not considered possible by all until Frisch and Peierls calculated that the quantity of the material necessary was much smaller than had been appreciated at first. Einstein communicated this in his famous letter to President Roosevelt in Aug. 1939, as reproduced in Fig. 2.4. If we needed to specify a date constituting the impetus for the Manhattan project that followed, then perhaps the date of this correspondence is it. However, the prospect of the weaponisation of this new source of energy divided scientists, with several declining to work on the development. Others in contrast served the weapons programme from their laboratories in their respective nation states or indeed moved to Los Alamos in the United States to work directly on the programme. Some Nazi sympathisers, most famously Werner Heisenberg, joined Germany's attempt to get the bomb before the allies, stimulating many years of debate and intrigue in the process.

While the focus of the nuclear weapon development in the 1940s from a nuclear engineering perspective was on achieving a reliable, rapid, high-yield criticality excursion as opposed to the controlled, sustained yield of energy needed to produce power, the vast investment in terms of expertise, infrastructure and resource accelerated our knowledge radically in terms of peaceful applications of nuclear energy. For example, techniques for uranium recovery from the ground, uranium enrichment, nuclear fuel manufacture, computing and radiation transport development all advanced significantly as a result. Indeed, the latter also set the foundations for the development of the architecture of the first computer systems—one of many by-products of this era.

2.8 NATIONAL TRENDS IN POWER REACTOR DESIGN

The development of the atomic bomb and its successors have been described in detail in many other texts, and this is not the objective of this work. However, an important influence of the entanglement of civil nuclear power development and atomic weapons research is that of war-time secrecy. This continued beyond World War II extending throughout the Cold War that followed until the fall of the Union of Soviet Socialist Republics (USSR) in the late 1980s. This led to reactor design preferences that were apparent from country to country and which, in many cases, are still evident today because of the long-lived operation of nuclear power facilities.

For example, the United States had a well-evolved electricity supply system with which to drive enrichment facilities primarily for the weapons programme, but also for related military nuclear activities that included the development of nuclear-powered vessels such as submarines and aircraft carriers. For these applications, a compact propulsion unit was required and the pressurised water reactor (PWR), with its combined use of water as both coolant and moderator, provided the answer; particularly, given the commitment of its pioneer Admiral Hyman G. Rickover to the PWR philosophy because of its inherent characteristic safety benefits. However, this was only possible if both uranium enrichment facilities (to cater for neutron absorption on hydrogen in light water) and the supply of high-grade pressure vessels (to maintain the water in its liquid state) were available. Ready access to the fuel in such reactors in order to extract plutonium for use in nuclear weapons was not required as alternative facilities had already been constructed in the war to cater for this requirement. Hence, the United States committed to the light-water reactor fleet on which its commercial nuclear power sector still relies today, a reliance that has spread to much of the world in terms of power reactor design. This

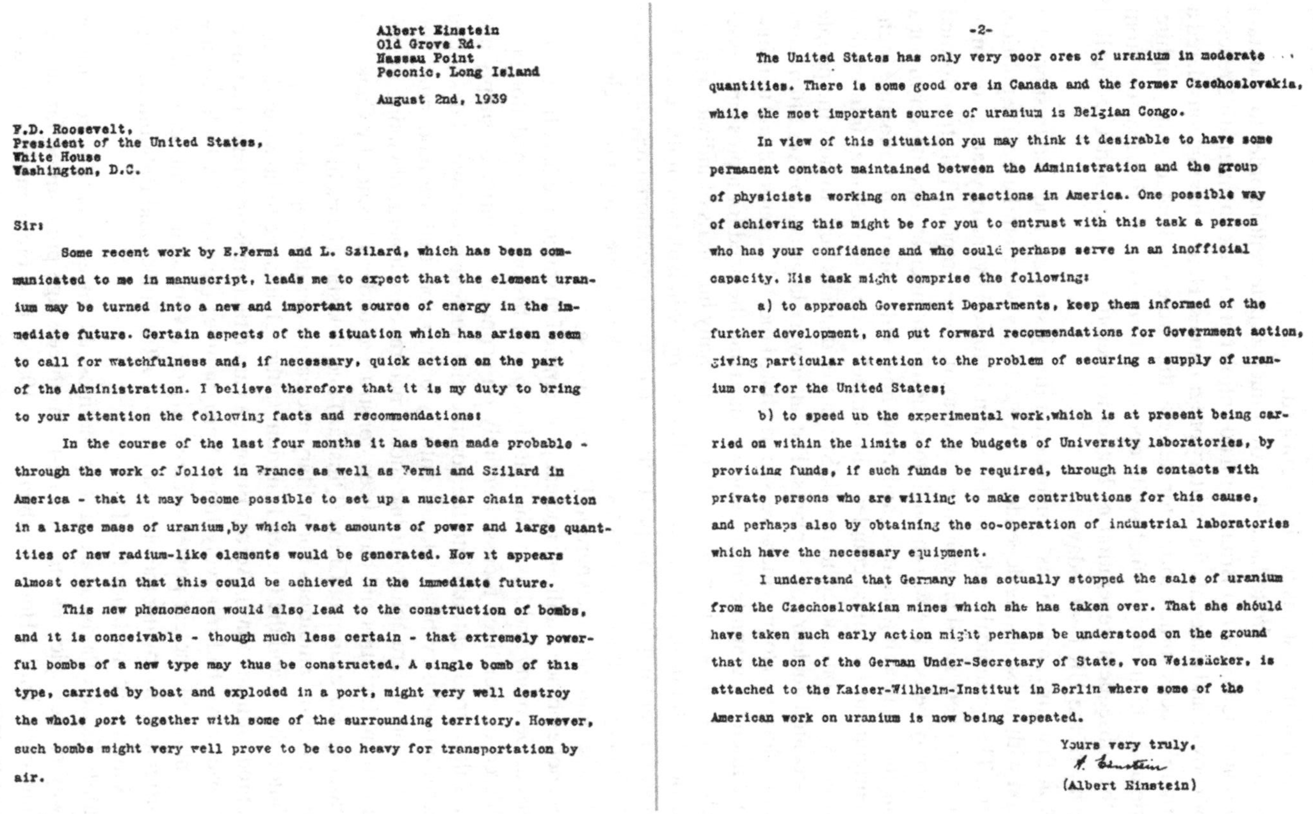

Albert Einstein
Old Grove Rd.
Nassau Point
Peconic, Long Island

August 2nd, 1939

F.D. Roosevelt,
President of the United States,
White House
Washington, D.C.

Sir:

Some recent work by E.Fermi and L. Szilard, which has been communicated to me in manuscript, leads me to expect that the element uranium may be turned into a new and important source of energy in the immediate future. Certain aspects of the situation which has arisen seem to call for watchfulness and, if necessary, quick action on the part of the Administration. I believe therefore that it is my duty to bring to your attention the following facts and recommendations:

In the course of the last four months it has been made probable - through the work of Joliot in France as well as Fermi and Szilard in America - that it may become possible to set up a nuclear chain reaction in a large mass of uranium,by which vast amounts of power and large quantities of new radium-like elements would be generated. Now it appears almost certain that this could be achieved in the immediate future.

This new phenomenon would also lead to the construction of bombs, and it is conceivable - though much less certain - that extremely powerful bombs of a new type may thus be constructed. A single bomb of this type, carried by boat and exploded in a port, might very well destroy the whole port together with some of the surrounding territory. However, such bombs might very well prove to be too heavy for transportation by air.

-2-

The United States has only very poor ores of uranium in moderate quantities. There is some good ore in Canada and the former Czechoslovakia, while the most important source of uranium is Belgian Congo.

In view of this situation you may think it desirable to have some permanent contact maintained between the Administration and the group of physicists working on chain reactions in America. One possible way of achieving this might be for you to entrust with this task a person who has your confidence and who could perhaps serve in an inofficial capacity. His task might comprise the following:

a) to approach Government Departments, keep them informed of the further development, and put forward recommendations for Government action, giving particular attention to the problem of securing a supply of uranium ore for the United States;

b) to speed up the experimental work,which is at present being carried on within the limits of the budgets of University laboratories, by providing funds, if such funds be required, through his contacts with private persons who are willing to make contributions for this cause, and perhaps also by obtaining the co-operation of industrial laboratories which have the necessary equipment.

I understand that Germany has actually stopped the sale of uranium from the Czechoslovakian mines which she has taken over. That she should have taken such early action might perhaps be understood on the ground that the son of the German Under-Secretary of State, von Weizsäcker, is attached to the Kaiser-Wilhelm-Institut in Berlin where some of the American work on uranium is now being repeated.

Yours very truly,
A. Einstein
(Albert Einstein)

FIG. 2.4

A copy of the letter sent by Albert Einstein to President Roosevelt in 1939 ('Einstein-Roosevelt-letter'. Licensed under Public Domain via Commons—https://commons.wikimedia.org/wiki/File:Einstein-Roosevelt-letter.png#/media/File:Einstein-Roosevelt-letter.png).

comprised primarily PWRs, but was joined a little later in the 1950s by boiling water reactor (BWR) systems, when the stability of boiling systems had been grasped.

In the United Kingdom, a political desire to have their own nuclear deterrent in order to maintain parity with the United States and the USSR left Britain with no option but to seek an alternative reactor design that would yield both power and materials for military programmes. Hence, having considered a range of reactor types, Britain's focus turned to reactor designs that were moderated with graphite, cooled with carbon dioxide and which could refuel while operating (*on-load*). British reactors from this era would influence the subsequent generation of higher-temperature, gas-cooled UO_2-fuelled reactors that are the mainstay of the UK fleet today.

In what was the USSR at that time, ambitions to maintain military strength and also to support a vast, modernising population with an escalating dependence on electricity were similar but the political context completely different. The tyrannical leadership by Stalin followed by many years of communism pushed the USSR through a very rapid programme of economic development. This transformed what was a largely rural economy before the war through to a significant nuclear superpower in but a few years. However, lacking the engineering facilities and know-how with which to manufacture high-grade pressure vessels or enrichment facilities, coupled with limited access to technical information from the West, 'Atom towns' were constructed in the USSR; entire districts built to host the people working at the associated co-located nuclear power plants. The latter were large complexes often comprised of several large reactors designed to provide access to irradiated fuel while on-load and to utilise uranium fuel of natural enrichment. A graphite moderator was used and the need for a pressure vessel was avoided by routing light-water coolant throughout the core via many hundreds of individual pressure tubes. Such characteristic reactor designs were accomplished largely in isolation of the rest of the world. This brought with it significant implications, particularly in terms of operational response characteristics and stability. The legacy of this continues today and is evident in the design of nuclear facilities influenced by these developments elsewhere in Eastern Europe.

For those nations that did not embrace a military nuclear programme but sought the benefits of civil nuclear power such as Canada, Germany and Japan, their nuclear power history developed along perhaps what might be considered clearer lines focussed entirely on economy, security of supply and safety. For example, as a result, Canada entered into the commercial nuclear power world designing and building a power generation industry based almost entirely on reactors that are moderated and cooled by heavy water. These reactors are able to utilise fuel of natural enrichment, obviating the requirement for uranium-enrichment facilities and have been constructed under a licence by Canada in several other countries including, for example, China (Qinshan) and Pakistan (Karachi-I).

Japan embarked on one of the most ambitious nonmilitary nuclear power developments, importing reactor designs from the United States and the United Kingdom (although it is the light-water designs from the United States that predominate today). Across Europe, a varied picture is apparent in terms of reactor designs. Some countries have had relatively limited programmes (several of which are now defunct), while others have embraced nuclear power strategically on a very significant scale via extensive fleets of reactors particularly following the oil crisis in the 1970s; France and Belgium perhaps being the most prominent examples of the latter trend.

In the developing world, India and Pakistan are among the most advanced countries in terms of nuclear power. India, by way of example, adopted a long-term, three-stage programme formulated by the preeminent Indian nuclear scientist Homi Bhabha in the 1950s. This programme is based on the realisation that India's indigenous thorium reserves were some 10 times that of uranium. Thus, it was thorium rather than uranium that offered the potential to serve the long-term ambitions of India in terms of its burgeoning economy and security of supply, given India's large population and rapid pace of progress towards developed status. To utilise thorium however requires breeding and refinement of ^{233}U from ^{232}Th. This requires several independent reactor stages as the ^{233}U is not present naturally in contrast with ^{235}U; hence the latter is needed to drive the production of the former, as discussed in Chapter 11.

2.9 THE FIRST REACTOR: CHICAGO PILE 1

Perhaps the most prominent example where military expediency and benefit to the wider civil power programmes came together is the development of the Chicago Pile 1 (CP-1) reactor. This was developed by Enrico Fermi and colleagues at the University of Chicago in 1942, and was the first nuclear reactor to achieve criticality and demonstrate a sustained chain reaction. As part of the Manhattan project, the focus of this activity was to determine if a controlled, self-sustaining nuclear chain reaction could be achieved, and hence to explore the properties of such a reaction in terms of reactor design, control and safety.

The choice of materials for the CP-1 reactor was of key significance. For the moderator (discussed in detail in Chapter 6), graphite was selected due to its chemical stability, low atomic mass, material rigidity and low neutron absorption cross section. Of critical importance for graphite-moderated reactors that followed, it was realised that the impurities in ordinary graphite, especially boron, would have a detrimental effect on the potential success of the reactor. Hence, approximately 360 tonnes of refined graphite in the form of bricks that were free from boron was used. For the fuel, a combination of pure uranium metal (~6 tonnes) and some ~30 tonnes of uranium dioxide was used. Uranium metal had only recently become available at the time that the reactor was constructed via recently developed manufacturing processes. The fuel was formed into short rods with spherical-shaped ends, each referred to as a nodule, that were inserted into the bricks of two of every three of the graphite layers of the CP-1 pile.

The CP-1 reactor was designed adopting a spherical form in order to maximise the multiplication of neutrons for a given physical quantity of uranium and to minimise neutron leakage. Control rods made from cadmium (due to its ability to absorb thermal neutrons very significantly) were configured to enter the reactor horizontally from the side, under electrical control. Despite the 'pile' as Fermi coined it (due to it representing a heap of graphite bricks) being the first of its kind, several primitive but effective safety measures were put in place. An emergency control rod (now referred to as a *scram rod* more widely) was withdrawn and fixed in its withdrawn position by a rope. This was to be lowered into the reactor in the event of an emergency. To cater for the possibility that the operator responsible for this might be incapacitated for some reason, another operator was armed with an axe to cut the rope. A cadmium salt solution was also available with which to flood the pile to halt the chain reaction, again based on neutron absorption. An arrangement of boron trifluoride neutron detectors were integrated into the assembly to monitor the flux level of neutrons to be used as an indicator of when the chain reaction became self-sustaining.

The pile was assembled under the stands of a sports field in a space formerly used as a squash court at the University of Chicago. Construction took several weeks of careful preparation and eventually, on the afternoon of 2nd Dec. 1942, Fermi and his team observed the characteristic escalation in neutron flux characteristic of a self-sustaining chain reaction when the control rods were gradually withdrawn,

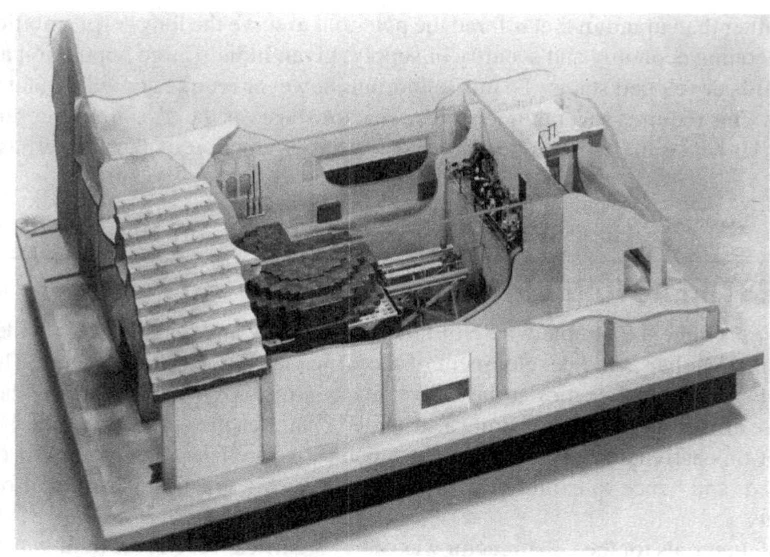

SCALE MODEL OF THE FIRST NUCLEAR REACTOR, DECEMBER 2, 1942
CHICAGO PILE NO. 1
THE UNIVERSITY OF CHICAGO

FIG. 2.5

A photograph of a scale model of the CP-1 reactor.

Reproduced with permission, Special Collections Research Center, University of Chicago Library.

little by little. The first controlled critical reaction lasted for just 4.5 min generating a mere 0.5 W. Interestingly, because of its low power in contrast with the majority of reactors that followed, CP-1 did not need any shielding or coolant. A photograph of a scale model of CP-1 is reproduced in Fig. 2.5.

2.10 ADVANCED REACTORS AND ALTERNATIVES TO ^{235}U

A further impact of the significant investment in time and effort that was made in the period 1945 onwards is that many alternatives to the uranium fuel cycle were considered. The evidence of this is present in the form of relics from this era in many countries that considered nuclear power as an option, even among those that have reduced or stopped their involvement since, and also among the many associated reports and scientific papers. Such relics include disused experimental reactors, exotic wastes and moribund reprocessing facilities.

One motivation for these developments was the anticipation that, given continued growth in nuclear power generation at the rate witnessed in the 1950s, naturally abundant reserves of ^{235}U would be exhausted, possibly as early as by the end of the 20th century. This stimulated significant interest in the production of the fissile plutonium isotope ^{239}Pu from ^{238}U, primarily because of the vast resource that ^{238}U constitutes relative to ^{235}U. Fast-breeder reactor systems designed to produce both power and ^{239}Pu were constructed and operated for many years in several countries including the United States,

United Kingdom and the former Soviet Union. At the same time, scientific interest also peaked in the context of nuclear fusion for similar reasons because it promised the benefits of nuclear power with an unlimited fuel supply drawing on deuterium and tritium as fuels; experimental fusion systems were developed in several countries across the world and continue to be the focus of this objective today. Furthermore, an interest in the use of fission reactor technology for propulsion evolved with several countries developing nuclear-powered submarines, aircraft carriers, ice breakers and freighter vessels. For example, the Otto Hahn nuclear-powered ship operated for many years based on a small PWR before being decommissioned. Even prototype nuclear-powered aircraft engines were developed but never commercialised.

As a consequence of these developments and investment, the mainstream low-enriched fission-based ^{235}U fuel cycle that constitutes the backbone of nuclear power generation today is complemented by extensive experience and knowledge of allied applications of nuclear engineering systems. These include the use of ^{239}Pu and ^{233}U (thorium) as fuels, fast and thermal breeder systems, molten salt reactor designs, fusion reactors and small, modular reactor designs.

There are several reasons that these alternatives to the ^{235}U-based fission approach have not found a mainstream use for power production. First, the forecast shortage of uranium did not materialise partly because natural gas became very competitive in terms of cost and offset the requirement for nuclear fuel. This effectively removed the urgency associated with alternatives to uranium and halted several fast breeder programmes. Second, the accident at Three Mile Island in 1979 and also those at Windscale, Chernobyl and more recently Fukushima have highlighted the potential for accidents at nuclear fission plants. In some cases, this has halted developments and delayed expansion plans for new nuclear power stations. Although extremely unlikely in terms of the amount of power produced by nuclear power plant, these incidents highlighted the need for much more stringent regulation to be in place for the construction, operation and decommissioning of nuclear plants. Also, it was clear that although statistically unlikely, such accidents were extremely expensive and could have long-lasting, often unanticipated consequences.

As a result, nuclear power was viewed more conservatively as a technology, centred on tried-and-tested systems and operating principles. Some reactor builds became expensive where the gap between investment and return lengthened due to project overruns and inflation rates escalated in between, becoming much less attractive for financiers: nuclear energy contrasts significantly with fossil fuel-based electricity production. Fossil fuel plants are relatively unsophisticated and cheap to construct and regulate, with the chief cost often being their fuel requirements. Furthermore, their operation can be adapted to exploit expected opportunities in the market values of fuel stocks. Nuclear plants, by contrast can present a significant, upfront construction cost and can take a long time to build. This can result in significant investment uncertainty, particularly when there are unforeseen delays in completion during a time when there are, by definition, no generation revenues. The cost of nuclear fuel in contrast is relatively insignificant (very little is used relative to fossil fuel alternatives) and reactors usually operate for a long time resulting in extensive, postconstruction revenue-generating periods.

2.11 REACTOR CLASSIFICATION BY GENERATION

Relatively recently, stimulated by the adoption of the classification 'Generation IV' to simplify reference to advanced reactor concepts that are under development and anticipated for adoption c.2030–40, a classification has been developed for all commercial nuclear power reactors currently in use and those

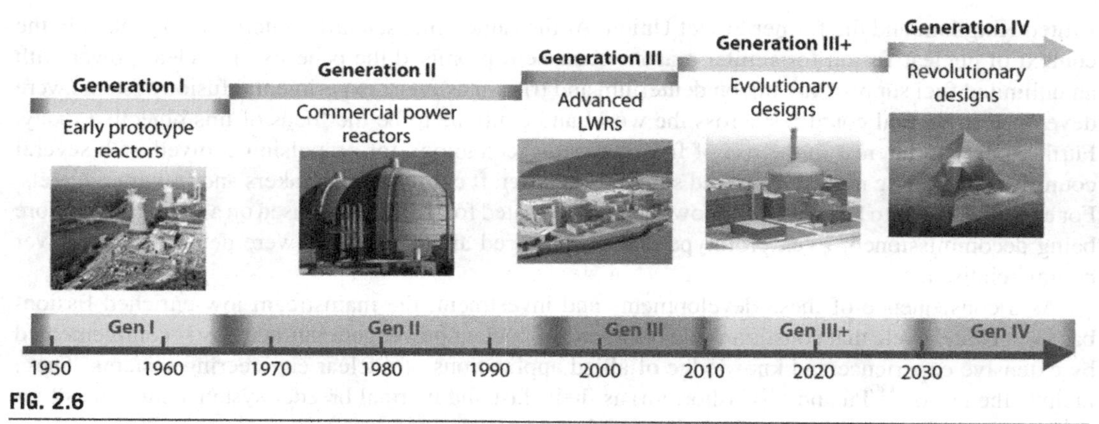

FIG. 2.6

The generations of nuclear power reactors; the time ranges indicated correspond to the first deployment of each generation ('Technology roadmap update for Generation IV nuclear energy systems', Jan. 2014, OECD Nuclear Energy Agency, https://www.gen-4.org/gif/jcms/c_60729/technology-roadmap-update-2013).

Reproduced from https://www.gen-4.org/.

that are now defunct. Future developments classified under *Generation IV* will be summarised in detail in Chapter 11. The classification for existing designs is as follows (Fig. 2.6):

Generation I: This generation of reactors refers to the earliest, prototype power reactors to be built and operated. It includes: the British gas-cooled 'Magnox' reactor type (discussed further in Chapter 10); the similar French 'Uranium Naturel Graphite Gaz' (UNGG) design; the first Soviet Reaktor Bolshoy Moshchnosti Kanalnyy (RBMK) graphite-moderated water-cooled designs; the American Shippingport PWR; type-1 BWR designs (the first being Dresden 1, Illinois, US, as discussed further in Chapter 10); and early prototype fast breeders (such as the liquid sodium cooled design of Fermi 1, Michigan, US).

The common features of this generation are that they all stem from the same era, having been designed and commissioned in the late 1950s and early 1960s, and the power output is usually small by the standards of the reactor generations that followed (later Magnox and the RBMK-1000 being exceptions to this trend). This generation of reactors is now reaching the end of its functional life with most plants having ceased operations due to either having been superseded in terms of economic viability, changes in a nation's energy policy or not meeting current regulatory requirements (e.g. as a result of operational wear and tear and often because renovation is not a practicable option). By way of example, the last and largest Magnox reactor at Wylfa in Wales in the United Kingdom closed down in Dec. 2015. Among the first reactors to produce electricity for civilian use of this generation, there is a smaller precursor to the RBMK design constructed near Moscow in the former Soviet Union—Obninsk 1—the first reactor in the world to produce commercially available electricity in 1954, the Calder Hall Magnox station in Britain (comprising four reactors), which was commissioned in 1956 and the Shippingport PWR in the United States that followed in 1957.

Generation II: This generation constitutes the majority of reactors operating in the World today. It encompasses both the PWR and BWR designs (developed and in widespread use both in the United States and France), the CANada Deuterium Uranium (CANDU) reactors, Advanced Gas-cooled Reactors (UK) and the Water-Water Energetic Reactor (VVER) pressurised water design (developed and in use in the former Soviet Union). The transition from Generation I to Generation II is typified in general terms by Generation II reactors being of larger power generation capacity, utilising oxide-based fuels capable of higher burn-up, thus giving more sustained operation free from the operational constraints associated with military requirements. Almost all Generation II reactor designs dispensed with the use of fuel of natural enrichment opting instead for low-enriched fuel typically 3–5% wt. ^{235}U instead. In contrast with Generation I, which in a minority of cases had utilised rather exotic fuel types including elevated enrichments and isotopic mixtures, the extent of the use of different fuel types in Generation II tends to have been limited to mixed-oxide fuels, with these plutonium-containing fuels used in some plants, particularly in France and Japan.

Generation III: This generation of reactor designs are advanced LWRs conceived as the natural follow-on from Generation II, utilising advancements in operational life expectancy and reductions in anticipated core damage frequency. These developments were inspired by research building on lessons learned from operational experience in the intervening years with Generation II systems. Generation III reactors are also linked by having been licenced in the 1990s era. However, the economic competitiveness of other sources of electricity (particularly gas) and changes in energy policy at that time (particularly in the United States and parts of Europe) has led to few of this reactor generation having been built thus far relative to earlier generations. Examples of those that have been constructed include the Sizewell B PWR in the United Kingdom, and the advanced boiling water reactors (ABWR) in Japan (Kashiwazaki and Ohma).

As the world's approach to generating base load electricity has developed, associated with strategies to decarbonise supplies and to cater for the rapid industrialisation of developing countries, an advancement of this generation of reactor designs has been evolved. This subgeneration is often referred to as Generation III+ although the distinction between this and Generation III is a little arbitrary. Reactor designs under this classification have been improved relative to Generation III in terms of, for example, resilience to aircraft strike (arising from concerns following the September 11 attacks); reduced cost (to achieve greater economic competitiveness particularly in terms of upfront construction cost); modular construction (to achieve better correspondence with investment forecasts and compliance with construction schedules); and both passive and active safety developments (e.g. to reduce calculated probabilities of severe accidents resulting in damage to the reactor core and to be consistent with regulatory requirements in the modern era).

Generation III+ reactors include the EPR, which is a developmental combination of the prior N4 (France) and Konvoi (Germany) designs. At the time of writing, several EPRs are under construction in China and applications are in progress for construction in the United States and in Great Britain. Perhaps the most high-profile developments to date are the EPRs currently under construction at Flamanville (France) and Olkiluoto (Finland). Other Generation III+ designs include the AP1000 (a Westinghouse PWR design with a focus on modular construction and passive safety currently under construction at Haiyang in China); an advancement of the ABWR referred to as the economic simplified boiling water reactor (ESBWR) and the advanced CANDU referred to as the ACR-1000.

REVISION GUIDE

On completion of this chapter you should:

- understand the physical requirements that were necessary for the formation and operation of natural nuclear reactors

- have learned of the various uses of uranium and its related minerals that preceded the discovery of fission and drove early mining and exploration activities
- know about the pioneering scientific discoveries, the approximate chronology by which they occurred and the people who made them, leading to the discovery of fission and the development of the first artificial nuclear reactor
- appreciate the influence of World War II on the development and peaceful use of nuclear power and also the distinction between systems of military origin and those designed and constructed purely for electricity production
- understand the context and the operation of the first operating nuclear reactor: Fermi's CP-1
- appreciate the distinction of advanced reactors and alternative fuels, ahead of discussion in Chapter 11
- understand the classification system used for current operating reactors, ahead of the discussion of Generation IV designs in Chapter 11

PROBLEMS

(1) List the four conditions that are considered to have been necessary to enable natural nuclear reactors to operate.

(2) Name three early uses of uranium that required it to be extracted from the ground before nuclear fission was discovered. Why have most such uses now declined?

(3) Both the former Soviet Union and Great Britain sought fission-based power reactor designs after the Second World War that were compatible with uranium of natural enrichment and that facilitated access to the nuclear fuel while the reactor was operating. Explain why this was required and why, in general, such requirements are not desirable for current reactor designs.

(4) In Hahn and Strassman's experiments that led to the discovery of nuclear fission, what would they have observed as an outcome to their fractional crystallisation procedures if the radium hypothesis had been correct and fission was not the explanation?

(5) What was the original justification for breeder reactors and why has their use waned significantly on a global scale? Summarise the long-term possibilities for fast reactor applications.

REFERENCES

[1] F. Gauthier-Lafaye, P. Holliger, P.-L. Blanc, Natural fission reactors in the Franceville basin, Gabon: a review of the conditions and results of a "critical event" in a geologic system, Geochim. Cosmochim. Acta 60 (23) (1996) 4831–4852.

[2] C. Degueldre, C. Fiorina, The proto-Earth geo-reactor: reassessing the hypotheses, Sol. Ear. Sci. 1 (2) (2016) 49–63.

[3] G.K. Gillmore, P.S. Phillips, G. Pearce, A. Denman, Two abandoned metalliferous mines in Devon and Cornwall UK: radon hazards and geology, in: International Radon Symposium, 2001, pp. 94–105.

[4] O. Hahn, F. Strassman, Über den Nachweis und das Verhalten der bei der Bestrahlung des Urans mittels Neutronenentstehenden Erdalkalimetalle, Die Naturwissen. 27 (1939) 11–15.

CHAPTER

FUNDAMENTALS OF RADIOACTIVITY

3

3.1 SUMMARY OF CHAPTER AND LEARNING OBJECTIVES

In this chapter the fundamentals of radioactivity are introduced as a basis for the subsequent, more specialised nuclear engineering chapters that follow. Radioactivity is the salient feature that separates nuclear engineering from the more general engineering disciplines. Its discovery a little more than 100 years ago was the stimulus for all beneficial applications of nuclear technology that have arisen since. However, it also distinguishes why nuclear systems need to be managed carefully along with the utmost respect for safety, health and the environment. In this chapter the radioactive decay law and its relationship with the decay constant are introduced; the various modes of radioactive decay are considered with particular reference to those of interest to nuclear energy; the characteristic radiations from these decays are discussed along with their interaction properties and secondary radiations; the relevance of the various radiation types is highlighted in terms of their implications for shielding and stopping power.

The objectives of this chapter are to:

- introduce the fundamental concepts of *radioactive decay* and its relationship with the amount of substance
- discuss the *radioactive decay law* and the definitions of the *decay constant*, *mean lifetime* and *half-life*
- explain the development of the decay law to cater for several isotopes constituting a *decay chain* and thus the concept of equilibrium between radioactive isotopes that decay simultaneously
- introduce the variety of radiation types of relevance to the reader in the context of nuclear energy
- discuss the specific radioactive decay phenomena of photon emission, α decay, β decay and internal conversion
- compare and contrast the interaction properties of *photons*, *heavy charged particles*, *electrons* and *neutrons*, including the implications of penetration depth and the density of ionisation for shielding requirements and radiation exposure
- highlight the possibility of secondary radiations such as *bremsstrahlung* and *annihilation photons*

Nuclear Engineering. https://doi.org/10.1016/B978-0-08-100962-8.00003-2

3.2 HISTORICAL CONTEXT: MARIE CURIE 1867–1934

Marie Curie was inspired to study uranium for her doctoral thesis by the earlier discoveries of X-rays and the observation of radiation emanating from uranium (Fig. 3.1). Her husband Pierre had invented the electrometer some years earlier and she had pioneered separations procedures by which it would become possible to extract new, radioactive elements. Together they observed that the intensity of the radiation emitted by radioactive substances is proportional to the amount of substance emitting it and thus is associated with processes in individual atoms, rather than being a larger-scale phenomenon. For this she received the Nobel Prize in Physics with Pierre and Henri Bequerel in 1903. Subsequently, she noticed that uranium compounds emitted more radiation than the pure, equivalent quantity of uranium and hypothesised that this was due to the presence of a much more radioactive constituent. She thus proved that thorium was radioactive and also discovered the elements radium and polonium. For this she received the Nobel Prize in Chemistry in 1911. She was also instrumental in setting up the first mobile medical X-ray units used in the First World War.

Marie Curie was the first woman to win a Nobel Prize; the first person and only woman to win two such prizes in different fields. She invented the term *radioactivity* and lends her name to a standard unit of radioactivity, the Curie, that is still used widely today. Perhaps her most significant legacy is the discovery of radium; this provided the generations of scientists in future with a means by which the structure of the atom could be probed leading to, for example, the discovery of the atomic nucleus. Indeed, she once expressed her belief that 'the way was now open for a truly great advance' [1].

3.3 INTRODUCTION

Radioactive decay is a stochastic process: every nucleus of an unstable isotope has a finite probability that it will decay at some point in the future. However, it is not possible to specify exactly when in time a particular nucleus will decay. Fortunately, atoms are so small that even minute quantities of matter

FIG. 3.1

Madame Curie.

comprise many billions of nuclei, thus constituting an extensive statistical group on which to base a stochastic analysis. Assuming that such a quantity is comprised of the same isotope, all the nuclei present will have the same propensity for decay as one another. Thus, while the destiny of an individual nucleus is impossible to predict, a large statistical sample makes it possible to draw very accurate conclusions about the likelihood of radioactive decay for all but the longest-lived possibilities. As a consequence, it has been possible to measure the probability of radioactive decay for most isotopes to a significant degree of accuracy. This is often inferred from measurements of the radiations that are emitted by the substance as a result of radioactive decay. Over the past hundred years or so, extensive research has provided a detailed picture of the properties of most of the isotopes of relevance to nuclear energy.

3.4 THE RADIOACTIVE DECAY LAW

Returning to the concept of a quantity of substance comprising many individual nuclei of the same isotope (as a mildly hypothetical example), we might consider 1 μg of the isotope cobalt-60 which would comprise approximately 10^{16} ^{60}Co nuclei. If the mass and thus the number of ^{60}Co nuclei present in such a sample were doubled, clearly each individual ^{60}Co nucleus would retain the same probability of decay as before because nothing has been changed about the individual properties of a given nucleus. However, since there are now twice as many of them, the rate at which this particular amount of substance decays will also double. This is directly analogous to the effect on the death rate of doubling a population, assuming the demographic, age distribution and so forth remain unchanged, since we have twice the sample size of individuals with each of them subject to the same probability of demise.

This concept can be generalised to state that the rate of decay[1] with time t of a given sample of number N nuclei of the same unstable nuclide is proportional to the number of nuclei present, thus,

$$-\frac{dN}{dt} \propto N \tag{3.1}$$

The relationship symbolised by Eq. (3.1) is important because it states mathematically that the rate of radioactive decay in a substance is proportional to the number of unstable nuclei in that substance; hence the greater the quantity of a given radioactive substance, the greater the rate of decay, as per Marie Curie's discovery. Since radioactive decay is often associated with the emission of radiation that can constitute a risk to health and the environment, this implies that the greater the amount of substance then the greater the risk (notwithstanding the influence of the specific properties of the isotope in question and of the radiation that is emitted). Thus, in terms of radioactivity the quantity of substance matters.

To obtain the expression for the number of unstable nuclei, N, remaining at some time, t in the future, Eq. (3.1) is re-arranged and integrated to yield Eq. (3.2) (see problem 3.1). The following limits are chosen for the integration: time $t=0$, corresponding the initial point when the number of unstable nuclei present is N_0, and time t at some point afterwards when the number is N,

$$N(t) = N_0 e^{-\lambda t} \tag{3.2}$$

[1]Since the term *decay* results in a reduction in the original quantity of N, the rate of decay dN/dt is by definition negative.

The constant of proportionality λ follows from Eq. (3.1). This equates the rate of decay to the number of unstable nuclei present as per Eq. (3.3) and is known as the *decay constant*,

$$\frac{dN}{dt} = -\lambda N \tag{3.3}$$

Further, the *activity*, A, is the product of the decay constant and the number of unstable nuclei in a given substance,

$$A = \lambda N \tag{3.4}$$

and thus the decay constant is equal to the activity divided by the number of unstable nuclei present; λ is a constant for a given mode of decay of a given isotope and is measured in units of reciprocal time, s^{-1}. The *lifetime* τ of a given isotope can be defined,

$$\tau = \frac{1}{\lambda} \tag{3.5}$$

which represents the mean time that an unstable species exists; at $t = \tau$ then N_0/e of the nuclei (or ~37% of the substance) remain (see problem 3.2).

Historically, the point in time at which *half* the amount of radioactive substance has decayed away is usually considered. This has been adopted universally to characterise the lifetimes of radioactive species. The time at which this occurs is known as the *half-life*, $t_{1/2}$, where

$$t_{1/2} = \frac{\log_e 2}{\lambda} \tag{3.6}$$

follows from a re-arrangement of Eq. (3.2) for the case that $N(t = t_{1/2}) = N_0/2$ (see problem 3.1). The use of either half-life or mean lifetime is advantageous because, even for very long-lived isotopes, a measurement of the change in decay rate with time can usually be made (e.g. via measuring the decline in the rate at which the emitted radiation is detected). This can invariably be done over time periods that are short relative to the lifetime of the nuclide, removing the need to measure over impractical periods of time. It is usually possible to extrapolate from these data, according to the exponential relationship of Eq. (3.2), and to extract $t_{1/2}$, or alternatively τ if preferred, and hence λ. For pure samples of relatively short-lived isotopes, the half-life might even be approximated visually from a plot of activity versus time by estimating the time at which the activity falls to half its initial level.

Given that activity, as defined in Eq. (3.4), is related to the number of unstable nuclei, there is often a need for this to be normalised in terms of mass so that we can compare the relative activities of different substances; this is known as the *specific activity* as per Chapter 1. In addition to being able to compare across isotopes with different half-lives, this normalisation is particularly useful when dealing with radioactive waste because wastes are separated and disposed of differently according to, in part, activity per unit mass. The specific activity of a given isotope can be extracted knowing the atomic mass and the half-life as the latter is, by definition, independent of mass.

3.5 MULTIPLE RADIOACTIVE DECAY PROCESSES AND EQUILIBRIUM

Where a radioactive isotope decays to another radioactive isotope, rather than to one that is stable, a cascade is formed. Such a cascade is known as a *decay chain* and can comprise several different radioactive isotopes connected by a variety of different modes of radioactive decay. The range of decay possibilities can exhibit a corresponding range of radiation emissions and a wide variety of half-lives. As introduced in Chapter 1, the isotope that undergoes the process of radioactive decay is usually denoted the *parent* isotope and that which the parent decays to is denoted the *daughter* (or *daughter product*); in a decay chain this specification can be repeated at several interconnected decays.

The simplest example of the situation described above is that which comprises two radioactive isotopes, the decay of which leads to the formation of a stable third such that the decay sequence comprises two consecutive steps. Each step is characterised by a separate decay constant denoted, for example, λ_P and λ_D for parent and daughter, respectively. The number of nuclei of the parent isotope decays at a rate given by Eq. (3.3), that is only a function of its decay constant and the number of parent nuclei present. In contrast, the number of unstable daughter nuclei changes at a rate dependent on the combination of their rate of *formation* (which is equal to the decay rate of the parent) and their rate of *disintegration* (the rate of decay of the daughter). Thus, the rate of change of the number of parent and daughter nuclei with time can be written as, respectively,

$$\frac{dN_P}{dt} = -\lambda_P N_P$$

$$\frac{dN_D}{dt} = \lambda_P N_P - \lambda_D N_D \tag{3.7}$$

The number of parent and daughter nuclei as a function of time follows, respectively, as per Eq. (3.8),

$$N_P(t) = N_P(0)e^{-\lambda_P t}$$

$$N_D(t) = \frac{\lambda_P N_P(0)}{(\lambda_D - \lambda_P)}\left(e^{-\lambda_P t} - e^{-\lambda_D t}\right) + N_D(0)e^{-\lambda_D t} \tag{3.8}$$

where the latter is known as the *Bateman equation* [2] (see Case Study 3.2), and $N_P(0)$ and $N_D(0)$ are the number of parent and daughter nuclei at time zero, respectively.

If a state is reached where the ratio of parent and daughter activities remains constant, they are said to be in *equilibrium*. Two scenarios are possible in this context, assuming we start our consideration at a point in time when none of the daughter exists, as follows:

- If the half-life of the parent is very long relative to that of the daughter isotope, the activity of the former is not observed to change noticeably in comparison with that of the latter. Meanwhile, the activity of the daughter is observed to increase until its rate of production is equal to its rate of disintegration. At this point the parent and daughter are said to be in *secular equilibrium* with one another. This is characterised by their activities being observed to be constant and equal in the period over which they are observed.
- If the half-life of the parent is longer than the daughter, but sufficiently short for decay in the parent to be evident, the activity of the daughter will increase to a maximum. It then declines at the same rate as the parent. At this point, the isotopes are said to be in *transient equilibrium*; their activities are observed to change but the ratio of the two remains constant in the period over which they are observed.

If the half-life of the daughter is long relative to the parent, equilibrium is never established. The parent decays away while the activity of the daughter increases, reaches a maximum and then that decays away too.

3.6 RADIATION TYPES

In nuclear engineering the focus is frequently on radiation arising from changes in the nucleus (nuclear radiation) as was introduced in Chapter 1. It is this that can pose a significant risk to health and the environment if not managed appropriately but which also enables processes and materials to be identified, managed and controlled remotely; often to very exacting degrees of sensitivity.

There is also a frequent emphasis on radiation that has sufficient energy and interaction strength to disrupt the electron shell structure of atoms and molecules, which is classified as *ionising radiation*. Often several different types of radiation are emitted together, such is the complexity and variety of the processes in the nucleus; as Chadwick wrote in 1931: 'It is clear that the nucleus of a heavy atom must be an exceedingly complicated structure...' [3]. In this text a similar convention is adopted and the discussion is focussed on nuclear, ionising radiation except for the case where radiation emissions arise sympathetically with nuclear processes or as a secondary effect of the radiation produced, such as for the case of X-rays arising from bremsstrahlung that is discussed later in this chapter.

In addition to providing information associated with contrasting nuclear processes, ionising radiation interacts with matter differently depending on its energy, mass (or lack of) and electrical charge (or lack of). Consequently, different radiation types deposit their energy in matter differently. To illustrate this fact, the energy deposited by a variety of different radiations (relative dose) versus depth in living tissue is shown in Fig. 3.2 taken from a study of the use of charged-particle radiations for medical applications[2] [4]. This variety of interaction behaviour has important implications for radiation protection particularly with respect to dose, dose measurement and shielding. It also influences the use of nuclear materials in nuclear power engineering applications, particularly in terms of control, fatigue and the lifetime of components.

3.6.1 PHOTON RADIATIONS: X-RAYS AND γ RAYS

X-rays and γ rays are high-energy components of the electromagnetic spectrum. In common with all electromagnetic photons, X-rays and γ rays have zero mass and travel at the speed of light in a vacuum.

X-rays and γ rays can have the same energy and are the same type of radiation exhibiting the same interaction properties with matter. However, they are distinguished from each other on the basis of the difference in their origin: X-rays are emitted as a result of transitions by electrons between energy levels in the structure of the atom. This can be either to establish the lowest-energy arrangement (in the case that an electron has left leaving a vacancy that needs to be filled), due to the decay of

[2]Given this comparison is from a medical perspective, the radiations and the energies involved are not quantitatively relevant to nuclear energy. Rather, they serve to illustrate, qualitatively, the variety of interaction possibilities that exist for different types of radiation.

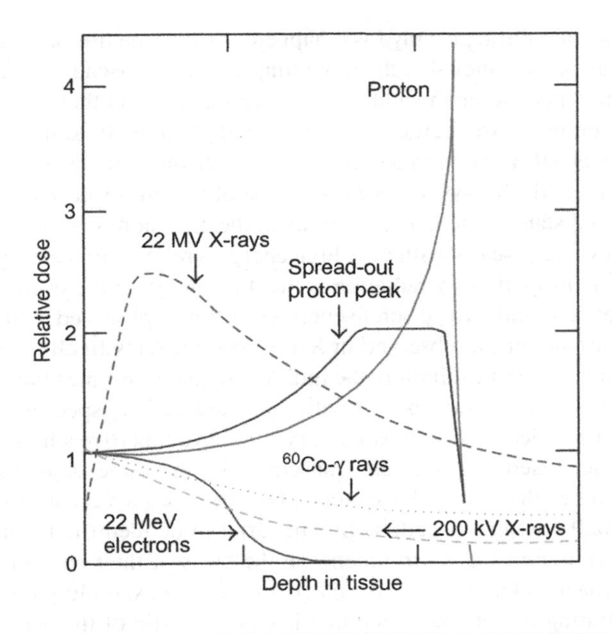

FIG. 3.2

Relative dose as a function of depth in tissue for X-rays, γ rays, electrons, unmodulated and modulated protons (the latter spread in energy by the use of absorbing filters), after Chianchetti and Amichetti [4].

an excited arrangement of the atom, or as *bremsstrahlung* (braking radiation), arising when a charged particle is decelerated by another (this is discussed further in Section 3.6.3). In contrast, γ rays arise from changes in the excited state of the nucleus.

The distinction between X-rays and γ rays might appear somewhat semantic but it is important because the presence of one variant of these radiations, as opposed to the other, indicates that a very different physical phenomenon has occurred. Further, the ways in which each of the radiations arise and are controlled in use can also be different. X-rays arising from the decay of atomic states in a nuclear engineering context tend to (although not always) occupy the lower portion of the photon energy spectrum typical from the decay of radioactive isotopes, extending through to ~150 keV. By contrast, γ rays can extend through to 2 MeV and beyond. There are exceptions to these generalisations with significant overlap in the energy of both types of radiation. For example, high-energy X-rays are produced routinely for therapeutic purposes with medical linear accelerators for cancer treatment while some isotopes, such as ^{241}Am, are characterised by γ-ray emissions of relatively low energy, ~60 keV.

Perhaps the most common source of X-rays in nuclear energy applications (aside from those that exploit particle accelerators from which stray X-ray fields might be evolved) is the variant of a radioactive process that is known as *internal conversion*. This phenomenon occurs occasionally as a variant of γ decay where, instead of emitting a γ ray (as it happens most of the time), an inner orbital electron is ejected instead. This leaves a vacancy that the remaining electrons cascade to occupy, emitting X-rays as they do so. Clearly this phenomenon should not be confused with other types of radioactive decay that might result in the emission of energetic electrons and photons. In terms of the photon spectrum that arises when both *internal conversion* and the *γ-ray decay* of excited states of an unstable nucleus are evident, both characteristic X-rays and γ rays can be observed, respectively.

The very tightly bound state of the nucleus results in the emission of γ radiation as a result of transitions between excited states, usually with very high energies relative to the range of X-rays described above. A further, specific property of γ radiation is that the energy of the transition between quantum states in the nucleus is often manifest as discrete energies or 'lines' observed in the spectrum, analogous to the discrete lines of atomic origin observed in X-ray spectra. As with all quantum mechanical systems, at a sufficiently high state of excitation, the individual quantum states blend into a continuum. In such a state so do the associated γ-ray spectra with the characteristic spectra; the latter observed for low-lying states where the variety of decay paths a nucleus can adopt to reach the ground state becomes limited to only a few, quantised possibilities. Consequently, the same sequence of γ-ray energies is always observed, specific to that particular isotope. This provides a very useful analytical benefit as it enables extremely small amounts of radioactive material to be identified isotopically, for example in process streams, waste and in the environment, on the basis of their characteristic γ-ray emission. Such an analytical approach is known as *γ-ray spectroscopy*. An example γ-ray spectrum for ^{137}Cs is given in Fig. 3.3, illustrating the 662 keV line that is characteristic of this isotope.

The high energy and lack of mass and electrostatic charge of γ-ray photons as entities renders them highly penetrating. They interact principally with the electrons that surround atomic nuclei and which constitute the conduction bands in metals. It is for this reason that γ-ray attenuation exhibits a strong dependence with the atomic number Z of the material in question and a rapidly changing, inverse,

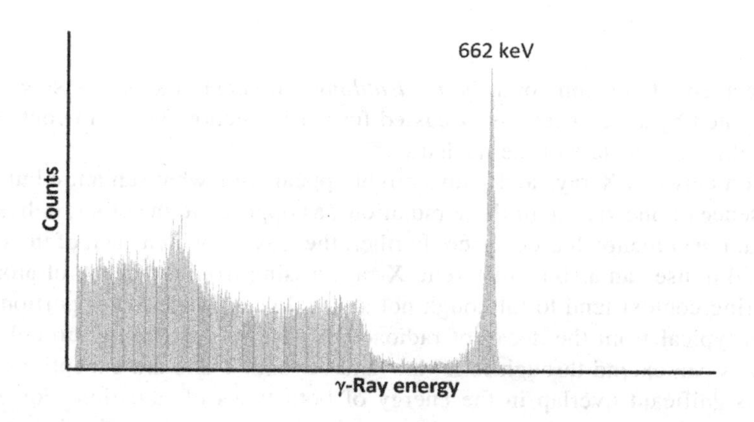

FIG. 3.3

A photon spectrum of counts versus energy for a ^{137}Cs source. Caesium-137 emits one prominent, characteristic γ ray at 662 keV via the decay of its ^{137m}Ba daughter and an associated X-ray due to internal conversion at 32 keV. Note the photopeak arising from the photoelectric effect at 662 keV and the continuum at lower energies to the left of the photopeak arising from Compton scattering of photons in the detector crystal. These data were taken with a cadmium zinc telluride (CZT) scintillation detector and a relatively small, sealed, isotopic source.

dependence with energy particularly at low γ-ray energies, that is $E_\gamma \lesssim 300$keV. γ-ray photons are attenuated most effectively by heavy atoms, such as lead and tungsten, and consequently these materials are often used for shielding and as collimators.

Unlike some types of radiation we shall consider later in this text, photons do not have a specifically defined penetration depth in a given material at a given energy. Rather, we observe the intensity (number of photons) to fall exponentially with the thickness of the material. To quantify this, the specific influence of a given material is incorporated via a parameter which describes the propensity of the material to attenuate the photons; this is known as the *linear attenuation coefficient* μ. Given the influence of energy referred to above, this coefficient is also a function of the γ-ray energy, E_γ, although this is rarely indicated explicitly, such that,

$$I(x) = I_0 e^{-\mu x} \tag{3.9}$$

where $I(x)$ is the intensity after a thickness x of material and I_0 is the intensity without any material present, that is $x = 0$. The linear attenuation coefficient, μ, is relatively simple to determine experimentally; for example, by measuring the intensity of a specific γ ray for a selection of thicknesses of a given material and then deriving μ from a fit to the data. However, because the interaction of photons is dependent on the electron density (which is usually proportional to the bulk density ρ), a version of the coefficient normalised with respect to bulk density is usually provided by convention in data tables.[3] This is known as the *linear mass attenuation coefficient*, μ_m, such that,

$$I(x) = I_0 e^{-\mu_m \rho x} \tag{3.10}$$

This allows for more flexible calculations to be made assuming the thickness and density of the material is known, along with the energy of the photon radiation of interest. It is a particularly useful parameter when selecting the material and scale of radiation shielding, collimation and also in specifying the requirements (i.e. size, composition) of photon radiation detection systems.

Photon radiation interacts via three main processes to an extent dependent on its energy:

- The photon can be absorbed completely, ejecting an orbital electron from an atom in doing so. This process known as the *photoelectric effect*.
- Alternatively, some of the energy of the incident photon can be imparted to eject an electron from an atom. The residual energy takes the form of a photon that is scattered relative to the angle of incidence and of a lower energy than the incident photon. This process is known as *Compton scattering*.
- For photons with energies above 1.022 MeV, they can undergo *pair production*. This is a process in which the incident photon is replaced by an electron–positron pair.

The distinction between the photoelectric effect and Compton scattering is significant because the former is strongly favoured for high atomic numbers and low energies, whereas for the latter the dependence on these properties is much reduced. It is also possible to see the individual components arising from these interaction processes in the corresponding spectra of photons in terms of their energy. For example, the photoelectric effect results in the entire energy of the photon being absorbed thus contributing to a prominent peak in the spectrum (known as the *photopeak*). In contrast for the case of

[3] http://www.nist.gov/pml/data/xraycoef/.

Compton scattering, where only a fraction of the photon energy is absorbed and the scattered photon is lost from the detector, a lower-energy, broad region is manifest in the spectrum that is known as the *Compton continuum*. This is comprised of many fractions of the total energy of a γ ray absorbed as a result of Compton scattering interactions and the partial deposition of the photon energy, as is also shown in Fig. 3.3.

3.6.2 HEAVY CHARGED PARTICLES: α DECAY AND α-PARTICLE RADIATION

There are a variety of decay pathways by which radioactive isotopes can achieve greater stability. One route is via the emission of energy in the form of electromagnetic radiation in terms of the emission of γ rays, as has been described above in the preceding section. Other routes exist in which greater stability is achieved, relative to the prior state of the parent isotope, via the emission of both mass *and* energy, as opposed to the emission of energy alone in the case of photon emission. It can be advantageous in the pursuit of greater stability for heavy isotopes to dispense with excess positive charge, leading to a reduction in the number of protons, to reduce nuclear instability arising from Coulomb repulsion. However, these pathways must be energetically favourable if they are to occur naturally and spontaneously; pathways that do not result in greater binding of the daughter nuclei will not result in greater stability and thus will not proceed without an external stimulus. One such pathway that is often energetically favourable is that of α decay. In this transformation an unstable nucleus ejects a ^{4}He nucleus or, as it is more commonly known, an α particle.

To illustrate this concept, it is commonplace to consider the variety of scenarios that might exist, albeit hypothetically, for a given isotope in terms of the energy consumed or evolved as a result of a nuclear transformation. To do this it is necessary to introduce the concept of *Q value*: this is the amount of energy that is exchanged as a result of a reaction particularly, in this context, a nuclear reaction. If energy is evolved by a reaction then the reaction is said to be *exoergic*, and the Q value is positive, while if energy is absorbed it is *endoergic* and Q is negative. One such example is presented in Table 3.1 below for the decay of ^{238}U. For each feasible reaction the energy (Q value) either expended (positive) or necessary (negative) as a stimulus has been calculated by summing the total nuclear binding energy (as defined in Chapter 1) of the parent and daughter isotopes and then calculating the difference. Since

Table 3.1 A Selection of Hypothetical Decay Scenarios for ^{238}U and Their Q Values

Reaction	Q (MeV)
$^{238}U(^{237}U,n)$	−6.27
$^{238}U(^{237}Pa,\ ^{1}H)$	−7.78
$^{238}U(^{236}Pa,\ ^{2}H)$	−11.16
$^{238}U(^{235}Pa,\ ^{3}H)$	−10.01
$^{238}U(^{234}Th,\ ^{4}He)$	**+4.19**
$^{238}U(^{233}Th,\ ^{5}He)$	−2.87
$^{238}U(^{232}Th,\ ^{6}He)$	−5.83
$^{238}U(^{232}Ac,\ ^{6}Li)$	−6.01

The exoergic case corresponding to α decay is highlighted in bold.

binding energy is manifest as the mass defect between an isotope and its constituent nucleons, the Q value is the difference in mass of the constituents of a reaction before and after it occurs.

Of the decay pathways presented in Table 3.1, only one of those considered for this example results in a positive Q value; that which results in the emission of a ^{4}He nucleus i.e. α decay. The other scenarios would result in a reduction in the binding energy of the products relative to ^{238}U whereas an increase, resulting from a decrease in mass overall reflecting increased binding and therefore greater stability, is necessary for the decay to be energetically favourable. Aside from the relative stability of the transuranic products (the isotopes of thorium, protactinium and actinium), α decay is often favoured because of the significant binding energy of the ^{4}He nucleus (7.07 MeV) over its neighbouring possibilities (cf. ^{3}H: 2.82 MeV and ^{5}He: 5.48 MeV). The α particle constitutes the lowest mass possibility by which the disruptive Coulomb repulsion in heavier nuclei can be abated most effectively, whilst being a very efficient means by which nuclei can shed protons and mass to achieve greater stability.

The Q value represents the energy released in an exoergic reaction, such as spontaneous radioactive decay. This takes the form of the kinetic energy T of the reaction products. For the energetically favoured example above,

$$^{238}\mathrm{U} \rightarrow {}^{234}\mathrm{Th} + \alpha$$

and assuming the parent isotope (^{238}U in this case) is at rest prior to the decay, then the momenta p of the α particle and the ^{234}Th nucleus (p_α and p_{Th}, respectively) must be equal since momentum is conserved. Consequently,

$$p_\alpha = p_{\mathrm{Th}} \tag{3.11}$$

and re-arranging with the expression for kinetic energy in terms of momentum $T = p^2/2m$,

$$Q = p_\alpha^2/2m_\alpha + p_{\mathrm{Th}}^2/2m_{\mathrm{Th}} \tag{3.12}$$

where m_α and m_{Th} are the mass of the α particle and ^{234}Th nucleus, respectively. Since we are usually interested specifically in the properties of α radiation, the kinetic energy of the α particle, T_α is

$$T_\alpha = Q\left(\frac{m_{\mathrm{Th}}}{m_\alpha + m_{\mathrm{Th}}}\right) \tag{3.13}$$

where the substitution for $p_{\mathrm{Th}}^2 = 2m_\alpha T_\alpha$ has been made, further to Eq. (3.11). In this case, and for all actinides that exhibit α decay, $m_{\mathrm{Th}} \gg m_\alpha$ and hence $T_\alpha \sim Q$. Thus, in all such cases, it is the α particle that acquires the majority of the kinetic energy in α decay. For the case of the reaction ^{238}U(^{234}Th, α) applying Eq. (3.11), $T_\alpha > 98\%$ of Q, giving $T_\alpha \sim 4.12$ MeV.

This amount of energy imparted to an α particle, it being a low-mass (relative to the parent isotope), tightly bound projectile, yields it a speed of ~5% the speed of light. α particles are ejected from atomic nuclei with no electron shell structure of their own and thus they constitute highly charged, highly energetic and massive projectiles at the nuclear scale. They belong to a class of radiation types often referred to as *heavy charged particles* to differentiate them from electrons, positrons and photons. As a result, α particles interact very significantly and directly with their surroundings and are highly ionising. Their interaction is comprised of two mechanisms: inelastic scattering with the electrons of the atoms in the material and elastic scattering with the constituent nuclei.

It is instructive to consider the maximum energy lost by a heavy charged particle to an electron ΔE in one interaction, assuming head-one elastic scattering, via the expression,

$$\Delta E \cong \frac{4m_e}{m_\alpha}E \tag{3.14}$$

for nonrelativistic energies where m_e is the mass of the electron, m_α is the mass of the ion (the α particle in this case) and E is the incident energy of the ion (see problem 3.5). The significant difference in mass between an electron and an α particle, that is $m_\alpha \sim 7300\, m_e$, results in very inefficient transfer of energy from the latter to the former at each collision. Thus, the α particle undergoes many interactions losing a little energy each time before losing all of its energy in a substance. Because of its relatively significant mass, the path of the α particle is not perturbed significantly by this cascade of interactions. Consequently, α particles and other heavy-charged particles can be approximated as travelling in straight lines in matter on this basis.

The intensity of the interaction of α particles in matter is usually confined to a relatively small volume, as depicted in Fig. 3.4. Here the interaction of 50,000 α particles, each of 4.12 MeV energy (as per α-particle emissions from ^{238}U estimated above) have been simulated in human skin and aluminium in Fig. 3.4A and B, respectively; note the use of different scales. This illustrates that, despite having relatively high incident energies, the α particles do not penetrate further than ~17.5 μm in the metal and ~25 μm in tissue. Three useful insights can be drawn from this illustration:

- The yield of ionisation from a relatively small number of α particles, each having significant energy relative to that needed to ionise an atom, is deposited in a very small volume.
- Small quantities of aluminium (along with thin layers of most other non-gaseous materials) serve as an effective shield against relatively energetic α particles emitted by radioactive substances.
- α particles are unlikely to penetrate beyond the protective dead layer of human skin (the thinnest epidermis typically of thickness ~50 μm on the eyelids). This highlights that α particles do not pose

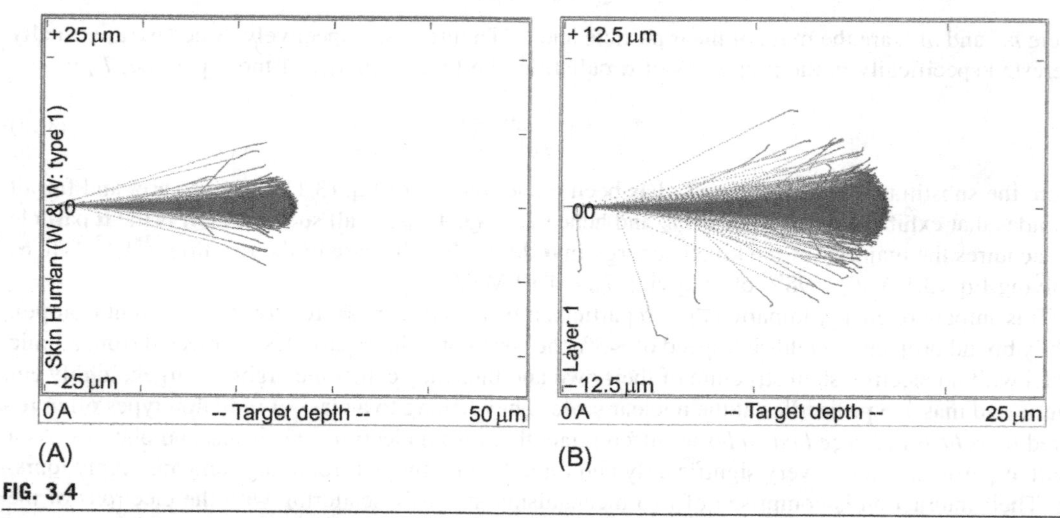

FIG. 3.4

Two simulations performed with the SRIM (Stopping and Range of Ions in Matter) package (www.srim.org) of 50,000 α particles, of energy 4.12 MeV, incident on (A) human skin exhibiting a penetration depth of ~25 μm (left) and (B) on aluminium with a depth ~17.5 μm (right), note the different axes scales.

an external risk of radiation exposure to living tissue. However, they can be a concern if exposure were to occur internally where the epidermis is not present for example as a result of inhalation or ingestion.

The mean rate of energy loss E with distance x (often referred to as the *stopping power*, dE/dx) is a function of several parameters; Bethe [5] derived the following equation to describe it,

$$-\frac{dE}{dx}=\frac{4\pi k_0^2 Z^2 e^4 n}{m_e c^2 \beta^2}\log_e\left[\frac{2m_e c^2\beta^2}{I(1-\beta^2)}-\beta^2\right] \tag{3.15}$$

where $k_0 = 1/4\pi\varepsilon_0$ and ε_0 is the permittivity of free space, Z is the atomic number of the interacting ion, e is the charge on an electron, n is the number of electrons per unit volume in the medium in which the interaction is taking place, β is the ratio of the speed of the ion to that of light and I is the excitation energy. I is a property of the medium and is often measured experimentally via the measurement of stopping powers [6].

Stopping powers can be calculated readily using databases available via the Internet or sourced from data tables. The conceptual significance of Eq. (3.15) is that only the charge of the ion, Ze, and the electron density of the medium, n, are important in terms of the stopping power of heavy, charged particles in matter. The stopping power describes the energy lost per unit distance travelled by an ion in a medium as a result of a sequence of interactions that yield charge from the ionisation of atoms. This can result in the disassociation of molecules that can, in turn, influence the biological function of cells in living tissue. Consequently, stopping power can be very important in terms of assessing radiation dose and the relative biological effectiveness of radiations. In this context dE/dx is closely related to the amount of energy deposited in a material by an energetic ion per unit distance which is known as the *linear energy transfer* (LET).

Perhaps most significantly, the process of heavy charged particles losing a little energy with each collision causes the ion to slow down a little with each collision and thus for it to become slightly more susceptible to further interactions as the process continues, as per the influence of the β parameter. This is reflected in the gradual decrease in β in Eq. (3.15) culminating in a maximum in dE/dx as $\beta \to 0$. This feature in terms of the depth of interaction is known as the *Bragg peak*. At very low energies, energy loss decreases sharply as the ions come to rest. This can also be appreciated from Fig. 3.4A and B as the ionisation comes to an abrupt end at a particular distance in both of the examples of aluminium and tissue. This property has three important implications:

- Heavy-charged particles have a defined maximum penetration depth for a given particle, material and energy.
- This depth can be varied in a given material by varying the energy of the ion, where an increase in energy leads to an increase in depth or range.
- We can be reasonably certain of the depth of penetration for a given particle type, material and energy.

Consequently, shielding and processes involving the stopping of charged particles in matter can be designed with relative precision. This is especially relevant, for example, in medical radiotherapy treatments using protons or heavy ions where the depth of penetration can be a critical benefit of the therapy over other alternatives and can be controlled by adjustment of the energy of the ion.

Returning to α decay itself, the effect on the parent nucleus is that it sheds two neutrons and two protons bringing about an overall gain in binding energy of the daughter product. The daughter is rarely stable and can be vulnerable to further decay, particularly via β-particle and γ-ray emission. Decay via α-particle emission is often part of what can be extensive decay chains involving several isotopes especially for the case of the heavier, actinide series of isotopes. In terms of the $Z-N$ plane of the chart of the nuclides, α decay is characterised by a shift diagonally, down and to the left, where the daughter M relative to the atomic mass A and atomic number Z of the parent is $^{A-4}_{Z-2}M$. This is characterised by the specific cell of the chart of the nuclides reproduced below in Fig. 3.5.

3.6.3 β RADIATION: ENERGETIC ELECTRONS AND POSITRONS

β radiation arises from β decay of which there are two main variants: β^- and β^+ decay. Various terminologies are used to describe these types of radioactive decay including *negative β decay* and *positive β decay*, respectively. The latter is also sometimes referred to as *positron decay*. There is a third variant, *electron capture*, in which an orbital electron is acquired by the nucleus which is effectively the inverse of β^- decay.

In the case of β^- decay a neutron n becomes a proton p with the emission of a β^- particle (e^-) and an antineutrino, $\bar{\nu}$. This is described by the reaction equation,

$$n \rightarrow p + e^- + \bar{\nu} \tag{3.16}$$

whereas for β^+ decay a proton becomes a neutron with the emission of positron e^+ (the antimatter equivalent of an electron) and neutrino ν,

$$p \rightarrow n + e^+ + \nu \tag{3.17}$$

that can only occur within the nucleus with the energy available from the decay transition.

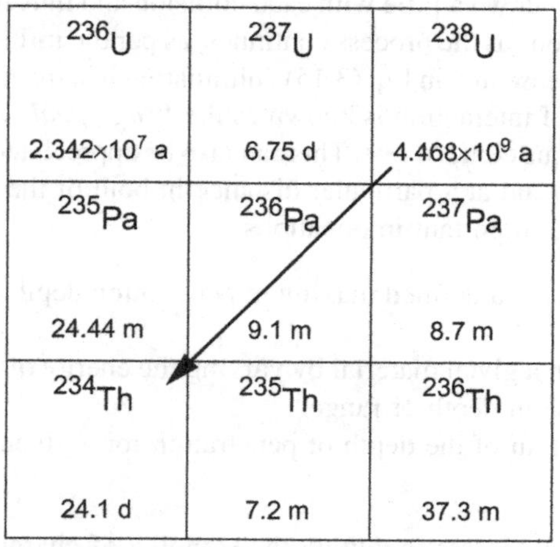

^{236}U 2.342×10^7 a	^{237}U 6.75 d	^{238}U 4.468×10^9 a
^{235}Pa 24.44 m	^{236}Pa 9.1 m	^{237}Pa 8.7 m
^{234}Th 24.1 d	^{235}Th 7.2 m	^{236}Th 37.3 m

FIG. 3.5

A 3 × 3 cell of the Z versus N plane illustrating the effect of α decay on the parent nucleus, in this case for the decay of ^{238}U.

Shortly after the discovery of radioactivity by Henri Bequerel in 1896, the properties of β^- particles were probed (e.g. their deflection in a magnetic field, charge-to-mass ratio, etc.) and they were found to be fast electrons. It is important to emphasise that the electron emitted in β^- decay arises from the nucleus of the parent isotope and has nothing to do with the atomic electron structure of the corresponding element.

An inconsistency in the energy emitted in β^- decay was identified in early investigations of the process which was subsequently associated with a second particle, the antineutrino (in the case of β^- decay). Neutrinos and antineutrinos are highly penetrating, electrically neutral, very low-mass forms of radiation that are difficult to detect. They escaped the detection systems of early β-particle experiments and it would be quite unusual to detect them in industrial applications today, more than a hundred years on. However, they are being explored for some emergent forms of safeguards monitoring of nuclear reactors [7].

This phenomenon is important, however, since the kinetic energy of the products of β^- decay (β^+ decay) is shared amongst the three products of the decay and thus, not knowing the energy of the antineutrino (neutrino), we cannot be certain of the β-particle energy. As a result, β particles are observed with energies spanning a continuum. As for α decay, the energy emitted in the decay arises from the difference in mass of the reaction constituents before and after the decay. However, discrete energies corresponding to this associated with the emitted β particles are *not* observed because the neutrino carries away a share that is not easily measurable in most engineering applications and is thus difficult to account for. The continuum extends up to what is termed the *end-point energy*; this is the point at which the antineutrino energy approaches zero. Consequently the maximum energy a β particle of a given decay can be inferred from the energy at this point. An example β-particle spectrum [8] is given in Fig. 3.6 for the case of ^{14}C.

Isotopes that decay via either β^- decay or β^+ decay are generally either rich in neutrons or protons, respectively, relative to the balance of nucleons associated with stable isotopes in the surrounding mass region. β decay is a route by which unstable isotopes can acquire greater stability as a result of a more balanced configuration of protons and neutrons. Where perhaps α decay is of interest in nuclear energy

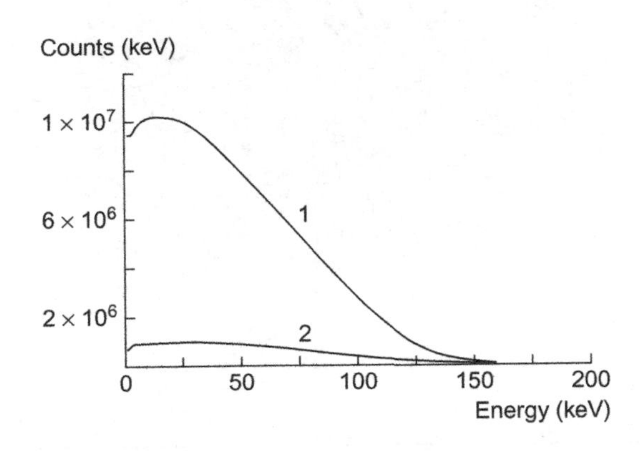

FIG. 3.6

The β-particle spectrum for ^{14}C (1) and the spectrum of random coincidences (2), after Kuzminov et al. [8], indicating an end-point energy in this case of ~150 keV.

because it is often associated with the actinides that have a central role in the fuel used, β decay is of relevance because it often arises as a decay path for the nuclei of the fission fragments. These nuclei, formed as a result of a fission event, tend to be neutron-rich and thus β^- decay is a prominent decay pathway by which this bias is resolved to bring about greater stability and nucleon symmetry. That said, β decay is observed to a greater or lesser extent throughout the isotopic landscape.

While similar in concept, the physical evidence for each variant of β decay is rather different. The β^- particles arising from β^- decay (being high-energy electrons) interact by transferring their energy to the bound electrons in the materials of their surroundings, ionising the associated atoms as they do so. While similar to the interaction of other charged particles, there are two important distinctions associated with the behaviour of electrons: First, they can be travelling at relativistic speeds and, second, they are much less massive, exhibiting mass symmetry with the electrons they interact with. Consequently, they exhibit large deflections with each interaction, describing erratic paths and have much less well-defined penetration depths than their slower, more massive counterparts.

By way of example, a simulation of β^- particles up to the average energy of emission from ^{14}C incident on 0.1 mm of aluminium is shown in Fig. 3.7. This illustrates the greater eccentricity in path relative to heavy-charged particles and the absence of a well-defined penetration depth. It is also noteworthy that, in this case, the electrons are much less energetic than was illustrated for ^{238}U α particles. Conversely, the penetration depth is of the same order of magnitude, highlighting the greater penetration capabilities of β particles over α particles.

The ionisation arising from the interactions of electrons is manifest as significant quantities of free charge carriers that can be detected directly as an electrical pulse or current. This might be achieved, for

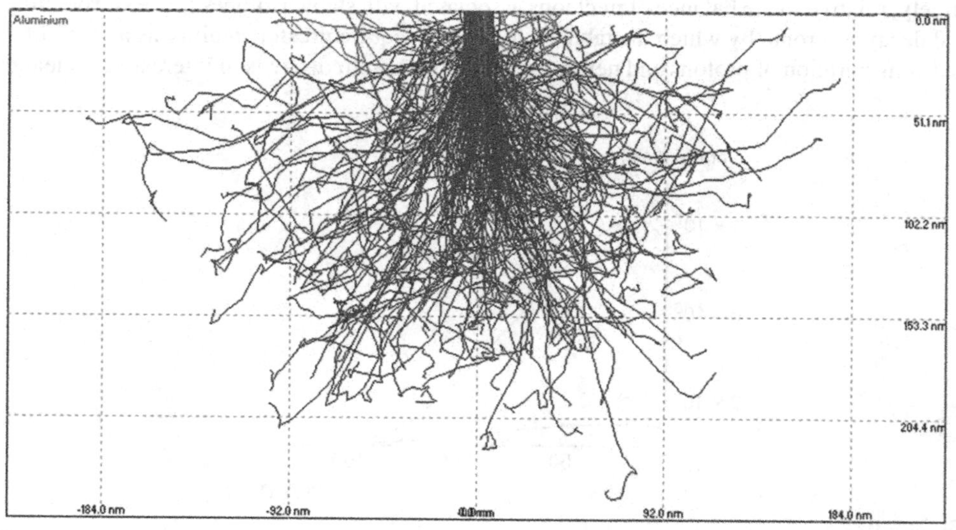

FIG. 3.7

A two-dimensional Monte-Carlo simulation of 100,000 electrons per keV up to the average energy for β^- decay in ^{14}C, incident at 0° onto aluminium of 0.1 mm thickness (http://www.gel.usherbrooke.ca/casino/). The red tracks denote backscattered electrons and the blue denote absorbed electrons.

example, between two electrodes held under a high voltage in a gas where the ionisation of the gas is observed as an electrical signal and is manifest as a brief drop in the voltage across the electrodes as the charge is conducted away. Also, as the β^- particles lose energy in matter, they can yield bremsstrahlung in the form of X-rays. This constitutes an indirect indicator of the presence of the interaction of high-energy electrons, often symptomatic of β^- decay. This phenomenon can also present a source of radiation dose that needs to be managed and understood that has different implications to that associated with β^- particles, particularly in terms of range and shielding requirements. However, this is only prominent for high-energy electrons (>1 MeV) interacting in high-Z materials as might arise, for example, from the interaction of high-energy electrons in the surroundings for medical radiation therapy.

β^+ particles or, as they are more widely referred to, *positrons* are positively charged but have the same mass and magnitude of charge as their electron counterparts. After any residual kinetic energy has been lost via interactions with the electrons in matter, the positron and an electron annihilate with each other, ceasing to exist and leaving behind a quantity of energy equivalent to their rest mass energies. Assuming the pair are at rest when this occurs, this constitutes 1.022 MeV which is manifest as two γ rays of 511 keV, each travelling at 180° to one another thus conserving momentum. Therefore, rather than observing direct ionisation as for β^- particles as an electrical signal or bremsstrahlung, the presence of β^+ decay can be indicated by characteristic 511 keV lines in associated γ-ray spectra; a potential source of confusion are the 511 keV photons that can arise from the annihilation process following pair production due to an unrelated γ-ray interaction. It is also possible for the nucleus of a proton-rich isotope to absorb an atomic electron from an inner electron shell, particularly when β^+ decay is not favoured energetically. This process is known as *electron capture*. As for β^+ decay, via electron capture, a proton becomes a neutron and a neutrino is emitted.

As discussed for the case of bremsstrahlung, the exchange of energy with mass and the transfer from one form of radiation to another can constitute a change of radiation protection focus, particularly since 511 keV photons are highly penetrating and thus are not as easily shielded as β^- particles. The back-to-back angular correlation of the photons arising from positron annihilation is exploited widely in an important medical imaging modality, *positron emission tomography* [9], which has also been used in industry for process analysis and flow monitoring. The corresponding changes in terms of the $Z-N$ chart of the nuclides for β^- decay, β^+ decay and electron capture are given in Fig. 3.8A–C, respectively.

3.6.4 NEUTRON RADIATION

Neutrons can arise from a variety of sources and nuclear processes associated with nuclear energy. One example is *stimulated fission* associated with operating reactors; this is discussed further in Chapter 4. Neutrons are also emitted in *spontaneous fission*, typically of the even-number transuranic isotopes, and also from the action of α particles on light isotopes (which yields what are known as [α,n] reactions).

Neutron production is feasible throughout the nuclear fuel cycle: the actinide isotopes are particularly susceptible to the modes of decay referred to above as constituents of spent fuel and of materials in the back-end of the cycle that can comprise storage, reprocessing, recycle and disposal activities. However, in terms of the general environment (aside from nuclear energy) neutrons are much less prevalent in comparison with the other radiations and decay modes discussed earlier in this chapter. There are other ways in which neutrons can be produced besides fission and [α,n] reactions, for example via photonuclear reactions and proton-induced mechanisms carried out with particle accelerators, but these are not widespread in nuclear energy generation and are thus beyond the scope of this text.

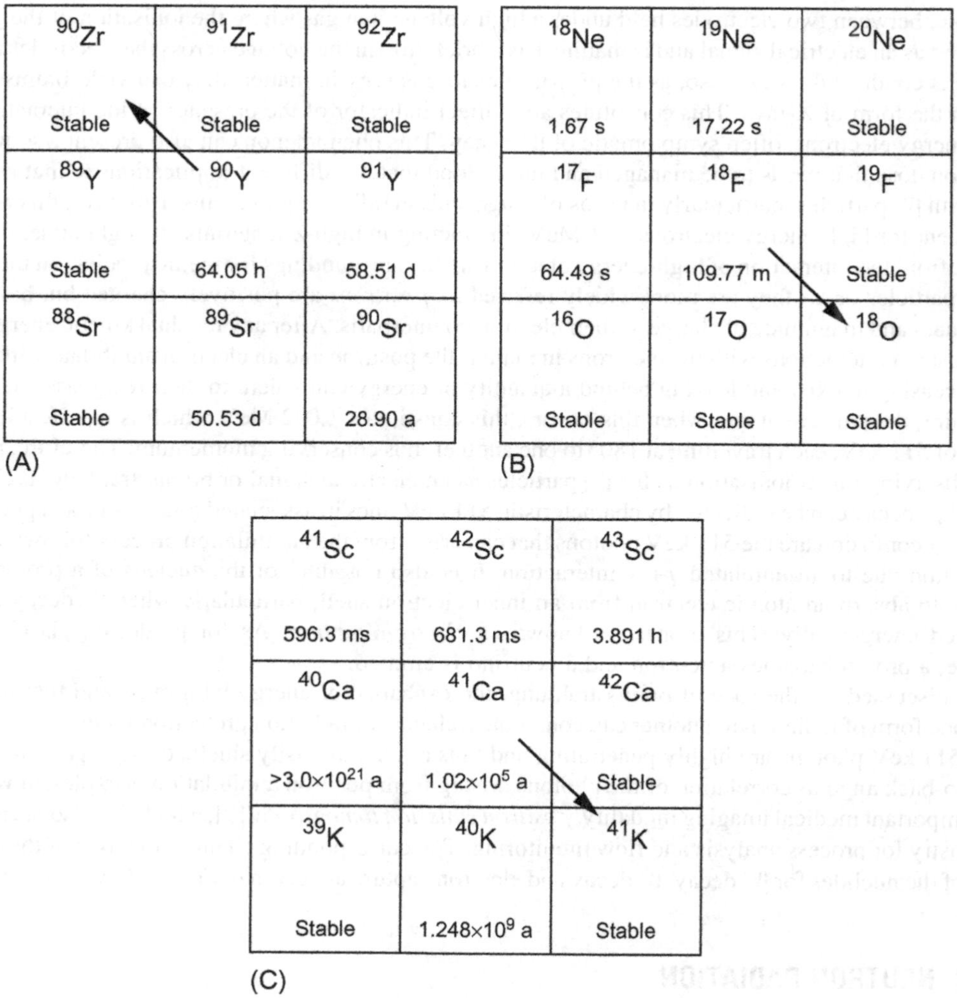

^{90}Zr Stable	^{91}Zr Stable	^{92}Zr Stable
^{89}Y Stable	^{90}Y 64.05 h	^{91}Y 58.51 d
^{88}Sr Stable	^{89}Sr 50.53 d	^{90}Sr 28.90 a

(A)

^{18}Ne 1.67 s	^{19}Ne 17.22 s	^{20}Ne Stable
^{17}F 64.49 s	^{18}F 109.77 m	^{19}F Stable
^{16}O Stable	^{17}O Stable	^{18}O Stable

(B)

^{41}Sc 596.3 ms	^{42}Sc 681.3 ms	^{43}Sc 3.891 h
^{40}Ca >3.0×10^{21} a	^{41}Ca 1.02×10^{5} a	^{42}Ca Stable
^{39}K Stable	^{40}K 1.248×10^{9} a	^{41}K Stable

(C)

FIG. 3.8

A 3 × 3 cell of the *Z* versus *N* plane from the chart of the nuclides illustrating the effect of (A) β^- decay, (B) β^+ decay and (C) electron capture on the parent nucleus, in this case for the decay of ^{90}Y, ^{18}F and ^{41}Ca, respectively.

Examples of stimulated fission, spontaneous fission and an $[\alpha, n]$ process are given below in terms of their typical reaction equations,

$$^{235}_{92}\mathrm{U} + n \rightarrow {}^{236}_{92}\mathrm{U}^* \rightarrow {}^{91}_{38}\mathrm{Sr} + {}^{143}_{54}\mathrm{Xe} + 2n \quad \text{(stimulated fission)}$$

$$^{240}_{94}\mathrm{Pu} \rightarrow {}^{89}_{38}\mathrm{Sr} + {}^{148}_{56}\mathrm{Ba} + 3n \quad \text{(spontaneous fission)}$$

$$^{241}_{95}\mathrm{Am} + {}^{9}_{4}\mathrm{Be} \rightarrow {}^{237}_{93}\mathrm{Np} + \alpha + {}^{9}_{4}\mathrm{Be} \rightarrow {}^{237}_{93}\mathrm{Np} + {}^{13}_{6}\mathrm{C}^* \rightarrow {}^{237}_{93}\mathrm{Np} + {}^{12}_{6}\mathrm{C} + n \; (\alpha, n)$$

Neutrons have no electrical charge but have significant mass relative to a number of other forms of radiation. They can possess energy at levels through from being in thermal equilibrium with their surroundings (0.0253 eV at 293 K) to beyond 1 MeV in nuclear energy applications; a full eight orders of magnitude. Their lack of charge limits the interaction of neutrons to atomic nuclei via the *strong nuclear interaction*; energy and momentum imparted to a nucleus by fast neutrons in such an interaction causes the nucleus to recoil. Being pushed out of its surrounding atomic electron structure, the recoil nucleus is highly ionised and thus causes ionisation as it recoils. Such an interaction is highly stochastic as there are a range of interaction possibilities.[4] For example, neutrons might interact strongly, transferring a lot of energy to an atomic nucleus, or weakly not transferring much at all. The variety of isotopes that they might interact with further influences the efficiency of energy transfer, and all of these variables influence the quantity of ionisation that is produced. The energy of the neutron influences whether the interaction is biased towards capture/inelastic scattering or elastic scattering. This has two significant implications:

- Neutron detection methods are often different to those used for radiations that interact electromagnetically and our interpretation of the detected response can be much more involved.
- The energy deposited per unit mass (dose) to living tissue arising from exposure to neutrons varies nonlinearly with energy.

Conventionally, neutrons are termed differently dependent on their energy (or occasionally termed *detection temperature*), as given in Table 3.2.

Despite fission being a significant source of neutron radiation, this is generally not encountered widely by people working in nuclear power facilities because access to the environments where it would arise (such as the primary containment of an operating nuclear reactor) is usually heavily regulated and access is restricted. Further, in some reactor designs, the facilities are so large that the neutron emission is heavily shielded by containment structures and the moderator.

Table 3.2 The Widespread Terminology Associated With Neutron Radiation in Terms of Ranges in Energy of Relevance to Nuclear Energy Applications

Neutron Energy	Neutron Terminology	Notes
~0.0253 eV @ 293 K, <0.2 eV	Thermal	In thermal equilibrium with their surroundings
>0.2 eV	Epithermal	With energies greater than thermal but less than a few eV
<0.4 eV	Cadmium	Strongly absorbed by cadmium
>0.6 eV	Epicadmium	Not strongly absorbed by cadmium
<1 eV through to 10 eV	Slow	Sometimes to 1 keV
1 eV → 300 eV	Resonance	Vulnerable to absorption resonances in ^{238}U
1 keV < E_n < 0.5 MeV	Intermediate	Between slow and fast
>0.5 MeV	Fast	Up to ~10–20 MeV

[4]The probability of a neutron interaction is quantified in terms of the *microscopic cross section* and the conventional symbol σ_i where i denotes the specific interaction concerned. This is considered in more detail in Chapter 4.

Neutrons are important in nuclear engineering for a number of reasons: They are the agent that sustains the critical chain reaction from which power is derived in fission reactors; measurements of the neutron flux in an operating fission reactor are often used to infer the power level; in prototype fusion reactor systems, measurement of the neutron output is the primary route by which the power output is estimated; measurements of the number of neutrons emitted by nuclear materials can also be used to infer the quantity present to help guard against materials being diverted for illicit purposes in *nuclear safeguards* and *accountancy* practices.

In contrast to photons, heavy-charged particles and electrons, the interaction of neutrons with matter is not dependent on the electron density of the material in which they interact. Rather, at low neutron energies, their attenuation is generally more significant for low atomic masses due to the more effective energy transfer to nuclei with masses closer to that of the neutron. There are also isolated peaks in attenuation for materials that have particularly high probabilities for the absorption of neutrons, such as isotopes of gadolinium and cadmium for the case of thermal neutrons.

Perhaps most importantly there is not a continuous, increasing trend of attenuation with atomic number in the way there is, for example, for photons. Aside from the specific isotopes that exhibit anomalously high neutron absorption characteristics referred to above, low-mass materials containing a lot of hydrogen such as water, plastic and other organic substances have the best shielding

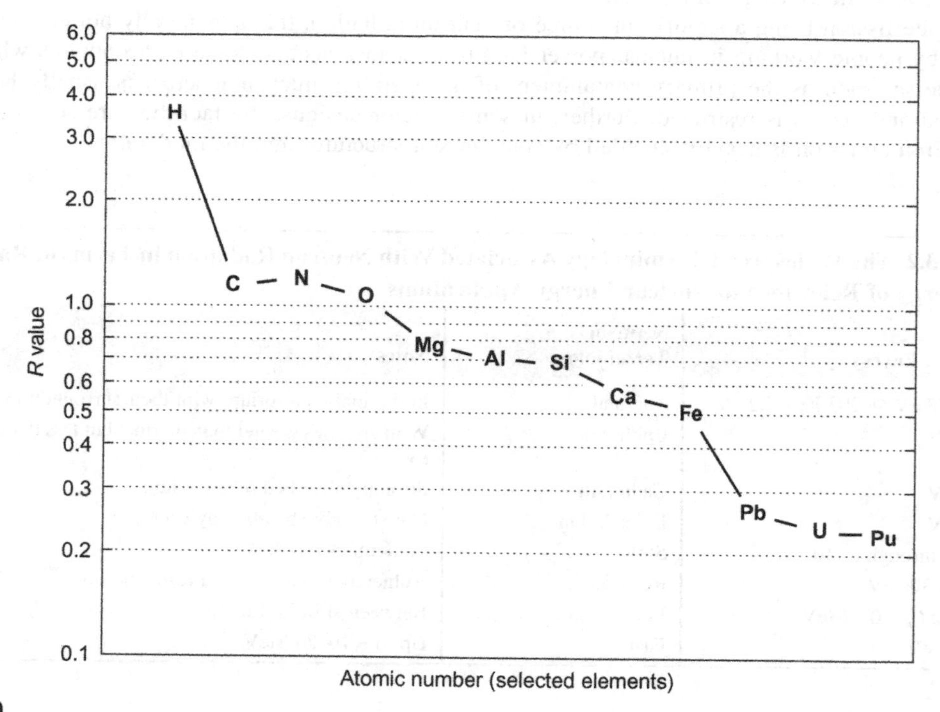

FIG. 3.9

The ratio of the attenuation coefficients (R) of 14 MeV neutrons to γ rays from ^{60}Co as a function of atomic number, after Eberhardt et al. [10].

characteristics for neutrons. Conversely, high-Z materials can be largely transparent to them although some of the more dense materials can exhibit reflective properties. One route to depicting this contrast between the interaction properties of neutrons and photons is to consider the difference in abilities of substances to attenuate neutrons over photons. This is shown in Fig. 3.8 in terms of the ratio of the attenuation of neutrons to photons, denoted R, as a function of the atomic number of the selected elements concerned [10]. This depicts the significant absorption characteristics for neutrons at low masses and the gradual decline with increasing atomic number (Fig. 3.9).

The indirect process by which neutrons interact with matter results in a nonlinear dependence of energy deposition with energy. This has significant implications for the measurement of radiation dose from neutrons that are discussed further in Chapter 14.

CASE STUDIES

CASE STUDY 3.1: CALCULATING THE SPECIFIC ACTIVITY OF CAESIUM-137.

To demonstrate a calculation of specific activity, we might consider the case of ^{137}Cs: it has a half-life of 30.17 years and an isotopic mass is 136.907 u, where u is the atomic mass unit, $u = 1.66 \times 10^{-27}$ kg.

By definition, an isotopic mass of 136.907 u corresponds to a molar mass for ^{137}Cs of 136.907 g mol^{-1} where a mole is equal to Avogadro's number $N_A = 6.023 \times 10^{23}$.

Therefore, 1 g of ^{137}Cs comprises:

$$N = 6.023 \times 10^{23}/136.907 = 4.40 \times 10^{21}\,\text{atoms}$$

and the activity can be calculated from this using Eq. (3.4):

$$A = \lambda N$$

Since, from Eq. (3.6),

$$t_{1/2} = \frac{\log_e 2}{\lambda}$$

the decay constant λ:

$$\begin{aligned}\lambda &= \log_e 2/(30.17 \times 86400 \times 365.25)\\ &= 0.69/9.52 \times 10^{8}\\ &= 7.28 \times 10^{-10}\,\text{s}^{-1}\end{aligned}$$

And hence the activity for 1 g ^{137}Cs:

$$\begin{aligned}A &= 7.28 \times 10^{-10} \times 4.40 \times 10^{21}\\ &= 3.20 \times 10^{12}\,\text{Bq/g}\end{aligned}$$

$$A = 3.20\,\text{TBq/g} \quad \text{or} \quad 86.5\,\text{Ci/g}$$

CASE STUDY 3.2: DERIVING THE BATEMAN EQUATION FOR A PARENT–DAUGHTER DECAY.

This study illustrates the definition of the Bateman equation, Eq. (3.8), and provides the subsequent step that relates it to the activity of the daughter isotope.

The numbers of unstable nuclei of the parent and daughter isotopes at time t are denoted $N_P(t)$ and $N_D(t)$, respectively, and the corresponding decay constants are denoted λ_P and λ_D. From Eq. (3.2). $N_P(t) = N_P(0)e^{-\lambda_P t}$ which can be substituted into a re-arranged form of Eq. (3.7) of:

$$dN_D = \lambda_P N_P\,dt - \lambda_D N_D\,dt$$

to yield

$$dN_D = \lambda_P N_P(0)e^{-\lambda_P t}dt - \lambda_D N_D\,dt$$

Re-arranging to collect like terms:

$$\lambda_D N_D\,dt + dN_D = \lambda_P N_P(0)e^{-\lambda_P t}dt$$

Multiplying through by $e^{\lambda_D t}$,

$$e^{\lambda_D t}\lambda_D N_D\,dt + e^{\lambda_D t}dN_D = \lambda_P N_P(0)e^{(\lambda_D - \lambda_P)t}dt$$

Noting that via the chain rule:

$$d\left(N_D e^{\lambda_D t}\right) = e^{\lambda_D t}dN_D + \lambda_D N_D e^{\lambda_D t}dt$$

and hence this can be substituted for the left side so that,

$$\int_0^t d\left(N_D e^{\lambda_D t}\right) = \int_0^t \lambda_P N_P(0)e^{(\lambda_D - \lambda_P)t}dt$$

$$N_D e^{\lambda_D t}\Big|_0^t = \frac{\lambda_P N_P(0)e^{(\lambda_D - \lambda_P)t}}{(\lambda_D - \lambda_P)}\Bigg|_0^t$$

$$N_D e^{\lambda_D t} - N_D(0) = \frac{\lambda_P N_P(0)}{(\lambda_D - \lambda_P)}\left(e^{(\lambda_D - \lambda_P)t} - 1\right)$$

Dividing through by $e^{\lambda_D t}$,

$$N_D(t) - N_D(0)e^{-\lambda_D t} = \frac{\lambda_P N_P(0)}{(\lambda_D - \lambda_P)}\left(e^{-\lambda_P t} - e^{-\lambda_D t}\right)$$

$$N_D(t) = \frac{\lambda_P N_P(0)}{(\lambda_D - \lambda_P)}\left(e^{-\lambda_P t} - e^{-\lambda_D t}\right) + N_D(0)e^{-\lambda_D t}$$

Multiplying through by λ_D to give parent and daughter activities A_P and A_D, respectively:

$$\lambda_D N_D(t) = \frac{\lambda_D \lambda_P N_P(0)}{(\lambda_D - \lambda_P)}\left(e^{-\lambda_P t} - e^{-\lambda_D t}\right) + \lambda_D N_D(0)e^{-\lambda_D t}$$

$$A_D(t) = \frac{\lambda_D A_P(0)}{(\lambda_D - \lambda_P)}\left(e^{-\lambda_P t} - e^{-\lambda_D t}\right) + A_D(0)e^{-\lambda_D t}$$

If there is no pre-existing daughter activity, that is, $A_D(0) = 0$, then,

$$A_D(t) = \frac{\lambda_D A_P(0)}{(\lambda_D - \lambda_P)}\left(e^{-\lambda_P t} - e^{-\lambda_D t}\right)$$

CASE STUDY 3.3: THE TIME TAKEN FOR ^{238}U AND ^{234}Th TO REACH SECULAR EQUILIBRIUM AND THE EQUILIBRIUM MASS OF ^{234}Th

^{238}U decays to ^{234}Th with a half-life of 4.47×10^9 years and then ^{234}Th decays with a half-life of 24.1 days. We wish to calculate how long is necessary for 1 μg of ^{238}U, that is isotopically pure at time zero, to reach secular equilibrium and to determine the mass of ^{234}Th at this point. The isotopic masses of ^{238}U and ^{232}Th are 238.0507 u and 234.0436 u, respectively.

At secular equilibrium (denoted time t_m), the activities of the parent and daughter are equal, $A_D(t_m) = A_P(t_m)$. Substituting this into the Bateman equation in terms of activity, as per Case Study 3.2:

$$A_P(t_m) = \frac{\lambda_D A_P(0)}{(\lambda_D - \lambda_P)}\left(e^{-\lambda_P t_m} - e^{-\lambda_D t_m}\right)$$

Since $A_P(t_m) = A_P(0)e^{-\lambda_P t_m}$ and cancelling the $A_P(0)$ on each side,

$$e^{-\lambda_P t_m} = \frac{\lambda_D}{(\lambda_D - \lambda_P)}\left(e^{-\lambda_P t_m} - e^{-\lambda_D t_m}\right)$$

Dividing through by $e^{-\lambda_P t_m}$ and re-arranging,

$$\frac{(\lambda_D - \lambda_P)}{\lambda_D} = 1 - e^{-(\lambda_D - \lambda_P)t_m}$$

$$e^{-(\lambda_D - \lambda_P)t_m} = \frac{\lambda_P}{\lambda_D}$$

Taking natural logarithms and re-arranging,

$$t_m = \frac{1}{(\lambda_D - \lambda_P)} \log_e \frac{\lambda_D}{\lambda_P}$$

Since $\lambda_D \gg \lambda_P$, an approximation can be made if desired. Substituting for $\lambda_D = \lambda_{Th}$ and $\lambda_P = \lambda_U$,

$$t_m \sim \frac{1}{\lambda_{Th}} \log_e \frac{\lambda_{Th}}{\lambda_U}$$

The decay constants are as follows:

$$\lambda_U = \frac{\log_e 2}{86400 \times 365.25 \times 4.47 \times 10^9} = 4.91 \times 10^{-18}\,\mathrm{s}^{-1} \quad \text{and} \quad \lambda_{Th} = \frac{\log_e 2}{86400 \times 24.1} = 3.33 \times 10^{-7}\,\mathrm{s}^{-1}$$

To give $t_m \sim 2.37$ years.
From Eq. (3.7), at equilibrium $dN_D/dt = dN_{Th}/dt$ and thus $\lambda_{Th}N_{Th} = \lambda_U N_U$. Hence,

$$N_{Th} = \frac{\lambda_U N_U}{\lambda_{Th}}$$

By definition, 238.0507 g comprises 1 mole of ^{238}U atoms. Further, $N_U \sim N_U(0)$ since the half-life of ^{238}U is long. Hence,

$$N_U = \left(6.023 \times 10^{23}/238.0507\right) \times 10^{-6} = 2.53 \times 10^{15} \text{ atoms}$$

and so from the equation for N_{Th} at equilibrium above,

$$N_{Th} = \frac{4.91 \times 10^{-18} \times 2.53 \times 10^{15}}{3.33 \times 10^{-7}}$$

$$= 3.73 \times 10^4 \text{ atoms or } 6.19 \times 10^{-20} \text{ mole}$$

In terms of mass of ^{234}Th at equilibrium, m_{Th}, as per the question: $m_{Th} = 6.19 \times 10^{-20} \times 234.0436$
Hence, $m_{Th} = 1.45 \times 10^{-17}$ g or ~ 14.5 ag

CASE STUDY 3.4: CALCULATING Q VALUES FOR α DECAY

The energy exchanged in a nuclear reaction (the Q value defined in Section 3.6.2) indicates whether a reaction is exoergic (releasing energy) or endoergic (requires energy to occur). In the latter case, a stimulus is necessary to make the reaction happen.

For the case of spontaneous radioactive decay that is the focus of this case study, there is a single primary component to the reaction: the unstable parent isotope. As a result of the disintegration process that follows, decay *products* are formed.

The energy that is exchanged in a nuclear reaction comes from the change in mass of the system before and after. This is manifest as the kinetic energy and possibly the excitation energy of the reaction products that often constitute radiation(s) emitted as a result of the reaction. The corresponding difference in binding energy holding the nucleus together can be determined by comparing the masses of the components in the reaction, before and afterwards. Returning to the ^{238}U example discussed in the text, two scenarios associated with the α-decay of ^{238}U are considered in this study for the purposes of illustration, as follows:

$$^{238}\text{U} \rightarrow {}^{234}\text{Th} + \alpha$$

and

$$^{238}\text{U} \rightarrow {}^{235}\text{Th} + {}^{3}\text{He}$$

Isotope	^{238}U	^{235}Th	^{234}Th	^{4}He	^{3}He
Mass/u	238.0507	235.0475	234.0436	4.0026	3.0160

The Q value can be obtained from the difference in mass before and after the reaction.

For the ^{238}U(^{234}Th, α) reaction,

$$\begin{aligned} Q &= M(^{238}\text{U}) - M(^{234}\text{Th} + {}^{4}\text{He}) \\ &= 238.0507 - (234.0436 + 4.0026) \\ &= 0.0045\,\text{u where, since u} = 931.5\,\text{MeV}/\text{c}^2 \\ Q &= +4.19\,\text{MeV (note it is positive i.e. exoergic)} \end{aligned}$$

For the ^{238}U(^{235}Th, ^{3}He) alternative,

$$\begin{aligned} Q &= M(^{238}\text{U}) - M(^{235}\text{Th} + {}^{3}\text{He}) \\ &= 238.0507 - (235.0475 + 3.0160) \\ &= -0.0128\,\text{u where, since u} = 931.5\,\text{MeV}/\text{c}^2 \\ Q &= -11.92\,\text{MeV (endoergic)} \end{aligned}$$

Hence the former occurs spontaneously while the latter is not favoured as a natural decay pathway in the absence of an external stimulus with energy > 11.92 MeV.

CASE STUDY 3.5: TWO CONTRASTING EXAMPLES OF β DECAY

Tritium (^{3}H) is neutron rich, its nucleus comprising twice as many neutrons as protons and is thus unstable to β^- decay with a half-life of 12.3 years. It follows the decay path given below,

$$^{3}_{1}\text{H} \rightarrow {}^{3}_{2}\text{He} + e^- + \bar{\nu}$$

The antineutrino is not usually observed. The emitted β^- particle has a very low energy (a mean energy of 5.7 keV). Detecting such low-energy β particles can be a challenge for many detector systems and this is the main reason why the quantity of tritium in the environment is often difficult to assess. The tritium decay pathway also constitutes one route by which ^{3}He gas is manufactured which is an important agent for thermal neutron detection due to its high neutron capture cross section at thermal energy levels. Tritium is also an important component of the fuel in fusion reactors.

Carbon-14 (^{14}C) is another example of a neutron-rich isotope which decays via

$$^{14}_{6}\text{C} \rightarrow {}^{14}_{7}\text{N} + e^- + \bar{\nu}$$

It has a half-life of 5730 years and an average β^{-}particle energy of 49 keV. Carbon-14 is not easily detected for the same reason discussed above in the case of tritium, but clearly it resides for significantly longer in the environment.

Interestingly, the natural formation of ^{14}C in the atmosphere via cosmic neutron irradiation of ^{14}N (via the [n,p] reaction) results in living organisms maintaining a constant abundance of ^{14}C (replenished by what they consume) until they expire. When their remains have been recovered many years later by archaeologists the quantity of ^{14}C remaining has been assessed and used to infer how long ago the organism died. This is the basis of *radiocarbon dating*.

REVISION GUIDE

On completion of this chapter, you should:

- understand that radioactivity is a property arising as a result of changes in the nucleus
- that radioactivity is often evidenced by radiation emitted from the nucleus as a result of radioactive decay and that it occasionally exerts an influence on the surrounding atomic electron shell structure to yield secondary radiation effects, such as bremsstrahlung or annihilation photons
- appreciate that, as a property of the atom, the amount of radioactivity correlates directly with the amount of substance
- be able to write down the mathematical relationship linking radioactivity and the amount of substance, and from this derive the following parameters: *decay constant*, *lifetime* and *half-life*
- understand the concept of *equilibrium* in a qualitative context for decays of at least two isotopes, specifically the definitions of *secular* and *transient* equilibria, and be able to relate this to the corresponding extension of the differential equation describing radioactive decay
- appreciate the variety of radiation types of relevance to the nuclear power community, their origins and the differences in the way they interact with matter
- understand the implications of penetration depth for the radiation types referred to above, and the implications this has for shielding and dose, at a qualitative level
- appreciate that neutrons are rather different from other radiation types because they have mass but are not charged and thus they interact very differently with matter

PROBLEMS

3.1 Given that

$$\frac{dN}{dt} = -\lambda N$$

derive the relationship between N and t in terms of the decay constant λ.
Hence, derive the relationship between the decay constant λ and the half-life $t_{1/2}$.

3.2 Derive the expression for the mean lifetime, τ, of a single radioactive species, and hence determine the relationship between mean lifetime and half-life, $t_{1/2}$.

3.3 Write down the decay equation for the emission of ^{12}C from ^{238}U, and hence calculate the Q value. Comment, on the basis of your result, as to whether you would expect to observe this decay experimentally. By considering the differences between the decay products of this and the corresponding scenario for α decay, would you expect to see ^{12}C emission as often as the associated α decay?

(The atomic mass for ^{238}U is 238.0507 u, for ^{12}C is 12 u and for ^{226}Rn is 226.0309 u.)

3.4 Write down the equation for α decay of ^{222}Rn. Hence calculate kinetic energy of the α particles you would expect from ^{222}Rn.

(The atomic mass for ^{222}Rn is 222.0176 u, for an α particle it is 4.0026 u and for ^{218}Po it is 218.0090 u.)

3.5 Consider a single interaction of an α particle with an electron: before the interaction the α particle has a kinetic energy T and the electron is at rest. After the interaction the α particle will have lost ΔT to the electron.

By considering the principles of the conservation of energy and momentum, before and after the interaction, derive the relationship between ΔT and T, stating any approximations you make.

Note: you should derive an expression for the speed of the α particle after the interaction to infer its kinetic energy afterwards and use the fact that $m_\alpha \gg m_e$ to simplify the result.

REFERENCES

[1] B.H. Orndoff, An interview with Madame Curie: an historical note, Radiology 71 (5) (1958) 750–752.

[2] H. Bateman, Solution of a system of differential equations occurring in the theory or radioactive transformations, in: Proc. Cambridge Phil. Soc. IS, 1910, pp. 423–427.

[3] J. Chadwick, J. Chadwick, Radioactivity and Radioactive Substances, Pitman, London, 1931.

[4] M. Chianchetti, M. Amichetti, Sinusoidal malignancies and charged particle radiation treatment: a systematic literature review, Int. J. Otolaryngol. (2012) art. ID 325891.

[5] H. Bethe, Zur Theorie des Durchgangs schneller Korpuskularstrahlen durch Materie, Ann. Phys. 397 (3) (1930) 325–400 (Translated in 'Selected works of Hans A. Bethe: with commentary', World Scientific Series in 20th Century Physics, vol. 18, pp. 77–154 (1997). ISBN 9810228767).

[6] M. Inokuti, J.E. Turner, Mean excitation energies for stopping power as derived from oscillator-strength distributions, in: 6th Symposium on Microdosimetry, vol. 1, Harwood Academic Publ. Ltd., 1978, pp. 675–687

[7] J. Carroll, J. Coleman, M. Lockwood, C. Metelko, M. Murdoch, C. Touramanis, G. Davies, A. Roberts, Monitoring nuclear reactors for safeguards purposes using anti-neutrinos, J. Phys. Conf. Ser. 598 (1) (2015) 012024.

[8] V.V. Kuzminov, N.Ja. Osetrova, Precise measurement of ^{14}C beta spectrum by using a wall-less proportional counter, Phys. At. Nucl. 63 (7) (2000) 1292–1296.

[9] A.K. Shukla, et al., Positron emission tomography: an overview, J. Med. Phys. 31 (1) (2006) 13–21.

[10] J.E. Eberhardt, S. Rainey, R.J. Stevens, B.D. Sowerby, J.R. Tickner, Fast neutron radiography scanner for the detection of contraband in air cargo containers, App. Rad. Isot. 63 (2005) 179–188.

CHAPTER

THE FISSION PROCESS

4

4.1 SUMMARY OF CHAPTER AND LEARNING OBJECTIVES

The objectives of this chapter are to:

- introduce the reader to the concept of *nuclear fission*
- provide a basis for the understanding of the *fission process* in terms of neutron interactions
- define the concepts of *microscopic cross section*, *macroscopic cross section*, *mean free path*, *neutron interaction types* and the *dependence of cross section with energy*
- provide a summary of the trends in behaviour of the nuclei of massive isotopes and particularly the dependence of *binding energy* and *binding energy per nucleon* with atomic mass
- demonstrate the advantage gained by massive isotopes that decay via fission, in terms of the greater binding energy that the products of fission achieve
- explore the properties of the products of fission and particularly the energetics of fission including common trends in *neutron energy spectra* and *multiplicity* of the emitted neutrons
- demonstrate how to calculate most *probable neutron energies* from the corresponding neutron energy spectra, the average multiplicity for a given fission reaction and to provide an introduction to the interpretation of the corresponding distributions

4.2 HISTORICAL CONTEXT: LISE MEITNER 1878–1968

Following Chadwick's discovery of the neutron, many researchers started to investigate what might be found by irradiating substances with this new, neutral form of radiation. Their focus at the time was the anticipation of the synthesis of new, heavy isotopes beyond the then heaviest, naturally occurring elements: uranium and radium. Meitner (Fig. 4.1), having studied physics under Boltzmann, had worked with Otto Hahn for 30 years studying radioactivity at the Kaiser Wilhelm Institute for Chemistry. Her physics knowledge complemented his chemistry abilities and had enabled them to make several important discoveries prior to that of fission. Following the formation of radioactive isotopes by Fermi with neutron bombardment, Meitner, Hahn and Strassman attempted to identify whether any might be examples of the elusive transuranic elements, yet to be discovered.

Meitner fled to Sweden when Austria was annexed by Nazi Germany in 1938. It was from here that she corresponded with Hahn and Strassman by letter, meeting in secret with Hahn in Copenhagen. Following this, Hahn and Strassman focused on the radioactive products of neutron interactions on uranium that they suspected to be isotopes of radium. However, they were to conclude instead that they were precipitates of barium and lanthanum due to their identical chemical properties. What they had

Nuclear Engineering. https://doi.org/10.1016/B978-0-08-100962-8.00004-4

FIG. 4.1

Lise Meitner.

discovered were radioactive isotopes of barium and lanthanum formed as fission fragments from fission induced in uranium with neutrons and observed for the first time. It was Meitner and Robert Frisch (her nephew) who first successfully explained the phenomenon. They drew on predicted behaviour based on the liquid-drop description of the nucleus, named the phenomenon 'fission' (Kernspaltung) borrowing the term from the process of cell division in biology and predicted the exceptional energy output from the curve of mass defect dependence with atomic mass.

4.3 INTRODUCTION

Nuclear fission—the process by which the nucleus splits into two (occasionally three) individual nuclei—presents us with a many-body problem whose properties are highly stochastic and that vary significantly across enormous dynamic ranges in terms of probability, half-life and energy. Perhaps then it is not surprising that understanding the fission process remains a very difficult problem which lacks a complete theoretical treatment. Fission is nonetheless profound in its effect on mankind. It defines, at one extreme, the upper limit of the periodic table and the extent of the matter which we experience in the world around us. Conversely, fission provides what is at present the easiest route by which the transformational amounts of energy that are bound up in the atomic nucleus can be exploited by man. In Fig. 4.2 the original apparatus used to prove the existence of fission is shown reconstructed.

The accepted qualitative description of the stimulated fission process is that an already large nucleus absorbs energy; for example, a neutron is absorbed by the nucleus. This causes the resulting compound nucleus to deform to a state from which, conceptually at least, its destiny can follow in one of two paths. It can either continue to deform further, leading to a complete separation of the parent nucleus into (usually) two smaller fragments, or it can undergo an alternative route of decay by which to dissipate its excitation energy. This point in the process is often called the *transition state*.

Ignoring *single-particle* influences on fission for the time being (i.e. the effect of the individual energy states of neutrons and protons in the nucleus), there are two *collective* competing influences

FIG. 4.2

A photograph of a reconstruction of the original apparatus with which nuclear fission was first observed.

on a nucleus' behaviour at the transition state: *Coulomb repulsion* and the *strong nuclear force*. The former arises from the repulsion of the constituent protons due to their like charges. The latter is one of the four fundamental interactions that acts at short range, at dimensions less than the nuclear scale typically $<10^{-15}$ m, to bind the nucleons together. The strong force is often introduced conceptually by analogy with a cohesive surface tension that acts to keep a droplet of fluid together in what is often referred to as the *liquid-drop model* of the nucleus. Conversely, the Coulomb force acts to elongate the nucleus to its transition state. The surface tension component acts to maintain the cohesion of the nucleus in contrast to the elongation.

However, a critical distinction between the two influences described above is that the force of Coulomb repulsion is proportional to the square of the number of protons, Z^2, where by contrast and as shown in Fig. 4.3, nuclear binding energy is proportional to the atomic mass, A. Hence, as we move from left to right on the chart of the nuclides with increasing mass, a point is reached where the force of Coulomb repulsion exceeds the cohesive binding of the nucleus, because Z^2 increases more quickly than A. The early symptoms of this are that the nucleus deforms and elongates. As the distance between the constituent protons increases due to the elongation, the Coulomb energy decreases (being inversely proportional to the separation between the protons, r). Conversely, the surface energy increases due to the increase in surface area with elongation. This leads to the *scission point* at which the parent nucleus essentially comprises two highly-charged nuclei (fragments) in contact with one another.

From the scission point the progression to complete fission is dramatic. The fragments accelerate away from each other due to the Coulomb repulsion of the constituent groups of protons, to 90% of their final kinetic energy in 10^{-20} s; they reach a speed of 5% speed of light and an acceleration corresponding to $\sim 10^{26}$ g. Although deformed at the instant of scission, the fission fragments relax to sphericity their surface energy decreasing but Coulomb energy increasing due to the reduction in r of each fragment (albeit much less than in the parent nucleus state since there are much fewer protons in each

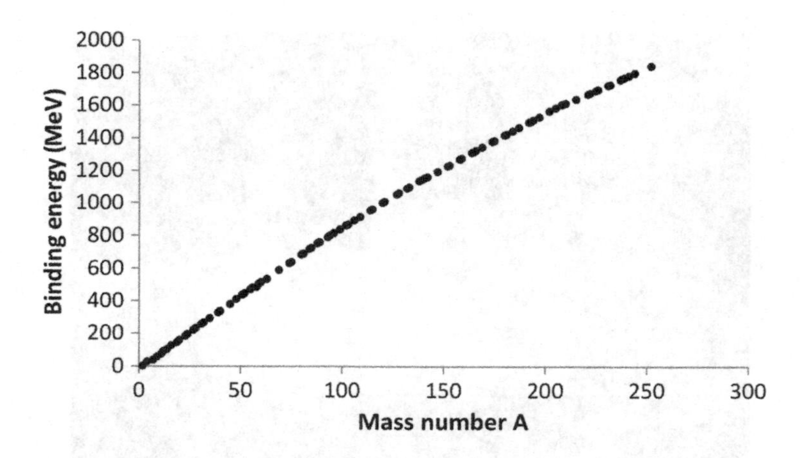

FIG. 4.3

Nuclear binding energy versus mass number.

fragment). Consequently, the excitation of each fragment increases. This energy is dissipated firstly by the emission of prompt neutrons somewhere between 10^{-18} and 10^{-13} s after scission (a process often referred to as *neutron evaporation*) and then the emission of γ rays (10^{-14} to 10^{-7} s).

The fission fragments at this stage are usually neutron-rich and radioactive and are thus susceptible to disintegration via β^- decay. Following β^- emission, the daughter products of the fragments can be in highly excited states with sufficient energy to emit further neutrons. However, these neutrons are emitted at a time later, that is, they are *delayed*, relative to the prompt emission of neutrons by the intervening stage of β^- decay; these are discussed further in Chapter 8. The majority of the energy released in fission takes the form of the kinetic energy of the fission fragments.

Finally, alongside the collective effects, the influence of the individual nucleons on fission can also be significant in terms of the probability fission will occur (discussed below) and the mass of the fission products. With regard to the latter, while fission is understandable via the liquid-drop model on the basis of the collective or overall behaviour of the nucleus, this predicts the mass of the fission fragments to be similar to one another yielding a symmetric mass distribution reflecting the expectation that on average the nucleus will separate into two halves of equal mass.

However, this is not what is observed. Rather, the mass of the fragments is distributed asymmetrically comprising a light fragment at $A \sim 90$ and a heavy fragment at $A \sim 140$. This is shown for ^{252}Cf, ^{235}U and ^{239}Pu in Fig. 4.4. In particular, the single-particle structure of the nucleus favours a fragment at $Z=50$ and $N=82$ since this is one of the most stable arrangements of the nucleons, comprising magic numbers for both protons and neutrons. Further, spontaneous fission half-lives are only predicted consistently with what is observed experimentally when corrections are included to account for the influence of the individual nucleon behaviour on the formation of the fragments. Interestingly, there is not a great preference in nature as to how the protons and neutrons are divided as a result of fission. Thus, since the parent nucleus has more neutrons than protons evidenced by the curve of the chart of the nuclides discussed in Chapter 3, the fission fragments that arise from fission are also usually neutron-rich and exhibit an immediate tendency to emit neutrons followed by a predisposition to β^- decay as a result.

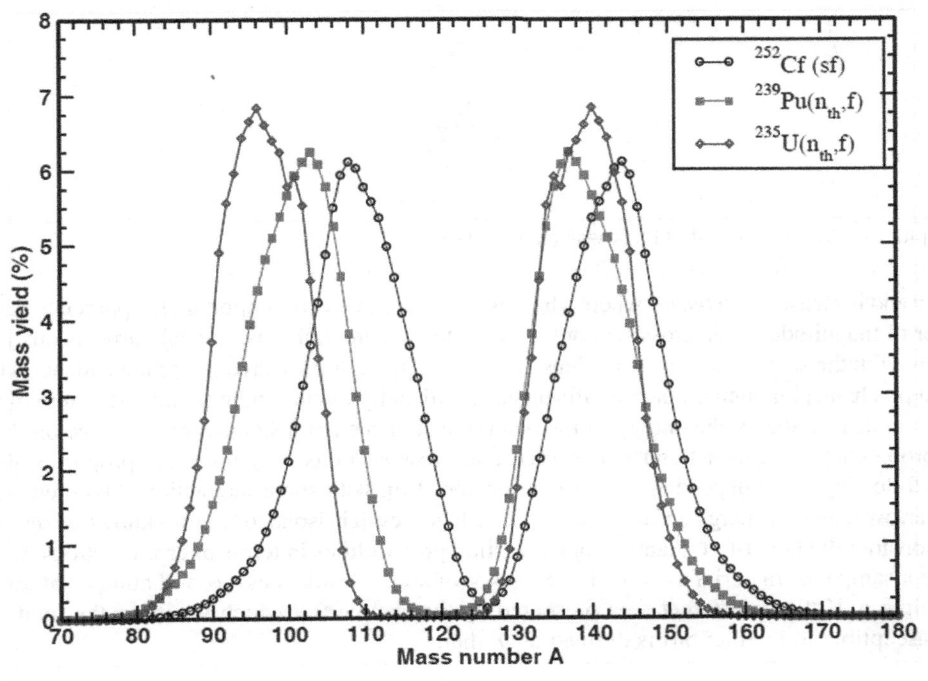

FIG. 4.4

Mass yield of fission fragments (%) versus mass number A for spontaneous fission in ^{252}Cf, and fission stimulated with thermal neutrons in ^{235}U and ^{239}Pu.

After O. Litaize, O. Serot, L. Berge, Fission modelling with FIFRELIN, Eur. Phys. 51 (2015) 177–191 with kind permission of the European Physical Journal *(*EPJ*).*

4.4 NEUTRON INTERACTIONS

4.4.1 DEFINITION OF MICROSCOPIC CROSS SECTION

The heavier atomic nuclei that are relevant to fission and nuclear energy are complicated entities that exhibit properties that are influenced by the behaviour of the nucleus as a whole (referred to as *collective*) and that of the individual nucleons of which they are constituted (referred to as *single-particle*). One of these properties with which we are particularly concerned within the context of fission is the probability that a neutron will interact with a nucleus. Given that there is a significant variety of interaction types, and that the probability of interaction is dependent on the specific nuclide in question and the energy of the incident neutron, there is a wide range of probabilities to consider.

In Chapter 1, the concept of there being a probability in order to quantify this was introduced where a formalism is adopted with the likelihood of the interaction symbolised as a cross-sectional area. This is analogous to a target used in archery, such that the bigger the target (area) the greater the interaction probability, originating from the pioneering scattering experiments carried out to elucidate the size of the nucleus. The corresponding probability is usually termed the *cross section* with the mathematical

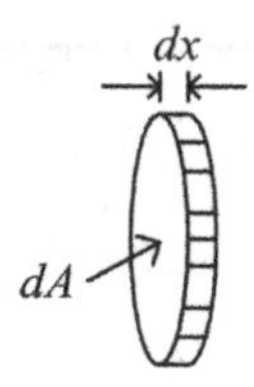

FIG. 4.5

A conceptual sample of material of thickness *dx* and area *dA*.

symbol σ and is measured in *barns*, where a barn is 10^{-24} cm^2. It is also important to appreciate that while the order of magnitude of the cross sections we tend to encounter, that is, several barns, is qualitatively consistent with the size of the atomic nucleus, interaction probabilities have strong dependencies that are quantum mechanical in nature that are influenced significantly by the single-particle structure of the nucleus. As indicated above, the energy of the neutron can also exert a significant influence on the interaction probability: a general trend is observed that cross sections vary inversely proportionally with energy (following a $1/v$ dependence with neutron speed v), with some interactions favoured intensely within narrow ranges in energy and not in others. The latter result in isolated features known as *resonances*.

To consider the benefit of quantifying interaction probabilities in terms of cross sections, we might consider a sample of material as shown in Fig. 4.5, of area dA, thickness dx and number of atoms per unit volume n. If the total target area presented by the sample (by which we mean the total nuclear 'area' susceptible to interaction) is defined as a, then

$$a = n\sigma dA\, dx \tag{4.1}$$

The interaction probability, p, being the ratio of the area susceptible to interaction to that of the physical area presented by the sample, is

$$p = n\sigma dx \tag{4.2}$$

This assumes implicitly that the atomic number density is uniform throughout the sample and that the sample is 'thin', that is that successive interactions are not plausible to the extent that they would have an influence on the incident energy of the neutron. It follows that the cross section relates the interaction probability to the density of nuclei encountered by a neutron along its path of interaction. For the sake of completeness as defined in Chapter 1, σ is often referred to specifically as the *microscopic cross section* as it refers to the property of a single nucleus; where 'cross section' is referred to implicitly but is measured in units of barns then it refers to the microscopic case, σ.

4.4.2 NEUTRON INTERACTION TYPES

As suggested earlier, there are a range of interaction types and a corresponding range of cross sections describing the associated probability that each might occur. For example, the result of the interaction of a neutron with a nucleus might be that the neutron transfers some of its energy to the nucleus and that the trajectory of the neutron is changed such that it is *scattered*. Alternatively, the neutron may be *absorbed* by the nucleus and cease to exist as a lone entity altogether. The variety of interactions can be grouped into two general classes associated with *scattering* and *absorption*, which can be represented schematically as in Fig. 4.6, and have the corresponding constituent microscopic cross sections σ_s and σ_a, respectively.

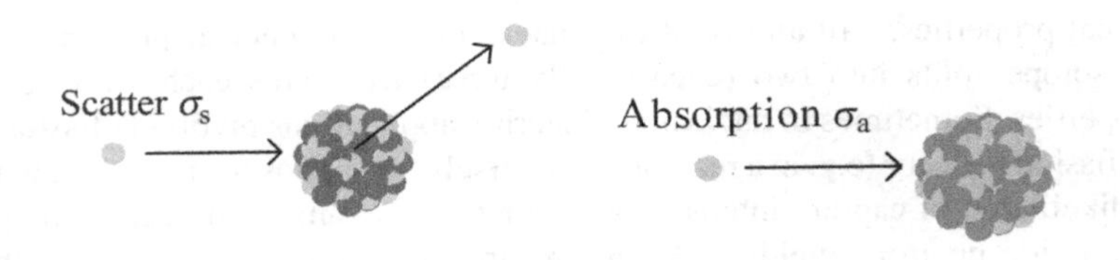

FIG. 4.6

Schematic representations of the processes of scatter (left) and absorption (right).

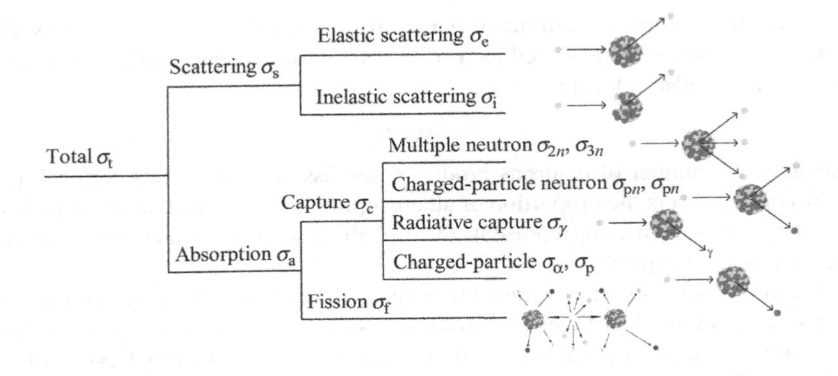

FIG. 4.7

Schematic representations of neutron interaction subclassifications.

The distinction of a scattering interaction is that the neutron exists in isolation afterwards whereas after absorption it does not. There are a further number of sub-classifications of both scattering and absorption, as depicted in Fig. 4.7. In nuclear engineering, we are particularly interested in the following sub-classifications of scattering: *elastic scattering*, where the kinetic energy of the system both before and after the interaction is the same and *inelastic scattering*, where some of the kinetic energy of the neutron is absorbed by the nucleus. We shall return to these when we discuss moderation in Chapter 6. With regard to fission, the following sub-classifications of neutron absorption are of most interest: that associated with *capture*, that is the process in which the interaction is deemed ended when the nucleus with the additional neutron is formed, and *fission* where the absorption of a neutron is followed rapidly by the nucleus separating into two fragments and the release of radiations of a wide variety of forms.

The total microscopic cross section can be defined thus,

$$\sigma_t = \sigma_e + \sigma_i + \sigma_c + \sigma_f \tag{4.3}$$

where σ_e, σ_i, σ_c and σ_f are the cross sections for elastic scattering, inelastic scattering, capture and fission, respectively. The sub-processes of multiple-neutron, charged-particle emission, etc. tend to be less significant in most nuclear engineering activities.

4.4.3 NEUTRON PRODUCTION PARAMETERS

Following the absorption of a neutron, the original nucleus ceases to exist in the form it did prior to the interaction. As a result of neutron capture a new isotope is formed which, while having almost identical chemical properties, will almost always have contrasting nuclear properties. In fission, the newly-formed isotope splits into two (occasionally three) fragments each with its own chemical and nuclear properties. Sometimes in nuclear engineering applications involving fission, our objective is to encourage fission to occur (e.g. in a reactor). Conversely, the opposite can be true where we wish to accentuate the likelihood of capture interactions instead—for example to quench the possibility of a reaction or to provide neutron shielding. In this regard, it is instructive to define the ratio of the capture-to-fission cross sections, α,

$$\alpha = \sigma_c / \sigma_f \tag{4.4}$$

As the potential for the supply of neutrons to maintain subsequent fission reactions is also often of interest, the number of neutrons produced as a result of fission per the number of nuclei destroyed in the process, η (eta), is also relevant,

$$\eta = \overline{\upsilon}\sigma_f / \sigma_a \tag{4.5}$$

where $\overline{\upsilon}$ is the average number of neutrons produced per fission event. Considering Eq. (4.5) more closely; the ratio σ_f/σ_a reflects the proportion of all absorption interactions that result in fission, where it is implicit that $\sigma_a = \sigma_\gamma + \sigma_f$ (here σ_γ represents the probability of absorption interactions not resulting in fission, that is radiative capture).

However, the discussion above is not meant to suggest that a small ratio need imply that the number of fissions is too small to yield sufficient neutrons to sustain a chain of subsequent fission reactions, given that the number of neutrons produced in fission (as per $\overline{\upsilon}$) is sufficiently large to offset this. The converse is also true. Here we observe the first indications of the multi-parameter dependence of fission which is reflected by the variety of material, fuel and enrichment properties on which it can depend. It should be borne in mind, however, that we are rarely faced with the highly simplified system implied by Eq. (4.5) associated with a *homogeneous* arrangement of fissile atoms. The use of cross-sectional data as inferred in Eq. (4.5) implies single values, that is, a single isotope at one neutron energy. However, a realistic system comprising of a diversity of nuclides and neutron energies together with the essential engineering features and interfaces (e.g. cooling, cladding, control rods, etc.) is usually considered via a sophisticated computer-based simulation of a heterogeneous reactor system.

By way of example, the cross-sectional data for a variety of actinide isotopes, together with data for α, $\overline{\upsilon}$ and η, are provided in Table 4.1. Note that:

- ^{233}U has the highest value of η, indicating that it is the most efficient producer of neutrons at thermal energies, relative to the other isotopes considered here.
- ^{233}U, ^{235}U, ^{239}Pu and ^{241}Pu all have significant values of η, but in the case of the plutonium isotopes this is because $\overline{\upsilon}$ is significant whereas in the case of uranium isotopes it is because the capture-to-fission ratios are less significant.
- ^{232}Th and ^{238}U have relatively low neutron absorption cross sections.
- ^{240}Pu has a high neutron absorption cross section but a relatively low fission cross section.

4.4.4 THE DEPENDENCE OF CROSS-SECTION WITH ENERGY

Neutrons can exist across a vast range in energy: in thermal reactor systems, their energy range can span some eight orders of magnitude from 0.0253 eV through to 5 MeV and beyond, as discussed

Table 4.1 Microscopic Cross Sections for Absorption σ_a, Capture σ_c and Fission σ_f (at Thermal Neutron Energies) and the Corresponding Capture-To-Fission Ratios σ, Mean Number of Neutrons Per Fission $\overline{\upsilon}$, and Number of Neutrons Per Nuclide Destroyed η for a Variety of Relevant Actinide Isotopes (https://www-nds.iaea.org/sgnucdat/a5.htm)

Nuclide	σ_a/b	σ_c/b	σ_f/b	α	$\overline{\upsilon}$	η
^{232}Th	7.35	7.35	–	–	–	–
^{233}U	574.6	45.5	529.1	0.087	2.50	2.30
^{235}U	681.4	98.8	582.6	0.169	2.44	2.09
^{238}U	2.68	2.68	–	–	–	–
^{239}Pu	1017.4	269.3	748.1	0.361	2.90	2.13
^{240}Pu	290	289.05	0.05	–	–	–
^{241}Pu	1373.2	362.1	1011.1	0.357	3.00	2.21

in Chapter 3. The probability of neutron interactions changes dramatically across this range evident in the dependence of the microscopic neutron cross sections with energy. Further, since the atomic nucleus exhibits properties that are both collective and single-particle in origin, we observe features corresponding to these traits in the dependence of cross-sections with neutron energy.

As an example, the neutron cross-section data are plotted as a function of energy in Fig. 4.8 for ^{238}U (the sum of all possible interactions referred to as the *total*) and ^{235}U (for fission only). Considering ^{238}U in the first instance, it is clear that the dependence with energy is very significant (note: the data are plotted on a $\log_{10}$–$\log_{10}$ plot covering cross sections across six orders of magnitude). It is apparent from the dependence with energy that, in general, the cross section follows approximately $1/\sqrt{E}$, where E is the neutron energy. This trend is indicative of the collective properties of the nucleus and is observed across many isotopes.

Further, there is a highly transient section in the energy region 10^{-5} to 10^{-2} MeV. These are the cross-section *resonances* referred to earlier, associated with specific regions in energy where the interaction likelihood is high. The significance of cross-section resonances in practical terms is that significantly higher interaction probabilities exist (some four orders of magnitude greater in this example) in this energy region than is the case either side. The region in energy where the individual resonances can be discerned is often referred to as the *resolved-resonance region*, while at higher energies in the *unresolved resonance region* the width of the resonance peaks is larger than the space between them and they are therefore no longer discernible as individual features.

In the context of ^{238}U, note that fission does not constitute a significant contribution to the total in Fig. 4.8 because the cross section for induced fission, σ_f, is negligible, as indicated in Table 4.1.

By way of comparison and contrast, the cross-section dependence with energy for induced fission in ^{235}U is also given in Fig. 4.8. This cross section is distinguished by being much larger at low energies than is the case for ^{238}U, consistent with the data for thermal energies in Table 4.1 being $\times 100$ times greater for ^{235}U than ^{238}U. The ^{235}U data also exhibit a more significant decline with energy than is the case for ^{238}U, but it is noteworthy that the resonances for ^{238}U, while in broadly the same energy region, are much more significant (with a few isolated exceptions) by at least an order of magnitude than in the case of ^{235}U. From the comparison of these cross-section data, it is clear that the probability of stimulated fission in ^{235}U is optimised at low neutron energies. The comparison of ^{238}U and

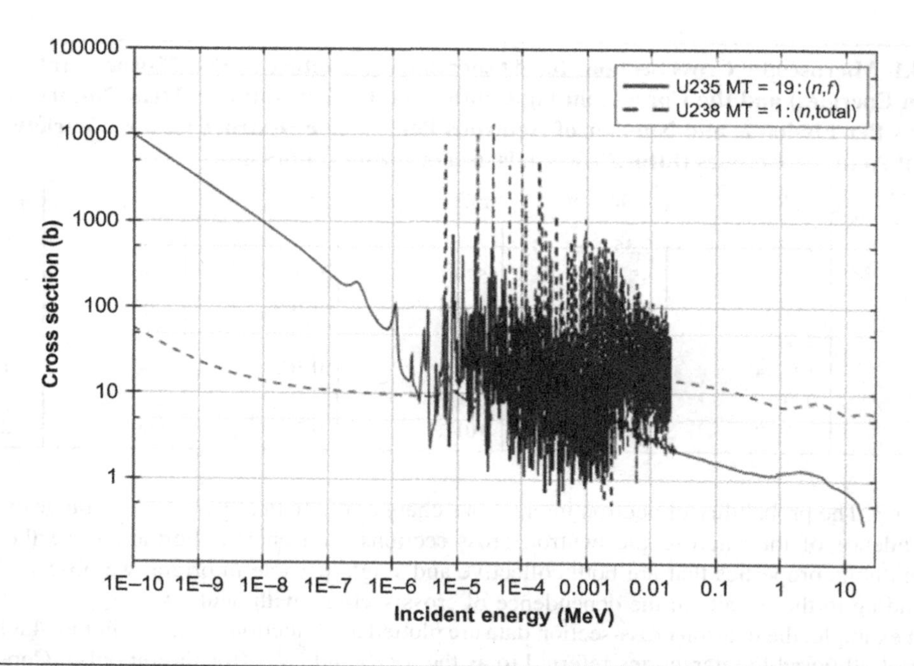

FIG. 4.8

Microscopic neutron cross sections versus energy, σ_f in ^{235}U (neutron-stimulated fission, solid line) and σ_t in ^{238}U (total, dashed line), produced with data from the Data Bank of the Nuclear Energy Agency and JANIS, http://www.oecd-nea.org/janisweb.

^{235}U cross-section data can be particularly useful because, while these isotopes are always present together in fission reactor systems, the data illustrate why their behaviour in a neutron field can contrast with one another.

4.4.5 INTERACTION RATES

In addition to considering the probability that a reaction might take place, we can also consider the rate at which a reaction is likely to occur as a function of time. This is important because too low a rate in a nuclear power context will not deliver sufficient energy for the production of electricity to be useful, whereas too high a rate will risk damage to the reactor. The accepted goal of commercial nuclear power generation systems is one of sustained production at as constant a level as possible to achieve optimum fuel utilisation and high levels of availability. There are other important requirements, such as reactor control, that are also reliant on effective reaction rates in materials that absorb neutrons.

The interaction rate is dependent on the physical properties of the material in which the reaction takes place, and those of the neutron population. Considering, for example, the conceptual material sample introduced in Section 4.4.1 and depicted in Fig. 4.5, we can also describe a beam of neutrons in terms of its density N and its speed v incident on such a sample. Recalling that the sample was of area dA and thickness dx, a volume dV of the beam passes through the sample in time dt, such that

$$\text{Number neutrons per unit time} = N\,dV/dt \qquad (4.6)$$

The distance the neutrons travel in time dt is $v\,dt$; expressing volume in terms of this and the area dA, gives

$$\text{Number neutrons per unit time} = NvdA \tag{4.7}$$

Given the expression for the interaction probability (Eq. 4.2), the reaction rate is equal to the number of neutrons (Eq. 4.7) multiplied by the interaction probability to yield,

$$\text{Interaction rate} = Nvn\sigma dV \tag{4.8}$$

which is a little cumbersome. However, considering the component of this expression that is the product of the atomic number density n and cross section σ, it is plausible to think of this conceptually as the sum of all the cross sections for a given quantity of substance, divided by the volume. A re-arrangement of Eq. (4.2) implies that the product $n\,\sigma$ is related to the probability of interaction per unit distance, which can be defined as the *macroscopic cross section* Σ,

$$\Sigma = n\sigma \tag{4.9}$$

Then the reaction rate can be expressed in terms of the product of the macroscopic cross section, neutron flux ϕ and the volume of the sample, thus,

$$\text{Reaction rate} = \Sigma\phi dV \tag{4.10}$$

where the neutron flux is the number of neutrons per unit area per unit time and Σ has units of inverse distance, as per its definition above.

As indicated earlier, our primary discussions of the microscopic cross section ignored the fact that transport through a material will influence the energy of a neutron and thus we must consider microscopic cross sections to be functions of energy, that is $\sigma = \sigma(E)$, for all but thin samples. Similarly, the assumption of a neutron flux that is unaffected by interaction in the sample only holds for thin samples; for all other scenarios where this is significant, $\phi = \phi(x)$, where x is the sample thickness and,

$$\phi(x) = \phi(0)e^{-\Sigma x} \tag{4.11}$$

as the neutron flux falls off exponentially with increasing distance with the attenuating properties of a given material described by the macroscopic cross section.

This can be further defined *as per*,

$$\Sigma = {}^{1}/\lambda \tag{4.12}$$

introducing λ which, in this context, is the *mean free path* of the neutron. Qualitatively, we can think of λ as being the average distance a neutron travels before it undergoes an interaction, where the type of interaction is specified by the nature of the corresponding microscopic cross section used in the specific definition of Σ, for a particular interaction type, e.g. scattering, fission etc. Quantitatively, λ is the distance over which the incident neutron flux $\phi(0)$ falls by a factor e, or by ~63%.

In industrial applications, we are often faced with materials that are compounds of several elements, particularly when considering substances that comprise fuels, moderators, coolants and shielding. In such cases, the macroscopic cross section can be calculated for each isotopic component of the compound via the product of the microscopic cross section and the corresponding number density, as per Eq. (4.9). These are then summed to yield the corresponding macroscopic cross section for the composite substance.

4.5 MASSIVE ISOTOPES AND THE CONCEPT OF BINDING ENERGY PER NUCLEON

The only isotopes in which fission can be induced by neutrons to a practically significant level and that, subsequently, yield sufficient numbers of neutrons to sustain fission in neighbouring nuclei are some of those of uranium and the transuranic elements. As discussed in Chapter 1, very few of these elements remain as primordial constituents of the Earth's crust since they are all radioactive and most have decayed away to a very significant extent; only ^{235}U remains to a degree that is extractable on an industrial basis and even in this case significant purification and refinement is required to render it of practical use in energy generation.

The nuclei of the transuranic isotopes are amongst the most massive yet identified and this has a direct influence on their stability and consequently their predisposition to decay via a variety of mechanisms, particularly fission. Earlier in this chapter the concept of nuclear binding energy scaling in proportion with mass was introduced, as depicted in Fig. 4.3. However, it is also important to compare the behaviour of nuclei alongside one another, normalised in terms of nucleon number. This can be calculated relatively easily by deriving the nuclear binding energy, via a comparison of the atomic mass of each isotope with the mass of its constituent nucleons, and dividing this through by the number of nucleons in each case as is illustrated in Case Study 4.3 for ^{235}U. If similar calculations are performed for all of the isotopes in the chart of the nuclides (or at least a representative sample of those lying along the valley of stability), and the results are then plotted as a function of the atomic mass number A, the dependence for the *binding energy per nucleon* as a function of mass number, is obtained, as shown in Fig. 4.9.

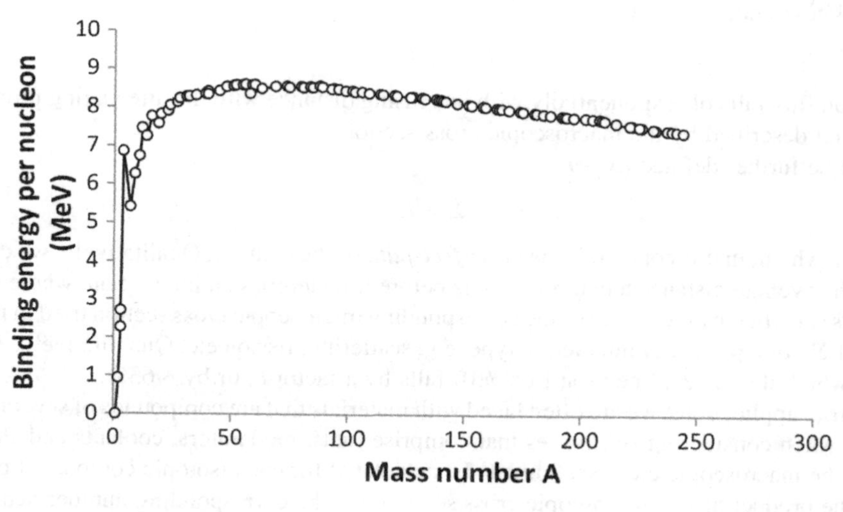

FIG. 4.9

Binding energy per nucleon as a function of atomic mass number, for ^{2}H through to ^{252}Cf.

Fig. 4.9 is insightful because it demonstrates that the binding energy per nucleon is not independent of mass. Such an independence might be expected if the strong nuclear force were simply some additive property escalating in proportion each time a nucleon were added to a given nucleus. Rather, it increases very rapidly for light isotopes to a maximum in the region $A \sim 56$ (associated with iron) and then declines much more gradually through to the highest mass numbers. This latter trend reflects, in part, the increasing perturbation on nuclear stability of the repulsion of the protons with increasing Z.

Two features of the binding energy per nucleon dependence with atomic mass are of particular interest: the energy that might be made available by separating an isotope into smaller, constituent parts as occurs in fission, and that that might be evolved by amalgamating light isotopes into heavier ones as occurs in fusion. To illustrate the first of these possibilities, let us consider a ^{252}Cf nucleus undergoing *spontaneous fission*. Californium-252 has a binding energy per nucleon of 7.30 MeV and hence a total binding energy of ~1838 MeV. It is susceptible to spontaneous fission, that is, it fissions without any external stimulus, as a natural mode of radioactive decay. While the specific products of a given fission event cannot be predicted with any certainty, we can anticipate some possibilities in terms of the mass regions of the fragments based on the mass asymmetry of fission known from previous measurements, such as depicted in Fig. 4.4. Almost always there is a light fragment and a heavy fragment even in the case of ternary fission,[1] viz. the following example,

$$^{252}_{98}\mathrm{Cf} \rightarrow {}^{139}_{55}\mathrm{Cs} + {}^{109}_{43}\mathrm{Tc} + 4n$$

The energy released in this event, by way of example, is the total binding energy of the fission products minus the binding energy of ^{252}Cf; since the total binding energy of the fragments is 1135 MeV plus 902 MeV for ^{139}Cs and ^{109}Tc, respectively, and that of ^{252}Cf is 1838 MeV, the difference is ~200 MeV. This, at some eight orders of magnitude, is much greater than the energy released in the breaking of an atomic bond in combustion. This is the principal reason why such excitement erupted with the pioneering observation of fission in 1938.

In addition to calculations of the energy released in fission, it is relatively straightforward to estimate the approximate way in which the energy is apportioned between the products of the reaction. A reasonable starting point is to assume that the majority of the kinetic energy of the fragments is due to the Coulomb repulsion between them; this is a reasonable assumption given the significant mass of the fragments relative to everything else that is ejected in a fission event. When this is done, it is clear that the kinetic energy of the fission fragments accounts for approximately 85% of the total energy available, with the rest apportioned relatively evenly amongst the prompt and delayed emissions of radiation (prompt and delayed neutrons, prompt and delayed γ, delayed β and capture γ rays). A typical distribution for a fission event in which approximately 200 MeV is released is given in Table 4.2. Clearly, at least one neutron (plus a few more on average to account for absorption and leakage) is needed to sustain the critical chain reaction. The majority of the rest of the products impart their energy to the surroundings, usually in a way that is optimised by design to the reactor coolant for the case of power generation applications. In addition to the leakage of some neutrons, other highly penetrating radiations (e.g. high-energy γ rays, anti-neutrinos, etc.) are lost from the plant.

[1]The third fragment in ternary fission is almost always the triton (^{3}H) and thus does not affect the mass of the other fragments significantly in terms of their mass region. Symmetric, spontaneous fission for ^{252}Cf has a much smaller mass yield than asymmetric fission.

Table 4.2 The Approximate Distribution of Prompt Energy Deposition Amongst the Products of a Fission Event Induced by Thermal Neutrons in ^{235}U

Product	Energy (MeV)	% Total
Fragments	170	94
Prompt neutrons	5	3
Prompt γ	6	3
Total	181	100

After D.G. Madland, Total prompt energy release in the neutron induced fission of ^{235}U, ^{238}U and ^{239}Pu, Nucl. Phys. A 772 (2006) 113–137

4.6 DIFFERENT MODES OF FISSION

If the more massive isotopes have a predisposition toward fission, this begs the question as to why some of these isotopes fission spontaneously where as others require a stimulus to do so; indeed why is it that the heavy nuclei beyond a limiting mass do not just fission leaving nothing else behind? The answer to these questions is complicated but our understanding of it benefits from a closer consideration of the energy that is needed to overcome the cohesion of the nucleus. The cohesion constitutes a threshold commonly referred to as the *activation energy*; this is the energy necessary to overcome what is termed the *Coulomb barrier*.

It is useful in this context if we think of a fission event happening in reverse, albeit hypothetically. In this scenario, as the two charged fragments are brought closer together, the Coulomb potential between them increases (along with the force of repulsion) as the separation r gets smaller. This is illustrated by the expression for Coulomb potential, V, and its inverse proportionality to the separation r of the fragments of atomic number Z_a and Z_b, in Eq. (4.13),

$$V = \frac{1}{4\pi\varepsilon_0}\frac{Z_a Z_b e^2}{r} \tag{4.13}$$

where e is the charge on the electron and ε_0 is the relative permittivity of free space, $8.85 \times 10^{-12}\ \mathrm{Fm}^{-1}$. The potential energy increases due to the inverse proportionality with separation until the parent nucleus coalesces. The binding energy of this nucleus is represented by a well that is incorporated into the trend with r for $r < r_0$ depicted *as per* Fig. 4.10 where r_0 is the radius of the coalesced nucleus. Based on this illustration, three scenarios are possible:

1. A nucleus might possess sufficient energy to place it above the Coulomb barrier. In this case, the activation energy is zero and there is nothing to prevent the nucleus proceeding immediately to fission, notwithstanding the influence of any quantum mechanical effects of the energy level structure of individual nucleons, which we shall ignore for the purposes of this illustration. For the very massive isotopes that might exist in this category, that is, $A > 300$, none exist for sufficiently long to have been identified, at least by current methods, due to their instability toward spontaneous fission.
2. A nucleus can have an excitation energy that places it near to the top of the Coulomb barrier but it does not have sufficient energy to pass over it. However, as a subatomic system behaving according to quantum mechanics, there is an appreciable probability that it might *tunnel* through the barrier to fission. This is the mechanism behind *spontaneous fission*, such as is observed in the case of ^{252}Cf, ^{244}Cm, ^{240}Pu and ^{238}U, for example. Generally speaking, the level of

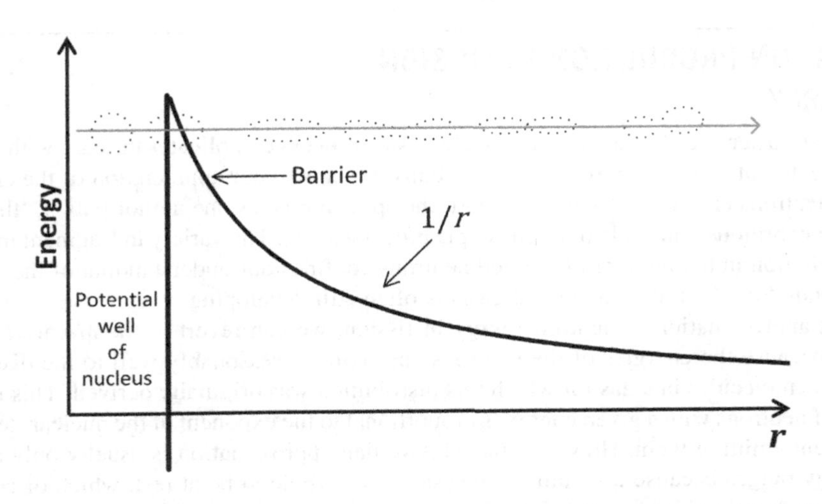

FIG. 4.10

A schematic depiction of energy versus separation of the fragmentation of a nucleus in fission.

excitation energy and thus the relative thickness of the Coulomb barrier has a significant influence on the probability of fission and consequently the observed half-life.

3. Or a nucleus might reside in the potential well with an excitation energy that is too low to yield any noticeable level of spontaneous fission because the probability of tunnelling through the relatively thick barrier at this level is too small. However, the absorption of a neutron (or photon) might place the nucleus so formed high relative to the barrier and thus cause it to be advantageously predisposed to fission. Hence *stimulated fission* is observed as in the case of ^{235}U and ^{239}Pu when these isotopes are effectively lifted by the absorption of a thermal neutron to the excited forms of ^{236}U* and ^{240}Pu*, respectively. For the case of those isotopes that are not vulnerable to stimulated fission with thermal neutrons, such as ^{238}U, the probability of fission can be raised by stimulation with higher energy neutrons. This is termed *above-threshold fission*; the higher incident energy is sufficient to yield the intermediate nucleus formed as a result of the neutron absorption to be higher in energy where the thickness of the barrier to fission is reduced relative to that at thermal energies.

While it is instructive to think of the parent nucleus as comprising two fragments preformed within the potential, this is unlikely to be the case in reality. Nonetheless, this does not undermine the benefit of the treatment based on the Coulomb barrier described above to understanding this complicated phenomenon; it is similar to the widely accepted theoretical treatment of α decay, with the α particle considered to be in some way preformed. The barrier methodology also enables half-lives to be predicted reasonably well and it supports the concept of activation energy, although this analysis is beyond the scope of this text. It also assists our understanding of the distinction between the spontaneous and stimulated fission decay pathways. Thus, the predisposition of matter to fission can be considered in terms of those isotopes susceptible to spontaneous fission, and those susceptible to stimulated fission; the former exhibit significant range in half-life but fission often only accounts for a very small fraction of the total decay probability of the isotope with the majority usually being confined to other modes of disintegration, principally α decay. Stimulated fission (also known as induced fission) affords the opportunity to establish and control a fission chain reaction. A discussion of this, and its industrial use, forms the focus of the subsequent chapters of this text.

4.7 NEUTRON PRODUCTION IN FISSION

4.7.1 ENERGY

As highlighted earlier, the stochastic nature of the fission process complicates the ease with which a full theoretical treatment of it is achieved. This difficulty extends to our appreciation of the energy spectrum of the neutrons emitted in fission; in short, the spectrum is, as one author puts it, 'the statistical average of an enormous multitude of microscopic processes' [1]. The variety in fragment mass, energy and the distribution in the number of emitted neutrons confines our understanding of the spectrum to empirical trends for which the theoretical basis is often still developing.

As a first approximation to neutron energy in fission, we can revert to the *Maxwell–Boltzmann distribution* because the energies of the neutrons approximate reasonably well to the distribution of energies of the molecules in a gas for which this distribution was originally derived. This implies that the number of neutrons with a given energy is proportional to the exponent of the nuclear 'temperature' of the fragment emitting them. However, the Maxwellian approximation is usually only adequate in limited energy ranges because it assumes the system as a whole to be at rest, which of course is not the case. Consequently, this approach provides a reasonable description of the distribution of energies in thermal energy ranges associated with the thermalized neutron distribution in a reactor utilising neutrons with thermal energies. The situation in which neutrons are boiling off a given fission fragment is rather different. As described earlier, the fragments are moving very rapidly indeed when neutrons are emitted from them and this needs to be included in a treatment of the neutron energy spectrum.

Early studies, including unpublished work by N. Feather, approximated the spectrum N of neutrons energies E_n to follow:

$$N(E_n) \propto E_n e^{-E_n/Q} \tag{4.14}$$

Eq. (4.14) assumes that the neutrons are emitted isotropically in the frame of reference of the fragment and that the associated fragment has accelerated to its maximum kinetic energy before emission, consistent with the neutrons being emitted $>10^{-15}$ s after scission. These assumptions have been confirmed experimentally [2]. In common with the Maxwellian approach, Q corresponds to the excitation state of the fragment or its 'temperature' (albeit temperature in a quantum mechanical context), giving

$$N(E_n) = 2\sqrt{\frac{E_n}{\pi Q^3}} e^{-E_n/Q} \tag{4.15}$$

An advancement of the Maxwellian approach that includes the motion of the fragment is the *Watt spectrum* [3]. In this case, the Maxwellian expression is transformed from the centre-of-mass frame of reference via a Galileo transformation to the laboratory frame (relativistic corrections only becoming necessary >20 MeV) [4],

$$N(E_n) = \frac{e^{-E_W/Q}}{\sqrt{\pi E_W Q}} e^{-E_n/Q} \sinh\left[2\sqrt{E_n E_W/Q}\right] \tag{4.16}$$

where E_W is the kinetic energy of the fragment per nucleon. A variety of data have been tested against this fit to provide values for the parameters E_W and Q culminating in specific expressions for a given isotope, such as in the case for ^{252}Cf where[6] $E_W = 0.359$ MeV and $Q = 1.175$ MeV,

$$N(E_n) = 0.64 e^{-0.85 E_n} \sinh\left[1.02\sqrt{E_n}\right] \tag{4.17}$$

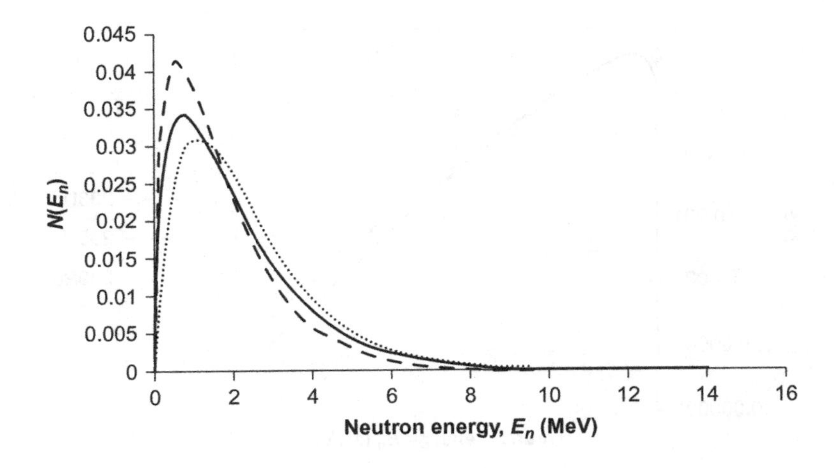

FIG. 4.11

Distributions of $N(E_n)$ versus neutron energy E_n for the Feather expression (dotted line, Eq. 4.14), Maxwellian (dashed line, Eq. 4.15) and the Watt spectrum for ^{252}Cf (solid line, Eq. 4.16) produced with data from the Data Bank of the Nuclear Energy Agency and JANIS, http://www.oecd-nea.org/janisweb. The data have been normalised to the sum being unity that is $\Sigma N(E_n) = 1$ to provide a consistent comparison showing the evolution of the empirical fit.

A comparison of the expressions given for Eqs (4.14), (4.15) culminating in the Watt spectra (Eqs 4.16, 4.17) is given in Fig. 4.11. It is noteworthy that the Watt spectrum provides a satisfactory model that is comparable with more sophisticated approaches based on microscopic models. Often, the various coefficients are subsumed into three parameters, that is, a, b and c, with a different group of these corresponding to each isotope, such that,

$$N(E_n) = ce^{-E_n/a} \sinh \sqrt{bE_n} \tag{4.18}$$

A comparison of spectra for thermal-neutron induced fission in ^{233}U, ^{235}U and ^{239}Pu is given in Fig. 4.12. Here it can be appreciated that the spectra are very similar to one another, with the most significant diversity becoming apparent at the higher energies. A similar dataset for neutron energies arising from spontaneous fission in ^{240}Pu, ^{244}Cm and ^{252}Cf is given in Fig. 4.13. In this case, the relative hardness of the ^{252}Cf spectrum over that of the others, especially ^{240}Pu, is evident but otherwise widespread similarity is apparent. A variety of nuclear databases provide the parameter basis for the calculation of a wide variety of fission neutron energy spectra.

Two parameters often extracted from fission neutron energy spectra are the *most probable neutron energy* and the *average neutron energy*. For example, for ^{252}Cf these are 0.7 and 2.1 MeV, respectively.

4.7.2 MULTIPLICITY

The term multiplicity is not specific to the consideration of fission neutrons. We might, for example, refer to the multiplicity of γ rays resulting from a cascade as part of the decay scheme of an excited nucleus. Multiplicity tends to refer to the number of events emitted within a specified window in time,

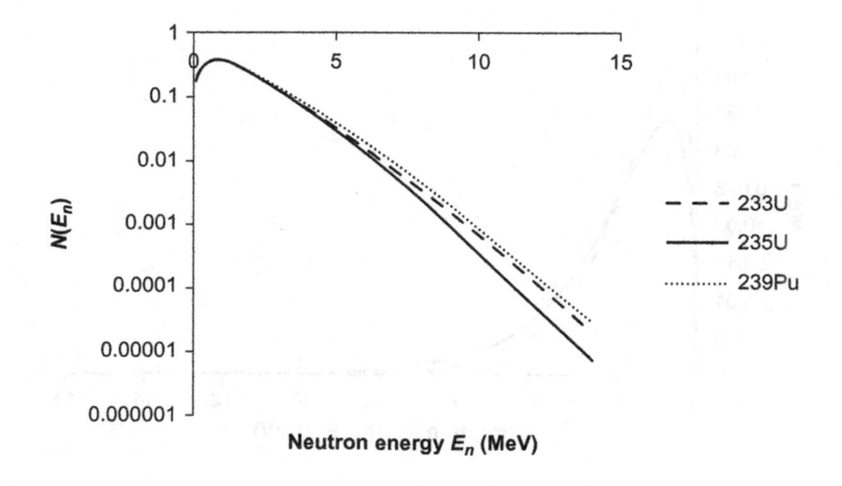

FIG. 4.12

Distribution $N(E_n)$ versus neutron energy E_n for thermal-induced fission in ^{233}U, ^{235}U and ^{239}Pu plotted in the more conventional way as $\log_{10} N(E_n)$ versus E_n.

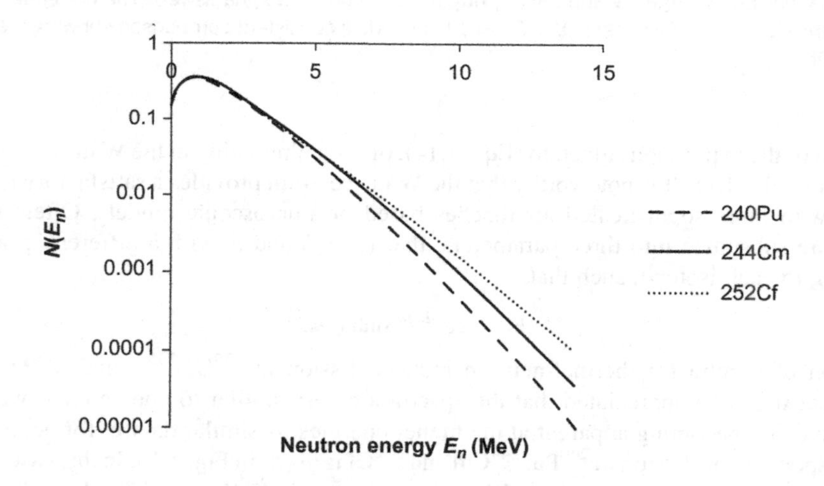

FIG. 4.13

Distribution $\log_{10} N(E_n)$ versus neutron energy E_n for spontaneous fission in ^{240}Pu, ^{244}Cm and ^{252}Cf.

where the width of the window is usually short relative to the inverse of the emission rate. We might try to detect such radiation in such a very small space of time because a correlated cascade of discrete transitions usually occurs very rapidly indeed (notwithstanding the presence of meta-stable states that do not concern us here). In restricting our detection to a brief, synchronised interval in time, we can filter in favour of radiations that are correlated in time and that are thus connected by virtue of their role in interconnected nuclear processes, while excluding those that are not. Uncorrelated sources of radiation include background radiation and scattering in the environment in which the measurement is made, for example.

Each fission event constitutes one of the most significant upheavals in the structure of matter at a nuclear scale and yields an associated number of prompt neutrons. As introduced earlier, because fission is a stochastic process we cannot predict exactly how many neutrons will be emitted from a specific fission event. However, we observe that the distribution of the number of neutrons emitted in fission is somewhat specific to a given isotope although not entirely distinct from that of others. For the case of induced fission, the distribution is also dependent on the energy of the stimulating neutron, exhibiting a trend in which the average of the multiplicity distribution scales linearly with energy in most cases. While there are much fewer isotopes with a susceptibility to fission relative to those emitting correlated γ-ray cascades, the measurement of neutron multiplicity is arguably less straightforward because neutrons can be more challenging to detect.

For a specific type of fission event we can define a probability (with the conventional notation of P_v) that v neutrons are emitted. The distribution of P_v versus the multiplicity v is called the *neutron multiplicity distribution*. In Fig. 4.14, a comparison of two multiplicity distributions for thermal-neutron induced fission is provided, for the cases of ^{235}U and ^{239}Pu. On this figure, as is customary with multiplicity distributions, the corresponding Gaussian fits are also provided for each isotope. This enables the average number of neutrons emitted in fission for each isotope to be calculated, usually given the notation $\overline{v}$. Notice how, on average, ^{239}Pu emits a significantly greater number of neutrons than ^{235}U, as reflected by their entries in Table 4.1, and their $\overline{v}$ values of 2.9 and 2.5, respectively. In Fig. 4.15, the corresponding multiplicity distributions for ^{235}U at four elevated neutron energies are given, illustrating the effect of this causing $\overline{v}$ to increase in correspondence.

One of the principal uses of neutron multiplicity arising from fission reactions is in the management of nuclear material to prevent it being used for illicit purposes; this known as *materials accountancy* in

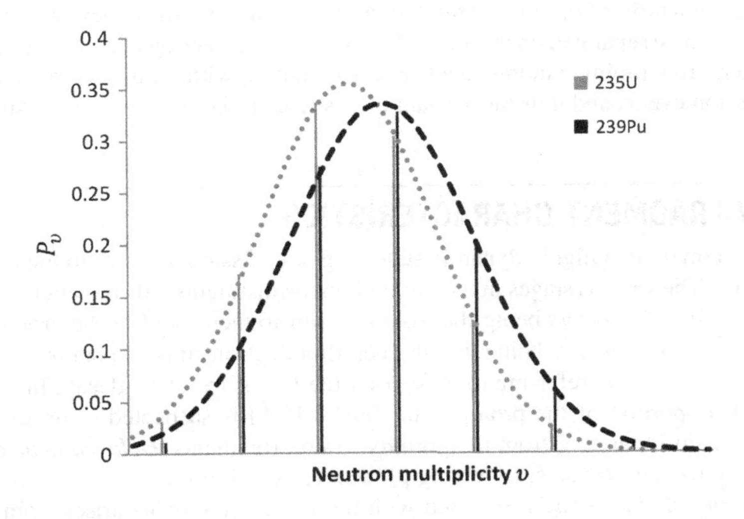

FIG. 4.14

Fission neutron probabilities P_v versus multiplicity v for thermal-neutron–induced fission in ^{235}U and ^{239}Pu, with Gaussian fits to the data (dotted and dashed lines, respectively).

After J.P. Lestone, Energy and isotope dependence of neutron multiplicity distributions, Report No. LA-UR-05-0288.

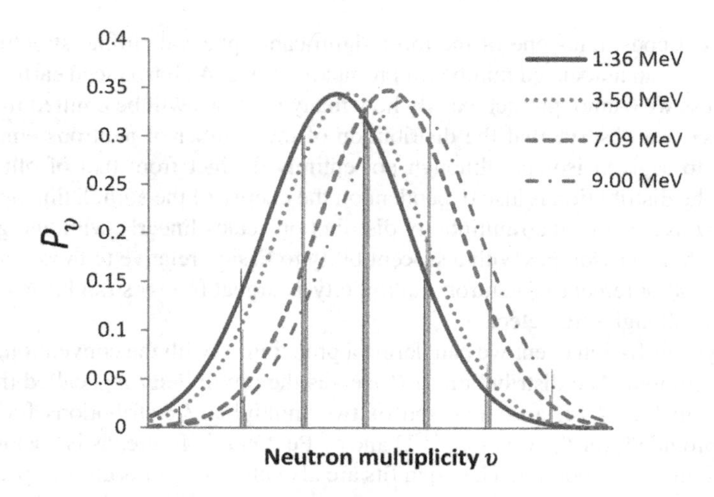

FIG. 4.15

Fission neutron probabilities P_υ versus multiplicity υ for fast-neutron induced fission in ^{235}U for various energies [2] of 1.36, 3.50, 7.09 and 9.00 MeV with corresponding $\overline{\upsilon}$ of 2.55, 2.81, 3.43 and 3.74, respectively, after Lestone.

the wider field of *nuclear safeguards*. The correlation in time of neutrons emitted from a fission event enables nuclear materials with a susceptibility to fission to be isolated from other sources of neutron radiation, such as the production of neutrons by the reaction of α particles on light isotopes. These sources of neutrons are not indicative of fission and thus not of fissile material; since they are not correlated with fission events they can be separated on this basis. In practice, the detection of a 'start' event constitutes a trigger opening a narrow timing window; neutron events falling within this window can be correlated specifically to fission events and thus the amount of fissile material present can be estimated.

4.8 FISSION FRAGMENT CHARACTERISTICS

Fission fragments remain in a highly dynamic state long after fission, relative to the timescales of the prompt phenomena. The various stages in their evolution are distinguished conventionally by name [5] with the *primary fission fragments* being the isotopes prior to their yield of the prompt neutrons and γ rays, the *primary fission products* being the isotopes after neutron emission but prior to β^- decay and the *secondary fission products* referring to those after the first stage of β^- decay. In terms of neutron emission, a small proportion of the prompt emission ($<10\%$) is suspected to occur at the point of scission, with the majority arising from the primary fission fragments. *Delayed neutrons* are emitted from the secondary fission products, following β^- decay, by definition.

The most striking characteristic associated with the fission fragments arises from radiochemical analysis and mass spectroscopy measurements. These indicate that the yield of fission fragments as a function of mass is asymmetric, as was introduced in Section 4.3 and depicted earlier in Fig. 4.4. Rather than a symmetric distribution centred around a mass of approximately half that of the original fissionable nucleus, two maxima are observed, one associated with a light (low-mass) fragment and another associated with a heavy fragment. The peak-to-valley ratio of the minimum between the fragment maxima can be as great as ~750 while the symmetric fission product yield (e.g. for isotopes such as ^{121}Sn in the case of the spontaneous fission of ^{244}Cm) can be as low as ~0.001%. To illustrate the extent of the mass asymmetry of fission fragments the yields as a function of mass for ^{244}Cm and ^{252}Cf are shown in Fig. 4.16.

The asymmetry in fission fragment yield with mass is observed to be qualitatively similar for most cases of fission, but some differences are observed with changes in mass of the fissionable isotope and excitation energy. The mass asymmetry is generally most pronounced at low energies (e.g. in the case of thermal neutron-induced fission of ^{235}U) and the yield of symmetric products increases with the

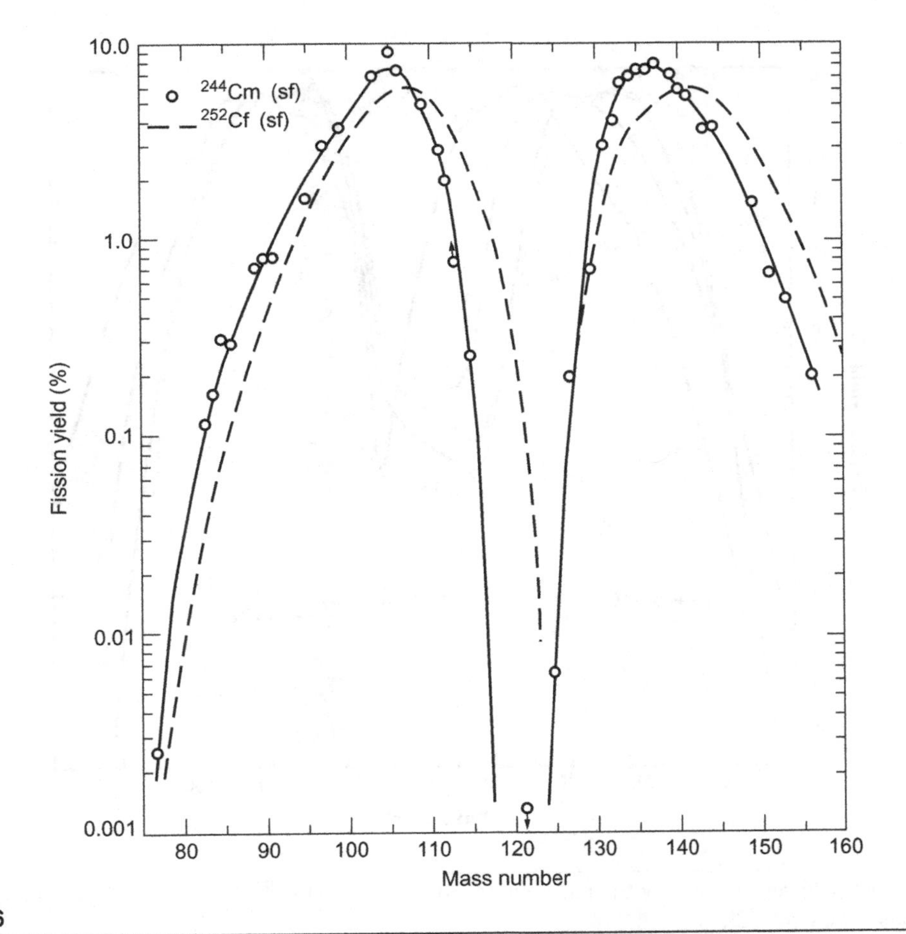

FIG. 4.16

Fission product yield versus mass number for spontaneous fission of ^{244}Cm and ^{252}Cf.

After K.P. Flynn, B. Srinivasan, O.K. Manuel, L.E. Glendenin, Distribution of mass and charge in the spontaneous fission of ^{244}*Cm, Phys. Rev. C 6(6) (1972) 2211–2214.*

energy of the stimulating particle [6]. Most often the stimulus is a neutron but other particles such as α particles, protons and photons can also stimulate fission.

Although a range of transuranic isotopes are susceptible to fission (either spontaneous or stimulated), scaling their mass does not result in a proportionate change in the mass of both fragments as might be expected. The heavy fragment (in the $A \sim 139$ region) is bounded in terms of nuclear structure by closed nucleon shell at 82 and 50 for neutrons and protons, respectively. This is manifest as a lower-mass boundary at $A \sim 130$ beyond which the mass of the heavy fragment is confined. Consequently, a variation in mass of the parent isotope is often observed more significantly as a change in mass of the light fragment while the mass of the heavy fragment changes significantly less. This is shown in terms of the fission fragment yield versus mass for a variety of isotopes in Fig. 4.17.

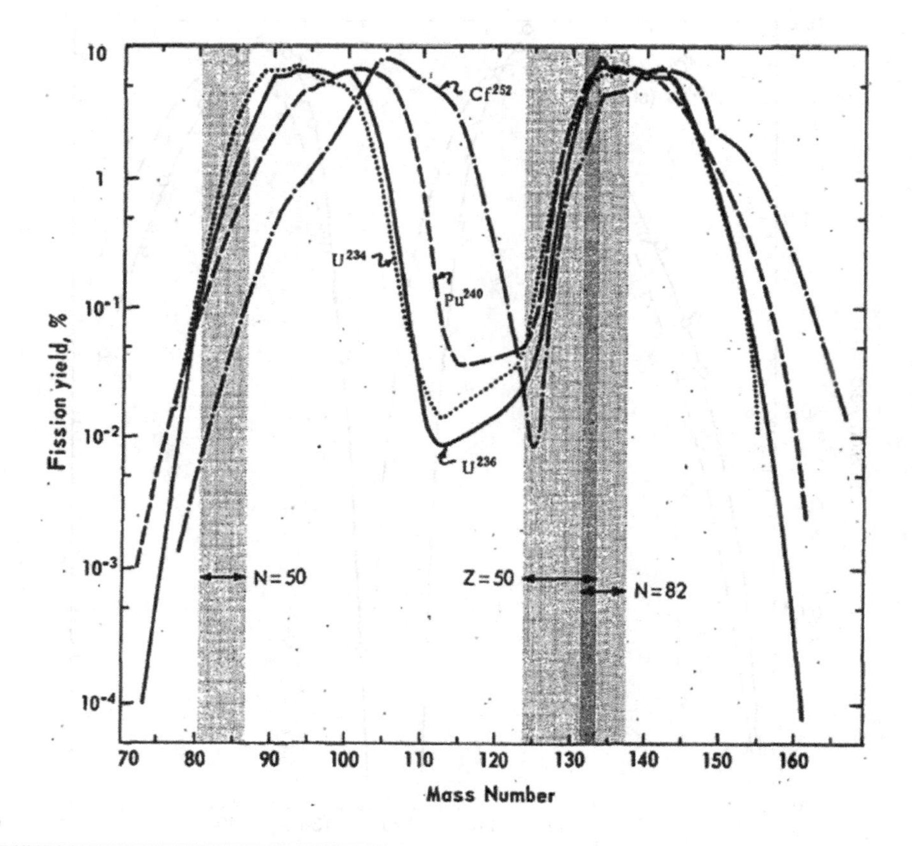

FIG. 4.17

Fission product yield versus mass for thermal neutron-induced fission for ^{233}U, ^{235}U and ^{239}Pu and spontaneous fission in ^{252}Cf.

With permission from the IAEA International Atomic Energy Agency, A.C. Wahl, Mass and charge distribution in low-energy fission, in: Physics and Chemistry of Fission, Proceedings of the Symposium on Physics and Chemistry of Fission Held by the International Atomic Energy Agency in Salzburg, 22–26 March 1965, Vienna, 1965, pp. 317–331.

The mass asymmetry observed of the fission fragments is not merely an interesting subtlety of the underlying nuclear physics. The control of power reactors, and the treatment and disposal of nuclear waste is complicated by the presence of a relatively small number of long-lived isotopes. A significant proportion of these are associated with fission products that show a strong correlation with the maxima in the fission product yield distribution with mass. These include, for example, ^{85}Kr, ^{90}Sr, ^{104}Ru, ^{135}Xe, ^{137}Cs and ^{149}Sm.

CASE STUDIES

CASE STUDY 4.1 NEUTRON SHIELDING PROPERTIES OF URANIUM

It is instructive to consider a case associated with a calculation of the mean free path of neutrons in matter, as this demonstrates the highly penetrative properties of the neutron.

Consider a 1-cm-thick plate of uranium metal that is depleted in ^{235}U, and hence for the sake of this example we can assume to comprise 100% ^{238}U. If a flux of neutrons is introduced from a ^{252}Cf source of $10^6\ s^{-1}\ cm^{-2}$ to one side of the plate, what flux would we expect to witness on the other side of the plate, ignoring any moderation by the plate and scatter in the surroundings?

Firstly let's consider Eq. (4.11), reproduced here for clarity,

$$\phi(x) = \phi(0)e^{-\Sigma x}$$

Given that $\phi(0) = 10^6\ s^{-1}\ cm^{-2}$ and $x = 1$ cm, we need to find Σ, where $\Sigma = n\sigma$ (Eq. 4.9). Since the density of uranium metal is 18.95 g cm^{-3} with a molar mass of 238.03 g, then $n = 4.80 \times 10^{22}\ cm^{-3}$.

We can approximate $\sigma = \sigma_t$ for ^{238}U at the most probable energy of neutron emission for ^{252}Cf, which is 0.7 MeV. From Fig. 4.8 it is clear that at 0.7 MeV, $\sigma_t \sim 10$ b. Thus,

$$\begin{aligned}\phi(x = 1\,\text{cm}) &= 10^6 \times e^{-4.80\times10^{22}\times10^{-23}\times1} \\ &= 6.2 \times 10^5\ s^{-1}cm^{-2}\end{aligned}$$

From this result, we can conclude that 62% of the neutrons incident on the uranium pass through it and hence, while uranium is not as good at stopping neutrons as other materials we might select (as discussed in Chapter 6), it is also clear that neutrons are extremely penetrating. Further, we have assumed the uranium does not affect the energy of the neutrons, and thus that the cross section is constant, which it will.

CASE STUDY 4.2: MEAN FREE PATH OF 1 MEV NEUTRONS IN URANIUM DIOXIDE OF 5 % WT. ENRICHMENT

To find the neutron mean free path for uranium dioxide, UO_2, it is necessary to know the density (10.97 g cm^{-3}) and that $^{235}UO_2$ has a molar mass of 267.04 g and $^{238}UO_2$ one of 270.03 g. Therefore 1 cm^3 UO_2 has a mass 10.97 g of which 0.55 g is $^{235}UO_2$ and 10.42 g is $^{238}UO_2$. Thus $n_{235} = 1.24 \times 10^{21}$, $n_{238} = 2.32 \times 10^{22}$ and $n_O = 4.90 \times 10^{22}\ cm^{-3}$. At 1 MeV, the total microscopic cross sections for ^{235}U, ^{238}U and ^{16}O are 6.8 b, 7.1 b and 8.2 b, respectively.

Given that: $\Sigma_{total} = \Sigma_1 + \Sigma_2 + \Sigma_3 + \cdots$, in this case

$$\Sigma_{UO_2} = \Sigma_{235} + \Sigma_{238} + \Sigma_{O_2}$$
$$= n\sigma_{235} + n\sigma_{238} + n\sigma_{O_2}$$
$$= (1.24 \times 10^{21} \times 6.8 \times 10^{-24}) + (2.32 \times 10^{22} \times 7.1 \times 10^{-24}) + (4.90 \times 10^{22} \times 8.2 \times 10^{-24})$$
$$= 0.58\,\text{cm}^{-1}$$

Hence, $\lambda_{UO_2} = 1.74\text{cm}$

CASE STUDY 4.3 CALCULATING BINDING ENERGY OF ^{235}U

The atomic mass of ^{235}U is 235.04 u and it comprises 92 protons and 143 neutrons. Using atomic mass units: $m_{235} = 235.04$, $m_p = 1.007$ and $m_n = 1.009$u.

Hence,

$$\begin{aligned} \text{Binding energy} &= (143 \times m_n) + (92 \times m_p) - m_{235} \\ &= 144.29 + 92.64 - 235.04 \\ &= 1.89\,\text{u and since } 1\,\text{u} = 931.5\text{MeV/c}^2 \text{ then} \\ &= 1760.54\,\text{MeV} \end{aligned}$$

CASE STUDY 4.4: FISSION FRAGMENT ENERGY DISTRIBUTION

Consider the following spontaneous fission example of ^{240}Pu, at the point just following scission:

$$^{240}_{94}\text{Pu} \rightarrow {}^{94}_{38}\text{Sr} + {}^{146}_{56}\text{Ba}$$

Assuming that at the point of fission the entirety of the energy released in fission is apportioned as kinetic energy to the fission fragments that is, there are no scission neutrons, let us estimate how much energy this is. ^{94}Sr has 38 protons and ^{146}Ba has 56. The kinetic energy in MeV due to the Coulomb repulsion of the fission fragments of charge Z_a and Z_b can be written as

$$T_f = \frac{Z_a Z_b e^2}{4\pi\varepsilon_0 r_0 \left(A_a^{1/3} + A_b^{1/3}\right)}$$

where r_0 is equal to 1.2 fm, arising from the expression for the nuclear radius $r = r_0 A^{1/3}$, A_a and A_b are the atomic mass numbers of the fragments and ε_0 is the permittivity of free space, 8.85×10^{-12} Fm^{-1}.

So in this case,

$$T_f = \frac{(38 \times 56)\left(1.6 \times 10^{-19}\right)^2}{\left(1.1 \times 10^{-10}\right)\left(1.2 \times 10^{-15}\right)\left(94^{1/3} + 146^{1/3}\right)} = 263\,\text{MeV}$$

In this analysis, we have assumed $r_0 = 1.2$fm which assumes the fission products are spherical whereas at the point of scission they are deformed. Often this is reflected by $r_0 = 1.8$ fm being used instead, giving an energy of 175 MeV. Interestingly, if we compare the binding energy difference between the parent and fission fragments, we obtain an estimate of 190 MeV, with the kinetic energy accounting for ~90% as anticipated.

REVISION GUIDE

On completion of this chapter you should:

- understand the process of fission, the introductory theoretical basis for its existence and the explanation this provides for what is observed
- appreciate the significance of neutron interactions in the discussion of fission, particularly the concepts of micro- and macroscopic cross sections, and the related concept of the neutron mean free path
- be able to sketch the dependence of the microscopic cross-section as a function of energy, the fission fragment mass distribution, fission neutron energy spectra and fission neutron multiplicity distributions
- be able to calculate nuclear binding energies, mass defects and the kinetic energies of primary fission fragments
- understand what is meant by the average and most-probable fission neutron energies and mean number of neutrons produced per fission

PROBLEMS

1. Consider a 1-cm-thick plate of high-density polyethylene of density 0.95 g cm^{-3} and atomic number density ratio of 33% carbon and 66% hydrogen. If we introduce a flux of neutrons from a ^{252}Cf source of 10^6 s^{-1} cm^{-2} with most probable energy of 0.7 MeV to one side of the plate, what proportion of the uninterrupted flux would we expect on the other side?

You should ignore any moderation of the neutrons by the plate and scatter in the surroundings. The total microscopic neutron cross sections at 0.7 MeV are 5.13 b for hydrogen and 3.03 b for carbon. The molar mass of polyethylene is 28.05 g mol^{-1}.

2. What is the mean free path for 1 MeV neutrons in light water at room temperature and pressure?

You should assume the density of water is 1 g cm^{-3}, its molar mass is 18.02 g and that the total microscopic cross sections for hydrogen and oxygen are 4.25 b and 8.15 b, respectively.

3. Consider the following spontaneous fission example just after the point of scission:

$$^{238}_{92}\text{U} \rightarrow {}^{147}_{57}\text{La} + {}^{91}_{35}\text{Br}$$

Assuming that the entirety of the energy released in fission is apportioned as kinetic energy to the primary fission fragments, estimate how much energy this is and what proportion of the binding energy gain it represents. Given that the kinetic energy is $<$ binding energy gained as a result of fission, where does the excess energy go?

The atomic masses of ^{238}U, ^{147}La and ^{91}Br are 238.0507 u, 90.934 u and 146.928 u, respectively. The atomic numbers of ^{147}La and ^{91}Br are 57 and 35, respectively, and $r_0 = 1.8$ fm.

4. Given the following data for neutron multiplicity of thermal-neutron–induced fission in ^{235}U and ^{239}Pu, calculate the average number of neutrons emitted per fission, $\overline{\nu}$, for these isotopes.

v	P_v	
	^{235}U	^{239}Pu
0	0.0317	0.0109
1	0.172	0.0995
2	0.3363	0.275
3	0.3038	0.327
4	0.1268	0.2045
5	0.0266	0.0728
6	0.0026	0.0097
7	0.0002	0.0006

5. Given that the energy spectrum N is a function of neutron energy E_n as given below, where a, b and c are constants, derive the expression for the most probable neutron energy. For the specific case of ^{252}Cf, where $a = 1.176$ MeV, $b = 1.02$ and $c = 0.64$, determine the most probable energy from your result.

REFERENCES

[1] F.H. Fröhner, Evaluation of ^{252}Cf prompt fission neutron data from 0 to 20 MeV by Watt spectrum fit, Nucl. Sci. Eng. 106 (1990) 345–352.

[2] C. Budtz-Jørgensen, H.H. Knitter, Simultaneous investigation of fission fragments and neutrons in ^{252}Cf (SF), Nucl. Phys. A 490 (1988) 307.

[3] B.E. Watt, Energy spectrum of neutrons from thermal fission of ^{235}U, Phys. Rev. 87 (6) (1952).

[4] D.G. Madland, Theory of neutron emission in fission, in: Workshop on Nuclear Reaction Data and Nuclear Reactors—Physics, Design and Safety, Trieste, Italy, LA-UR-98-797, 1998.

[5] R. Vandenbosch, J.R. Huizenga, Nuclear Fission, Academic Press, New York, 1973.

[6] Y. Aritomo, S. Chiba, K. NishioIndependent, Fission yields studied based on Langevin equation, Prog. Nucl. Energy 85 (2015) 568–572.

FURTHER READING

[1] O. Litaize, O. Serot, L. Berge, Fission modelling with FIFRELIN, Eur. Phys. J. A 51 (2015) 177–191.

[2] J.P. Lestone, Energy and isotope dependence of neutron multiplicity distributions, Report No. LA-UR-05-0288.

[3] K.P. Flynn, B. Srinivasan, O.K. Manuel, L.E. Glendenin, Distribution of mass and charge in the spontaneous fission of ^{244}Cm, Phys. Rev. C 6 (6) (1972) 2211–2214.

[4] International Atomic Energy Agency, A.C. Wahl, Mass and charge distribution in low-energy fission. Physics and chemistry of fission, in: Proceedings of the Symposium on Physics and Chemistry of Fission Held by the International Atomic Energy Agency in Salzburg, 22–26 March 1965, Vienna, 1965, pp. 317–331.

[5] D.G. Madland, Total prompt energy release in the neutron induced fission of ^{235}U, ^{238}U and ^{239}Pu, Nucl. Phys. A 772 (2006) 113–137.

CHAPTER

THE ACTINIDES AND RELATED ISOTOPES

5

5.1 SUMMARY OF CHAPTER AND LEARNING OBJECTIVES

While there are many radioactive isotopes with a wide range of physical properties, a thorough knowledge of all of them is not necessary in the context of nuclear energy. Many are too short-lived to warrant much engineering attention, notwithstanding the continuing need for curiosity-driven research. Further, many of those isotopes with half-lives long enough to pose a hypothetical concern are not formed in proportions in nuclear reactors to be significant in the context of nuclear power; rather, they are only formed by the dedicated bombardment of materials with particle accelerators. This leaves a relatively small number of radioactive species (approximately 20), formed in reactors or as daughter products, that are sufficiently long-lived to require management to keep people safe and to avoid pollution of the natural environment.

The central family of elements that concern us first in this chapter are the actinides and to a lesser extent the associated isotopes that follow in their decay chains. Derived from these there is also a selection of very light isotopes (arising either as a result of ternary fission or via neutron capture) and a mid-range group that arise as fission products of the nuclear chain reaction (the latter are referred to formerly as secondary fission products as we are usually interested in those that follow the primary β^- decays). The fission products subdivide further between a light and a heavy fraction as reflected in the asymmetry in yield with mass discussed in Chapter 4. In this chapter, we explore this range of isotopes to establish a foundation on which to base the subsequent chapters.

The objectives of this chapter are to:

- introduce the reader to the relatively limited and yet diverse range of radioactive species that arise as a result of processes in thermal power reactors, and the properties of these isotopes
- encourage the reader to divide their consideration of these species in terms of the main isotopic groupings
- identify the natural origin of these isotopes where applicable and the artificial origin of these substances in the context of nuclear power
- provide details of the isotopic properties in each case, including modes of decay, half-lives, emissions, and reference to important daughter products where appropriate
- highlight, where applicable, applications that these isotopes have been used for, or for which they might be used in the future

Nuclear Engineering. https://doi.org/10.1016/B978-0-08-100962-8.00005-6

- introduce concerns where they exist associated with the behaviour of an isotope in the natural environment, in living species and in prospect concerning radioactive waste disposal

5.2 HISTORICAL CONTEXT: GLENN THEODORE SEABORG 1912–99

If Albert Einstein is *the* theorist of the 20th century, then perhaps Glenn Seaborg is the equivalent experimentalist (Fig. 5.1). His achievements include the discovery of numerous elements and their isotopes leading to a Nobel Prize before he was 40 years old. His scientific legacy is to have challenged the perceived rigidity of the group structure of the periodic table, and to have introduced a new family to the table: the *transuranic elements*.

Seaborg is arguably most famous for his discovery of plutonium and particularly the plutonium-239 isotope made infamous by its use in atomic weapons. With co-workers he also discovered iodine-131, which has important uses in medicine, and americium that is used in millions of smoke detectors throughout the world. Plutonium (in the form of plutonium-238) is also the source of thermal energy that has been used to power some of the furthest-reaching exploratory space probes ever invented. Seaborg synthesised and identified both primordial isotopes, long since absent in the Earth's crust as a result of radioactive decay, and also artificial isotopes made for the first time.

5.3 INTRODUCTION

The prominent distinction of nuclear energy over all other forms of power generation is that a nuclear phenomenon is exploited with which to derive it, rather than an atomic or molecular process. This physical property is integral to the mechanisms by which nuclear energy is harnessed and those by which

FIG. 5.1

Glenn Seaborg.

the waste from its use is managed. At the start of the fuel cycle, the central component in this context is the actinide series of elements and particularly uranium due to its unique role as the only naturally occurring primary source of nuclear fission, in the form of the isotope ^{235}U. The subsequent radioactive inventory arises from fission, radioactive decay and neutron capture and results in the formation of many other radioactive isotopes. Once the short-lived isotopic inventory has decayed away, what remains is a relatively small number of isotopes that warrant further interest. The discussion of these substances in terms of their radioactivity and their role in nuclear power generation is the focus of this chapter.

5.4 THE ACTINIDES

5.4.1 COMMON PROPERTIES OF THE ACTINIDE SERIES OF ELEMENTS

The sequence of elements that follow actinium in the periodic table is collectively referred to as the *actinides*, as depicted in the Periodic Table given in Fig. 5.2. The isotopes of these elements are numerous and all are radioactive; together they constitute the most massive known nuclei, as elements they are all very dense metals. While they share some nuclear properties as a sequence there is also significant diversity. The complex electron structure that is common to all is manifest as a wide variety of valence possibilities and an associated range of contrasting chemical properties. This is reflected perhaps most tangibly by the wide range of vivid colours that their compounds can exhibit.

At a nuclear level, the range of properties that the actinides exhibit is similarly diverse. The variety of decay pathways by which many of the isotopes are connected, the energies of their radioactive emissions and the range of their half-lives makes for perhaps the most complicated sequence of radioactive materials known. Furthermore, while some of the actinides occur naturally, many exist today as a result of synthesis, both by design and default, because the primordial abundance of these has long since been depleted naturally via radioactive decay; plutonium is a case in point. Conversely, others have been synthesized for the first time such as protactinium. In total, actinium, thorium, uranium, neptunium, plutonium, americium, curium, *berkelinium*, californium, *einsteinium*, *fermium*, *mendelevium*, *nobelium* and *lawrencium* constitute the actinide series, with those elements that are heavier than uranium constituting the transuranic elements. Those in italics in the preceding list are not produced in sufficient quantities in reactors to concern us in a nuclear engineering context. Conversely, thorium, uranium, neptunium, plutonium, americium and curium are the main focus of applications in the nuclear industry and are among the longest-lived actinides that are encountered.

Some of the actinides also exhibit significant fission cross sections, as depicted in the chart of the nuclides reproduced in Fig. 5.3. However, with the exception of uranium and plutonium, practical uses of the actinides are not widespread. Several were used in medical applications early in the 20th century; for example, thorium was used as a contrast agent in early uses of X-rays in the form of a fluid suspension of thorium dioxide. In most cases improved knowledge of the radiotoxicity of these elements has caused many such practices involving them to cease as more benign alternatives have been identified to replace them. Several of these elements and the properties of their isotopes had not been identified until the Manhattan project and the pioneering work of Seaborg and co-workers in the late 1930s and early 1940s. For some of these substances it was only possible to make minute amounts while others exhibited such short half-lives as to not to serve any useful purpose. Making large amounts of the anthropogenic variants of the actinides can render them very expensive indeed and, besides their radiotoxicity, as heavy metals many of these elements are chemically toxic too. The manufacture of

1	2	3	4	5	6	7	8	9	10	11	12	13	14	15	16	17	18
1 **H** 1.008																	2 **He** 4.0026
3 **Li** 6.94	4 **Be** 9.0122											5 **B** 10.81	6 **C** 12.011	7 **N** 14.007	8 **O** 15.999	9 **F** 18.998	10 **Ne** 20.180
11 **Na** 22.990	12 **Mg** 24.305											13 **Al** 26.982	14 **Si** 28.085	15 **P** 30.974	16 **S** 32.06	17 **Cl** 35.45	18 **Ar** 39.948
19 **K** 39.098	20 **Ca** 40.078	21 **Sc** 44.956	22 **Ti** 47.867	23 **V** 50.942	24 **Cr** 51.996	25 **Mn** 54.938	26 **Fe** 55.845	27 **Co** 58.933	28 **Ni** 58.693	29 **Cu** 63.546	30 **Zn** 65.38	31 **Ga** 69.723	32 **Ge** 72.630	33 **As** 74.922	34 **Se** 78.97	35 **Br** 79.904	36 **Kr** 83.798
37 **Rb** 85.468	38 **Sr** 87.62	39 **Y** 88.906	40 **Zr** 91.224	41 **Nb** 92.906	42 **Mo** 95.95	43 **Tc** (98)	44 **Ru** 101.07	45 **Rh** 102.91	46 **Pd** 106.42	47 **Ag** 107.87	48 **Cd** 112.41	49 **In** 114.82	50 **Sn** 118.71	51 **Sb** 121.76	52 **Te** 127.60	53 **I** 126.90	54 **Xe** 131.29
55 **Cs** 132.91	56 **Ba** 137.33	57-71 *	72 **Hf** 178.49	73 **Ta** 180.95	74 **W** 183.84	75 **Re** 186.21	76 **Os** 190.23	77 **Ir** 192.22	78 **Pt** 195.08	79 **Au** 196.97	80 **Hg** 200.59	81 **Tl** 204.38	82 **Pb** 207.2	83 **Bi** 208.98	84 **Po** (209)	85 **At** (210)	86 **Rn** (222)
87 **Fr** (223)	88 **Ra** (226)	89-103 #	104 **Rf** (265)	105 **Db** (268)	106 **Sg** (271)	107 **Bh** (270)	108 **Hs** (277)	109 **Mt** (276)	110 **Ds** (281)	111 **Rg** (280)	112 **Cn** (285)	113 **Nh** (286)	114 **Fl** (289)	115 **Mc** (289)	116 **Lv** (293)	117 **Ts** (294)	118 **Og** (294)

*** Lanthanide series**	57 **La** 138.91	58 **Ce** 140.12	59 **Pr** 140.91	60 **Nd** 144.24	61 **Pm** (145)	62 **Sm** 150.36	63 **Eu** 151.96	64 **Gd** 157.25	65 **Tb** 158.93	66 **Dy** 162.50	67 **Ho** 164.93	68 **Er** 167.26	69 **Tm** 168.93	70 **Yb** 173.05	71 **Lu** 174.97
# Actinide series	89 **Ac** (227)	90 **Th** 232.04	91 **Pa** 231.04	92 **U** 238.03	93 **Np** (237)	94 **Pu** (244)	95 **Am** (243)	96 **Cm** (247)	97 **Bk** (247)	98 **Cf** (251)	99 **Es** (252)	100 **Fm** (257)	101 **Md** (258)	102 **No** (259)	103 **Lr** (262)

FIG. 5.2

The periodic table of the elements showing the actinides and the lanthanides (http://www.chem.qmul.ac.uk/iupac/AtWt/table.html).

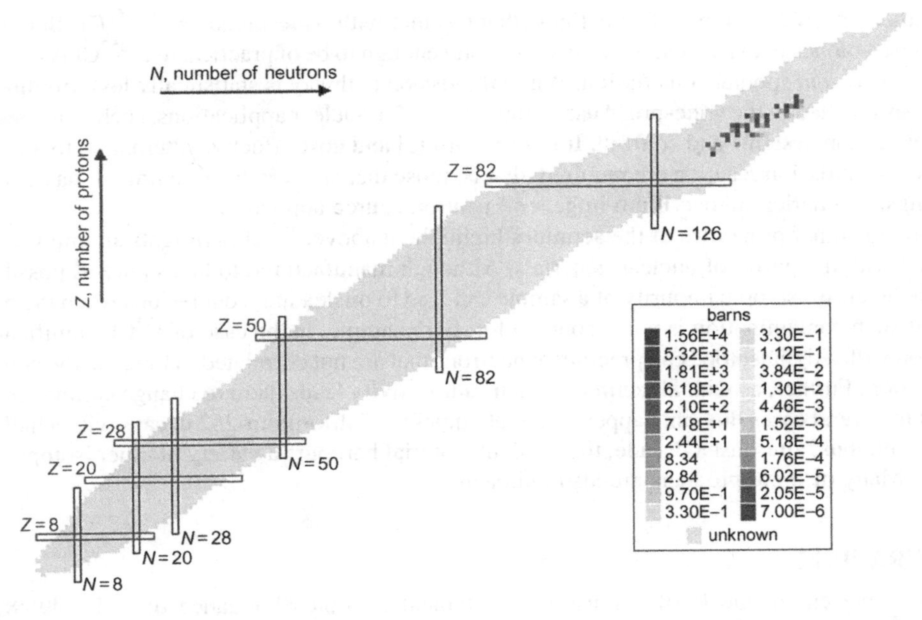

FIG. 5.3

The chart of the nuclides comprising the plot of proton number versus neutron number. The fissile isotopes are highlighted in terms of their thermal neutron fission cross sections, colour-coded as per the legend ('Chart of Nuclides—Thermal neutron fission cross sections' by National Nuclear Data Center/Brookhaven National Laboratory, reproduced with their permission).

plutonium and particularly the large-scale enrichment of uranium in the Second World War and following in the Cold War resulted in some of the largest industrial facilities ever constructed.

Aside from the energy derived from uranium and plutonium, several other actinide isotopes benefit our day-to-day lives significantly. Americium-241, formed from the β^- decay of ^{241}Pu, decays via α-particle emission. When this is mixed with a light isotope that has a susceptibility to yield neutrons due to the stimulus by α particles, it can constitute a neutron source. The example reaction for a mixture with beryllium is as follows:

$$^{241}\text{Am} + {}^{9}\text{Be} \rightarrow {}^{237}\text{Np} + \alpha + {}^{9}\text{Be} \rightarrow {}^{237}\text{Np} + {}^{(9-x)}\text{Be} + xn \quad (5.1)$$

This provides a flexible and long-lived source of neutrons based on a miniscule quantity of americium. The americium is acquired via separation from plutonium that arises following the irradiation of uranium with neutrons. Such sources, including mixtures with beryllium and lithium (referred to as americium–beryllium—AmBe and americium–lithium—AmLi, respectively), have proven very useful in applications in neutron metrology and oil exploration. Given its relatively long half-life of 432 years, ^{241}Am also provides an essentially constant source of α particles that can be used to monitor air for perturbations caused by smoke particles in a vast number of smoke detector systems. Conversely, so little of it is needed that it does not pose a waste disposal problem in domestic applications of this type and it is relatively easily protected against dispersal in the environment in the event of fire.

A further, specific example of an actinide that has met with widespread use is ^{252}Cf. Being one of the heaviest isotopes in existence, with a half-life long enough to be of practical use, ^{252}Cf is susceptible to both α decay *and* spontaneous fission. While the fission pathway is statistically less prominent it is this that has resulted in the widespread use of this isotope for nuclear applications, such as the testing of neutron detection systems and so forth. It is a very useful and cost-effective alternative to a reactor if neutrons from a fission reaction are required, that is, those that are correlated in time. It has also found uses alongside americium–beryllium in general neutron source applications.

Among the other properties of the actinides highlighted above, ^{252}Cf highlights an important issue associated with the purity of nuclear materials. Although manufactured to be as pure as possible, unavoidable levels of isotopic impurity of a sample can lead to misleading conclusions when the material is used, if such contamination is not accounted for. For example, in the case of ^{252}Cf, significant contamination with a light isotope can precipitate neutrons that are not correlated in time via the competing (α,*n*) reaction. Further, as with all actinides, their radioactivity leads them to change composition with time and in several cases this can happen relatively quickly. Californium-252 decays with a half-life of 2.6 years and thus, after just a decade, the resident material harbours a variety of other isotopic fission products. Many of these products are also radioactive.

5.4.2 URANIUM

Uranium is present in the Earth's crust with a typical isotopic abundance of ^{238}U: 99.28% wt., ^{235}U: 0.71% wt. and trace quantities (~0.0057% wt.) of ^{234}U. It was discovered in 1789, long before the discovery of nuclear fission in 1938 which caused it to attract attention for energy production and military applications. Uranium is approximately 40 times as abundant as silver, and its metal form has a melting point of 1132°C and a boiling point of 3818°C. In its solid, metallic form it is only mildly radioactive (arising predominantly from α decay in ^{238}U). This usually requires that it is handled with gloves to limit contact with the skin. Like many metals, when finely divided in the form of a powder it burns readily and is pyrophoric on violent impact. In the form of a powder or dust it is more hazardous than as a solid as the risks of ingestion and internal α-particle exposure via respiration are more significant. In environments where uranium dusts arise (and for that matter any actinide powders that have the potential to become airborne), the loose material is often immobilised with an adhesive spray; a method termed *fogging*. This greatly reduces the risk of ingestion. Also, air-fed suits are often used if people are required to access an environment heavily contaminated by loose, dry debris containing uranium or other actinide species. A comprehensive review of the exposure pathways and health effects of natural uranium is that of Brugge et al. [1].

Alongside the fission-related applications of uranium that we shall come to shortly, it has found relatively little use more widely. Its high density has led to it being considered for ballast and its high-*Z* composition renders it useful for radiation shielding. Such applications exploit its depleted form (see below) where its relative worth in terms of its use as a primary resource for the production of energy and potential risk as a nuclear material are effectively removed by the reduction of the ^{235}U content. Its high density and pyrophoric properties have led to its highly publicised use in conventional weapons, that is for use in bullets and shells.

In its naturally occurring form the isotopic abundance of uranium as recovered from the Earth's crust is usually termed as being of *natural enrichment*. This corresponds to the major isotopic constituents in proportion: 99.28% wt. ^{238}U, 0.71% wt. ^{235}U and 0.0057% wt. ^{234}U. Any artificial increase in the ^{235}U composition (^{235}U usually being the focus of interest by virtue of its susceptibility to stimulated fission) results in the material being termed *enriched*. There is a great distinction between the inference posed and the relevant use of *low-enriched uranium* (LEU), which is <20% wt. ^{235}U, and *highly-enriched* material, defined as being >20% wt. The world's operational power reactors exploit lightly enriched uranium fuel cycles, typically <5% wt. ^{235}U, since high levels of enrichment are expensive to achieve and highly-enriched material poses a more significant criticality risk and safeguards concern. With its ^{235}U content reduced to less than that of natural enrichment, the uranium by-product of the enrichment process is termed *depleted*. This typically has an isotopic composition in the range 0.25–0.3% wt. ^{235}U.

Examples of two enrichments are shown schematically for a pool of 140 atoms in Fig. 5.4 by way of illustration, while data for the most prominent four uranium isotopes are given in Table 5.1. As referred to above, depleted uranium does not have a great diversity of uses. However, its major isotopic constituent, ^{238}U, might be utilised for the production of plutonium in the future for energy applications. This offers enormous implications for low-carbon power production not limited by the relatively low natural abundance of ^{235}U, as discussed further in Chapter 11.

As can be appreciated from Table 5.1, all of the uranium isotopes of interest to the nuclear engineering community as listed have very long half-lives. With the exception of perhaps in the modelling of what are at present largely hypothetical, long-term waste disposal scenarios, these isotopes do not reduce in composition significantly with time as a result of radioactive decay. The corresponding

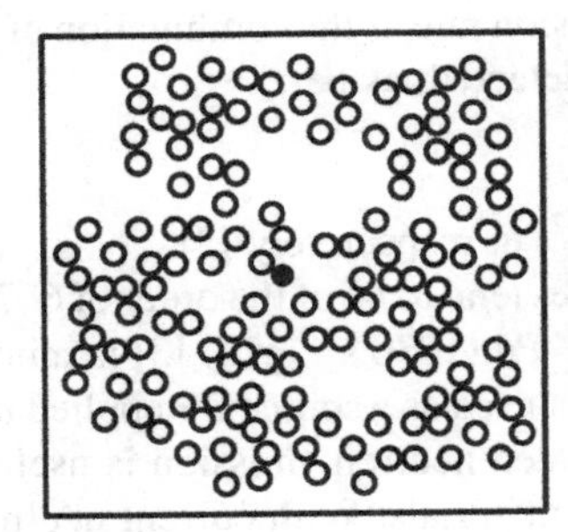
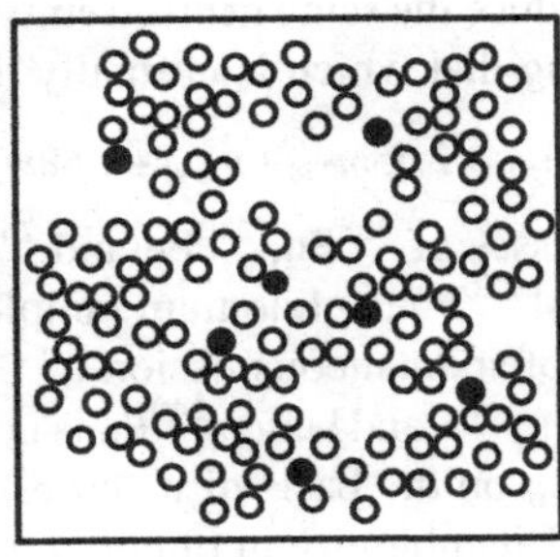
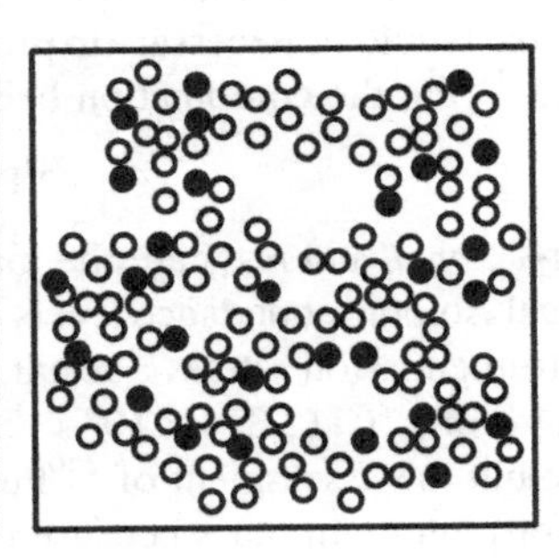

FIG. 5.4

A schematic representation of uranium enrichment in terms of relative number of atoms of each isotope. 0.71% wt. or 1:140 atoms ^{235}U (left), 5% wt. or 1:19 atoms ^{235}U (middle) and 20% wt. or 1:4 atoms ^{235}U (right). A similar rationale yields a ratio of 1:330 for depleted uranium of 0.3% wt. ^{235}U (not shown).

Table 5.1 Isotopic Abundances and Half-Lives of the Isotopic Constituents of Natural and Depleted Uranium

Isotope	% wt. (Natural)	% wt. (Depleted)	Half-Life/Years
^{238}U	99.28	99.8	4.51×10^9
^{236}U	–	0.0003	2.39×10^7
^{235}U	0.71	0.2	7.10×10^8
^{234}U	0.0057	0.001	2.47×10^5

specific activity of low-enrichment forms of uranium is therefore relatively low also. The radiation that is emitted comprises α, γ and a miniscule amount of neutron radiation; neutron emission arises predominantly from a small amount of spontaneous fission (SF) in ^{238}U. There are other uranium isotopes of relevance, such as ^{232}U and ^{233}U, which are generally products of reactions involving thorium. These are discussed later in this chapter.

5.4.3 PLUTONIUM

Plutonium occurs in the world almost entirely as a result of the nuclear processes developed over the last 60 years. In elemental form, it is a metal with a density almost twice that of lead and is formed as a result of neutron capture on uranium-238. There are several isotopes of plutonium but only ^{238}Pu, ^{239}Pu, ^{240}Pu and ^{241}Pu usually occur with a combination of an abundance that is significant and with half-lives that are long enough to be a concern for waste management and decommissioning activities.

The even-numbered isotopes of plutonium tend to undergo spontaneous fission thus emitting neutrons; ^{240}Pu is the most prominent in terms of the emitted neutron flux. However, this is small per unit mass relative to the more prominent exponents of spontaneous fission such as ^{252}Cf or ^{244}Cm. Plutonium-240 can be used in analysis methods providing a nonintrusive assessment of plutonium content to guard against proliferation of this material. Taken together, plutonium isotopes exhibiting spontaneous fission are termed the *^{240}Pu effective mass* '$^{240}Pu_{eff}$' to account for the total even-numbered contribution to neutron emission via spontaneous fission. The ^{240}Pu effective mass is equal to the mass of ^{240}Pu that would be necessary to produce the same neutron emission rate as the combination of ^{238}Pu, ^{240}Pu and ^{242}Pu, the combination being that which is normally detected, as per,

$$^{240}Pu_{eff} = 2.55m_{238} + m_{240} + 1.68m_{242}$$

where m_{238}, m_{240} and m_{242} are the masses of ^{238}Pu, ^{240}Pu and ^{242}Pu, respectively [2].

Typical isotopic abundance levels of ^{240}Pu in plutonium samples tend to be of the order of 6–7% wt. The neutron emission rate as a result of spontaneous fission of ^{240}Pu ($1030\ s^{-1}\ g^{-1}$) [3] is much less than that [4] of ^{252}Cf ($2.314 \times 10^6\ s^{-1}\ \mu g^{-1}$), and hence ^{240}Pu is not used as a source for applied uses of neutrons, but the assessment of $^{240}Pu_{eff}$ on the basis of its correlated neutron emission is useful as a marker for plutonium mass because of its ubiquity in plutonium materials. With current neutron detection systems it is relatively straightforward to measure the characteristic, time-correlated neutron emission from $^{240}Pu_{eff}$ passively, that is, without an external source. This can then be related to the quantity of material present.

All of the plutonium isotopes, with the exception of ^{241}Pu, decay via α decay with some attendant β^- and γ decay. The energies of the α-particles are in the 5 MeV energy region while the γ-ray emissions are generally of low energy, <500 keV. Plutonium-241 decays via β^--emission to ^{241}Am, which subsequently undergoes α decay. The main radiological risk from plutonium is internal because of the possibility of its α activity damaging the internal tissues of the body. Exposure to these isotopes can be eliminated by sealing material at risk of dispersal in a plastic bag and via the use of suitable personal protective equipment, for example, the use of a respirator, whole-body coverage, and so on.

The primary internal radiological concern with plutonium is via inhalation because it is not as readily absorbed into the bloodstream via ingestion, although any suspicion of significant contamination of the environment (particularly foodstuffs) would be a major concern. Uptake data are largely similar to that of americium with similar risks, such as retention in the liver and bone. Although often linked with uranium due to their joint association with the Manhattan project, plutonium is quite a different proposition in most respects to uranium. Discovered over 150 years later than uranium, at the height of the race for the atomic bomb in the Second World War, plutonium present in the world today is almost entirely synthetic in origin. The exceptions are the trace amounts that occur, occasionally together with uranium ores. As with uranium and the other actinides, the chemistry of plutonium is complex due to its extensive and complicated electron structure; the same is true of the radioactivity of its isotopes of which there is an extensive sequence. Plutonium metal melts at 640°C and boils at 3235°C. Unlike uranium, its intense α decay (particularly of ^{238}Pu) results in even a relatively small amount generating sufficient thermal heat to be noticeable. Anthropogenic sources of plutonium in the environment include nuclear accidents and fallout from nuclear weapon tests carried out before the Comprehensive Test Ban Treaty in 1996.

Plutonium is highly toxic because its long-term retention in bone can cause it to interfere with white blood cell production in the body. Incidents involving plutonium where internal contamination is suspected to have occurred are treated extremely seriously for this reason; a summary of the decays and half-lives of the most prominent plutonium isotopes is given below (Table 5.2).

Plutonium is produced unavoidably in nuclear reactors running on the thermal uranium/plutonium (U/Pu) fuel cycle; indeed, a significant fraction of the day-to-day electricity produced by thermal-spectrum, uranium-fuelled nuclear reactors is derived from the fission of ^{239}Pu produced in the reactor. Plutonium-239 is formed via neutron capture on ^{238}U and subsequent double β^- decay via

$$^{238}_{92}\mathrm{U} + n \rightarrow {}^{239}_{92}\mathrm{U} \xrightarrow{\beta^-} {}^{239}_{93}\mathrm{Np} \xrightarrow{\beta^-} {}^{239}_{94}\mathrm{Pu} \quad (5.2)$$

In addition to production in nuclear power reactors, significant quantities of plutonium were produced throughout the arms race of the mid-to-late 20th century for use in nuclear weapons. At the time of the race to perfect the atomic bomb, plutonium (and specifically ^{239}Pu) provided a route to a fissile isotope without the energy-intensive requirement for high levels of isotopic enrichment that is essential with uranium. Being elementally distinct from uranium, the separation process with plutonium becomes one of chemistry rather than one based on minute differences in mass, as it is for ^{235}U. The end of the Cold War has led to agreements to reduce nuclear weapon stockpiles and specifically to dispose of excess weapons plutonium; most recently the US–Russia Plutonium Management and Disposition Agreement (PDMA).

The fissile nature of ^{239}Pu yields benefits every day in the world's reactors with approximately 1% of electricity derived directly from fission in plutonium without any requirement for its separation or retrieval. There is an alternative fuel route that utilises plutonium in the fresh fuel, that is, it is inserted at

Table 5.2 Half-Lives and Decay Pathways for the Most Prevalent Plutonium Isotopes

Isotope	Decay Pathways	Half-Life/Years
^{242}Pu	α	373,300
^{241}Pu	β	14.3
^{240}Pu	α, SF	6561
^{239}Pu	α	24,110
^{238}Pu	α	87.7

the point of start up as a *mixed-oxide* (MOX) of uranium and plutonium, but this is not currently widespread aside from its use in France and Japan. Given the significant stockpiles of plutonium that exist, a technically feasible avenue remains for this material to be disposed of in power-generating reactors, which is arguably preferable to its long-term geological disposal. However, at present the economics of nuclear fuel production and use tends to favour once-through cycles based on uranium. This is discussed further in Chapter 16.

Alongside ^{239}Pu, several other plutonium isotopes have found specific applications other than as a source of energy in nuclear reactors. Plutonium-238 exhibits α emission of relatively high specific activity that yields thermal heat. This has been exploited to provide a source of electrical power and heat for applications in space, especially those associated with long-distance exploration probes such as Voyager, Cassini and several Mars rovers in the form of radioisotope thermoelectric generators (RTGs). In these applications ^{238}Pu fulfils the requirements for there to be no moving parts, it does not require a source of oxygen, is long-lived and has an extremely high energy density; electricity is usually derived via the Seebeck effect. The hazards associated with plutonium are not usually a significant problem in this context because such probes are generally not designed to return to Earth, assuming that necessary precautions are taken to ensure that, should an accident occur, the RTG housing is able to prevent the dispersion of radioactivity on re-entry; extensive research has focussed on this particularly following the Challenger disaster in 1986 and earlier incidents. Thermal batteries based on this technology have been responsible for much of the knowledge we have of the furthest reaches of the galaxy; this application constitutes perhaps the most profound, peaceful known application of plutonium aside from nuclear energy.

5.4.4 THORIUM

Having considered uranium and plutonium from a nuclear energy perspective, there are relatively few actinides left to consider that present an opportunity for fission. For those isotopes that are susceptible to fission and therefore offer the potential for electricity generation, most are either too low in abundance, too difficult to extract in sufficient quantities to render them economically viable or their thermal neutron fission cross section is too low to be exploited. It is worth emphasising that all isotopes of this type that are synthesised in a reactor usually need to be separated from the host spent-fuel stream of which they are part if they are to be used as a fuel in subsequent reactor operations. This is usually achieved via a liquid–liquid extraction process as discussed in Chapter 13 and is a major undertaking if done on an industrial scale, predominantly because of the high levels of radioactivity. Thorium provides an alternative route to a fissile uranium isotope and can be attractive in parts of the world where alternative fuel reserves that might otherwise offer security of supply, for example, uranium, oil, gas, coal, and so on are scarce by comparison.

Thorium is relatively abundant in the Earth's crust and in some cases is present in significant quantities where uranium is a less viable candidate for extraction, particularly in India. Despite being a primordial resource, a thorium-based fuel cycle operating independent of the U/Pu cycle requires the separation step described earlier to yield thorium-derived fissile fuel for use in reactors. The use of thorium is well understood since it was also explored for use in nuclear applications before being largely disregarded in favour of uranium and plutonium. To breed ^{233}U from ^{232}Th the U/Pu cycle is necessary because not enough ^{233}U exists to prime the thorium cycle independently of ^{235}U.

Recently, partly due to the resurgence in nuclear power development as a result of the worldwide desire to decarbonise electricity supplies, thorium has become of topical interest. India, due in part to its extensive thorium reserves, has had long-term plans based on thorium for several decades. It is a slight misnomer to state that it is thorium that is to be used, however, since the thorium component is actually not fissile but rather provides a fertile pathway by which we are able to access the fissile uranium isotope, ^{233}U. ^{233}U is not present to any great degree in natural uranium deposits as it is short-lived, relative to the isotopes of ^{235}U and ^{238}U, and neither is it produced in significant quantities in the U/Pu fuel cycle. Therefore, it cannot be refined from uranium ores as is the case for ^{235}U. Rather, ^{233}U can be produced from ^{232}Th in direct analogy to the production of ^{239}Pu from ^{238}U, via neutron capture and double-beta decay, as per the reaction as follows:

$$^{232}_{90}\mathrm{Th} + n \rightarrow {}^{233}_{90}\mathrm{Th} \xrightarrow{\beta^-} {}^{233}_{91}\mathrm{Pa} \xrightarrow{\beta^-} {}^{233}_{92}\mathrm{U} \tag{5.3}$$

Isotopes with this property (^{232}Th for the production of ^{233}U and ^{238}U for the production of ^{239}Pu) are known as *fertile*. While not presenting thermal fission possibilities themselves these isotopes offer a route by which fissile material might be obtained indirectly. This is of great significance, as we shall discuss further in Chapter 11, since the abundances of ^{238}U and ^{232}Th are much greater than ^{235}U. Fissile material accessed via the conversion of these isotopes to ^{239}Pu and ^{233}U, respectively, offers the potential for a source of nuclear energy for several thousand years. This contrasts significantly to a reliance on the relatively limited abundance of ^{235}U where most estimates of the lifetime of this resource are <100 years.

Thorium is occasionally heralded as a ‘safe’ alternative to uranium and plutonium because, at least in the modern era, it is rarely associated with weaponisation and it has often been integrated into reactor system designs (at least in principle) via accelerator-driven systems. These systems are envisaged to operate such that the critical chain reaction dies away when the accelerator is switched off. However, ^{233}U was explored extensively in terms of weapons use and, while viable, offered too awkward a preparation approach because of its resident radioactivity to warrant further investigation. Further, prominent nuclear reactor accidents, such as Three Mile Island and Fukushima Daiichi, highlight that reactor safety not only requires that the critical chain reaction be shut down as effectively as possible in the event of a problem, but also that the energy evolved by the radioactive decay of fission product activity in the fuel is dissipated safely once the reactor is shutdown. In this regard, a thorium-fuelled reactor would have the same vulnerability since decay heat is unavoidable in fission reactor systems, as will be discussed in Chapter 14.

The origin of the radioactivity associated with ^{233}U referred to above are the products of the decay chain that follows from the disintegration of ^{232}U; the latter is formed via several neutron-stimulated reactions on ^{232}Th. These products yield several intense, high-energy γ-ray emissions associated with the decay to their stable, ground states. These penetrating γ rays present a significant exposure risk associated with the handling of most materials comprising ^{233}U, contaminated with ^{232}U. This is in stark contrast to ^{235}U, which can be handled relatively easily in the form of low-enriched, fresh (i.e. un-irradiated) uranium-based fuel. From one point of view this provides an additional, in-built degree of proliferation resistance in that anyone handling ^{233}U materials would risk harm to themselves. Conversely, the additional sophistication required in handling the material for bona fide fuel preparation purposes greatly complicates the process and causes costs to rise. The key attributes associated with the use of thorium are discussed further in Chapter 11.

In summary, ^{233}U has benefits in terms of its neutron emission that enables breeding at thermal and epi-thermal neutron energies; the melting point of thorium is higher than uranium offering the potential for greater accident tolerance in related fuel materials and, finally, wastes derived from its use will have much reduced levels of plutonium and minor actinide content (see below) in comparison with the U/Pu cycle. This property offers potential benefits in terms of long-term waste management. For nations with significant thorium reserves, the ^{233}U fuel route via thorium continues to present attractive possibilities, albeit as part of a long-term >50-year energy strategy.

5.4.5 CURIUM, AMERICIUM AND NEPTUNIUM

Curium, americium [5] and neptunium are three of the less prominent transuranic elements of the periodic table; all are synthetic and in terms of wastes fall into the category of 'minor actinides' (MA). They are perhaps of most interest when designing, modelling and operating nuclear reprocessing plant, and in managing and disposing of spent nuclear fuel. They are all radioactive and often they are worked with remotely in a shielded cell. Some research has been done to explore their extraction from nuclear waste to reduce the overall activity before disposal, which is discussed in Chapter 15.

In the case of curium, the specific isotope of interest tends to be ^{244}Cm as the other curium isotopes, such as ^{242}Cm, are relatively short-lived while ^{243}Cm and ^{245}Cm are only produced in very small amounts. Curium-244 is formed as a result of neutron capture reactions in nuclear reactors and thus arises in environments and processes associated with spent nuclear fuel. While its main decay pathway is α decay, in common with ^{240}Pu, ^{244}Cm is susceptible to spontaneous fission at a level of $1.4 \times 10^{-4}\%$. This is some 25 times more significant than the SF channel in ^{240}Pu. Thus ^{244}Cm yields a significantly greater neutron flux as a result (1.08×10^7 SF s^{-1} g^{-1} for ^{244}Cm compared with 1.02×10^3 SF s^{-1} g^{-1} for ^{240}Pu). This can present significant shielding and criticality protection issues when dealing with high-activity raffinates in reprocessing and handling spent fuel, but also provides routes by which fissile content in these media might be verified too [6]. In terms of long-term disposal, the half-life of ^{244}Cm with respect to α decay is relatively short at 18.1 y (compared with the half-life for SF of 1.25×10^7 y [7]) [8]. Consequently, ^{244}Cm only warrants consideration in terms of waste management in the near term, that is <100 y. Curium has relatively few uses but has been considered for use in RTGs, especially in the event there is a shortage of ^{238}Pu for this purpose in the future (see also the earlier discussion on ^{238}Pu).

Americium does not occur naturally in the Earth's crust but is formed via the capture of neutrons on plutonium, usually in reactor systems in which ^{239}Pu captures two neutrons to become ^{241}Pu, which then β^- decays to ^{241}Am. Similar reaction schemes result in the ^{242}Am and ^{243}Am isotopes but they are only usually represent a small proportion of a typical nuclear americium inventory arising from thermal spectrum reactors. Amercium-241 is the most abundant of these isotopes and decays via α-particle emission and associated γ decay to ^{237}Np with a half-life of 432 y; the principal γ-ray line from ^{241}Am is at 60 keV.

Unlike curium, ^{241}Am has several important uses as discussed earlier. The attenuation of its low-energy γ-ray emission has also led to it being used in thickness gauges. The relatively high specific activity of ^{241}Am requires that only very minute amounts are needed in such applications.

Some of the americium that resides in the environment has arisen from nuclear weapons tests, while there is the possibility of localised contamination near those few reactor sites where serious accidents have occurred. Typically, americium forms an oxide and is relatively insoluble, instead binding very strongly to soils if it were to arise. In plutonium the β^- decay of ^{241}Pu results in the 'in-growth' of ^{241}Am with time, which can be used as a basis for the assessment of the age of this material, via either wet-chemistry α-spectrometry analysis or γ-ray spectroscopy of the 60 keV ^{241}Am line. Americium is generally only a hazard where it is taken up into the body because of the short range of the α-emission and the low-energy γ-ray emission. Taken up through respiration and ingestion, most americium will be excreted with only 0.05% entering the blood. About 10% of this will be excreted with the rest being deposited in the liver and the bone. As with the ingestion of most minerals, the amount deposited in these areas of the body depends on the age of the individual and several other factors.

Most neptunium isotopes are too short-lived to be of more than academic interest [9], with the exception being ^{237}Np. Neptunium-237 is formed from neutron capture on uranium and from the α decay of ^{241}Am and it has a very long half-life of 2.14×10^6 y. It is fissile but with a very low cross section and has a much higher critical mass than ^{239}Pu. Concern over the potential hazard from ^{237}Np to date has focussed on it arising from weapons fallout, effluents from damaged fission reactors, from reprocessing and plutonium production facilities and its incidence as long-lived component of high-level radioactive wastes. Neptunium-237 is produced in reactors with a yield of ~0.1% that of plutonium production, which increases with ^{235}U enrichment. It can be used as the feedstock for the production of ^{238}Pu via neutron bombardment. However, because of its long half-life, ^{237}Np is forecast to be a prominent remnant in radioactive wastes from about 10,000 to ~30 million years after disposal. Trace quantities of ^{237}Np are found in nature in uranium ore concentrates as a result of neutron bombardment from spontaneous fission of ^{238}U, for example in the ratio $^{237}Np/^{238}U$: 1.8×10^{-12}. The chemistry of neptunium is somewhat unique in the periodic table because it forms a stable pentavalent ion, NpO_2^+. Neptunium is considered to be more mobile than americium or plutonium in soils and stone. Uptake by plants is more significant than for plutonium, curium or americium, whereas uptake by animals is considered to be similar to plutonium.

5.5 PRODUCTS OF NEUTRON ACTIVATION IN REACTORS

5.5.1 TRITIUM

Early texts on light-water reactor systems [10] rarely discuss tritium despite its prevalence in these reactors arising from neutron activation and, to a lesser extent, ternary fission. This is largely because our knowledge as an engineering community has evolved significantly since the era associated with these reports (usually the late 1950s) along with the technology to provide a sufficiently low limit of detection for tritium; such is the elusive and pervasive nature of tritium contamination.

Tritium, ^3H, is the heaviest isotope of hydrogen and is formed primarily in reactor systems by neutron activation of the deuteron, ^2H, most often in the coolant and moderator of heavy water (D_2O, where $D={}^2H$) reactors [11]. It can also be formed via neutron irradiation of lithium, boron, by the bombardment of deuterons with deuterons (often referred to as the D-D reaction) and also as a product of fission. The D-D, lithium and boron reactions are given in Eqs. (5.4)–(5.6), respectively,

$$^2H + {}^2H \rightarrow {}^3H + n + 3.3\,MeV \tag{5.4}$$

$$^6Li + n \rightarrow {}^4He + {}^3H + 4.69\,MeV \tag{5.5}$$

$$^{10}B + n \rightarrow 2\,{}^4He + {}^3H + 0.335\,MeV \tag{5.6}$$

Tritium is formed naturally in very small quantities primarily in the atmosphere via cosmic irradiation of nitrogen and, to a lesser extent, via reactions with ^{6}Li, deuterium and fission in terrestrial material. It forms the corresponding oxide (HTO, where T = ^{3}H) rapidly and is dispersed like water, with there being evidence for a relative concentration in the condensed phase due to the lower vapour pressure in this form. Tritium is an important isotopic component of the fuel used in fusion reactors.

Deuterons exist in reactor systems either as a constituent of heavy water coolants and moderators or are themselves formed in light water (i.e. $^{1}H_2O$ to be explicit) coolants and moderators via neutron activation of hydrogen to deuterium, and subsequently to tritium. The deuterium pathway in heavy-water reactors tends to be more significant than in light-water systems, as might be expected; neutron flux levels, burn-up, and so on notwithstanding. However, the use of boron as a neutron absorber (both in control/shutdown rods and as an additive to coolants), and the presence of lithium as a contaminant and as a component of ion exchange resins, provide alternative pathways for the production of tritium in light water reactors. For this reason, a transition to ^{7}Li-containing resins, as opposed to those of natural lithium composition containing the susceptible ^{6}Li isotope used in early operations, has resulted in a favourable reduction in tritium production.

Tritium arising as a fission product in nuclear fuel can in principle migrate through fuel cladding, although this pathway is regarded as limited and certainly minor in heavy-water reactors compared to the deuterium route. Most tritium derived from fission in the fuel only risks widespread liberation in the wastes arising from reprocessing or as the result of a core-damage accident. As a potential substitute for hydrogen in hydrogenous media there is the possibility for tritium to reside in oils, plastics, rubbers, grease and living tissue. In the environment, while tritium radiotoxicity is low relative to the other sources of anthropogenic radioactive contamination, tritium can often be rendered airborne more easily.

Tritium undergoes β^- decay to become ^{3}He and has a relatively short half-life of 12.3 years. However, its half-life is not sufficiently short for it to be ignored in waste streams nor where it arises as an anthropogenic pollutant in the environment. The tritium β^- emission has an average energy of 5.7 keV with a maximum of 18 keV and is thus not sufficiently penetrating to pose an external radiation hazard, indeed it is often difficult to detect especially in the field. Conversely, the formation of tritiated water (HTO) is favoured energetically. Since the associated vapour can penetrate the skin in this form, tritium can constitute an internal source of exposure, albeit usually too low in concentration to constitute a concern. Of greater significance in relative terms, given the energetic preference for the formation of liquids (as opposed to HT, for example), is the exposure of workers to tritium contamination arising from system faults, that is, minor leaks associated with faulty gaskets and seals in coolant systems. Retention in the body is dependent on the chemical form the tritium acquires, with the difference in mass compared with tritium considered to have some biological significance in the function of cells, aside from the effect of the emission of β^- radiation on surrounding tissues. However, the salient challenge associated with tritium is not one of it being a high hazard *per se* but rather the difficulty of monitoring it in the environment due to the low-energy of its β^--particle emission.

5.5.2 SODIUM-24

Sodium-24 is formed via the neutron activation of natural sodium, ^{23}Na, according to,

$$^{23}_{11}\mathrm{Na} + n \rightarrow ^{24}_{11}\mathrm{Na} + \gamma \qquad (5.7)$$

Sodium-24 is a β^- emitter, emitting β^- particles with a maximum energy of 1.39 MeV and it has a half-life of 15 h. Although comparatively short, the high-energy nature of this emission has had implications for the design of fast-breeder reactors to isolate the activated eutectic sodium-potassium (NaK) coolant from latter coolant stages. As discussed in Chapter 11, sodium is favoured as a coolant in fast reactors because it has good heat transport properties while not being an effective moderator of neutrons; thus its use does not degrade the desired fast neutron spectrum in these reactor types.

The chemical reactivity of sodium, coupled with the products of this reaction that include, for example, sodium hydroxide, increases the hazard potential associated with ^{24}Na. This is because in the form of a hydroxide, or indeed a salt, there is the potential for it to be taken up by the body through unprotected contact with the skin, respiratory and digestive tracts. Thus, if internalised in this way it poses an internal radiation hazard via this route to living tissue. However, the biological half-life of ^{24}Na is short and thus it is not resident in the body for more than a few days. In the event of a nuclear accident where a significant exposure to neutrons is suspected, the formation of ^{24}Na via neutron activation of natural salts in the body can be measured and used as a means for inferring the degree of exposure to a given individual [12].

5.5.3 COBALT-60

Cobalt-60 has a half-life of 5.27 years and decays via the emission of two high-energy γ rays at 1.17 and 1.33 MeV. The relatively high energies of these emissions, coupled with the widespread distribution of this radioactive isotope throughout nuclear plants incorporating cobalt-based steels, result in this isotope being a common radiological feature in decommissioning the structural components of reactors. The short half-life of ^{60}Co, relative to decommissioning timescales, can justify postponing the removal of contaminated infrastructure until the ^{60}Co activity has decayed by a significant fraction, thus simplifying operations and lowering exposure risk. The isotopic reference to ^{60}Co is a slight misnomer (as is the case for a number of isotopes that emit γ rays after a radioactive decay transition) since it is actually an excited state of nickel-60 that emits the γ rays, following the β^- decay of ^{60}Co, via

$$^{59}_{27}\mathrm{Co} + n \rightarrow {}^{60}_{27}\mathrm{Co} \xrightarrow{\beta^-} {}^{60}_{28}\mathrm{Ni} + e^- + \bar{\nu}_e + \gamma \tag{5.8}$$

Cobalt-60 formation via neutron activation of ^{59}Co can follow a succession of similar neutron activation processes of isotopes of iron in the steel comprising the structure of a reactor. Cobalt-60 as an isotope manufactured specifically for the purpose is also used in the external beam radiotherapy treatment of cancer, because of its simple and well-understood γ-ray spectrum and the ease with which it can be produced with reactors and particle accelerators. It has also been the subject of a few isolated but nonetheless serious incidents arising upon disposal, usually in such instances because of negligence in the scrap metal industry. While ^{60}Co can pose an ingestion risk, the predominant radiation hazard is the external dose due to its γ-ray emission.

5.6 FISSION PRODUCTS

5.6.1 KRYPTON-85

Krypton-85 has a similar half-life at 10.76 years to ^{60}Co and decays via a combination of β^--decay (0.69 MeV) and γ decay (0.51 MeV) to rubidium-85, which is stable. Krypton-85 arises as a fission product in spent nuclear fuel, although trace amounts arise naturally as a result of cosmic irradiation in the atmosphere. At room temperature krypton is a gas and, as a noble element, it is very unreactive. Along with tritium, ^{14}C and ^{129}I, ^{85}Kr is one of the four most important *volatile* fission products. These isotopes are not confined to the liquid waste stream in reprocessing and, instead, concentrate in off-gas systems as a result of their gaseous state at room temperature. Similarly, volatile fission products can escape fuel cladding in the event of a core damage accident. As a noble element, ^{85}Kr complexes with relatively little, making its chemical extraction very difficult. Conversely, therefore, rarely is it sequestered by biochemical processes in living tissues and rather it is dispersed rapidly and uniformly in the environment if released [13]. If ingested ^{85}Kr will pass through the body at a rate determined by the biological half-life of the ingestion route (e.g. breathing, consumption), exposing the body internally while it is resident. It is readily excreted. Thus, ^{85}Kr usually poses a relatively small radiological hazard internally.

The main risk from ^{85}Kr is an external skin dose from immersion in a cloud of the gas that might arise from a plume in the event of a serious reactor accident in which volatile fission products are vented to the atmosphere. Views are often divided as to whether ^{85}Kr should be sequestered and disposed of along with other, nonvolatile radioactive wastes. However, the main scenario by which its release would come about is as a result of reprocessing where, while calculations might indicate that dispersed doses to the population would be significantly less than background, a precautionary strategy of immobilisation is usually pursued.

In nuclear accidents where there have been volatile fission product releases but ostensibly no other emission due to the containment (consider, for example, Three Mile Island discussed further in Chapter 14), ^{85}Kr often constitutes a significant component of the emitted radioactivity. Other volatiles are usually either too short-lived in terms of waste disposal to be a concern (such as the xenon isotopes) or too low in abundance (such as ^{81}Kr). Since filtering and immobilisation of ^{85}Kr can be difficult and complicated, in some emergency situations it has been stored (in the vessel or building in which it has gathered) until it can be released under controlled conditions; control of such releases enables exposure from the γ-ray component (often referred to as γ shine) in what would otherwise be a relatively concentrated plume containing this isotope to be dispersed to acceptable levels.

5.6.2 STRONTIUM-90

Strontium exists as a mixture of four stable isotopes and sixteen radioactive isotopes have been synthesised and identified. Strontium-90 has a half-life of 28.8 years and a relatively high-energy β^- emission of 0.546 MeV. It has been a radiological concern since the dawn of the nuclear industry associated primarily with the fallout from atmospheric nuclear weapons tests. The other sources of ^{90}Sr in the environment are releases from major nuclear accidents (Chernobyl and Mayak as discussed in Chapter 14) and releases from nuclear plants to the environment (the latter being the smallest contributor).

Strontium-90 arises as a fission product in the low-mass portion of the fission product distribution with a high yield relative to other fission products. Its decay leads to yttrium-90, which is also radioactive emitting a high-energy β^- emission,

$$^{90}_{38}\text{Sr} \xrightarrow{\beta^-} {}^{90}_{39}\text{Y} + e^- + \overline{\nu}_e + \gamma \tag{5.9}$$

Releases of strontium isotopes from nuclear reactors operating under normal conditions are generally considered very unlikely because it is contained within the fuel cladding. The high-energy radiation emitted by ^{90}Sr has been used as an energy source for portable instruments, especially for those used in space and in remote weather stations, where the heat generated by the radiation is converted to electricity [14].

Strontium-90 is also formed from the decay of the short-lived ($t_{1/2}=33$ s) fission product ^{90}Kr, via a similarly short-lived β^- decay of ^{90}Rb ($t_{1/2}=2.7$ m), viz.,

$$^{90}\mathrm{Kr} \xrightarrow{\beta^-} {}^{90}\mathrm{Rb} \xrightarrow{\beta^-} {}^{90}\mathrm{Sr} \tag{5.10}$$

Aside from ^{90}Sr, of the six radioactive strontium isotopes that are formed as a result of fission even the longest-lived (^{91}Sr and ^{89}Sr with half-lives of 10 h and 53 days, respectively) exist too briefly to warrant a concern.

The primary pathway for ^{90}Sr to enter the body [15] is via ingestion. This can be especially significant because the isotope can be concentrated in dairy products due to its elemental similarity to calcium. Such products are consumed widely, often close geographically to where they are derived from agriculture. Following a hypothetical release to the atmosphere arising from a serious accident, a plausible scenario is that ^{90}Sr is deposited on the ground following precipitation where it either remains on vegetation to be eaten by animals or is washed into the earth to be taken up by shallow-rooted plants. Its chemical similarity to calcium (notwithstanding being approximately double the mass) results in ~80% passing through the body and ~20% remaining. The latter can become deposited in place of calcium in bones and teeth; hence the widespread reference to ^{90}Sr as being a 'bone seeker'. If retained in such calciferous tissues the high-energy β^- emission of both ^{90}Sr and its daughter product, ^{90}Y, constitutes a source of exposure to those tissues. If ingested in quantities necessary to produce very high radiological dose rates (typically one thousand times greater than those received from background radiation), ^{90}Sr is known to increase the risk of bone cancer and leukaemia in animals, and is thus suspected to do so in people. Not surprisingly, releases associated with nuclear plants operating under normal conditions are generally too low to be measured and thus ^{90}Sr only poses a risk in the event of serious core-damage accidents [16].

5.6.3 YTTRIUM-90

Yttrium-90, the daughter product of ^{90}Sr, is also radioactive and decays to ^{90}Zr, which is stable. Yttrium-90 has a much shorter half-life than ^{90}Sr, at ~64 h, and it emits a very energetic β^- particle at 2.28 MeV (compare for example with the emission from strontium-90 described above). Thus, its radiotoxicity is more significant than ^{90}Sr, while being effectively transported to the site of exposure during the longer-lived state as ^{90}Sr. This corresponds to a very high dose per unit activity. As a decay product of ^{90}Sr, ^{90}Y exposes the same tissue types as ^{90}Sr following the calcium-related routes for uptake and hence it is the combination of these isotopes that is the main concern. The formation pathway of ^{90}Y is given as follows:

$$^{90}\mathrm{Sr} \xrightarrow{\beta^-} {}^{90}\mathrm{Y} \xrightarrow{\beta^-} {}^{90}\mathrm{Zr} \tag{5.11}$$

An unusual feature of ^{90}Y is that its β^- emission is so energetic that the associated β^- particles can penetrate further through the materials of personal protective equipment than other sources of β

contamination. This was of particular relevance in 1979 during the clean-up of the Three Mile Island accident. It has been dealt with by developing new instrumentation, remote handling capabilities and higher-density clothing not unlike fire-fighting equipment; such is the effect that a single isotope can exert on the clean-up of nuclear debris.

5.6.4 TECHNETIUM-99

Technetium-99, in common with several isotopes covered in this discussion, exists almost entirely as the result of nuclear activities with any primordial abundance arising only in trace amounts as a result of fission in uranium ores; even its elemental name harks from the Greek word for *artificial*: 'technetos'. There are no stable isotopes of technetium and, of the numerous radioactive variants that are known, only three exist for sufficiently long to warrant attention in the context of nuclear waste or decommissioning. These are ^{97}Tc, ^{98}Tc and ^{99}Tc. However, only ^{99}Tc is produced in sufficient quantities (via the decay of the fission product ^{99}Mo which has a fission yield of 6.06%), and with a half-life (2.1×10^5 years), to be of a concern in managing nuclear plants.

Technetium-99 is formed as a result of fission arising in the lighter-mass region of the fission fragment distribution. It can also be made with a particle accelerator but fission is the most common source, both as a by-product in power reactors and bred specifically for medical purposes in isotope-production reactors. Under normal operations, as with other nonvolatile fission products, the majority of ^{99}Tc resulting from fission is confined to the fuel element. Consequently, the main sources of ^{99}Tc in the environment are from nuclear reprocessing activities, nuclear accidents and weapons tests. In the environment technetium forms an oxide—technetium pertechnetate (TcO_4^-) and is thus readily water-soluble. This has the unusual property that the pertechnetate ion has negative valency despite containing the metal technetium, which complicates the ease with which technetium is extracted from the environment. However, the long half-life of ^{99}Tc renders it of low radio-toxicological risk relative to other products of fission with similar levels of yield.

In the body, ^{99}Tc constitutes an exposure hazard but externally it poses very little risk of exposure. It decays via β^--particle emission with energy 0.294 MeV to become the stable isotope of ruthenium, ^{99}Ru. It is taken up by the body via inhalation or ingestion and a significant proportion (50%–80%) is taken up by the bloodstream, following which it can be deposited preferentially in the thyroid, stomach and liver. However, technetium does not spend very long in the body; most of it will clear with a biological half-life of less than two days with the vast majority of it having left the body in ~20 days due to its high solubility in water.

5.6.5 RUTHENIUM-103 AND -106

Ruthenium-103 and -106, like ^{99}Tc, are two further examples of light-mass fission products that are radioactive and decay via β^--decay. In the case of ^{103}Ru, the half-life is so short (40 days, β^- emission at 0.2 MeV) that while being responsible for a significant proportion of the fission product radioactivity in spent fuel immediately postirradiation, this declines very quickly and falls well within the customary cooling period of irradiated fuel being several years in a spent fuel pond. For the case of ^{106}Ru, the half-life is longer (373 days, β^- emission at 0.04 MeV), which indicates that any contamination by this isotope would need to be left for approximately 10 years for it to decay to sufficiently low levels. The ruthenium isotopes pose a greater hazard internally than they do externally because of the nature of its radioactivity and the relatively low energies of the β^- emissions; there are numerous other unstable ruthenium isotopes but ^{103}Ru and ^{106}Ru have the longest half-lives and are hence usually the most abundant we encounter. ^{106}Ru decays to rhodium-106, which has a very short half also decaying via β^- particle emission. Early studies on animals [17] indicated that following internalisation ruthenium isotopes tend to accumulate in the major internal organs (kidney, spleen, lungs) and the bone.

5.6.6 THE IODINE ISOTOPES

The iodine isotopes, in common with the caesium isotopes discussed in the following section, originate as fission fragments in the heavier portion of the distribution of yield with mass. However, unlike most elements formed as fission products, iodine sublimes from the solid phase to the gaseous phase at 114°C; a temperature that is low relative to the operating conditions of nuclear plant. Consequently, iodine is usually in gaseous form in a nuclear context and hence can be liberated if fuel cladding is compromised in a core damage accident or when it is stripped to enable fuel to be reprocessed. Iodine, being a halogen, is highly reactive and therefore rarely exists in elemental form instead forming compounds readily both with metals and organics. In these forms (essentially as salts and organic complexes) present in the environment following a serious, hypothetical accident, the main pathway to the body for radioactive iodine atoms would be via contaminated foodstuffs and dairy products.

A further important property of iodine in a nuclear context is that it is concentrated preferentially in the thyroid gland. The thyroid gland is responsible for the production of peptide-based hormones that control the metabolism, growth and development. These hormones contain iodine and consequently the thyroid sequesters this on an elemental basis from the diet to synthesise these compounds irrespective of the isotopic variant of iodine or whether it is radioactive or not. Clearly, when radioactive iodine is present in the environment as a result of nuclear weapons fallout or a reactor accident, there is a risk that it will be concentrated in the thyroid and pose an increased risk because of the refinement via this biochemical pathway. This can result in the thyroid gland being exposed to iodine-borne radiation more significantly relative to the rest of the body. Since the thyroid gland in children is smaller and is growing with cell multiplication occurring at a higher rate than in adults, the risk of iodine exposure and the potential for cancer incidence in the young is more significant. A well-known scenario as occurred following the Windscale fire in 1957 (discussed in Chapter 14) is for milk to become contaminated following a serious accident via precipitation on grazing pasture, recognising milk usually constitutes a larger proportion of a child's diet than that of an adult. For this reason, countermeasures are often considered as part of the emergency response planning at nuclear sites including, for example, the distribution of potassium iodate tablets. Potassium iodate acts to dilute the iodine stream ingested by people in the event of a significant release and reduces the proportion of radioactive iodine retained by thyroid. The very short-lived iodine isotopes are strongly suspected to have been the source of most thyroid exposure following the Chernobyl accident, rendered airborne immediately after the primary explosion. However, their short lifetimes complicate the ease with which they are measured in the environment in such scenarios.

There are several radioactive isotopes of iodine. The isotopes ^{131}I, ^{132}I, ^{133}I, ^{134}I and ^{135}I are most significant as regards internal exposure but, while produced in significant yield from fission, their half-lives are sufficiently short (longest half-life being 8.1 days for ^{131}I) for their risk to negligible in decommissioning and waste effluent contexts. Iodine-129 is different since although it decays via a combination of β^- emission and γ-ray emission, in common with the other iodine isotopes, unlike them it has a half-life of 16 million years and emits very low-energy β^- particles (~150 keV) that are difficult to detect. This complicates the ease with which ^{129}I is measured but conversely reduces the radiological risk too (the γ-ray emission is also low in energy). Nonetheless, because of the long half-life of ^{129}I and its highly specific residence properties in the body, it is necessary to take its immobilisation into account during waste management and disposal.

5.6.7 THE CAESIUM ISOTOPES

The caesium isotopes arise in the nuclear fuel cycle as fission products and are part of the higher-mass peak in the fission yield mass distribution. There are many isotopes of caesium (32 in total). In a nuclear power context, the main focus tends to be on ^{134}Cs and ^{137}Cs. Caesium is very chemically active and thus it forms compounds readily, it is highly soluble in water and passes through the food chain and into the body easily. Caesium isotopes can migrate rapidly in the environment particularly in the presence of water. Both ^{137}Cs and ^{134}Cs can pose internal and external radiological risks.

Caesium-137 has a half-life of ~30 years and decays via β^- emission to the metastable state of ^{137m}Ba. It is the decay of the latter that is usually of principal interest because of its very high specific activity (characterised by a half-life of 2.6 min) and because of the associated single, energetic γ-ray photon at 662 keV. This feature has some benefits because it has led to the use of ^{137}Cs in many applications aside from the nuclear fuel cycle including the calibration of instruments and in medical radiotherapy. Consequently, although most ^{137}Cs present in the environment is primarily as a result of fallout from weapons tests and the Chernobyl accident, it can also arise as a result of discharges, accidents, misuse and inadvertent disposal of caesium sources manufactured for medical use, albeit to a much-reduced extent. The relatively high-energy γ rays from both ^{137}Cs and ^{134}Cs are ideal for assessment with high-resolution γ-ray spectroscopy systems.

Caesium-134 can present less long-term concern than ^{137}Cs because of its shorter half-life (2.1 y) but nonetheless decays via a similar combination of β^- decay and high-energy γ-ray emission (the latter with energy 1.6 MeV). Therefore, the risk from exposure to it is significant shortly following accidents and inadvertent releases but the isotope has usually decayed away as far as decommissioning and waste management is concerned. Given the difference in their half-lives, the relative fractions of these isotopes measured in the natural environment can be used to infer the origin of the caesium contamination on the basis of the time it has resided there. This is useful where contributions from weapons testing many years ago (that now comprise almost entirely ^{137}Cs) might otherwise be confused with more recent environmental contributions comprising both ^{137}Cs and ^{134}Cs.

5.7 SUMMARY

In the following Table 5.3, a summary is provided of the data for all of the examples of isotopic discussed in this chapter, including half-life, their principal decay pathways and associated risk factors.

Table 5.3 Half-Lives, Principal Decay Pathways (With Energies in MeV) and Risk Factors for the Isotopes Discussed in This Chapter

Isotope	Half-Life	Principal Decay Pathway (Energy of Emission in MeV)	Risk Factors
Tritium	12.3 years	β^- (0.019)	Internal ingestion/inhalation
Sodium-24	14.97 h	β^- (1.39)	Fast breeder reactors, internal dose risk
Cobalt-60	5.3 years	γ (1.33, 1.17)	All reactor systems where steel is used. External dose risk
Krypton-85	10.7 years	β^- (0.687)	As a gaseous release from a fission reactor. External skin dose
Strontium-90	29 years	β^- (0.546)	Bone and bone marrow
Yttrium-90	64 h	β^- (2.28)	Bone and bone marrow
Technetium-99	2.13×10^5 years	β^- (0.293)	Thyroid, stomach and liver
Ruthenium-103	40 days	β^- (0.04)	Main organs and bone
Ruthenium-106	372.6 days	β^- (0.0394)	Main organs and bone
Iodine-129	1.6×10^7 years	β^- (0.15)	Thyroid
Iodine-131	8.04 days	β^- (0.606)	Thyroid
Caesium-134	2.07 years	β^- (0.658), γ (1.6)	Internal and external exposure
Caesium-137	30.17 years	β^- (0.512), γ (0.662)	Internal and external exposure
Americium-241	432 years	α (5.49)	Internal exposure; liver and bone
Plutonium-238	87.7 years	α (5.49), SF	Internal exposure; liver and bone
Plutonium-239	2.4×10^4 years	α (5.15)	Internal exposure; liver and bone
Plutonium-240	6.6×10^3 years	α (5.16), SF	Internal exposure; liver and bone
Plutonium-241	14.35 years	β^- (0.0208)	Internal exposure; liver and bone

CASE STUDIES

CASE STUDY 5.1: PRIMORDIAL ESTIMATES OF ^{235}U ENRICHMENT

Current levels of natural ^{235}U enrichment are the result of an extensive period of radioactive decay. In the absence of a significant flux of neutrons to stimulate fission in ^{235}U, it decays via α decay to ^{231}Th with a half-life of 7.04×10^8 years. What was the level of enrichment 2 billion years ago?

The answer can be found by considering the half-life as considered in earlier chapters, but this time the abundance now corresponds to N (0.71% wt.) and that 2 billion years ago to N_0. Hence,

$N = N_0 e^{-\lambda t}$ and $\lambda = \log_e 2/t_{1/2}$

λ in this case is $3.1 \times 10^{-17}\ \text{s}^{-1}$, and thus we obtain: $0.71 = N_0 e^{-3.1\times10^{-17}\times6.3\times10^{16}}$

Re-arranging, $N_0 = 0.71/0.14 = 5.07$

However, we have to recall that ^{238}U (as the other major constituent of uranium) has also been decaying, independently of ^{235}U, via α decay with a half-life of 4.47×10^9 years. Repeating the calculation for ^{238}U,

$\lambda = 4.9 \times 10^{-18}\ \text{s}^{-1}$, and thus we obtain: $99.29 = N_0 e^{-4.9\times10^{-18}\times6.3\times10^{16}}$

Re-arranging, $N_0 = 99.29/0.73 = 136.01$

So, as percentages, the proportion of $^{235}\text{U} = \dfrac{5.07}{5.07 + 136.01} \times 100 = 3.6$ % wt. and similarly $^{238}\text{U} = 96.4\%$ wt.

CASE STUDY 5.2: RELATIVE PROPORTIONS OF ^{134}Cs AND ^{137}Cs

Faced with a soil sample that has been subject to fallout from weapons tests and Chernobyl, comprising caesium in proportions associated with ^{134}Cs and ^{137}Cs, what can we infer about the origin of the contamination?

Given that the original mean ^{134}Cs/^{137}Cs isotopic ratio for Chernobyl is 0.524 ± 0.006, if a ^{134}Cs concentration of 0.28 ± 0.04 Bq m^{-3} and a ^{137}Cs concentration of 5.0 ± 0.1 Bq m^{-3} were measured shortly after the accident then we can determine the ^{137}Cs contamination of weapons test origin, as follows:

Measured ratio $= 0.056 \pm 0.008$, as determined from the above measurements,

$$\text{Measured ratio} = \sum {}^{134}\text{Cs contributions} / \sum {}^{137}\text{Cs contributions}$$

In this example, all weapons test contributions to ^{134}Cs are zero because we shall assume all ^{134}Cs has decayed away, given the majority of atmospheric testing ceased a lot earlier and the half-life is 2.1 y. ^{137}Cs, with its half-life of 30 y, still makes a significant contribution to the denominator in the expression for the ratio above. Thus, this becomes,

$$\text{Ratio} = {}^{134}\text{Cs}_{\text{Chernobyl}} / \left({}^{137}\text{Cs}_{\text{Chernobyl}} + {}^{137}\text{Cs}_{\text{Tests}}\right) = 0.056$$

So we have two equations,

$$ {}^{134}\text{Cs}_{\text{Chernobyl}} = 0.524\,{}^{137}\text{Cs}_{\text{Chernobyl}}$$

$$ {}^{134}\text{Cs}_{\text{Chernobyl}} = 0.056\,{}^{137}\text{Cs}_{\text{Chernobyl}} + 0.056\,{}^{137}\text{Cs}_{\text{Tests}}$$

Subtracting, $0 = 0.468\,{}^{137}\text{Cs}_{\text{Chernobyl}} - 0.056\,{}^{137}\text{Cs}_{\text{Tests}}$

$$\text{so } {}^{137}\text{Cs}_{\text{Tests}} = 8.36\,{}^{137}\text{Cs}_{\text{Chernobyl}}$$

$$\text{since } {}^{137}\text{Cs}_{\text{Chernobyl}} + {}^{137}\text{Cs}_{\text{Tests}} = 5.0,\ {}^{137}\text{Cs}_{\text{Chernobyl}} = 0.534\ \text{Bq m}^{-3} \text{ and } {}^{137}\text{Cs}_{\text{Tests}} = 4.47\ \text{Bq m}^{-3}$$

REVISION GUIDE

On completion of this chapter you should:

- understand the range of specific isotopes that are of interest to the nuclear engineering discipline in terms of their decay pathways, retention in the body and radioactivity properties
- understand the distinction in terms of the keywords *actinides*, *transuranics*, *fission products* and *products of neutron activation*
- appreciate the influence that the range in *half-life*, *energy of radiation emission*, *fission yield* and *chemical properties* have on the properties of these isotopes in the environment and when ingested
- be aware of some of the relevant historical literature associated with the discovery, properties and behaviour of these isotopes in nuclear plant processes and in the environment

PROBLEMS

1. Calculate the specific activity of 1 g of tritium and 1 g of ^{137}Cs. Hence or otherwise, comment on the relative risk of exposure associated with these isotopes; in what ways are they similar and in what aspects are they different?
   The atomic mass of tritium is 3.016 u and of ^{137}Cs is 136.907 u. The half-lives are 12.3 y and 30.17 y, respectively.

2. The natural levels of ^{232}Th abundance we experience today have not always been the same because, in the absence of a significant flux of neutrons to stimulate the production of ^{233}Th, ^{232}Th decays via α decay with a half-life of 1.4×10^{10} years. If the average abundance in the Earth's crust is currently 9.6 mg ^{232}Th per kg what was it 2 billion years ago?
3. Iodine-133 can present a significant radio-toxicological hazard where it occurs. Research the properties of ^{133}I and hence describe why you might conclude this to be the case. Hence or otherwise, describe why it might present a cause for concern in the event of a serious accident, but also why we might not be able to conclude this easily from isotopic measurements of it in the environment.
4. Krypton-85 is generally not a significant cause for concern in terms of ingestion. Explain why this is the case with reference to its half-life, propensity for sequestration by the body and the radiation it emits.
5. Explain why neptunium-237 is an isotope of specific interest in the long term for nuclear energy.

REFERENCES

[1] D. Brugge, J.L. de Lemos, B. Oldmixon, Exposure pathways and health effects associated with chemical and radiological toxicity of natural uranium: a review, Rev. Environ. Health 20 (3) (2005) 177–193.

[2] F.H. DuBose, Calorimeter-based adjustment of multiplicity determined 240Pueff: known-α analysis for the assay of plutonium. N-TRT-K-00020, Savannah River Nuclear Solutions, 2012.

[3] Y. Kimura, M. Saito, H. Sagara, C.H. Yan, Evaluation of proliferation resistance of plutonium based on spontaneous fission neutron emission rate, Ann. Nucl. Energy 46 (2012) 152–159.

[4] R.C. Martin, J.B. Knauer, P.A. Balo, Production, distribution and applications of californium-252 neutron sources, Appl. Radiat. Isot. 53 (4–5) (2000) 785–792.

[5] J.D. Navratil, W.W. Schultz, The production, recovery, properties and applications of americium and curium, JOM J. Miner. Met. Mater. Soc. 45 (2) (1993) 32–34.

[6] N. Miura, H. O. Menlove, The use of curium neutrons to verify plutonium in spent fuel and reprocessing waste, LA-12774-MS, 1994.

[7] W.C. Bentley, Alpha half-life of ^{244}Cm, J. Inorg. Nucl. Chem. 30 (8) (1968) 2007–2009.

[8] D.M. Barton, P.G. Koontz, The spontaneous fission half-life of ^{244}Cm, J. Inorg. Nucl. Chem. 32 (3) (1970) 769–775.

[9] R.C. Thompson, Neptunium—the neglected actinide: a review of the biological and environmental literature, Radiat. Res. 90 (1982) 1–32.

[10] A.W. Kramer, Boiling Water Reactors, Addison-Wesley, New York, 1958.

[11] D.G. Jacobs, Sources of tritium and its behaviour upon release to the environment, ORNL/NSIC-37, US Atomic Energy Commission TID-24635, 1968.

[12] R.H. Mole, Sodium in man and the assessment of radiation dose after criticality accidents, Phys. Med. Biol. 29 (11) (1984) 1307–1327; H. Mizuniwa, et al., Dose evaluation to workers at JCO criticality accident based on the whole body measurement of sodium-24 activity and area monitoring, J. Atom. Energy Soc. Jpn. 43 (1) (2001) 56–66.

[13] P. Taylor, A survey of methods for separating and immobilising krypton-85 arising from a nuclear fuel reprocessing plant, AECL-10252, 1990.

[14] J.G. Morse, Energy for remote areas, Science 139 (3560) (1963) 1175–1180.

[15] B.L. Larson, K.E. Ebner, Significance of strontium-90 in milk. A review, J. Dairy Sci. 41 (12) (1958) 1647–1662.
[16] O.G. Raabe, Three-dimensional models of risk from internally deposited radionuclides, in: O.G. Raabe (Ed.), Internal Radiation Dosimetry, Medical Physics Publishing, Madison, WI, 1994, pp. 633–658 (Chapter 30).
[17] R.C. Thompson, et al., Physiological parameters for assessing the hazard of exposure to ruthenium isotopes, AEC Research and Development, [TID-4500, HW-41422], Hanford Atomic Products Operation, Richland, WA, 1956.

CHAPTER

MODERATION

6

6.1 SUMMARY OF CHAPTER AND LEARNING OBJECTIVES

In this chapter, the concept of neutron moderation is introduced in the context of reconciling favourable fission cross-section probabilities at low neutron energies with the contrasting high energies of neutrons that are emitted in fission. The most widespread moderator materials in use today are introduced and the qualitative and quantitative trends in the properties of these substances are discussed. The microscopic basis for neutron moderation is considered, which includes a discussion of the scattering kinematics of neutrons in moderator materials. The basis on which moderator performance is estimated is included and the properties of the main moderator materials in use in the world's nuclear reactors are compared and contrasted.

The objectives of this chapter are to:

- establish the general concept of *neutron economy*
- introduce the qualitative trends of scattering and absorption cross-section data of moderators with neutron energy
- introduce and derive the relationship between neutron energy after scattering, the atomic mass of the moderator and scattering angle
- introduce the relationship between neutron energy and the number of scattering interactions
- provide a comparison of the benefits and drawbacks of the moderator materials in mainstream use today
- provide a quantitative comparison of the principal properties of moderator materials
- define the *collision parameter*, *differential scattering cross section* and the *empirical scattering law*

6.2 HISTORICAL CONTEXT: JAMES CHADWICK, 1891–1974

James Chadwick worked with Ernest Rutherford first at the University of Manchester and subsequently, after the First World War, at Cambridge (Fig. 6.1). At that time, Rutherford was studying the disintegration of atoms with α particles to yield protons. Together, both had hypothesised the possibility of a neutral subatomic particle to explain several inconsistencies in the structure of the atom and early preliminary findings as to what the nucleus might comprise. Earlier, Walther Bothe and Herbert Becker had observed an unusual form of radiation when they bombarded beryllium with α particles that was considered to be of 'great penetrating power'. Meanwhile, Frédéric and Irène Joliot-Curie had observed that ionisation increased in a nearby vessel when the same radiation from beryllium was incident on a hydrogenous substance; observations consistent with protons being ejected from the substance by the mystery radiation.

Nuclear Engineering. https://doi.org/10.1016/B978-0-08-100962-8.00006-8

FIG. 6.1

James Chadwick.

Chadwick carried out experiments by using an ionisation chamber in which he observed that protons were ejected from gelatine by the beryllium radiation and, similarly, nuclei could be ejected from a variety of substances such as helium, lithium, carbon, and so on. This was evident from the significant degree of ionisation that they provoked in the gas of the chamber and because the ranges of the recoiling products were observed to be much greater than was feasible based on a hypothesis similar to Compton scattering by γ-ray photons. Chadwick realised that this implied a type of radiation comprised of relatively massive entities rather than, as was considered might be the case, γ-ray photons. He derived expressions for the maximum velocity of the protons and the other recoils in terms of the mass of the mystery radiation. By comparing the maximum ranges of the protons, among other factors, Chadwick was able to infer the mass of the mystery radiation to be approximately that of the proton. However, he also observed that it passed easily through a thick piece of lead and thus that it could not be comprised of protons nor consistent with any other known form of radiation. Thus, it had to be the sought-after 'neutron'.

Chadwick was awarded the 1935 Nobel Prize for physics for the discovery of the neutron. This had a profound effect on the understanding of the structure of atomic nuclei. Neutrons can be used to disintegrate the heaviest isotopes as is the case in fission. Chadwick's discovery resolved many theoretical difficulties associated with our understanding of the nucleus and completed the now universally accepted model of the nucleus comprised of neutrons and protons.

6.3 INTRODUCTION

An important design feature that distinguishes what are termed *thermal-spectrum* power reactors from *fast-spectrum* systems and fusion prototypes is the presence of a *moderator*. Not to be confused with something that might be perceived as 'moderating' the performance of these machines, a nuclear moderator raises the probability of fission by reducing the energy of neutrons. In thermal-spectrum reactors operating with fuel of low enrichments, a moderator is used to shift the initial energy of fission neutrons to the thermal part of the neutron spectrum where fission is more probable.

As was discussed in Chapter 4, in the consideration of the fission process, the majority of neutrons emitted in fission are born fast in contrast with the microscopic fission cross section that falls rapidly with increasing neutron energy, as per $1/v$ where we recall v is the speed of the neutron. To reconcile this, the neutron population is usually caused to interact with a substance to reduce the kinetic energy of the neutrons; the substance employed to do this is called a *moderator*. While there are other ways in which a chain reaction can be established in the absence of moderation (e.g. by raising the enrichment significantly), most modern power-generating reactors are required to generate large quantities of heat because this favours power generation efficiency, fuel utilisation and ultimately the economics of operating a plant. This brings with it important requirements in terms of thermal mass and materials robustness, which tend to favour low fuel enrichments. Further, as we shall learn in Chapter 12, enrichment is an expensive element of the nuclear fuel cycle that has a strong influence on economic viability. Also, concern over the proliferation of nuclear material is reduced if low-enriched fuels can be used.

The principal role of the moderator in a nuclear reactor is thus to shift the spectrum of fission neutrons to lower energies where the fission cross section is more favourable; in doing this the probability of each individual interaction is raised (moderation) rather than increasing the number of interactions that are likely to result in fission (as might be achieved through greater enrichment). The use of a moderator is thus very significant: it enables fuels of low enrichments to be exploited for power generation that would not be possible otherwise.

6.4 THE CONCEPT OF NEUTRON ECONOMY

The lifetime of a neutron in a reactor is considered along with other elementary reactor principles in Chapter 8. However, it is instructive to rehearse the salient aspects of this prior to the full consideration of moderation in this chapter. The concept of *neutron economy* is a useful and flexible idea that will arise on several occasions in this context, especially when comparing and contrasting different reactor designs. It simplifies our appreciation of several broad principles of the behaviour of the neutron population on which the function of nuclear reactors depends.

There is effectively only one source of neutrons in a fission reactor: the nuclei comprising the fuel that are undergoing nuclear processes in the core. These processes can be subdivided in terms of four specific phenomena:

1. *Induced fission* of fissile isotopes such as ^{235}U and ^{239}Pu.
2. *Spontaneous fission* of, for example, isotopes such as ^{238}U and ^{240}Pu.
3. [α,n] reactions driven by the α-particle emissions from minor actinides. The abundance of the minor actinides increases with burn-up; one example is the curium isotope, ^{244}Cm, as discussed in Chapter 5.
4. The interaction of high-energy γ rays (typically >8 MeV) that stimulate the emission of neutrons (*photo-neutrons*) either directly or as a result of the associated phenomenon; a process known as *photo-fission*.

In an operating reactor the prominent source of neutrons is induced fission; this contributes both *prompt* and *delayed* neutrons, that is, those arising immediately following scission of a given nucleus and those that follow at some time after the β^--decay of the primary fragments, respectively. In Chapter 4 it was discussed that the mean number of neutrons emitted per fission event, $\overline{\upsilon}$, is a property of the isotope that fissions; in the case of ^{235}U and ^{233}U we recall that this is approximately 2.5 while for ^{239}Pu it is 2.9, for fission induced by thermal neutrons.

If the propensity for fissile isotopes *not* to fission (which can be inferred by the number of neutrons emitted per nuclide destroyed, η) is also taken into account, then fissile isotopes might be ranked accordingly in terms of eta: $\eta(^{233}\text{U})=2.30$, $\eta(^{239}\text{Pu})=2.13$ and $\eta(^{235}\text{U}) = 2.09$, as per Table 4.1. This analysis demonstrates, albeit simplistically, that the specific isotope undergoing fission as a constituent of nuclear fuel can have a significant influence on the neutron economy of a given reactor design, especially since it is the principal *source* of neutrons. However, it is also important to appreciate that this contribution is not derived from a single isotopic species but from a variety, and that the nature of this variety changes with the amount of energy produced by the fuel during operation; this being the product of power and time that is termed *burn-up*. For example, with uranium-fuelled reactors there is also the unavoidable contribution from ^{239}Pu that arises soon after start-up due to neutron capture on ^{238}U. There will also be a relatively small amount of *fast fission*, predominantly from ^{238}U.

Having considered *sources* of neutrons in nuclear fuel we also need to consider the salient neutron *sinks* to continue our discussion of neutron economy. There are several processes by which neutrons are lost from a reactor that can be characterised as either *absorption* or *leakage*. We have already considered an example of the former implicitly associated with the scenario that a fissile nucleus absorbs a neutron but does not fission as a result; the neutron is absorbed and is lost from the system. The formation of an excited state of ^{236}U* from neutron absorption on ^{235}U, rather than fission, is an example of this. There are other sinks: neutron absorption in moderators and coolants are two principal examples. Since microscopic absorption cross sections change significantly with energy, the influence of this also needs to be borne in mind.

Leakage on the other hand is a consequence of real reactors not being infinite, that is there are physical boundaries that define the physical extent of the reactor core within which a chain reaction can be self-sustaining. Inevitably, some neutrons scatter beyond these boundaries never to return; thus, they too are lost from the reactor. Clearly, the interaction properties of the materials selected for use in a reactor core have a significant influence on the propensity for neutrons to leak and thus to be lost from the system, as does the ratio of the surface area of the core to its volume.

The following general guidelines can be adopted when considering the interplay of *reactor design* and *neutron economy*:

- The choice of fissile isotope and fuel enrichment influences neutron *income*. With the majority of power reactors to date being based on ^{235}U there has been limited flexibility as to the primary isotope of choice. This is likely to continue, notwithstanding ongoing developments associated with breeding from fertile isotopes, as discussed in Chapter 11. Raising the enrichment generally results in greater numbers of neutrons being generated overall (as a result of there being more fission interactions) and thus higher income. It is worthy of note, that harder neutron spectra i.e. higher energy, result in higher neutron multiplicities from a given fissile isotope. However, the counter effect of the decline in the microscopic fission cross section with increasing neutron energy needs to be borne in mind too.
- Reactor design, geometry and the choice of materials selected for moderator, coolant and structural components influence what we might term neutron *expenditure*. Where this selection results

in losses that are relatively low, neutron economy might be sufficient to provide some flexibility in terms of the level of fuel enrichment the reactor can accommodate, including low enrichments and fuels of natural enrichment. Where, conversely, expenditure is relatively high, the reactor might be constrained to fuels necessitating artificial enrichment.

It is worth re-stating that neutron economy is not the only factor that influences the level of fuel enrichment that is selected: increases in load factor, burn-up and optimising the periods between fuel loading are also factors that favour the enrichment of fuels above natural levels. The widespread availability of enrichment facilities has enabled the majority of the world's power reactors to exploit these benefits, notwithstanding the economic factors associated with enrichment that were highlighted earlier.

6.5 DESIRABLE PROPERTIES OF MODERATORS

Almost all substances exert some moderating influence on neutrons, to a greater or lesser degree (even ourselves). However, few alternatives meet the challenging and often conflicting engineering demands of a nuclear power reactor. The requirements of moderator materials span the essential requirements of having a *low atomic mass*, a *high neutron scattering cross section*, a *low neutron absorption cross section* and being *resilient in use* despite being exposed to high levels of radiation over long service periods. Fundamental to this selection are the interactions of neutrons that result in *scattering* and *absorption*; these are explored in the following sections.

6.5.1 NEUTRON SCATTERING INTERACTIONS REVISITED

The majority of neutrons pass through matter unperturbed because of the tiny scale of the nucleus relative to the size of the atom. However, a few pass sufficiently close to a nucleus to be scattered by the strong nuclear interaction. This causes the neutron to change direction and to exchange some of its energy with the nucleus, the latter which is usually at rest or near to it relative to the energetic state of the neutron before the interaction. If the sum of the kinetic energies before and after the interaction is the same, then the interaction is termed *elastic scattering*. If the interaction is significant, such that a significant quantity of energy is exchanged, this may result in the nucleus being elevated to an excited state and its composition might also be affected. In such a case, the sum of kinetic energy is reduced after the interaction relative to that before and the interaction is termed *inelastic scattering*. Elastic scattering is the most desirable interaction in moderators of thermal spectrum fission reactors because of the energy transfer that occurs.

6.5.1.1 Trends in microscopic neutron cross sections in moderators

Later in this chapter a quantitative comparison of the microscopic elastic scattering and absorption cross sections is provided. Before this is considered it is instructive to consider the qualitative trends in cross section of the principal moderating elements used in power reactors. These include *hydrogen* and *deuterium* (and *oxygen* as the other constituent of light water and heavy water), and also *carbon* (as the main constituent of graphite). A comparison of the microscopic elastic scattering cross sections and the corresponding absorption cross sections for these isotopes are given in Figs 6.2 and 6.3, respectively.

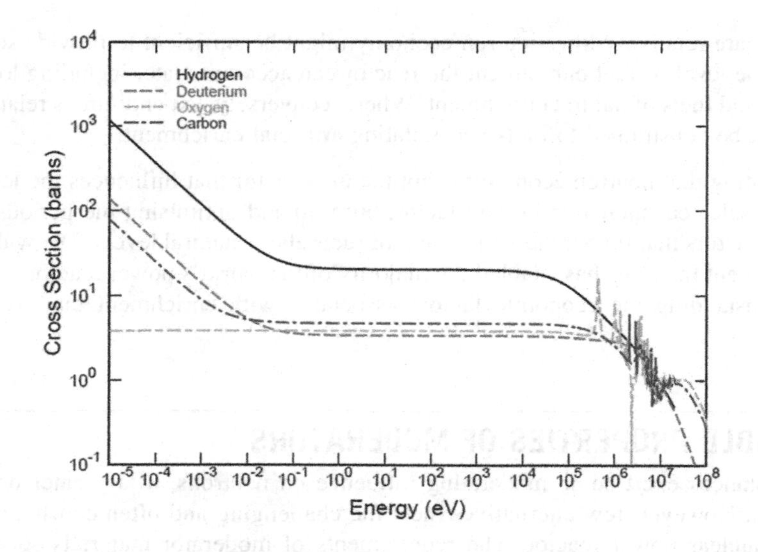

FIG. 6.2

Microscopic cross sections for elastic scattering of neutrons on hydrogen, deuterium, oxygen and carbon as a function of neutron energy.

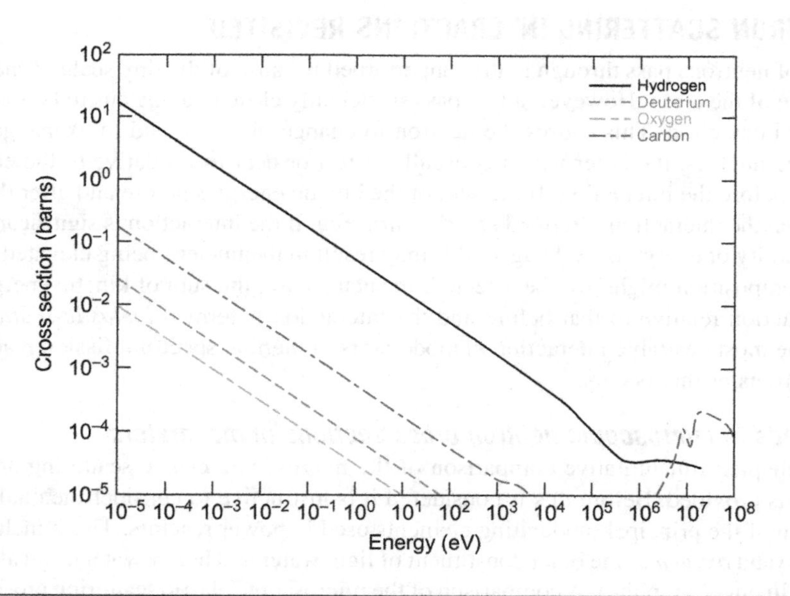

FIG. 6.3

Microscopic cross sections for neutron absorption on hydrogen, deuterium, oxygen and carbon as a function of neutron energy.

The dependence of the elastic scattering cross section with neutron energy, shown in Fig. 6.2, demonstrates that for the majority of the neutron energy range associated with fission neutrons (i.e. 0.0253 eV $\rightarrow \sim$0.1 MeV) the probability of scattering is largely independent of neutron energy. In this range, hydrogen exhibits the most significant probability of scattering with the other isotopes having similar cross sections to each other but approximately a factor of 10 lower than that of hydrogen throughout the range, up to 0.1 MeV. Beyond this level the cross sections fall off rapidly and converge with each other, with there being some resonance structure in the case of oxygen and carbon beyond 0.5 MeV.

The dependence of the microscopic absorption cross section with neutron energy in Fig. 6.3 is rather different to the corresponding case of scattering. A rapid decline is observed (note that the plots in Figs 6.2 and 6.3 are on $\log_{10}$–$\log_{10}$ axes) for all isotopes. The descending rank order, in terms of absorption cross-section magnitude throughout the energy range, is hydrogen, carbon, deuterium and oxygen. Hydrogen is two orders of magnitude greater than carbon, with carbon a factor of ten greater than deuterium.

It is clear from this consideration that it is not necessary to be overly concerned with the dependence of neutron scattering probability with energy in moderators as it is slowly varying. Conversely, the high absorption cross section for hydrogen does justify further consideration, especially at thermal neutron energies. It is also apparent that, as a result of moderation, neutron absorption in all cases is most significant at thermal energies. Consequently, this is where the most significant negative impact on neutron economy might occur, after moderation.

6.5.1.2 The dependence on atomic mass and the collision parameter

The most efficient transfer of energy occurs when there is symmetry between the mass of the nucleus and the neutron that is scattering from it. Symmetry is approached for those elements with the smallest number of neutrons and protons, and consequently *low-mass isotopes moderate neutrons most efficiently*.

The optimum energy-transfer scenario is that of a neutron colliding head-on with the nucleus of a hydrogen atom (a proton). The near perfect mass symmetry of a neutron with the proton for the optimum scenario results in the neutron being brought abruptly to rest while the proton is ejected from the hydrogen atom with an energy and momentum close to that of the incident neutron. Macroscopically, this is analogous to the abrupt arrest of the cue ball in pool or billiards following a head-on collision with the ball in play; the latter recoils at an angle π to the direction of the cue ball with the same speed (friction losses in this analogy notwithstanding). This scenario, before and after the interaction, is depicted in Fig. 6.4A and B.

Efficient moderation such as that provided by light isotopes is important for two reasons. First, the more efficient the energy transfer that is possible per interaction, the smaller the number of interactions necessary to thermalize a neutron. Secondly, it ensures that neutrons are thermalized in a small volume that is desirable for compact reactor designs. The former is critical to minimise absorption and associated neutron expenditure since every interaction brings with it the finite probability of the neutron being absorbed rather than scattering. Moderation in a small volume helps minimise the possibility of leakage.

The optimum, back-to-back scattering scenario described above is the most efficient in terms of energy transfer. However, it is also relatively unlikely and neither can the neutron scattering trajectory

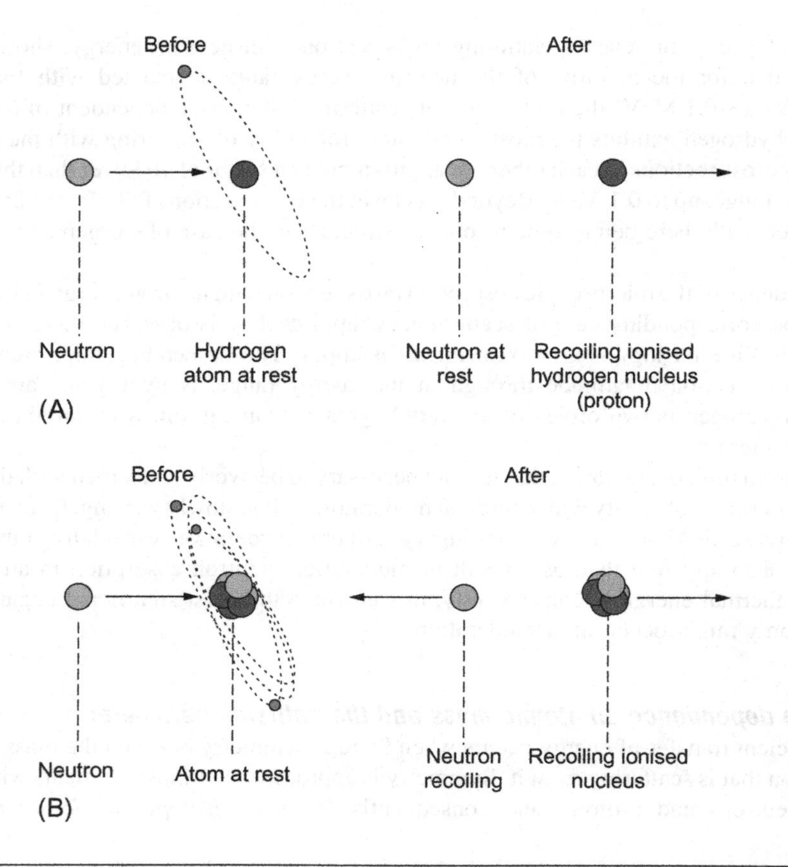

FIG. 6.4

Schematic representation of back-to-back scattering of a neutron (A) on a hydrogen atom (left) and (B) from a more massive nucleus (right), before and after elastic scattering (not to scale).

be constrained only to occur head-on. Equating the energy and momentum before and after scattering results in the following expression (see Case Study 6.1),

$$E' = \left[\frac{(A^2 + 2A\cos\theta + 1)}{(A+1)^2}\right] E \tag{6.1}$$

where E is the kinetic energy of the neutron before scattering, E' is that afterwards, A is the atomic mass of the scattering nuclide and θ is the scattering angle relative to the incident angle of the neutron in the centre-of-mass frame of reference. For the optimum case for energy transfer of back-to-back scattering, $\theta = \pi$ and thus $\cos\theta = -1$ so that Eq. (6.1) becomes

$$E' = E\left[\frac{(A-1)^2}{(A+1)^2}\right]_{\theta=\pi} \tag{6.2}$$

where $\alpha = [(A-1)/(A+1)]^2$ is referred to in some works as the *collision parameter*. If the scenario for hydrogen where $A=1$ is considered, it is clear that for back-to-back scattering, Eq. (6.2) reduces to $E' \sim 0$ since $\alpha \sim 0$ given the reasonable approximation that the mass of the neutron is equal to the mass of the proton. This implies that the neutron is brought to rest after just one interaction (note its energy cannot realistically be reduced below thermal equilibrium with its surroundings, i.e. $E' = 0.0253\,\text{eV}$ at 293 K). While liquid hydrogen is unsuitable as a moderator, having a boiling point ~20 K and being highly explosive, hydrogenous substances such as polyethylene and water[1] constitute excellent moderators due to their significant hydrogen content and the efficient scattering dynamics illustrated above.

6.5.1.3 The differential neutron cross section and the empirical scattering law

When the concept of the microscopic neutron cross section was introduced in Chapter 4, we considered a number of interaction types, for example, absorption, capture, fission, and so on, in which the incident neutron is consumed in the corresponding process. For elastic scattering, given its direct relevance to neutron moderation, the situation is a little different because a number of outcomes are possible in terms of the angle θ through which the neutron is scattered relative to its incident direction and the corresponding energy after scattering, E'. The microscopic cross section for elastic scattering that we have considered thus far, σ_s, is implicitly the integral across all of the possible final outcomes in terms of E',

$$\sigma_s(E) = \int_0^\infty \sigma_s(E \to E')\, dE' \tag{6.3}$$

where $\sigma_s(E \to E')$ represents the cross section for the specific component of $\sigma_s(E)$ describing the possibility that a neutron is scattered from the incident energy E to a final energy E'. Here, E' is defined as being between E' and $E' + dE'$. Interestingly, $\sigma_s(E)$ corresponds to what might be measured experimentally most easily in terms of an average of observations of many, individual, unconstrained scattering interactions. Since Eq. (6.3) can be expressed as

$$\sigma_s(E) = \int_0^\infty \frac{d\sigma_s}{dE'}\, dE' \tag{6.4}$$

$\sigma_s(E \to E')$ is often referred to as the *differential scattering cross section*. This is an important concept because we are often interested in the likelihood of scattering in a particular direction or to a particular final energy.

To express $\sigma_s(E \to E')$ in terms of the more tangible variable of scattering angle, θ, it can be expressed further to Eq. (6.4),

$$\frac{d\sigma_s}{dE'} = \frac{d\sigma}{d\Omega} \cdot \frac{d\Omega}{dE'} = \frac{d\sigma}{d\Omega} \cdot \frac{\sin\theta\, d\theta\, d\varphi}{dE'} \tag{6.5}$$

where $d\Omega = \sin d\theta d\varphi$ is the differential of the solid angle in the centre-of-mass frame of reference defined in spherical polar coordinates, and φ is the angle of azimuth. The energy after scattering E' is independent of φ, as per Eq. (6.1). In this context $d\theta$ defines the width of a ring on the surface of a

[1]It is necessary to use a weighted average $\alpha = \bar{\alpha}$ for hydrogenous compounds to take account of the elements present for which $A > 1$.

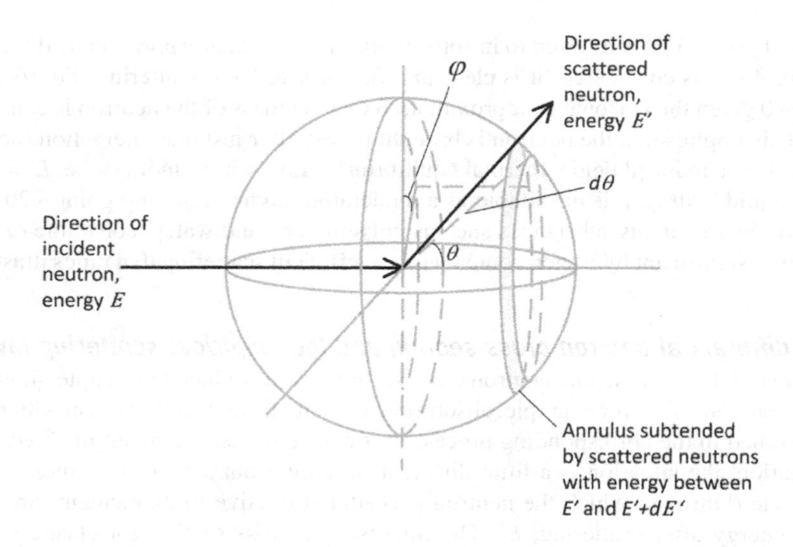

FIG. 6.5

A schematic representation of neutron scattering in spherical polar coordinates. Here, the scattering angle relative to the angle of incidence is θ and the differential scattering cross section are defined in terms of the neutron describing an angle of between θ and $\theta + d\theta$ and hence having an energy after scattering of E' and $E' + dE'$.

sphere for $0 \le \varphi \le 2\pi$ that defines the directions of the scattered neutron that are possible, as shown schematically in Fig. 6.5.

Differentiating Eq. (6.1) with respect to θ we obtain

$$\frac{dE'}{d\theta} = -\frac{2A\sin\theta}{(A+1)^2}E \tag{6.6}$$

and thus substituting into Eq. (6.5) gives

$$\frac{d\sigma_s}{dE'} = -\frac{d\sigma_s}{d\Omega} \cdot \frac{\pi(A+1)^2}{AE} \tag{6.7}$$

In the centre-of-mass frame of reference there is no preferential direction in which neutrons are scattered that is, it is isotropic. This is a property that is often referred to as the *empirical scattering law*. Hence, $d\sigma_s/d\Omega$ as a function of E is constant and thus so is $d\sigma_s/dE'$, as per Eq. (6.7).

If $d\sigma_s/dE'$ is sketched as a function of E', then we obtain the dependence shown in Fig. 6.6, specifically in this example for the case of hydrogen. This comprises a straight line of zero gradient that extends in energy from the maximum transfer possible, at $\theta = \pi$ and $E' = E$ ($E' = 0$ for the case of hydrogen), through to the minimum possible energy transfer (corresponding to angles of incidence $\theta \to 0$ and $E' \to E$). From Fig. 6.6 it can be appreciated that the average neutron energy after scattering over many scattering interactions is half the sum of the maximum and minimum energies; for the case of hydrogen shown in Fig. 6.6 this is $\overline{E'} = E/2$. Consequently, when a neutron interacts with the proton constituting a hydrogen nucleus it will, *on average*, lose half its energy.

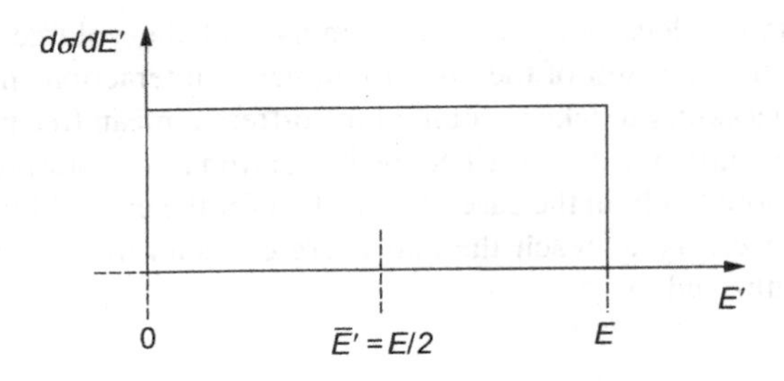

FIG. 6.6

A schematic representation of neutron scattering probability as a function of scattering energy (angle) in the centre-of-mass frame of reference, for hydrogen.

The scenario in which a neutron interacts head-on with a scattering nucleus and is thermalized by a single scattering interaction is a rare circumstance relative to the range of other possible scattering trajectories, for the reasons considered earlier in this section. Most neutrons thermalized by scattering in moderators will do so via a significant number of successive interactions that, despite being a sequence of interactions, still occurs very rapidly. It is instructive to consider the approximate number of interactions necessary to thermalize a neutron, again on an *average* basis.

To do this, consider the following scenario: Since on average with each scattering interaction on hydrogen a neutron loses half its kinetic energy, then the average kinetic energy $\overline{E'_n}$ of a neutron after n such scattering interactions with an incident energy of E can be written as

$$\overline{E'_n} = \left(\frac{1}{2}\right)^n E \tag{6.8}$$

Re-arranging to obtain the average number of interactions necessary to reduce a neutron with to an energy of $\overline{E'_n}$ in this case gives the following expression:

$$n = 1.44 \log_e \frac{E}{\overline{E'_n}} \tag{6.9}$$

Assuming an initial energy of $E = 1$ MeV and the final energy is $\overline{E'} = 0.0253$eV gives $n \sim 25$. This illustrates how efficient hydrogen is at moderating neutrons, attenuating them from speeds of 1.4×10^7 to 2200 m s^{-1} in just 25 interactions.

For the purposes of comparison, it is useful to consider ^{238}U as both a relevant and contrasting case: The significantly greater mass of the ^{238}U nucleus, relative to that of the neutron, results in the latter only transferring a small fraction of its energy (<2%) at each scattering interaction. This is depicted schematically in Fig. 6.4B where the neutron retains a significant proportion of its energy, recoiling in the opposite direction to the more massive nucleus in this example.

Substituting for $A = 238$ in Eq. (6.2) as per the case for ^{238}U gives a minimum scattered neutron energy for this case of $E' = 0.98\,E$, since in this case we are far from a symmetry in terms of the mass of the neutron and the uranium nucleus it is interacting with. This gives and average energy after scattering of $\overline{E'} = 0.99\,E$. Solving similarly to Eq. (6.9) with a value of 0.99 for the average collision parameter indicates that ~1740 interactions are required to thermalize the same 1 MeV neutron as per the final and incident energies used in the previous example, respectively.

The example given above does not necessarily imply that the volume required to moderate a neutron is drastically larger as a result of the higher number of interactions necessary for thermalisation; to estimate this it is necessary to take account of the different mean free paths for scattering of the different materials and the random walk described by the neutron as it is slowed. However, leakage and absorption are generally more likely in the case of ^{238}U than for the case of hydrogen due to the higher number of interactions necessary to reach thermal energies, neutron absorption cross sections and specific geometries notwithstanding.

6.5.2 PROPERTIES OF SPECIFIC MODERATORS

A summary of the principal nuclear properties of the relevant constituent isotopes in the most widely used moderator materials is given in Table 6.1. These comprise: hydrogen as a constituent of light water, deuterium as per heavy water, oxygen as a constituent of both and carbon for the case of graphite.

6.5.2.1 Light water

Light water is abundant, cheap and has the benefit of there being over 200 years of engineering knowledge of its thermophysical properties of heat transfer and flow. Where it is used as a moderator in thermal-spectrum power reactors, it is almost always used on a dual-purpose basis with it also being the primary coolant. In this arrangement it offers a combination of favourable neutron scattering cross section throughout the required energy range (in the case of both of its constituent elements hydrogen and oxygen), high specific heat capacity and Newtonian flow behaviour. The use of light water has enabled reactor designs that are compact and with predictable control characteristics, particularly in the event bubbles form in the water as a result of unanticipated increases in power; the latter feature is an important safety attribute and is discussed in more detail in Chapter 8.

The principal disadvantage of light water as a moderator is that the capture cross section of hydrogen at thermal neutron energies is high, relative to alternative moderator elements such as carbon (graphite) or deuterium (heavy water), as the data in Table 6.1 demonstrate. Neutron absorption on hydrogen results in the formation of deuterium and constitutes an undesirable sink for neutrons. A further operational aspect associated with the use of water is that the irradiation of mineral impurities in it can lead to contamination. It is thus necessary to use light water in a highly purified form to minimise the extent to which this occurs.

Table 6.1 Microscopic Cross Sections for Neutron Capture, σ_c (n,γ), and Elastic Scattering for Thermal and 1 MeV Energies, With α, $\overline{E'}/E$ and n for Constituent Isotopes of Moderators in Widespread Use (http://www.oecd-nea.org/janisweb and JENDL-4.0 via http://wwwndc.jaea.go.jp/)

Moderating Isotope	σ_c (n,γ)/b		$\sigma_s(n,elastic)$/b		α	$\overline{E'}/E$	n (To Thermalise 1 MeV Neutron)
	Thermal	1 MeV	Thermal	1 MeV			
Hydrogen (^{1}H)	0.332	0.00003	30.27	4.25	0	0.5	25
Deuterium (^{2}H)	0.00051	0.000006	4.26	2.87	0.11	0.56	30
Carbon (^{12}C)	0.00386	0.00002	4.94	2.60	0.72	0.86	116
Oxygen (^{16}O)	0.00019	0.0001	3.96	8.15	0.78	0.89	150

A significant advantage of light water, in addition to its widespread availability, is that it can be drained from a power plant, decontaminated and disposed of at the end of a reactor's useful life; these are procedures that are well-practiced and understood. Further, as a liquid, it is immune to issues associated with radiation-induced decline in material integrity that can be a concern for solid, graphite moderators.

6.5.2.2 Heavy water

From a conceptual perspective, replacing hydrogen in light water with deuterium to constitute heavy water results in a medium that has very similar thermophysical properties to light water (e.g. boiling point, critical point, density, fluid flow, etc.) and beneficial neutron interaction properties. In particular, the thermal neutron capture cross section of deuterium is several hundred times smaller than that of hydrogen while the important feature of mass symmetry between the neutron and, in this case the deuteron, is largely conserved relative to higher-mass alternatives. This has a direct benefit in terms of neutron economy as the removal of hydrogen removes a prominent absorption-based sink for neutrons. Consequently, reactor designs that are moderated with heavy water can exploit similar steam cycle principles to light water reactors but a wider range of fuel enrichments is open to them.

The principal disadvantage of reactors moderated with heavy water is that it is expensive and thus has to be conserved in use. Second, tritium is produced in the heavy water as a result of neutron activation and this can constitute an environmental concern due to its radioactivity, as discussed in Chapter 5. Most reactor designs that are moderated with heavy water use it as a primary heavy coolant that is separate from the moderator, with a secondary coolant cycle that is based on light water. This and other reactor designs are covered in more detail in Chapter 10.

6.5.2.3 Graphite

Graphite was the first moderator used in nuclear reactor systems, starting with Fermi's CP-1 pile through to both the early materials production and power-generating reactor designs. It remains the moderator in use across almost all of the power reactor fleet currently operating in Britain. Although it is obsolete with respect to the commercial power reactors currently under construction, it is relevant to several advanced concepts under consideration for the future.

In similarity to the case of deuterium in heavy water, carbon as the constituent element in graphite has the principal benefits of a high microscopic neutron scattering cross section and a low thermal neutron absorption cross section. It is relatively low mass, albeit less efficient as a moderator than hydrogen or heavy water in this respect. As in the case of heavy water, the low absorption cross section affords benefits in terms of neutron economy. This was important in the early nuclear power era, when enrichment capabilities were not widely available and the onus was often on fuels of natural enrichment and thus reactor systems of relatively low neutron economy.

Graphite being a solid has both benefits and disadvantages. From a positive perspective, it constitutes a matrix into which fuel can be loaded thus providing relatively easy access to fuel while the reactor is on load without the need to depressurise the entire coolant system. This was important when fuel burn-up periods were constrained, often as a result of radiation damage in metal fuels, which required nuclear fuel to be changed more frequently than is necessary today. It is relatively rare for graphite-moderated reactors to refuel on load as frequently today because most of them now use enriched oxide fuels that provide higher levels of burnup over longer periods. Further, the requirement to retrieve irradiated fuel with low burnup characteristics is less widespread. From a disadvantageous perspective, graphite is susceptible to radiation damage. In some of the first nuclear reactors (now obsolete) this was manifest as both deformation and the storage of potential energy (Wigner energy), the latter which required careful management to avoid uncontrolled temperature escalations. A further disadvantage is that graphite cannot be transported away as easily at the end of the life of a reactor, as is possible with liquid moderators, and many tonnes of graphite are necessary in a large power reactor. With time and reactor use, the graphite undergoes radiation damage, which degrades its structural integrity and effectiveness as a moderator, and it also becomes radioactive. This imposes a limit on the functional life of the plant because it is impossible to replace a large graphite moderator, in situ.

Despite these disadvantages, the use of graphite as a moderator has enabled reactors to be designed and operated with gaseous coolants, particularly carbon dioxide and helium. Such designs return high thermodynamic efficiencies relative to water-cooled/water-moderated reactors because the outlet temperatures are not limited by the critical point of a liquid coolant. For future designs in which high temperature, high-efficiency operation with gaseous coolants is desirable, graphite offers significant potential, albeit often in the form of a moderator material that is part of an integrated, composite fuel design.

CASE STUDIES

CASE STUDY 6.1: SCATTERING DEPENDENCE OF NEUTRON ENERGY WITH ANGLE

In the centre-of-mass frame of reference the elastic scattering of a neutron off a nucleus can be considered schematically, as per Fig. 6c.1,

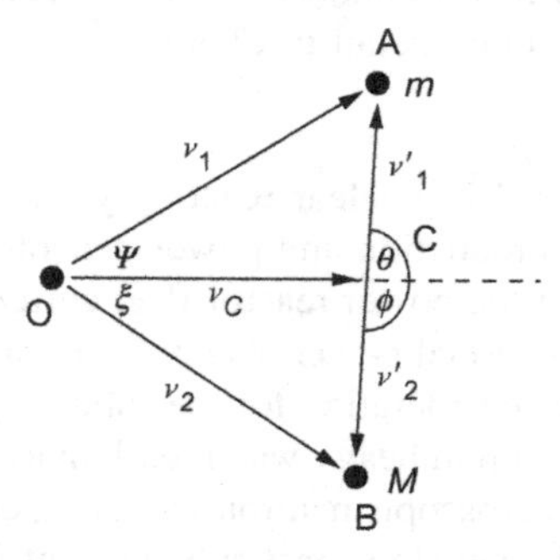

FIG. 6C.1

A schematic representation of the elastic scattering interaction of a neutron off a nucleus where the site of the collision in the laboratory frame is O, and the positions of the neutron and nucleus following the interaction are A and B, respectively.

where m is the mass of the neutron and M is the mass of the scattering nucleus, the velocity of the centre of mass frame is v_C, and that of the neutron and moderating nucleus in the laboratory frame are v_1 and v_2, respectively. The initial velocity of the neutron is u and the velocity of the neutron in the centre-of-mass frame is v'_1. For the purposes of this illustration isotropic scattering in the centre-of-mass frame of reference has been assumed, which is considered reasonable for most nuclear reactor applications. As vectors, it is implicit in this derivation that the direction of the velocities is defined relative to the direction of the centre of mass and, for example, the direction of the scattered neutron, as per v'_1, is defined as being at an angle θ.

Since the momentum of the neutron *before* the collision and that of the scattered neutron and recoiling nucleus *after* the collision are conserved, in the centre-of-mass frame of reference the two can be equated thus,

$$mu = (m+M)v_C$$

$$v_C = mu/(m+M)$$

where it is assumed that the nucleus is initially at rest.

The velocity of the neutron in the centre-of-mass frame is $u-v_C$ and hence,

$$v_1' = Mu/(m+M)$$

Considering the triangle OCA formed by the trajectory of the neutron, the cosine rule gives the velocity of the neutron after scattering,

$$v_1^2 = v_C^2 + v_1'^2 - 2v_C v_1' \cos(\pi - \theta)$$

Since $\cos(\pi-\theta) = -\cos\theta$,

$$v_1^2 = v_C^2 + v_1'^2 + 2v_C v_1' \cos\theta$$

and substituting for v_C and v_1' in terms of u from the expressions above,

$$v_1^2 = \left[\frac{mu}{m+M}\right]^2 + \left[\frac{Mu}{m+M}\right]^2 + \frac{2mMu^2\cos\theta}{(m+M)^2}$$
$$= (m^2 + M^2 + 2mM\cos\theta)\, u^2/(m+M)^2$$

Substituting $m \sim 1$ u and $M \sim A$ u and defining incident energy of the neutron as E and its energy after scattering as E', where $E = mu^2/2$ and $E' = mv_1^2/2$,

$$E' \sim \frac{A^2 + 2A\cos\theta + 1}{(1+A)^2} E$$

CASE STUDY 6.2: ESTIMATING THE NUMBER OF SCATTERING INTERACTIONS NECESSARY TO THERMALIZE A NEUTRON

Further to the discussion of Eq. (6.3), for moderators where $A > 1$, $\overline{E'}$ after n collisions is given by half the sum of the minimum and maximum energies after scattering, αE and E, respectively, where E is the energy of the incident neutron, as per,

$$\overline{E_n'} = E\left[\frac{1+\alpha}{2}\right]^n \qquad (6c.1)$$

and thus,

$$n = \frac{\log_e E/\overline{E_n'}}{\log_e 2/(1+\alpha)} \qquad (6c.2)$$

is the number of collisions necessary to reduce neutron energy from E to the average $\overline{E_n'}$.

Most moderators in use are compounds comprising several constituents, such as light water and heavy water, and hence it is necessary to derive the corresponding average value for the collision parameter $\overline{\alpha}$ based on α for each of the isotopic constituents weighted by the corresponding macroscopic cross section in each case. This takes the combined moderating influence of each constituent into consideration via the corresponding macroscopic cross sections. The results of such an analysis are presented in Table 6c.1 for light water, heavy water and also uranium dioxide for comparison purposes.

An alternative, widespread and generally more accurate approach is to consider the number of collisions necessary to yield the average value of $\log_e E/\overline{E_n'}$, where $u = \log_e E/E'$ is known as the *lethargy*. In this case,

$$n = \frac{u}{\xi} \qquad (6c.3)$$

and ξ is the *average logarithmic energy decrement per collision*. This is a property of the moderator of choice where [1] $\xi = 1 + [\alpha/(1-\alpha)]\log_e \alpha$. A comprehensive summary that provides further alternatives is that by Narita and Narita [2].

Table 6c.1 Average Collision Parameter $\overline{\alpha}$, Fractional Energy Loss Per Collision and the Number of Collisions Necessary to Thermalize a Neutron With an Energy of 1 MeV for Water, Heavy Water and Uranium Dioxide, Based on Eq. (6c.2) and Cross Sections Averaged Over a Fission Spectrum

Moderator	$\overline{\alpha}$	$\overline{E'}/E$	n
H_2O	0.03	0.51	26
D_2O	0.35	0.67	44
UO_2	0.87	0.94	270

REVISION GUIDE

On completion of this chapter you should:

- understand the need for moderation in thermal-spectrum nuclear power reactors on the basis of resolving the disparity between the low-energy advantages of the microscopic fission cross section with the high-energy emission of neutrons from fission
- be able to summarise the ideal properties of a neutron moderator in terms of natural abundance, the phase of the moderator material, neutron scattering properties, absorption cross sections and atomic mass
- know what the principal moderating materials used in thermal-spectrum commercial power reactors are and be able to summarise the benefits and disadvantages of each
- understand how to derive the dependence of neutron energy with atomic mass and scattering angle, and the formalisms for obtaining the average neutron energy and the approximate number of scattering interactions required to thermalize a neutron
- understand and be able to sketch the qualitative trends of scattering and absorption cross section with neutron energy, and be able to compare and contrast the relative scale of each according to the isotopes comprising each moderator substance
- be able to describe the physical basis for the dependence of moderation on atomic mass, the differential scattering cross section and the empirical scattering law

PROBLEMS

1. Concrete is a ubiquitous material in nuclear reactor systems that is used, for example, for the floor slab and often as part of the primary containment vessel.

   For the purposes of this example, concrete is defined as comprising: tricalcium silicate ($3CaO{\cdot}SiO_2$), dicalcium silicate ($2CaO{\cdot}SiO_2$), tricalcium aluminate ($3CaO{\cdot}Al_2O_3$), tetracalcium aluminoferrite ($4CaO{\cdot}Al_2O_3{\cdot}Fe_2O_3$) and also a varying quantity of both free and bound water depending on its state. Calculate $\overline{E'}/E$ for each of the principal isotopes of these compounds. You should assume the principal isotope is the stable, most abundant isotope in each case. Hence, comment on the moderating role of light water where it might be present as a result of hydration.

2. Draw the dependence of the rate of change of cross section with energy as a function of neutron energy after elastic scattering in the centre-of-mass frame for general cases, i.e. not just for hydrogen. What is the mathematical limit for the average neutron energy after a single scattering interaction for $\overline{E'}$ as $A \to \infty$?
3. Derive the general form of an expression for the number of scattering events n required to thermalize fast neutrons from an energy E to an energy $\overline{E'_n}$.

   Hence, calculate number of scatters necessary to thermalize neutrons with an initial energy of 2.1 MeV for the following variety of isotopes: ^{23}Na, ^{184}W and ^{208}Pb.

   What can you conclude about the moderating properties of these materials when used to cool fast reactors or as radiation shielding?
4. A neutron undergoes elastic scattering and describes a random walk with a constant distance between interactions in a sample of graphite.

   If the absolute distance from the point of the first interaction is $\lambda\sqrt{z\pi/6}$, where z is the number of interactions and λ is the mean free path of the neutron, what is this distance for the thermalisation of a neutron with an energy of 1 MeV, given density of graphite is 2.26 g cm^{-3}? What is (are) the critical assumption(s) in this calculation?
5. In Chapter 4 the reaction rate in a volume of material 1 cm^3 was defined as $\Sigma\phi$, where Σ is the macroscopic cross-section and ϕ is the scalar neutron flux. If the neutron flux is 10^{16} cm^{-2} s^{-1}, what would be the approximate rate of neutron capture in light water due to the hydrogen component as a percentage of the total flux, assuming the flux is fully thermalized?

   Compare this with the rate for the deuterium component of heavy water, and hence comment on the change in enrichment needed to offset the capture detriment of hydrogen.

   The densities of light water and heavy water are 1 g cm^{-3} and 1.11 g cm^{-3}, respectively, the molar masses of light water and heavy water are 18.015 g mol^{-1} and 20.028 g mol^{-1}, respectively, and the microscopic capture cross sections for hydrogen and deuterium at thermal energies are 0.33 b and 5.1×10^{-4} b, respectively.

REFERENCES

[1] S. Glasstone, M.C. Edlund, The Elements of Nuclear Reactor Theory, D. Van Nostrand Co., Inc., Princeton, NJ, 1952

[2] M. Narita, K. Narita, Average number of collisions necessary for slowing down neutrons, J. Nucl. Sci. Technol. 26 (9) (1989) 819–825.

CHAPTER

COOLING AND THERMAL CONCEPTS

7

7.1 SUMMARY OF CHAPTER AND LEARNING OBJECTIVES

In this chapter the fundamental concepts associated with cooling and the elementary thermal processes of relevance to nuclear energy are introduced. One of the most attractive features of nuclear power is that a reactor constitutes a high-density source of heat energy that only requires intermittent refuelling. When nuclear energy was first introduced, the first power reactors could be substituted almost directly for the coal- and oil-fired plant that had preceded them.

At this time, when electrification was expanding at a pace but the continued use of smog-producing, fossil-fuelled power stations close to city centres were becoming undesirable, a nuclear fission reactor offered a smog-free alternative that does not need daily deliveries of large quantities of fuel; this is an important benefit that continues to pay dividends not only in terms of infrastructure requirements associated with a particular station, but also to long-term operational cost forecasts.

Consequently, the first generations of nuclear power stations often used ancillary systems (such as heat exchangers, boilers, coolant cycles, etc.) that were similar to those used in earlier fossil-fuel power stations but adapted to accommodate the higher power density. Only as experience was accrued was it appreciated that there were benefits in using alternative coolants and different cycles. It is worthy of note, especially with reference to the preceding chapter, that in a significant number of cases the coolant performs the roles of both the heat transport agent and neutron moderator.

The objectives of this chapter are to:

- revise the laws of thermodynamics
- introduce the generic concept of the *heat engine* and the physical limitations of the theoretical and practical operation of these systems
- introduce the *Carnot cycle* as being the most efficient, hypothetical heat engine
- introduce the *Rankine cycle* as a practical simplification describing the operation of the heat transfer function in a nuclear reactor system
- review the use of the most common *liquid coolants* used in commercial nuclear reactor systems, including their *physical properties*, *advantages* and *disadvantages*
- review the use of the most prominent *gaseous coolants*
- review the use and ideas surrounding the more *exotic coolants*
- introduce the function of *steam turbines* as the most widespread conversion technology in use in nuclear power reactor systems

Nuclear Engineering. https://doi.org/10.1016/B978-0-08-100962-8.00007-X

- introduce the concepts of *thermal conductivity* and *heat transfer* in terms of maintaining nuclear fuel in its safe state
- introduce some elementary concepts of thermo-hydraulics, particularly with regard to boiling

7.2 HISTORICAL CONTEXT: SAMUEL UNTERMYER II, 1912–2001

Early light water nuclear reactors were pressurised to prevent boiling in the core and the formation of bubbles because it was anticipated that the associated, thermally stimulated change in density of a coolant that also served as the neutron moderator might cause the reactor to malfunction and overheat. Samuel Untermeyer's significant contribution (Fig. 7.1) in this regard was to realise that steam bubble formation might actually provide a stable and self-controlling environment. He instigated a series of boiling water reactor (BORAX) experiments and invented what he termed the *steam-forming neutronic reactor and method of operating it* [1].

Untermeyer realised that if bubbles and steam *did* form in an overheating light-water reactor, that had been designed to decline in reactivity as a result of a fall in moderator density, then the nuclear reaction would actually slow down. In 1953 he demonstrated that boiling in an operating reactor was a stable condition and that, furthermore, the property of increased boiling leading to a reduction in power could be used to control power increases. This affords several advantages over pressurised systems including being able to operate at lower pressures, removing the requirement for steam generators or a pressure vessel. It also reduces the risk of coolant leakage due to there being fewer pipes, fewer welds, fewer large diameter pipes and no steam generator pipes. There are currently 78 operable reactors based on this principle worldwide.

FIG. 7.1
Samuel Untermyer II.

7.3 INTRODUCTION

Cooling is essential to the operation of nuclear power plants for a number of reasons. The principal reason is one of safety: even a small fission reactor has the potential to generate a great deal of heat in a short period of time. Without cooling of one form or another significant and irreversible damage to the reactor fuel could occur, causing it to deform and to potentially breach the cladding, and thus risk the release of radioactivity; in the most extreme cases the fuel might even melt. Also related to the integrity of the reactor is the need to absorb and dissipate the heat generated by the residual radioactivity in the fuel when the reactor is shut down; this often requires active cooling for a period of days to weeks after a reactor has ceased operation. Clearly, a very important requirement in the context of reactor systems that are used to generate power, aside from the needs of safety, is the ability to transport the heat evolved by the reactor to a place where it can be converted to perform useful work. This usually involves the generation of electricity but might also be achieved via the direct use of high-pressure steam. In some reactor designs, such as pressurised water reactors and heavy water moderated reactors, cooling also ensures that the moderator remains in the liquid phase, as per the design basis. Cooling also keeps the fuel below temperatures at which significant changes to its crystal structure might occur and hence avoids deformation that might result otherwise. Effective cooling is also necessary to ensure that temperatures are kept below those at which oxidation of reactor components and cladding is known to occur.

Some early nuclear reactors were small and very low power, i.e. less than 1 MW, and therefore did not require *active* cooling in the sense that a medium was required to be transported throughout the reactor core by independently powered systems. However, this is not to suggest that cooling was in some way not required; these systems were invariably designed either to derive heat transfer from natural ventilation or for natural circulation of a fluid of some sort, with the advantage that their power output was limited to an extent that did not require a more sophisticated arrangement. As demands grew both in terms of materials production and power generation, it became necessary to assist the transport of the coolant through the reactor core. This enabled higher power outputs to be achieved, of the order of 10's of MW and above, as a result of greater heat transfer efficiency and, particularly when pressurised coolants were adopted, enabled higher thermodynamic efficiencies to be reached as a result of higher operating temperatures.

To enable these developments it has been necessary to integrate a number of related systems as part of the cooling circuits of most nuclear power plants; these were alluded to in Chapter 1, when the generic nuclear reactor was discussed, and further consideration will be given to them in Chapter 10 when specific, mainstream reactor designs are discussed. Although the specification and rating vary with a given reactor design, there are some features that are common to most nuclear power reactors:

- *Pressuriser*: This component establishes the elevated pressure in the coolant system, predominantly of pressurised water reactors, and enables changes in pressure to be accommodated when they arise without there being significant change or instability.
- *Steam generator or boiler*: This provides the interface between the coolant that is in contact with the core and the substance from which useful work is derived, where this is required for indirect cycle reactor designs.
- *Coolant pumps* (often referred to as circulators in gas-cooled designs): These are usually electrically powered with an independent, back-up power supply. The action they provide enables the coolant to be transported around the reactor coolant system, notwithstanding some designs considered for the future that exploit natural circulation.

- *Feedwater system*: This is the secondary side of the coolant system with which power is generated and provides the interface that enables the waste heat to be exhausted to the environment, either via outfall to the ocean or major reservoir, or via condensation in cooling towers.
- A variety of coolant treatment systems, driers, condensation systems and re-circulators are also often used to ensure the coolant is maintained in optimum condition and is compatible with the various parts of the reactor coolant system.

7.4 FUNDAMENTAL TERMINOLOGY

The apocryphal scene of James Watt observing the lid being lifted off a kettle by the pressure of steam may not have actually occurred but nonetheless illustrates the requirement that a substance is necessary to convert *heat* into useful *work*. This can be summarised as follows:

A *working substance* is necessary to receive and reject heat energy, from which a proportion can then be directed to perform *useful work*.

A system in which this conversion takes place is referred to as a *heat engine*. In Watt's kettle, the working substance was water. Water remains the substance of choice for the majority of the world's heat engines, big and small, including many nuclear power reactors. Water is cheap, abundant and benign. Very importantly, a significant amount of energy is required to raise its temperature relative to the same mass of other fluids; that is, it has a large *specific heat capacity*.

To extract energy from a supply of heat to perform work, Watt and others in the 18th and early 19th centuries observed that it was often necessary to convert energy from one form to another. They also observed that it was necessary to do this as efficiently as possible. In the context of nuclear energy, several stages of conversion are usually required:

- The primary form of energy arises from the fission event—*nuclear energy*.
- The nuclear energy is converted to *heat energy* predominantly via the interactions of the fission fragments but also those of the neutrons and the other products of fission with their surroundings. An important goal of the nuclear reactor designer is to ensure that as much of this energy as possible is absorbed by the working substance, in order that the transport of this to where it can be converted into *kinetic energy* is optimised.
- The conversion of kinetic energy into *electrical energy* then follows, usually via generators.

Electricity is usually the end point because this form of energy has the benefits of being transported easily and efficiently by cables, above or beneath ground. Electricity is also compatible with virtually all of the energy requirements of developed societies, save some aspects of transport and domestic heat. At the time of writing, transport and domestic heat are rapidly migrating to exploit low-carbon electricity too, rather than continuing to being reliant on fossil fuels; for example, the widespread electrification of motor vehicles will draw inevitably, in part, on electricity produced by nuclear reactors in the future, along with that from other sources.

In contrast, the heat evolved by the earliest nuclear reactor designs was a by-product of their primary purpose, which was to produce nuclear material. The role of the working substance at that time was primarily to ensure that the heat arising from fission in the core was directed away from the plant to prevent damage to the fuel and reactor infrastructure, and hence the primary design objective of reactor cooling was then, as it is now, safety. In addition to those that used water, some of the earliest reactor designs were cooled by air. This was ejected into the atmosphere without being exploited further. As will be described in this chapter, the environment (alternatively termed the *ambient*) plays an important role in the thermodynamics of heat engines.

The first power-generating nuclear reactors were pioneered at a time when heat engines falling into the *external combustion* category, that is, where the source of heat is external to working substance as opposed to internal combustion engines, had been subject to a great deal of development and optimisation over the preceding 200 years. These developments had culminated in the coal-fired, steam-based power stations of the 1930s and 1940s. Therefore, the first reactors to be harnessed to do useful work (as opposed to producing nuclear materials) were often mated simply with pre-existing steam-raising plant designs derived from fossil-fuelled power plant use and experience.

Subsequently, coolant circuit designs became more complicated and dedicated to the nuclear power plant as operating experience was gained with the higher energy density source of heat. A distinction of nuclear power coolant cycles is that there is not an equivalent to the hot flue gas evolved directly from the fuel, that would be analogous to the primary working substance in external combustion engines. Therefore, nuclear power reactors often use a combination of two working substances:

- The *primary* coolant that serves as the transfer medium for the heat evolved by the fuel via its cladding.
- The *secondary* that is in contact with the electrical conversion plant.

Heat is usually transferred from the primary to the secondary via a *heat exchanger* (more commonly termed a *boiler* especially in the case of gas-cooled reactors). Boiling water reactors are an exception to this approach.

The design and performance of the coolant cycle for a particular reactor system has a significant influence on its thermal efficiency. Since this, in turn, has a significant bearing on the station's long-term economic viability, the choice of cooling approach can constitute a critical feature in the viability of a reactor design. In many reactors the components of the cooling system are often the largest and most obvious aspects of the plant apparatus. This is due to the capacity necessary to transport the concentrated supply of heat away from the reactor core.

7.5 ELEMENTARY THERMODYNAMICS

7.5.1 THE LAWS OF THERMODYNAMICS AND THE GAS LAWS

In considering cooling in nuclear power reactor systems it is important to review the laws of thermodynamics. These represent the fundamental performance limitations of these systems and of all related heat engines. The first law of thermodynamics states that:

1. *Kinetic energy and heat energy are interchangeable but energy can neither be created nor destroyed.*
   The first law establishes an important boundary on the operation of all heat engines in terms of the net energy available for transfer between a heat source and a heat sink.

Of particular significance to the design of all power stations, based on the principle of the heat engine and particularly nuclear power reactors, the second law states:

2. *For two bodies in contact with one another heat will flow from the body at the higher temperature to that at the lower temperature until the temperatures of both are equal.*
The second law is effectively a statement concerning the disorder of the system. It is often stated alternatively that the disorder (or *entropy*) associated with a spontaneous natural process must increase. Consequently, if energy in the form of heat is to be used to do useful work (such as to compress a piston or cause a turbine to turn) this constitutes a localised *reduction* in entropy. Given there must always be a net increase in entropy (as inferred by the second law), this is assured via the increase in disorder associated with heat that is exhausted to the surroundings. This leads to an often more familiar phrasing of the second law:
It is impossible to use all heat energy in a heat engine to do useful work.
This interpretation of the second law implies that a heat engine can never be 100% efficient; indeed they are often a lot less efficient than this; if 100% efficiency were possible entropy would have to have decreased, akin to a hot reservoir getting hotter at the expense of energy from a connected cold reservoir, which is not possible. This limitation has important commercial and safety-related consequences for nuclear power reactors and indeed all power stations based on energy conversion with a heat engine.

The systems considered in this chapter exploit a change in the condition of a working substance that is usually a fluid, where heat can bring about changes in pressure, volume and temperature. Any heat supplied to the working substance will comprise a sum of the *external work done* on the surroundings plus any *change in the internal energy* of the substance. Heat supplied to a volume of the working substance brings about expansion; it is usually this that acts to do work on the surroundings. One illustration of the second law considered above is that, to do work, the working substance must remain expanded. Hence, it must retain some of the heat as internal energy to do this. However, this is not manifest as *work done*. The internal energy of the substance is inferred by its *temperature*. This describes the tendency of the substance to yield its internal energy to the surroundings as a function of entropy.

To explore this quantitatively, it is instructive to consider the behaviour of a gas under isothermal conditions (i.e. constant temperature). *Boyle's law* is a statement that the product of pressure P and volume V for a perfect gas[1] is constant,

$$PV = \text{constant} \tag{7.1}$$

This describes the scenario where a change in volume is brought about in an isolated volume of gas, as depicted in Fig. 7.2. This illustrates the change as an increase from V_1 to V_2 where there is a sympathetic change in pressure of P_1 to P_2 if the temperature of the gas is constant. By Boyle's law, the following is true,

$$P_1V_1 = P_2V_2 \tag{7.2}$$

and the *ratio of expansion* can be defined as r where

[1]A perfect gas is one for which the ideal gas law holds and that is in thermal equilibrium and free from intrinsic chemical reactivity. For a nonperfect gas the nonlinearity associated with the nature of the gas and the expansion conditions is reflected by an index n such that $PV^n = \text{constant}$.

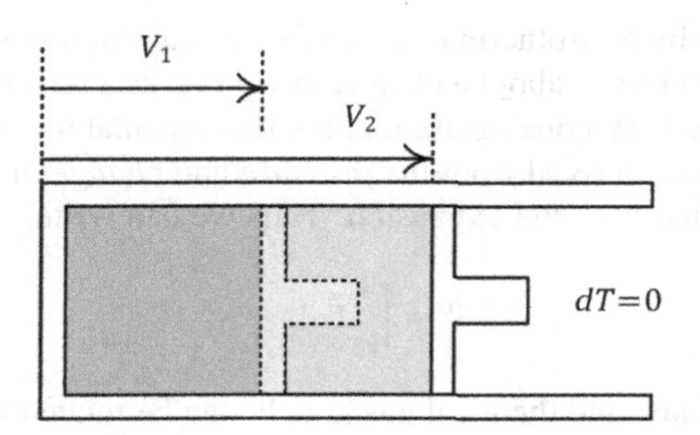

FIG. 7.2

A schematic illustration of isothermal expansion (i.e. at constant temperature) in a sealed, insulated cylinder of a volume of gas of V_1 to a volume V_2.

$$r = \frac{P_1}{P_2} = \frac{V_2}{V_1} \tag{7.3}$$

For the corresponding cases of constant pressure or constant volume, *Charles' law* applies where either the quotient of volume V to temperature T or that of pressure P to temperature T is constant, such that

$$\frac{V}{T} = \text{constant} \tag{7.4}$$

and

$$\frac{P}{T} = \text{constant} \tag{7.5}$$

respectively, where temperature is quoted in units of absolute temperature. Combining Boyle's and Charles' laws gives,

$$\frac{PV}{T} = \text{constant} \tag{7.6}$$

Eq. (7.6) is very useful because we can describe the state i of a fluid in terms of P_i, V_i and T_i where

$$\frac{P_i V_i}{T_i} = \text{constant} \tag{7.7}$$

However, we also know that a mole[2] of any perfect gas at standard temperature and pressure occupies the same volume. This associates quantitative significance to the constant in Eqs (7.6) and (7.7) and which can be related to the *gas constant*, $R = 8.31\ \text{J mol}^{-1}\ \text{K}^{-1}$. Hence, the *ideal gas law* can now be defined as

$$PV = nRT \tag{7.8}$$

where n in this case is the number of moles, P has the units of Pascals (Pa) and V those of m^3.

[2]The amount of substance that contains as many atoms or molecules as there are atoms in 12 g of ^{12}C; that is, Avogadro's number: $6.023 \times 10^{23}\ \text{mol}^{-1}$.

The external work done W by an isothermal stage of expansion of a working substance at temperature T (as depicted in Fig. 7.2) brings about a change in volume as a result of an increase in pressure. The *work done* can be obtained from consideration of the fundamental relationship with the product of *force* and *distance*. These can be derived from the *pressure* and *change in volume*, respectively, over the change in volume associated with the expansion. Thus we can write,

$$W = \int_{V_1}^{V_2} P dV \tag{7.9}$$

Via substitution from Boyle's law and the ideal gas law, W can be related to the ratio of expansion as follows:

$$W = nRT \int_{V_1}^{V_2} \frac{dV}{V} \tag{7.10}$$

$$W = nRT \log_e r \tag{7.11}$$

Hence, assuming we have an estimate for the expansion ratio, Eq. (7.11) provides a route to an assessment of the work done as a result of the expansion of a gas in response to supplied heat.

It is worthy of note, especially given the elementary nature of this introduction, that in the thermal cycle of a nuclear power station at power the steam produced is under significant pressure and does not behave as a perfect gas. The thermodynamic properties cannot therefore be derived on this basis and, instead, are often sourced from experimental data and interpolations between them.

7.5.2 HEAT ENGINES

Developing our definition of a heat engine a little further we might regard this as being a machine that converts *heat energy* into *kinetic energy* and vice versa. The kinetic energy is then available to do *useful work* where by *useful* it is implied that it can be directed to carry out some desirable function. There is a wide variety of heat engine designs in regular use that are familiar such as the internal combustion engines in cars and the compressors that bring about refrigeration in fridges and freezers. However, we shall focus our discussion steam turbines as they are in widespread use to transfer the heat from nuclear reactor cores to electrical generators with which to produce electricity.

It is worthy of note that, in some heat engines, the expanded working substance is often exhausted to the surroundings and is lost to the system; consider, for example, internal combustion engines in motor vehicles and steam locomotives. In this case, the hot gases are often exhausted directly from the cylinder in the former, and the expanded steam is usually lost from the piston in the latter. This satisfies the second law in terms of the need for there to be an increase in entropy and also illustrates two examples of the destiny of the heat energy that is not converted into useful work for these cases. It is possible to optimise the usefulness of this energy further via, for example, low-pressure turbines or as a supply of distributed process heat. However, this constitutes an increase in the efficiency with which heat is extracted from the working substance, rather than a contravention of the second law.

Steam generators exploit a physical process in which heat is transferred from a source at a higher temperature via a working substance (or as is more usually referred a coolant) to a sink that is at a lower temperature. This leads to changes in the pressure and volume of the coolant. These changes are usually exploited to cause a turbine to rotate, thus performing mechanical work via a shaft to a generator to produce electricity.

To mechanise the function of heat on the working substance to provide a sustained contribution of useful work over an extended period of time, i.e. power, it is usually necessary to repeat a number of operations in a *sequence*. A sequence of operations has the advantage that it can be repeated indefinitely whereas, otherwise, the supply of energy would be intermittent and its use highly inefficient. In this context, a sequence is referred to as a *cycle*. There are two main distinctions between types of cycles: those that are *ideal* and those that *actual*. Ideal cycles represent the theoretical optimum that might be achieved in the absence of friction and heat loss to the environment via, for example, conduction and convection. These cannot be realised in practise, examples include: the *Carnot cycle* and the *Otto cycle*. Conversely, actual cycles represent real processes that can be implemented in working engines. Ideal cycles provide a theoretical performance basis with which to compare the performance of actual cycles; actual cycles are, by definition, always less efficient than the corresponding ideal example.

7.5.3 THE CARNOT CYCLE

The *Carnot cycle* is the ideal cycle against which all *external* combustion heat engines are usually compared, at least in the first instance. The *Otto cycle* is the corresponding ideal cycle for comparison with *internal* combustion engine designs. The Carnot cycle describes the maximum theoretical efficiency achievable with a perfect coolant and insulation properties with optimum working conditions. As an ideal cycle its performance cannot be replicated in practise.

The Carnot cycle describes the transfer of heat from a source to a sink wherein some of this energy is directed to perform useful work. The cycle comprises four individual stages: two of expansion and two of compression. The heat source is conventionally assigned a temperature T_1 and the sink a temperature T_2, where $T_1 > T_2$. Although it represents a theoretical optimum, a number of practical examples can be used to illustrate the principle of the Carnot cycle, given the corresponding efficiency cannot be achieved in reality. The most common example is a piston operating on a gaseous working substance in a cylinder, as shown in Fig. 7.3. Carnot envisaged the piston being the *prime mover* connected to a crank with which to supply the rotational motion necessary to lift a specified mass. The four stages of the Carnot cycle are as follows:

Stage 1: Heat from the source is supplied to the working substance in the cylinder. This causes the substance to expand from V_1 to V_2 and hence to perform work on the surroundings. The internal energy of the gas increases by an amount Q_1. However, since the cylinder is open at this stage the temperature remains constant at T_1; the expansion is thus termed *isothermal* (Fig. 7.3A).

Stage 2: The cylinder is then sealed and perfectly insulated so no further heat is supplied or lost from the working substance. The substance continues to expand *adiabatically*, that is, energy is not lost or gained from the system during this stage. The volume increases from V_2 to V_3 resulting in a fall in temperature from T_1 to T_2 (Fig. 7.3B).

Stage 3: The surroundings now work on the system, acting to cause the piston to compress the gas from V_3 to V_4. Since the cylinder is once again open, energy Q_2 is exhausted to the sink *isothermally* at T_2 (Fig. 7.3C).

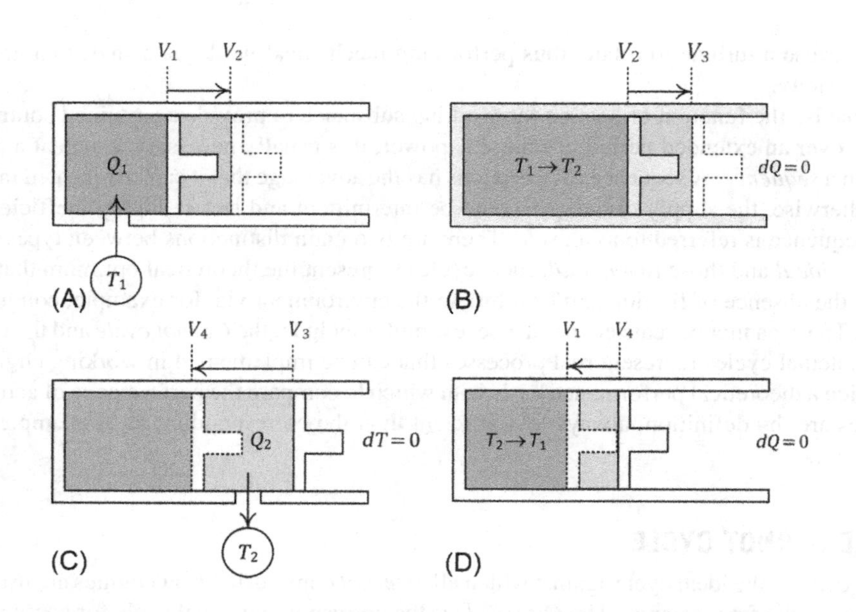

FIG. 7.3

The four stages in the Carnot cycle. (A) Stage 1: Isothermal expansion under heat input Q_1, (B) Stage 2: Adiabatic expansion accompanied by a fall in temperature T_1 to T_2, (C) Stage 3: Isothermal compression, Q_2 exhausted, (D) Stage 4: Adiabatic compression accompanied by an increase in temperature T_2 to T_1.

Stage 4: Finally, the cylinder is sealed once more and the surroundings continue to work on the working substance compressing it further from V_4 to V_1 with there being no exchange in energy, thus returning it to the starting temperature T_1 (Fig. 7.3D).

To calculate the efficiency of the Carnot cycle it is necessary to determine the ratio of the work done (by the piston under expansion) to the total heat energy supplied to the cylinder to make it expand. The latter is denoted by Q_1. The work done W is the difference between the heat supplied and that which is exhausted, $W = Q_1 - Q_2$. Using the derivation for the energy required to bring about the corresponding ratio of expansion, r, (Eq. (7.11)) the heat supplied Q_1 and that exhausted Q_2 can be defined as,

$$Q_1 = nRT_1 \log_e {}^{V_2}/_{V_1} \tag{7.12}$$

$$Q_2 = nRT_2 \log_e {}^{V_3}/_{V_4} \tag{7.13}$$

Since the ratios of expansion for stages 1 and 3 are the same, the efficiency η can be expressed as,

$$\eta = \frac{W}{Q_1} \tag{7.14}$$

where the numerator $Q_1 - Q_2 = nR(T_1 - T_2)\log_e r$. The common factors cancel with those in the denominator to give,

$$\eta = \frac{T_1 - T_2}{T_1} \tag{7.15}$$

Eq. (7.15) is of great significance in the design of heat engines because it specifies that the optimum, theoretical efficiency is directly proportional to the difference between the temperatures of the *heat source* and the *heat sink*. In most heat engines these correspond to the temperatures of the *inlet* (source) and *outlet* (sink), respectively. Thus, to optimise the efficiency of such an engine, it is necessary to construe as large a temperature differential between these as is possible. Further, Eq. (7.15) is compliant with the second law of thermodynamics since an outlet temperature of absolute zero would be necessary to obtain 100% efficiency (not allowed under the second law and which is not physically achievable).

There are several reasons why the Carnot optimum cannot be achieved in practise. First, it is impossible to contrive a cylinder that is totally insulating; some heating of the cylinder wall will always occur and thus some heat is lost. It is also inconceivable that the *opening* and *sealing* of the cylinder at the various corresponding stages might achieve perfect *isothermal* and *adiabatic* conditions, respectively. This will also degrade the practical efficiency achieved in contrast to the optimum derived above.

Many heat engines exploit a working substance that exists across the liquid–vapour phase boundary of that substance. In such cases the coolant no longer exhibits the properties of a perfect gas and it is necessary to allow for the energy associated with the phase change of the substance. This corresponds to the energy associated with the vaporisation and condensation of the coolant; the *latent heat*. Specifically, latent heat is taken in to vaporise the coolant in stage 1 and is rejected at stage 3. In addition, to be isothermal the pressure is constant for these stages too. For the ideal case, the efficiency is the same as derived for the Carnot cycle above (Eq. 7.15) but in designing the thermal cycle it is essential that the latent heat exchange is taken into account.

7.5.4 THE RANKINE CYCLE

The vapour-based adaptation of the Carnot cycle is improved when employed on a practical basis, usually by integrating a separate condenser. This has the effect of extending stage 3 by a stage of active condensation. Here, as much of the vapour is condensed as possible to the initial volume present at the start of the cycle, stage 1. Heat is then needed to raise the condensate to the start temperature T_1 of stage 1. This requirement lowers the efficiency of the cycle yet further below that of the ideal Carnot cycle. However, this ensures that the practical benefits of expansion from operating across the liquid-phase boundary are exploitable, particularly in driving mechanical systems such as pistons and turbines.

7.6 PROPERTIES OF WORKING SUBSTANCE AND COOLANTS

7.6.1 IDEAL REQUIREMENTS

Carnot remarked that a working substance must have the following properties:

1. It should be able to *receive and reject heat energy in large quantities*.
2. It should be *adaptable to changes of pressure and temperature*, within the performance envelope of the engine in which it is used.
3. It should be *chemically stable* within the performance requirements of the engine.

In addition, where the external heat source is a nuclear reactor that is in intimate contact with the core, we might add that the substance also needs to:

4. Have a *low susceptibility to neutron activation* by virtue of low neutron capture cross sections of the isotopes comprising the substance. This can require that trace contaminants susceptible to such activation are removed from the substance prior to use.
5. Be *integrated in sympathy with the moderator* of the specific reactor design in which it is used. This can be a complex requirement because in some reactor designs the coolant and the moderator are one and the same and thus the coolant also needs to have effective moderation properties. In others, inadvertent moderation by the coolant needs to be avoided as otherwise it can complicate reactor control and neutron economy (as we shall discuss in Chapter 8). In fast reactors, where moderation is to be avoided, coolants with very low moderating properties are usually desirable.
6. The *neutron absorption cross section should be as low as possible* so as to avoid neutron loss in the coolant and thus minimise any associated reduction in neutron economy in the reactor system.

Requirement 1 is met by selecting a substance with a high specific heat capacity, high thermal conductivity and stable flow characteristics. Requirement 2 is realised by selecting a substance that does not disassociate as a result of repeated pressure/temperature cycling. Requirement 3 effectively rules out most substances that are highly flammable, significantly non-Newtonian (that is, those that exhibit non-linear flow characteristics), susceptible to chemical decomposition or that cause extensive corrosion of reactor components. Requirement 4 is often ensured by requiring that demineralised coolants are used, such as deionised water, and also that isotopes susceptible to neutron activation have been removed from coolant additives that are used for long-term reactor control. Requirement 5 is often a concatenation of the earlier requirements in that low-flammability, chemically stable and noncorrosive coolants are usually favoured. Requirement 6 is difficult to avoid where the substance of choice is light water, especially because otherwise light water offers many significant advantages. In such cases, the effect on neutron economy is usually offset by increasing the enrichment of the fuel. Clearly, as is common in most engineering applications, a compromise is sought as not all the requirements can be met at the same time.

For indirect reactor systems that comprise two coolant loops exchanging heat energy via an intermediate steam generator (as introduced in Chapter 1), the primary coolant is almost always maintained in the same physical state throughout the primary circuit often by maintaining it under pressure. The heat transferred in the steam generator to the secondary coolant usually acts on light water to facilitate the generation of steam in the secondary circuit. This is usually directed to a multistage turbine system. There are exceptions to this arrangement where the steam from the primary is used directly, which will be discussed in Chapter 10.

An important distinction of nuclear systems in this context is that, where the medium is maintained in the liquid phase as a primary loop coolant, its principal functions are for the absorption of heat, heat transport and heat transfer. Thus we seek a *high specific heat capacity* as a prominent performance characteristic. A summary of the thermophysical parameters of a variety of coolants of relevance to nuclear energy is given in Table 7.1 and that of typical reactor operating conditions is given in Table 7.2.

Table 7.1 Typical Thermophysical Parameters of Various Coolants[a] [2–4]

Coolant	Specific Heat Capacity/kJ kg^{-1} K^{-1}		Density/ kg m^{-3}	Viscosity		Critical Point	
	c_p	c_v		Kinematic/ $\times 10^{-6}$ m^2 s^{-1}	Dynamic/ $\times 10^{-6}$ Pa s	T_C/K	P_C/ MPa
H_2O (PWR)	4.1843[b]	4.1565[b]	997.05[c]	0.89[c] (0.12)	82.57	647.096	22.064
D_2O (CANDU)	4.2277[b]	4.2152[b]	1104.36[c]	0.99[c]	1095[c]	643.847	21.671
CO_2 (AGR)	1.22	0.655[b]	23.09	1.68	38.78	304.250	7.390
Helium	5.19[b]	3.12[b]	–	–	–	5.195	0.228
Sodium	1.383[d]	1.262[d]	927	7.42[d]	688.27[d]	–	–
Sodium-potassium (NaK)[e]	1.277	–	845[d]	0.60	507	–	–
Lead	0.1473[f]	–	10,670[f]	0.25	2702.59[f]	–	–
Lead-bismuth (Pb-Be)	0.150[g]	–	10,300	0.32	3270[g]	–	–

Note: *Most parameters are subject to change at the temperatures and pressures of the corresponding reactor circuits (data for reactor type where available shown in parentheses).*

[a] *http://www.peacesoftware.de/einigewerte.*

[b] *293 K and 1 atm.*

[c] *298 K and 1.01 atm.*

[d] *371 K.*

[e] *Typically 22% sodium, 78% potassium.*

[f] *~600 K.*

[g] *400 K.*

Table 7.2 Typical Operating Conditions of Various Coolants for a Variety of Mainstream Reactor Types (Reactor Type Shown in Parentheses as Discussed in Chapter 10) for Primary Circuits

Coolant	Coolant Temperatures/K		Coolant Pressures/10^6 Pa		Typical Operating Mass Flow Rates/kg s^{-1}	Typical Operating Volumetric Flow Rates/m^3 s^{-1}
	Inlet	Outlet	Inlet	Outlet		
H_2O (PWR)	565	596	15.83	15.51	19,200 [5]	27.71
H_2O (BWR-5)	551	559	7.14	7.03	13,670	17.99
D_2O (CANDU)	539	583	11.25	9.89	7700	9.89
CO_2 (AGR)	612	912	4.10	16.70	4067 [6]	176.14

7.6.2 LIQUID COOLANTS: LIGHT WATER AND HEAVY WATER

Water (taken to imply in this context both light water, H_2O, and heavy water, D_2O) has what might be considered almost the ideal thermophysical properties for use as coolants in nuclear reactor power systems. Water is particularly relevant for nonsupercritical steam cycles where outlet liquid-phase temperatures and pressures in primary circuitry are less than its critical point. Perhaps most importantly, several hundred years' operating experience exists associated with water as a coolant, starting with the industrial revolution. This spans both liquid, vapour and superheated forms culminating in the high-efficiency, multistage steam cycles that are employed today. However, this experience was based on the use of water as the coolant in boilers, where the heat source was the supply of hot flue gases usually from coal- or oil-fired combustion processes, rather than nuclear reactors.

In correspondence with Carnot's recommendations (Section 7.5.1), water has a high specific heat capacity (4200 J kg^{-1} $°C^{-1}$) reflecting its ability to absorb and reject heat in significant quantities. It is chemically stable within the performance envelope of most current heat engine designs and is not flammable. Excepting heavy water, light water is available in significant quantities without great expense, notwithstanding the need to factor in the costs of purification and decontamination for most related reactor-based cycles. The critical points of both light and heavy water (given the current discussion is limited to nonsupercritical coolant cycles) are sufficiently high to enable water to be maintained in its liquid phase under pressure while yielding adequate thermodynamic efficiencies. This is an essential requirement for use in the primary coolant loops of pressurised water reactors. Very importantly, assuming a reactor plant is located sympathetically with semi-infinite sources of light water such as oceans, lakes or large rivers, water constitutes a naturally available heatsink for condenser systems. In this regard, it is the temperature of water constituting these heatsinks that sets the limiting ambient level for reactor coolant systems.

A subtler benefit of water as a coolant arises from its rheological properties and, specifically, the linear, Newtonian relationship that it exists between shear stress and shear rate in the liquid phase. This affords it great flexibility and predictability in use over a wide range of flow rates and circulation regimes because the viscosity is constant, notwithstanding the effect of temperature. This property also lends the use of water for natural circulation requirements being sought in several new and forecast reactor designs, to provide mitigation in accident scenarios on a passive basis where forced circulation is lost.

Water does, however, have a number of disadvantages as a reactor coolant. For example, heavy water and to a lesser extent light water are susceptible to the formation of tritium, as was discussed in terms of its moderating properties in Chapter 6. It is also usually necessary to use water in highly purified form to avoid its inadvertent contamination via neutron activation of impurities, especially in BWRs. Such impurities can enter the flow from the reactor coolant system itself with time, comprising either residual particulate debris from manufacturing processes or the products of corrosion. Water is also corrosive when in contact with some steels. To mitigate against this, corrosion-resistant materials, particularly those comprising the reactor pressure vessel and pipework, are used where possible. High levels of cleanliness are adopted when the coolant circuit is constructed to avoid unnecessary contamination by swarf and manufacturing debris.

In a related context, a significant thermophysical disadvantage of water is its critical point. This limits its use as the primary coolant for future reactor developments where the onus is often on obtaining greater thermodynamic efficiency, via higher primary outlet temperatures, unless it is used in supercritical form. The potential to optimise the Carnot efficiency by raising the turbine inlet temperature

(T_1 in Eq. 7.15) is limited with nonsupercritical water, as per Table 7.1. Supercritical water is known to exhibit significant properties of corrosion and, thus, either alternative primary coolants are necessary or corrosion-resistant materials will be required for the supercritical coolant circuit in these reactor designs. This will be considered in more depth in Chapter 11. For existing reactor plants, using liquid-phase water as a primary coolant, the critical point sets a hard physical limit on the thermodynamic efficiency that is achievable. However, it should be emphasised that while *thermodynamic efficiency* is only one of several influences on the *operational efficiency* of a nuclear plant, it is the latter that has a strong influence on the economic viability of the installation. Operational efficiency is a function of the time that a plant is on load at its optimum power level. This is reflected by a high *capacity factor*[3] and benefits from a combination of high levels of operational availability *and* thermodynamic efficiency. Capacity factor is a strength of light water reactors.

The production of steam by boiling and its use as a working substance in heat engines has caused some specific terminology to evolve associated with its use. When water is fully atomised into individual molecules it is invisible and termed *dry*. If not fully vaporised or allowed to cool, droplets of water exist, which renders it visible; in which case it is termed *wet steam*. Less heat energy is required to reach steam in a state where some water particles remain. Such a state is described quantitatively by the steam having a *dryness fraction*. This fraction is the mass of steam as a function of combined mass of the steam and water present. Steam that is dry but in contact with the water from which it has been produced is termed *saturated*. Continued heating of saturated steam beyond dryness causes its temperature to rise, in which case it is termed *superheated* since in this state its temperature is higher than the point at which it is vaporised.

7.6.3 GASEOUS COOLANTS

For reactor designs where the coolant and moderator are not the same unified volume of substance, in contrast to the case of light water reactors, there are a number of important coolant options to be considered. The requirements of a separate, dedicated coolant contrast significantly with the properties of a combined water-based coolant/moderator, because sympathetic moderation properties are no longer required of the coolant. Hence, the requirement for coolants for these reactor designs is often for them to exhibit low cross sections for both neutron scattering *and* absorption. This often turns a reactor designer's attention to coolants that are in the gaseous phase under typical reactor operating conditions.

There is an important historical context behind the development of gaseous reactor coolants. When it was necessary to remove nuclear fuel frequently from reactor cores to extract nuclear materials from it and because of the burn-up limitations of early metallic fuel materials, solid-phase moderators such as graphite were selected because they offered the flexibility to do this while a reactor was operating. This was much more convenient than the process of shutting a reactor down, waiting for the decay heat to subside, depressurising the system and removing pressure vessel components to get inside. However, this resulted in reactor cores that are much larger than most water-moderated alternatives. Routing

[3]The *capacity factor* of a power unit is the ratio of its output over a set period of time to its output if it were to operate full specified capacity continuously over the same period of time.

liquid coolants throughout such a core can be sophisticated, as demonstrated by perhaps the most extreme exponent of such a system; the Soviet RBMK reactor design (discussed further in Chapter 10).

Gaseous coolants offer a convenient alternative in such cases where, assuming the gas in question is cheap and abundant, the perfusion of the gas throughout the system can be achieved more easily and cost effectively. One further advantage of gaseous coolants, at the time reactor designs were at the height of their technical development in the 20th century, is that the heated gaseous outlet substituted almost directly in conventional boiler plants (albeit at lower temperatures) for the hot flue gases usually derived from fossil-fuel plants. This enabled established boiler designs to be used as the basis for early gas-cooled reactor steam generators.

Several requirements of reactor coolants are common to both liquids and gases: nonflammability, chemical stability, low susceptibility to neutron activation and good thermophysical properties (particularly high heat capacity and thermal conductivity). The lower density of gases usually requires that higher flow rates are necessary, relative to liquid phase coolants, to achieve similar heat transfer rates. Gaseous coolants are, by definition, already vaporised and consequently it is not possible to lose a high-capacity condensate from the system as a result of vaporisation as is the case with most liquid coolants, particularly water. This effectively removes the upper temperature limit in terms of Carnot efficiency associated with the critical point, so removing an important constraint in terms of thermodynamic efficiency. Naturally, there are other engineering constraints to be considered at higher operating temperatures, such as fuel-clad integrity and the action of the high-temperature gas on reactor componentry, and so on. Nevertheless gas-cooled nuclear reactors generally have higher thermodynamic efficiencies than their water-cooled counterparts principally because of their higher operating temperatures.

The move away from the use of reactors for both power and nuclear materials production has largely removed the imperative to refuel on-load. Rather, the economic focus for power reactor operation has moved towards higher burn-up with longer periods between fuel loading and shorter refuelling outages to optimise the proportion of time spent at power. This has led to the widespread adoption of light-water reactor technology throughout the world with gas-cooled reactors falling out of favour. The UK remains a significant exponent of the gas-cooled technology today. However, a central theme of several future reactor designs currently under consideration is a drive to achieve higher thermodynamic efficiencies. One route to this goal is to use gaseous coolants to increase outlet temperatures significantly beyond current levels. Gaseous coolants may offer several further advantages in the future that are not being exploited widely at present and these will be discussed in further detail in Chapter 11. For example, the absence of significant moderation lends the use of gaseous coolants to some fast reactor concepts.

The characteristically high upfront cost of nuclear plant and strong competition from natural gas with regard to cost of electricity generation are causing research attention to turn to the *direct* use of the heat generated by fission reactors [7]. This is in contrast to continuing to employ downstream electricity generation systems and incurring the conversion efficiency penalties that they introduce. An important application of this approach might be, for example, in the production of bulk chemical products used as the feedstock in pharmaceuticals production. The outlet gas stream from the reactor would be used as the source of process heat to drive the chemical processes directly. Used in this way, a reactor plant could replace an on-site, fossil-fuelled plant offering a combination of power and heat with the advantages of being low-carbon and rarely needing to be refuelled (if at all in the context of small modular reactor designs). A further advantage of gaseous coolants is that, in contrast with reactor designs that use the combination of graphite moderation and light-water coolant, gas-cooled reactors lack the additional moderation that can result in relatively high moderator-to-fuel ratios. This can result in operational instability in the event of nucleation in a liquid coolant, as will be discussed in Chapter 8 and also in Chapter 14. This represents an important design benefit, particularly in terms of passive safety requirements associated with unexpected changes in reactor power.

7.6.3.1 Carbon dioxide

Of the few options for gaseous reactor coolants that exist, given the caveats discussed above, carbon dioxide is perhaps the most common in use today albeit now being in decline in favour of light water. While several countries explored gas-cooled reactors with carbon dioxide in the 1960s and 1970s, the most common design is the *Advanced Gas-cooled Reactor* (AGR) using low-enriched uranium dioxide fuel, carbon dioxide coolant and a graphite moderator. These reactors replaced the previous generation of carbon-dioxide-cooled reactors in the UK that were based on metal fuel of natural enrichment (Magnox) clad in magnesium alloy. Both designs will be discussed further in Chapter 10.

With reference to Table 7.1, carbon dioxide has a significantly lower specific heat capacity and density relative to those of water, as might be expected given its gaseous state. While it has a lower dynamic flow rate in comparison to light water systems (for example, the Sizewell B pressurised water reactor in Table 7.1), the volumetric flow rate is approximately sixfold the flow rate of water-cooled plant (i.e. PWR with the comparison based on an AGR). Carbon-dioxide-cooled reactors usually require dedicated gas production facilities on site and gas circulators to achieve such flow rates. The latter are used effectively in place of the primary coolant pumps in water-cooled reactors.

To exploit the benefits of the vapour-phase Rankine cycle described earlier in this chapter, nuclear reactors cooled with carbon dioxide are almost always used in a dual-cycle, indirect configurations. The gas constitutes the primary coolant and light water is almost always used in the secondary, due to its abundance and the other beneficial attributes described earlier. The interface between the two sides of the coolant system is the reactor boiler. Typically, several boilers are configured per reactor core. While being relatively unreactive, at high temperatures carbon dioxide can react with graphite and some steels. The main source of induced radioactivity in carbon dioxide is that of trace impurities of ^{13}C (1.11% abundance and not radioactive) that becomes ^{14}C as a result of neutron activation; ^{14}C is a relatively long-lived β^- emitter [8].

7.6.3.2 Helium

Helium has also been considered for use as a reactor coolant [9,10]. It satisfies most of the requirements specified above and, as a noble gas, it has the benefit of being chemically unreactive. Further, as reflected in the differences in its thermophysical properties and flow characteristics, it offers a heat transfer performance that is superior to carbon dioxide. This is usually reflected by it requiring reduced pumping power. However, for mainstream use in gas-cooled reactors the important parameter that has prevented its widespread use thus far has been its cost as compared to the relative abundance and availability of carbon dioxide.

For future reactor design developments, where the emphasis is on raising operating temperatures to achieve better thermodynamic efficiencies, helium is featured in several possibilities. It offers greater chemical stability at higher temperatures (~850°C) than carbon dioxide in addition to its heat transfer advantages. Helium also offers the potential for direct, single-loop reactor circuit designs mated to a helium-drive turbine and the supply of process heat, as described above. It is indicated in theoretical analyses, however, that the potential exists for impurities to be released into the coolant from in-core components and materials, predominantly graphite, at such extreme temperatures. These will have to be removed from the helium in these systems [11].

7.7 EXOTIC COOLANTS

7.7.1 SODIUM

For coolant applications in fast reactors, where minimising the moderation of neutrons is of paramount importance, liquid metals have long been used due to their attractive heat transfer characteristics; liquid sodium and variants of it have been a widespread choice [12,13]. Sodium is a liquid at relatively low temperature (melting point 370 K), it has the effective thermophysical properties one would expect from a metal and is also relatively inexpensive. It has the requisite low neutron-scattering property required of fast reactor coolants (to ensure that any moderation of the neutron flux by such a coolant is minimised). It also has a low neutron absorption cross section thus limiting the possibility of neutron activation or deleterious effects on neutron economy. Its relatively high boiling point (1156 K) provides for a reasonably extensive temperature range of operation and thus does not require pressurisation in use. This greatly simplifies the extent of related equipment, pressuriser, instrumentation, and so on, that is necessary. The use of sodium is compatible with stainless steel, removing the need for exotic, corrosion-resistant process plant materials and maintenance regimes.

However, sodium is highly reactive with water and therefore significant measures are necessary to minimise the possibility that it might come into contact with moisture, as this could result in fire. From a radiological perspective, sodium activates under neutron irradiation to ^{24}Na. This isotope has a relatively short half-life of ~15 h, decaying via β^- decay to ^{24}Mg (as discussed in Chapter 5) but otherwise sodium is relatively free from activation concerns. Like many reactor coolants, sodium has to be used in high-purity form to avoid the activation of trace impurities. This can require a dedicated purification plant adding to the cost of its use. It is opaque, which requires that alternatives to visual inspection methods are used to assess processes in plant in place of traditional camera use. Liquid sodium has an intrinsic tendency towards a positive void effect which is discussed further in Chapter 14. This must be minimised by core design and by avoiding the possibility of boiling. Sodium has been used both in isolation and as a eutectic alloy (a mixture of metals with a combined lower melting point) with potassium, typically in proportions 22% and 78% for sodium and potassium, respectively.

7.7.2 LEAD

In comparison with sodium, lead has the significant advantage that it is compatible for use in contact with air and water, again for fast reactor applications [14,15]. For example, design options such as locating steam generators in close proximity with the primary vessel are feasible with lead that are not possible in the case of sodium. Both lead and a lead-bismuth eutectic alloy have been explored as fast reactor coolants; the latter predominantly via use in Russian nuclear submarines. However, significant neutron activation of bismuth yielding the highly radiotoxic ^{210}Po as a product, along with the scarcity of bismuth, has caused attention to revert to lead. Lead has a significantly higher melting point (~600 K) than the lead-bismuth alloy (typically 396 K) but its relatively high boiling point (2022 K) means that a sufficient range of operation is available without the risk of boiling.

Lead is chemically toxic and it is highly reactive with stainless steel. The corrosion that results can only be prevented by maintaining an iron oxide layer on steels in contact with the lead, together with strict temperature control regimes, limitations of flow and purification operations. The flow limitations alone result in a limit in power density and hence one on overall reactor capacity. As with sodium, lead is opaque necessitating special in-service techniques (especially since it is very dense and corrosive). The use of lead as a coolant also requires special attention to be given to accident scenarios associated with seismic resilience due to its very high density.

7.8 STEAM TURBINES

Most large-scale nuclear reactors used to generate electricity today use steam turbines. Although reciprocating engines based on the phase change of a vaporised working substance (low pressure steam) against a piston were arguably the first useful heat engines some 300 years ago, the conversion of heat to kinetic energy from *high-pressure steam* exploits the benefits of a turbine. Turbines offer many advantages including

- Reduced requirements for maintenance and repair.
- Direct drive of the generator shaft.
- A reduced requirement of oil and hence a much-reduced propensity for contamination of the exhausted steam by lubricating oil.
- Reduced vibration and thus capability for operation at very high speeds and thus high steam inlet pressures.
- The system is fully enclosable thus minimising the possibility of interference or accident.
- Large generating power capabilities are possible coupled with a small and compact form factor.
- The ability to exploit the full expansion of the steam over several stages (i.e. high- through to low-pressure turbine stages) and thus to achieve high efficiencies.
- Reduced steam consumption.

Conceptually, a turbine is a rotor attached directly to a shaft such that its rotation causes the shaft to rotate. The rotation of this shaft is used to generate electricity via a generator set that is located further along in the turbine hall. The working substance, travelling at high velocity and having been expanded following the application of heat (in this context from a reactor), is directed to impart its energy to the rotor. Energy is extracted from the steam in the forms of heat (thermal energy) and kinetic energy. The outlet is thus at a lower temperature than the inlet, as a result of the heat that is extracted, and the velocity of the steam at the outlet is reduced relative to the inlet as a consequence of the removal of kinetic energy.

A great deal of research and development has been applied to the design of rotors, particularly the performance of the materials of which they are made and the texture of the surface, to optimise the efficiency of the energy transfer from the working substance to the rotor. The design of the first turbines, such as the *de Laval turbine* [16], illustrates the fundamental concept. This employed a series of nozzles to direct the steam to sculpted surfaces in the perimeter of the rotor, as shown in Fig. 7.4. In such a design, the steam expands rapidly through the tapered openings of the nozzles. This ensures that the steam is fully expanded when directed to a series of blades machined into the rotor that are angled in sympathy with the direction of the incoming steam.

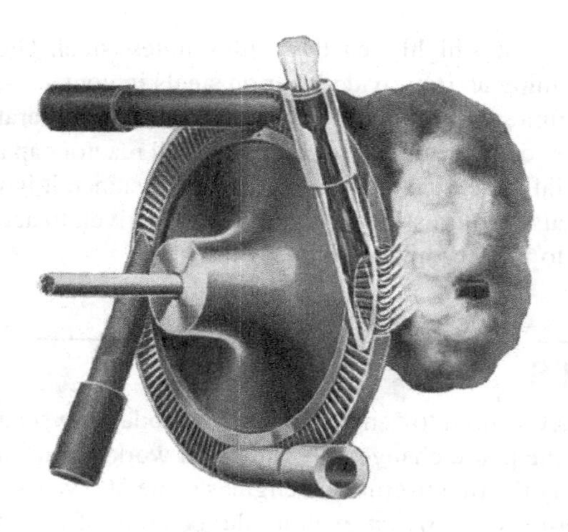

FIG. 7.4

The wheel and nozzle components of the de Laval steam turbine.

After E.S. Lea, E. Meden, The De Laval steam-turbine, Proc. Inst. Mech. Eng., 67 (1) (1904) 697–714.

Perhaps the most important advantage of a turbine is that, unlike reciprocating engines where the working substance is often exhausted to the surroundings after one cycle, the expanded steam passing through the blades of one rotor can be ducted to a selection of subsequent rotors; the outlet of one rotor constituting the input, or 'nozzle' for the next. A modern rotor comprises a sophisticated arrangement of vanes and blades, some of which are fixed and some which are moving. Most turbines employed in modern nuclear power stations are *impulse-reaction turbines*, easily identifiable by the convergent spacing of the turbine blades. Such turbines are so-called because they are caused to turn both by the impulse imparted by the *arrival* of high-velocity steam acting on the turbine blades at its input, but also by the reaction of the steam *leaving* the moving vanes. This imparts a reverse thrust similar to the way in which an untethered hose recoils from high-pressure water flowing from it.

Each of the stages in a multistage turbine comprises the rotor, stator, one fixed set of blades and a moving set. Each stage extracts a small amount of energy from the passing steam, which is then exhausted to the next stage. Absorbing energy from the expanded working substance gradually in this way minimises the loss of steam with the overall pressure drop across the whole turbine achieved by compounding a series of small drops at each stage.

Steam turbines are usually divided into a high-pressure portion and a low-pressure portion. The high-pressure portion comprises a set of turbine blades of relatively small diameter. The diameter increases gradually (evident as an increase in height of the turbine set) as the steam passes through before it is exhausted to the low-pressure portion. The blades in the low-pressure portion are of a much bigger diameter than the high-pressure stage because the pressure of the steam is decreasing as energy is extracted from it and it expands.

The choice of the materials and the shape of the blades in different parts of the turbine are highly-detailed design factors that are optimised to ensure that the turbine machine can withstand high temperatures and also to minimise the possibility of steam leaking past the blades without performing useful work. Of the few disadvantages associated with turbines, the fragility of the turbine blades and the requirement for a high vacuum essential to their operation are perhaps the most significant issues (an example of a relatively modern turbine system exposed during servicing is shown in Fig. 7.5).

FIG. 7.5

A photograph of the turbine systems at the Heysham II Advanced Gas-cooled nuclear power station in Lancashire, United Kingdom.

By Peter Joslin with permission of EDF.

7.9 ELEMENTARY THERMAL HYDRAULICS

The goal of nuclear thermal hydraulics analysis is to inform our understanding of the response of the fluid-based components of a nuclear reactor system to thermal stimuli to ensure that the right decisions can be made in terms of the operation, performance and, above all else, safety. The high power density of nuclear fuel, in comparison with conventional heat sources that are used in steam raising plants (usually combustion-based systems), means there is sufficient energy to exceed the temperature limits of fuel, cladding and coolant, which could result in serious consequences if not properly understood and managed.

As is described further in Chapter 14, reactor systems are designed and operated within what is termed a *multi-barrier philosophy*. Several components of such a system have a significant bearing on the thermal aspects of the plant. For example, the integrity of the fuel constitutes the first barrier, the fuel clad the second, the primary circuit the third barrier and so forth. In general terms, thermal hydraulic considerations assess the capacity for a given design to withstand anticipated but unplanned changes in reactor behaviour (i.e. *transients*), accidents which the reactor system is designed to withstand (thus being deemed to be *within design basis*) and severe, *beyond design basis* accidents. Such considerations enable us to determine whether engineering safety and performance margins are satisfactory. They also inform outage planning, inform decisions associated with uprating plants in terms of power and availability, and also with respect to extending the life of operating plants. Perhaps most importantly, given the imperative of reactor operation is to maintain cooling, thermal hydraulic understanding provides insight as to what conditions might lead to cladding surfaces losing contact with coolant, and hence the potential for a rapid temperature escalation of the fuel.

Power-generating reactors are usually large, heterogeneous systems. They comprise many different materials, a range of opening spaces, constrictions and flow paths. There are many interfaces with the potential to influence fluid flow and behaviour. This geometrical complexity, combined with a range of heat transfer scenarios, the potential for two-phase flow and a variety of flow regimes constitutes a complex system with a response that is not readily described by a small number of closed-form equations with just a few variables. Most often, computer-based simulations are necessary based on empirical relationships. A wide variety of codes have been developed corresponding to the range of scenarios and situations that might occur and for which a response is needed, including steady-state operation and accidents. It is worthy of note that several scenarios—for example, related to severe transients and beyond design basis accidents—cannot be replicated experimentally, certainly not with the accuracy desired, and rarely if ever at the same scale. Thus the simulation environment is critical to anticipating such responses. The thermal hydraulic situations that are usually of interest include:

- *steady-state operation* and hence estimating power output
- *transients* within the normal operational performance envelope, that is occurrences not anticipated as precursors to severe accidents
- *natural circulation* behaviour
- *severe accidents*, usually involving a loss of coolant particularly with respect to the performance of critical circuit components

7.9.1 THERMAL CONDUCTIVITY AND THE HEAT TRANSFER COEFFICIENT

Unlike most other power generation systems based on a heat engine, the fuel in a nuclear reactor remains largely unchanged after use as a macroscopic physical entity but is augmented isotopically and also as a material, particularly at a microscopic level. It is of paramount importance that nuclear fuel remains intact despite there being ample energy available in a power reactor with the potential to breach the cladding and to cause extensive damage to the fuel inside. The role of a coolant in this regard is simple: it must ensure that the heat transfer from the fuel is sufficient to keep the temperature of the fuel below its maximum permissible limit. The physical properties of the coolant and the way in which it is supplied can thus have a direct influence on the maximum fission rate and hence the power output of a reactor; forced convection, in which the coolant is pumped through a reactor core in a turbulent state is the most widespread arrangement that is used.

In Fig. 7.6 a schematic of a simplified portion of a heterogeneous reactor core is shown, with fuel, cladding and coolant channels depicting a vertical flow of coolant with an enlarged portion showing the fuel, the gap between it and the cladding, and the cladding itself.

In the fuel, fission generates heat and raises its temperature. This is highest at the centre of the fuel element and a minimum at the edge of the fuel where the heat is lost from the fuel most readily; the temperature typically describes a parabola as a function of distance from the centre to the outer boundary of the fuel, that is the distance from the centre of a cylindrical fuel element. The heat is conducted from the fuel to the cladding, the latter which has no role in the heat generation process under normal operation. The rate at which heat q is conducted via a given material per unit area in one dimension is given by the Fourier expression for *heat flux*, dq/dt (energy per unit area per unit time), viz.,

$$\frac{dq}{dt} = -k\frac{dT}{dx} \tag{7.16}$$

where dT/dx is the temperature gradient in terms of distance x from the upper temperature to the lower temperature, and k is the *thermal conductivity* of the material. This reflects the fact that the rate at which

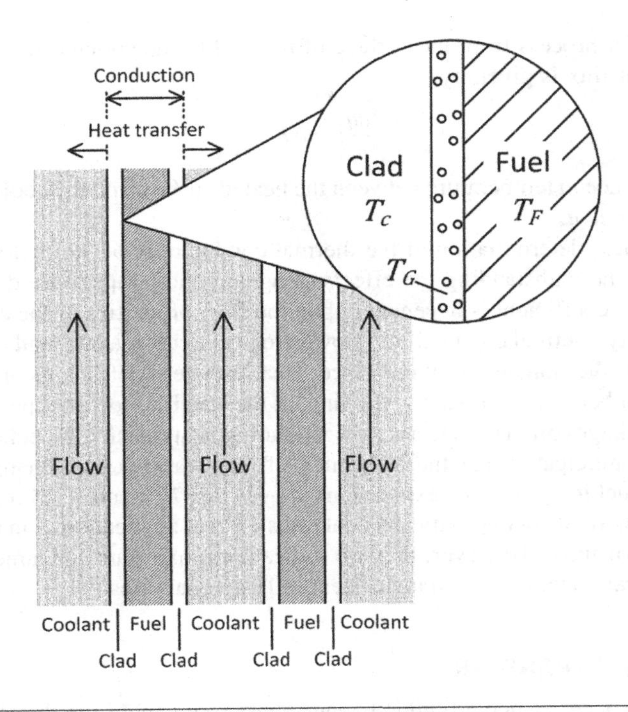

FIG. 7.6

A schematic of a portion of a nonboiling reactor core showing fuel elements in a vertical arrangement with cladding and coolant (liquid or gas) flowing up vertically through channels in between the fuel elements (greatly simplified and not to scale). The temperatures of the fuel, cladding and the gap between them are defined T_F, T_C and T_G, respectively.

energy is conducted is proportional to the temperature gradient and that the heat flows in the opposite direction to the temperature gradient, from hot to cold, consistent with the first statement of the second law of thermodynamics given in Section 7.5.1. In this case it has been assumed that k is a constant when, especially over the temperatures that a nuclear reactor operates, it will actually vary significantly with temperature; long-term changes in the fuel and its dimension over time will also influence it. As depicted in Fig. 7.6 there is also a very narrow gap (typically ~0.1 mm for a fresh fuel assembly) between the fuel and the cladding. Despite being narrow, conductivity is relatively poor across this gap and so its influence can be tangible. Furthermore, the extent of this spacing is known to reduce as the fuel expands with use and this can also cause it to lose its uniformity. Changes in composition also occur as gaseous fission products are evolved into the space; these factors also influence the conduction of heat from the fuel to the cladding and can render it difficult to forecast. If this is rationalised in terms of the volume of the fuel, and the surface temperature and thermal flux of the fuel are known, then the internal temperature can be determined in terms of the dimensions (usually the diameter) of the fuel; this enables maximum permissible fuel temperatures to be estimated by design in terms of the dimensions and composition (thermal conductivity) of the fuel material.

The thermal energy from the fission reaction in the fuel is transferred via the cladding to the coolant, the latter that is in the form of a fluid in order that the energy can be transported away efficiently. The convective heat transfer process from the surface of the fuel to the coolant follows Newton's law of cooling where the heat flux is given by,

$$\frac{dq}{dt} = h\Delta T \tag{7.17}$$

where ΔT is the difference in temperature between the heated surface and the cooling medium, and h is the *heat transfer coefficient*.

Whereas an empirical determination of the thermal conductivity of the fuel and cladding is relatively straightforward, notwithstanding the effects of ageing and temperature dependence described above, the heat transfer coefficient is dependent upon the flow properties of the coolant, for example, its density, heat capacity, thermal conductivity, flow velocity, viscosity, etc., and also the dimension of the flow path in which the transfer is taking place. Heat transfer coefficients are thus often derived empirically from a number of formalisms depending on the specific flow regime in question. Computational simulation packages often used to assist in estimating heat transfer, especially for extraordinary scenarios where it is anticipated that the efficiency of heat transfer may change significantly and potentially implicate fuel integrity. The expressions above (Eqs 7.17 and 7.16) represent the simplest, linear formalisms for thermal conductivity and heat transfer; real applications can require more sophisticated developments of them. However, they do indicate an important fundamental principle: high temperature gradients are necessary to transfer heat in large quantities.

7.9.2 BOILING HEAT TRANSFER

The process of *boiling* occurs when a liquid changes phase to a vapour via the formation of bubbles, usually due to heating. Boiling is important in nuclear power reactors because the majority of operable reactors in the world use water (either light or heavy water) as a coolant. Thus the potential for boiling to occur exists at temperatures that fall within design basis. Further, two designs (BWRs and RBMKs) use steam direct from boiling through contact with the reactor core while most designs provoke steam generation in feedwater (secondary) stages. Here we are focussed on the significance of boiling in reactor cores rather than in these subsequent parts of the plant due to its direct influence on the nuclear characteristics of the reactor system.

A common thread to the mitigation of accidents, particularly in the case of pressurised water reactors, is to avoid of a loss of coolant at the clad–coolant interface to ensure that cooling is maintained at all times. This can be complicated because the dependence of the transfer of thermal energy from a heated surface with temperature to a liquid that provokes boiling is not continuous. Rather, there are several phases of behaviour that a liquid exhibits that influence the efficiency of heat transfer, which are afforded a number of important definitions. This is described below in reference to Fig. 7.7, which illustrates five stages of boiling as a function of *heat flux*[4] increasing from left to right, and Fig. 7.8, which is a plot of heat flux versus difference in temperature ΔT of the heated surface and the saturation temperature of the liquid (known as a *boiling curve*), in this case for water and isopropanol for the purposes of comparison. The various stages of heat transfer are described generically by restating *Neutron's law of cooling* slightly differently, as per Eq. (7.17),

[4]*Heat flux* is the rate of *heat transfer* through a surface, measured in Watts per unit area.

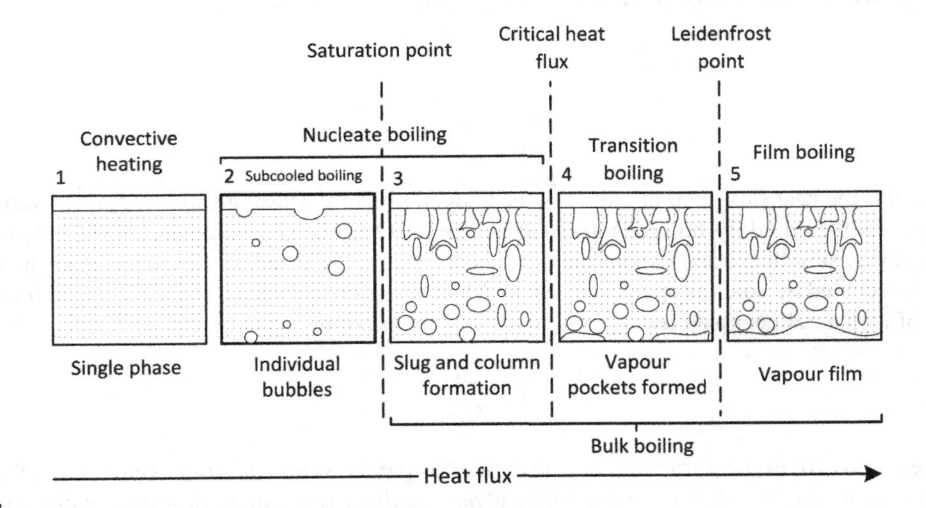

FIG. 7.7

The five stages of heat transfer to a liquid, assuming a heated surface at the base of a vessel, square in elevation with a liquid–vapour interface towards the top of vessel. Heat flux increases from left to right, from stages 1 to 5, with the critical heat flux shown between nucleate boiling and transition boiling, and the Leidenfrost point located between transition boiling and film boiling.

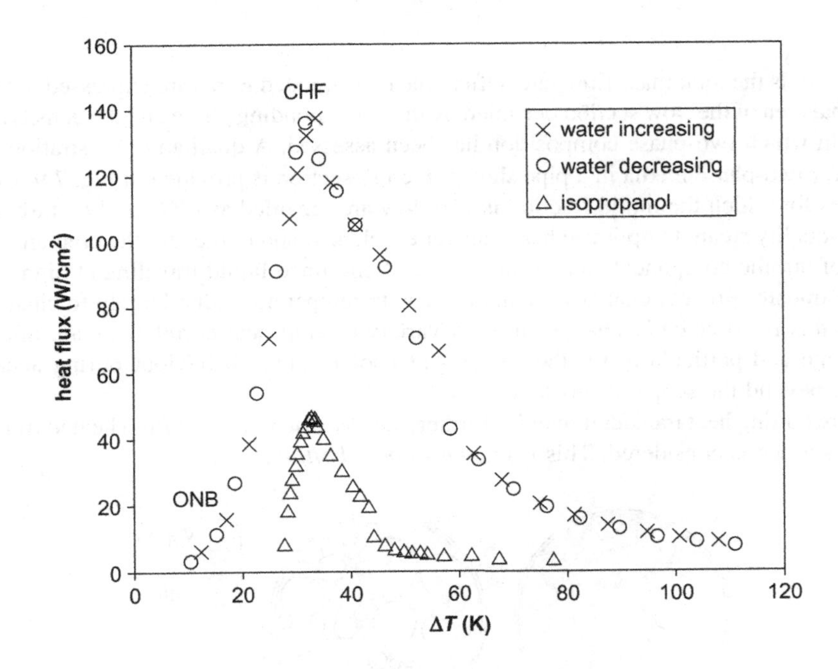

FIG. 7.8

A boiling curve (heat flux versus the excess temperature ΔT between the heat source and the saturation temperature), for distilled water at pressure 0.1 MPa and isopropanol at pressure 0.12 MPa.

After H. Auracher, W. Marquardt, Heat transfer characteristics and mechanisms along entire boiling curves under steady state and transient conditions, Int. J. Heat Fluid Flow 25 (2004) 223–242 with labels highlighting ONB and CHF by the author.

$$\frac{dq}{dt} = h(T - T_s) \tag{7.18}$$

where dq/dt is the heat flux per unit area, T is the temperature of the heated surface, T_s is the *saturation temperature* of the liquid (its boiling point) and h is the heat transfer coefficient as before. In two-phase mixtures comprising liquid and vapour there can be a range in the extent to which the space in a given channel is occupied by voids as opposed to liquid. Thus, the *void fraction, α*, is defined as the fraction of the area of a channel occupied by voids A_v to that of the mixture, $A_v + A_l$,

$$\alpha = \frac{A_v}{A_v + A_l} \tag{7.19}$$

where the *quality* (or dryness fraction) can be defined as per the *static quality* χ_{st} (the ratio of the mass fraction in vapour phase to that of the mixture) and *flow quality* χ (the vapour flow rate relative to that of the mixture) defined as,

$$\chi_{st} = \frac{m_v}{m_v + m_l} \tag{7.20}$$

$$\chi = \frac{\dot{m}_v}{\dot{m}_v + \dot{m}_l} \tag{7.21}$$

and $\dot{m} = \dot{m}_v + \dot{m}_l$ is the total mass flow rate. Often, the mass-related terms are expressed in terms of the cross sectional area of the flow section occupied by the corresponding phase and the density, dependent on the way in which two-phase composition has been assessed. A qualitative illustration of such an assessment for two-phase media in a pipe shown in cross section is provided in Fig. 7.9. Conventionally, mixtures for which the liquid fraction is very low are regarded as *high quality*; turbines usually require high-quality steam to optimise heat transfer efficiency, and to prevent the premature corrosion and failure of turbine components due to the action of entrained liquid traveling at high velocities.

The heat transfer process changes dramatically with temperature, due largely to changes in void fraction and this is reflected in changes in h. A variety of empirical correlations are often used for h at each stage and particularly for the analysis of coolant phase behaviour during accidents [17] but these are beyond the scope of this text.

To explore boiling heat transfer dynamics further, the case for a stagnant fluid heated from below by a horizontal surface is considered. This is known as *pool boiling*.

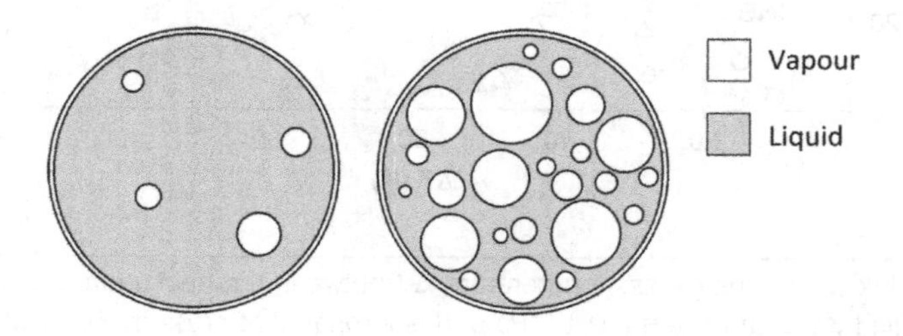

FIG. 7.9

An illustration of void fraction, low (left) and high (right) for the case of liquid–vapour flow in a pipe cross section.

7.9.2.1 Convective heating

When a liquid is at a temperature that is below its boiling point it is termed *subcooled*. When it is heated, it absorbs energy from a source of heat (usually a surface in intimate contact with the liquid in the context of boiling and nuclear energy). This energy is propagated via thermal currents throughout the liquid (*free convection*) to result in vaporisation of the liquid at the liquid–vapour interface. This gradually adds to the vapour pressure via an interface with the vapour phase that is usually at a point distant from the source of heat. This corresponds to stage 1 in Fig. 7.7 where the medium is present as a *single phase* only (liquid) and the heat flux increases with increasing temperature, as shown in Fig. 7.8, up until $\Delta T \sim 20$ K.

7.9.2.2 Nucleate boiling

The region beyond $\Delta T \sim 20$ K signifies the *onset of nucleate boiling*, that is from the point labelled ONB in Fig. 7.8, where a subtle increase in heat transfer is evident from the increase in the gradient of the graph. Here, the temperature of the heated surface in contact with the liquid has reached a temperature that is greater than the saturation temperature T_s of the liquid. First, given a relatively limited temperature excess of between $4\text{K} < \Delta T < 10\text{K}$ at this stage, bubbles form at *nucleation sites*. These might arise, for example, from pockets of gas trapped in microscopic features or cavities on the surface that arise due to the specific locations of these being slightly more hydrophobic than the surrounding surface. Continued heating causes the bubbles to become detached from the heated surface as they gain sufficient buoyancy; they then have the potential to rise to the liquid–vapour interface (if there is one) and to yield their vapour to the ambient or, alternatively, to collapse en route. Whether or not they collapse is dependent on the temperature of the surrounding bulk liquid relative to T_s. The separation of the bubble from the heated surface generates turbulence at the surface. This causes mixing of the fluid and enhanced heat transfer by convection, and consequently increased heat flux.

In this regime, *subcooled boiling* describes the state when vaporisation occurs close to the surface supplying heat (stage 2 in Fig. 7.7, known as *local boiling*). The liquid phase still has capacity to absorb heat and is characterised by regions of the liquid that have a higher void fraction than others, that is, comprising a heterogeneous distribution of voids. Bubbles formed in localised areas of nucleation migrate into the bulk and collapse due to the influence of the subcooled state. For bubbles to survive, the pressure inside them needs to be greater than the pressure from the liquid that surrounds them. Thus the vapour inside needs to be at a higher temperature than the liquid outside. When this is the case the vapour is referred to as *superheated*; superheating occurs at the heated surface to cause bubble formation but the subcooled bulk at this stage is sufficient to cause bubble collapse.

As heat transfer continues, causing the temperature of the liquid to increase, a limit is reached beyond which the addition of any further energy only results in a phase transition of the liquid to vapour, rather than an increase in the temperature of the liquid. At this point the liquid will have reached its boiling point. Hence, a *saturated liquid* is one that has absorbed the maximum energy possible and the *saturation temperature* and *saturation pressure* are the temperature and pressure at which boiling occurs. Boiling of a saturated liquid is known as *bulk boiling*. This is characterised by a homogeneous void distribution, as depicted by stage 3 in Fig. 7.7 and the region in temperature from $\sim 20\text{ K} < \Delta T < \sim 35$ K in Fig. 7.8. At this stage the boiling state contrasts with the subcooled boiling state in which separate bubbles rise individually and collapse in the bulk, since they no longer collapse.

7.9.2.3 Transition boiling

For surface temperature excesses of 10–30 K greater than T_s, the rate of bubble formation increases such that some of the bubbles coalesce. This yields larger conglomerations of several voids (often referred to as *slugs*) that rise in *columns* that reach to the liquid–vapour interface. These constitute effective pathways for vapour to reach the ambient but, conversely, they impede the mixing at the heated surface that had enhanced heat transfer at lower levels of heat flux. As heating continues pockets of vapour begin to form over some areas of the heated surface. Thus, while heat transfer continues to increase, the rate of increase is reduced, corresponding to stage 4 in Fig. 7.7. This stage is known as *transition boiling* and spans the stages of nucleate boiling and film boiling.

7.9.2.4 Departure from nucleate boiling, critical heat flux and boiling crisis

Nucleate boiling results in an increase in heat transfer due to the enhanced mixing caused by the turbulence near to the heated surface due to bubble formation, as described earlier. However, the pockets of vapour formed (as per stage 4) eventually coalesce to constitute a film across the entire surface, insulating it from the liquid (stage 5). The heat flux at the point that this happens is known as the *critical heat flux* (CHF) because at this point heat transfer via convection is staunched suddenly by the insulating effect of the vapour film. Unabated, this is accompanied by a sudden increase in the temperature of the heated surface because heat transfer is now reliant on a combination of convection in the vapour and radiative effects across the film, which is much less efficient than forced convection in the liquid phase. This corresponds to the maximum in Fig. 7.8 at $\Delta T \sim 35$ K, labelled CHF.

The associated threshold at this point is referred to as the *boiling crisis* and the state is one of *film boiling*. The formation of the film signifies the end of nucleate boiling or the *departure from nucleate boiling* (DNB). It is illustrated by, for example, the observation that droplets on a very hot surface elevated by a boiling film require longer to be vaporised than those on a surface at a lower temperature where the film is absent. The restriction on heat transfer is known as the *Leidenfrost effect*. The temperature at which the heat transfer is minimised, following the point of CHF, is the *Leidenfrost point*. In Fig. 7.8 for water this is $\Delta T > 100$ K.

The dynamics of boiling, as described above, have important implications with respect to nuclear reactor systems. For example, DNB can result in *dryout* at the surface of the fuel cladding; the state where there is no liquid in contact with the clad. This leads to a reduction in heat transfer (as per stage 5 in Fig. 7.7) at the coolant–clad interface. This loss of the heat sink at the clad–coolant interface would lead to a dramatic increase in fuel temperature. If this is not mitigated by the design of the fuel and/or clad or the response of emergency core cooling systems, damage could result.

7.9.3 FLOW

In addition to boiling phenomena, heat transfer in reactor systems is influenced significantly by both passive (natural) and active (forced) flow of the coolant; indeed an unplanned loss of flow can itself be a very serious cooling issue. The possibility of two-phase flow arising as a result of voiding, results in a number of flow regimes or *patterns*. These can influence the performance of the reactor coolant system; their behaviour is augmented by the elevation of the flow path (pipe), constrictions and openings.

Assuming a horizontal flow path, as depicted in Fig. 7.10, low flow rates are characterised by *stratified flow*. Here the liquid flows along the lower section of the vessel with the gaseous (vapour) phase flowing above it. A significant difference between the flow rate of the liquid and that of the vapour can

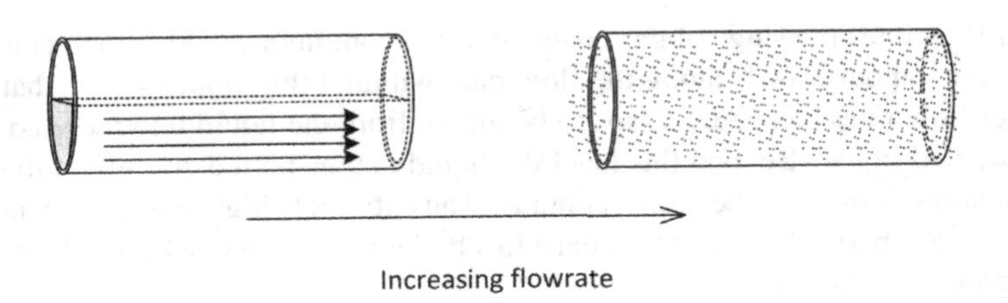

FIG. 7.10

A depiction of the extremes of two-phase flow patterns in a horizontal pipe section, from stratified flow (left) at very low flow rates through to dispersed flow (right) at high flow rates. Stratified flow is only feasible for horizontal flow paths.

result in turbulence at the interface between the two. At still higher flow rates a disturbance of the vapour–liquid interface is characterised as *wavy flow*. As flow rate is increased a variety of patterns are possible. These constitute a trend between stratified flow at the more quiescent, low flow-rate extreme, through to *dispersed flow* in which the liquid is finely dispersed in the vapour as a mist at the other.

7.9.3.1 Vertical flow regimes and maps

In the context of a reactor core, vertical coolant flow behaviour is often of greater interest with regard to heterogeneous LWR cores because of its relevance to the flow of the coolant up through the fuel matrix or along fuel channels. In this arrangement, heat is transferred not from a horizontal surface but from the vertical boundaries constituted by the clad of the fuel elements. Single-phase liquid flow (solid) in this arrangement is incompressible and subject to the phase transition from laminar to turbulent flow, being most often in the latter state during steady-state operation due to the large heat transfer rates that can be achieved. Two-phase, liquid–vapour flow can be widespread in steady-state scenarios due to nucleation and gaseous desorption from the bulk liquid. Boiling is a common accident scenario that is considered especially for pressurised water reactors whereas it is a steady-state scenario in boiling water reactors.

Vertical, two-phase flow is often differentiated in terms of whether it is *homogeneous* (or bubbly) flow and *heterogeneous* (or churn) flow. The former is associated with relatively low void fractions where there is little interaction between individual bubbles. As the void content increases, the interaction increases between bubbles, larger voids form and start to fill the pipe. Eventually only a thin layer of liquid remains between them and pipe wall and the voids are longer than the diameter of the pipe. When this occurs the bubbles are referred to as *slugs*; they typically have hemispherical caps and have also been referred to as *Taylor bubbles*. The uniform dispersion of the bubbles in the liquid phase present at lower void fractions has degenerated at this point.

Such *slug flow* tends to occur in small-diameter pipes. In wider channels when continuity is lost the gas and liquid are separated with voids of gas and plugs of liquid in between them. This flow pattern is referred to as *churn*. The media in this state is *heterogeneous*, often referred to as *churn-turbulent* flow, with the flow broken up by slugs of gas. At higher void levels this extends to what might be termed *semi-annular flow*. Here, the passage of the liquid is intermittent and separated by voids of vapour. Clearly at this level, there is a large variation in void fraction that yields fluctuations in pressure resulting in contrasting behaviour in mixing and heat transfer rates.

At higher gas-flow rates the slugs of gas coalesce into a continuous void, which constitutes *annular flow*. In this case the vapour travels along the flow path within a thin liquid sleeve that lines the walls of the pipe; higher flow rates can cause wisps to be drawn from the liquid into the void and droplets of the liquid are entrained in gas. Beyond this level the liquid is transported entirely in the form of a mist (*mist flow*), culminating in *dryout* where the vapour is of an extremely high quality with no steam remaining. All such phases described above might coexist in a boiler tube or channel should the requisite conditions arise, as depicted in Fig. 7.11.

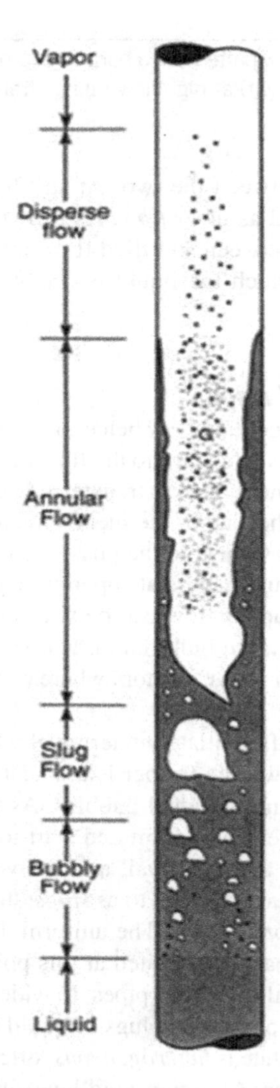

FIG. 7.11

The evolution of steam-water flow in a vertical boiler tube.

After H. Abdulmouti, Bubbly two-phase flow: Part I—characteristics, structures, behaviours and flow patterns, Am. J. Fluid. Dyn. 4 (4) (2014) 194–240

Given the significant influences of the different flow regimes described above on mixing and heat transfer, there is frequently a requirement to predict when and how transitions between flow transitions might occur. These transitions are not distinct, and they are dependent on a significant number of factors while the ease with which they are measured can vary significantly for different situations. Examples of measurement approaches are the use of x-ray and flash photography and, more recently, the attenuation of laser light. Experimental data tend to be scattered around perceived transition zones rather than defining hard boundaries. Nonetheless, such analysis enables maps of flow regimes to be constructed as a function of gas and liquid flow rates (typically mass per second per unit area). An example of one such map is given in Fig. 7.12.

7.9.4 RELEVANCE TO ACCIDENT SCENARIOS AND SAFETY

The analysis of two-phase flow scenarios introduced in the above sections is particularly relevant to safety and the assurance of defence-in-depth arguments (discussed further in Chapter 14). The three structural barriers—the *fuel clad*, *reactor vessel* and the *containment*—are vulnerable to stresses induced by thermal and pressure loads that can result from sudden phase transitions. Thermal hydraulic

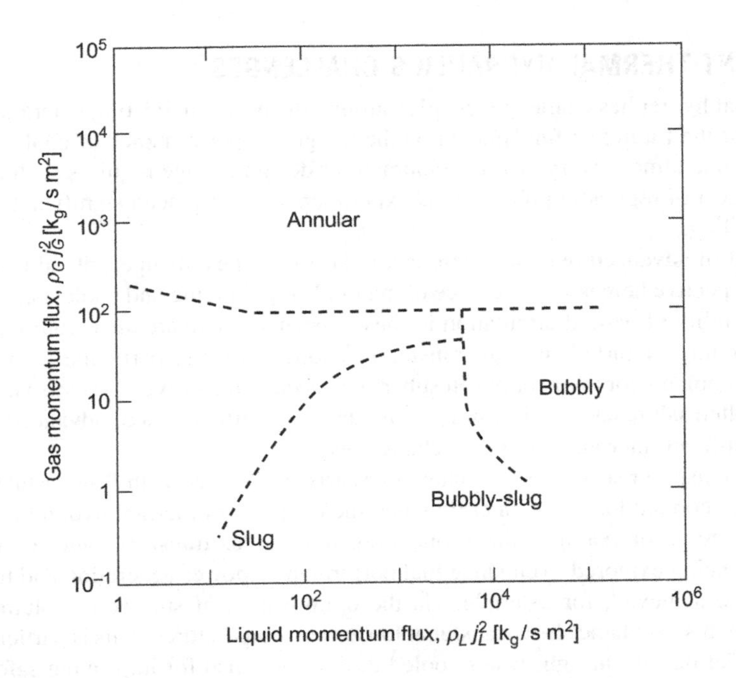

FIG. 7.12

A flow regime map for an air–water mixture in a vertical, 3.2-cm-diameter pipe showing the experimentally observed transition boundaries, validated for air–water flow at atmospheric pressure and steam-water flow at high pressure.

After G.F. Hewitt, D.N. Roberts, Studies of Two-Phase Flow Patterns by Simultaneous X-ray and Flash Photography, UKAEA Report Aere M2159, 1969. Adapted by H. Abdulmouti, Bubbly two-phase flow: Part I—Characteristics, structures, behaviours and flow patterns, Am. J. Fluid. Dyn. 4 (4) (2014) 194–240.

calculations are used to assess that the margins used to ensure that such loads do not exceed limits are sufficient to withstand such scenarios.

Early power-generating nuclear plants were designed in the absence of the computer-based, sophisticated modelling capabilities that exist today and with a much-reduced knowledge base derived from operational experience and experimental research. Consequently, the designs of these systems were significantly over-specified in terms of safety and operational performance criteria to accommodate the relatively primitive understanding of thermal hydraulics at that time. They were often first operated via gradual, stepwise start-up test and measurement procedures.

The current state-of-the-art exploits the existence of a much greater quantity of data derived from, for example, the dedicated testing of all critical components such as steam generators, coolant pumps, upper plenum, and reactor cores with which to infer the suitability of the designs of emergency core cooling systems. This is complemented by extensive computer-based models and simulations such as RELAP5.[5] These more detailed thermal hydraulic analysis capabilities enable the inherent conservatism in earlier safety analyses of circuit designs to be recovered, in order for proposals for operation beyond original estimates of plant life or uprating to be considered that would not have been possible previously.

7.9.5 CURRENT THERMAL HYDRAULICS CHALLENGES

Currently, thermal hydraulics challenges revolve around the needs of existing reactors and designs being developed for the future; so fundamental is the integrated performance of coolant behaviour and reactor response that almost every new development or design change requires such an analysis. The complexity and cost of replicating phenomena experimentally has placed significant emphasis on the benefits of modeling.

In the context of advanced reactors, there is a focus on studies of supercritical light water reactor designs. The perspective here is on the choice of materials for cladding and coolant circuits, and also to explore the possibility of natural circulation in these systems. There are also requirements for an improved understanding of turbulent heat transfer in exotic coolants, particularly for example lead-bismuth eutectic coolants for applications in subcritical accelerator-driven systems and in liquid metal fast reactors. Molten salt reactor design and performance is a further area of advanced reactor analysis that presents significant thermal hydraulics challenges.

In the area of existing reactor designs, challenges arise associated with flow instabilities in natural circulation boiling coolant loops, channel flow and the coupling of thermal hydraulics and neutronics phenomena to address, for example, the implications of complex transients, and so on. New fuel geometries are also being explored to enhance fuel performance, power generation and to increase burn-up. This might be achieved, for example, via the optimisation of surface-to-volume ratio of fuel-channel arrangements to enhance heat transfer at lower fuel temperatures. This is particularly important to uprate the power density in light-water cooled designs and also for improving safety margins.

A significant area of thermal hydraulics applications is the broad area of severe accident analysis, particularly associated with loss of coolant scenarios based on transient-stimulated problems, valve failure and leak rates in light water reactors. Most empirical models are developed on test facilities that are understandably smaller than real reactor systems. Consequently, the scaling of these models

[5]http://www4vip.inl.gov/relap5/.

on which computational simulations are based using data derived from such test facilities and hence validated for full-sized power plants is a further area of activity. Also related to safety, there is an ongoing need for the study of multiphase flow, particularly two-phase processes such as those associated with critical heat flux, departure from nucleate boiling, dryout and multiphase flow properties in reactor fuel channels.

CASE STUDIES

CASE STUDY 7.1: A COMPARISON OF SPECIFIC HEAT CAPACITY AND A SIMPLE EXPANSION PROCESS

In this elementary illustration we consider a simple thermal-expansion process based on a gaseous working substance. The energy necessary to raise the temperature of the gas is compared with that needed to yield isothermal expansion of the same gas against a piston. Consider a simple piston of mass 1.00 kg and cross-sectional area 0.630 cm^2, oriented in the vertical plane that is acting under gravity of 981 cm s^{-2}. This is shown schematically in Fig. 7c.1.

In this case the working substance contained in the cylinder below the piston is a 2.00 cm^3 volume of air. The density of air ρ is 1.23 kg m^{-3} and its heat capacity at constant pressure c_p is 1.00 kJ kg^{-1} K^{-1}. The air in the piston is caused to expand by heat being supplied to it which, for the purposes of this example, raises its temperature from 293 to 493 K. The thermal coefficient of expansion β for air is 3.43×10^{-3} K^{-1}. For the purposes of this illustration we assume there to be no friction between the piston and the wall of the cylinder, that the seal between the piston and the cylinder wall is perfect and that no heat is conducted through the cylinder wall or the piston by conduction. Let us determine the distance that such an expansion on such a piston might bring about, and compare the relative scales of the energy required to raise the temperature of the gas compared to that needed to do work.

To calculate the heat, ΔQ, supplied to raise the temperature of the air:

$$\Delta Q = mc_p\Delta T$$

$m = \rho \times V = 1.23 \times 2 \times 10^{-6} = 2.46 \times 10^{-6}$ kg and $\Delta T = 200$ K. Hence,

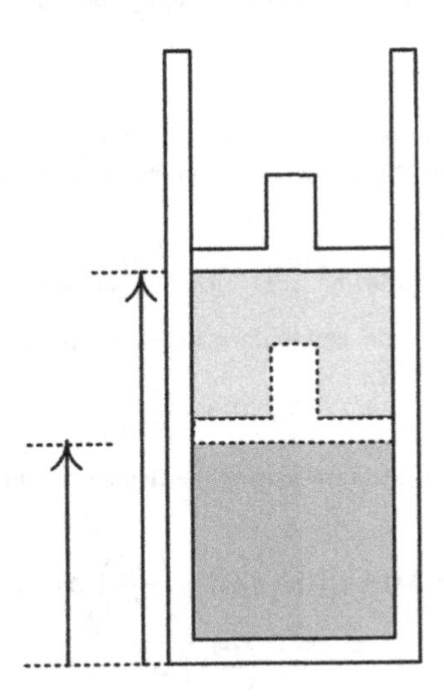

FIG. 7c.1

A schematic diagram of a piston operating in the vertical plane under gravity.

$$\Delta Q = 4.92 \times 10^{-4}\,\text{kJ}$$

To calculate the expansion ΔV brought about by this temperature increase we shall need to consider that at constant pressure the fractional increase in volume, $\Delta V/V$, is proportional to the temperature change, ΔT, where the constant of proportionality is the thermal coefficient of expansion, β,

$$\Delta V = V\beta\Delta T = 2 \times 10^{-6} \times 0.00343 \times 200 = 1.37 \times 10^{-6}\,\text{m}^3$$

This corresponds to a distance the piston has moved of 2.18 cm. So the work done W being the product of force and distance is,

$$W = 1 \times 9.81 \times 0.0217 = 2.14 \times 10^{-4}\,\text{kJ}$$

Thus the energy required to do work on the piston is ~2.3 times that necessary to raise the temperature of the gas, which illustrates the benefit of work being performed by the gas having been expanded by the application of heat.

CASE STUDY 7.2: THERMAL CONDUCTION EQUATION

To derive the thermal conduction equation we consider that the change per unit time (rate) in the internal energy, U, per unit volume of a substance is related to the specific heat capacity, c_v, and the density, ρ, of the substance as per,

$$\frac{dU}{dt} = \rho c_v \frac{dT}{dt}$$

The internal energy changes due to the production of heat in the substance, which tends to cause it to increase (in the current context due to fission) and due to heat loss which causes it to fall. In this case we shall adopt the convention that the rate of energy production per unit volume is denoted H and that this is uniformly distributed throughout the fuel. The quantity of heat, q, lost per unit time per unit volume is related to the heat flux, which is a vector quantity. Returning to Eq. (7.16) for one dimension (x), the heat flux is given by,

$$\frac{dq_x}{dt} = -k\frac{dT}{dx}$$

Therefore in three dimensions, writing $dq_x/dt = q'_x(x, y, z)$, and given that $\underline{\nabla} = \frac{\partial}{\partial x}\hat{\underline{i}} + \frac{\partial}{\partial y}\hat{\underline{j}} + \frac{\partial}{\partial z}\hat{\underline{k}}$,

$$\underline{q}' = -k\underline{\nabla}T$$

and $\underline{q}' = q'_x\hat{\underline{i}} + q'_y\hat{\underline{j}} + q'_z\hat{\underline{k}}$.

To determine the heat flow from a given volume we assume $\underline{q}' = \underline{q}'(x, y, z)$ varies slowly and thus can be approximated by a Taylor series, which for the benefit of recall enables a function, say $f(x)$ that is differentiable at $x = a$, to be approximated thus,

$$f(x) = f(a) + \frac{\dot{f}(a)}{1!}(x-a) + \frac{\ddot{f}(a)}{2!}(x-a)^2 + \cdots$$

Hence for the case of q'_x, by way of example and limiting to two terms around the origin, $a = 0$,

$$q'_x(x) = q'_x(0) + \left(\frac{\partial q'_x}{\partial x}\right)_0 x$$

So for the case depicted in Fig. 7c.2, the heat flow leaving the elemental volume in the x direction is the flux at $x + \Delta x$ minus the flux at x,

$$q'_x(x+\Delta x) - q'_x(x) = q'_x(x) + \left(\frac{\partial q'_x}{\partial x}\right)_x \Delta x - q'_x(x)$$

$$= \left(\frac{\partial q'_x}{\partial x}\right)_x \Delta x$$

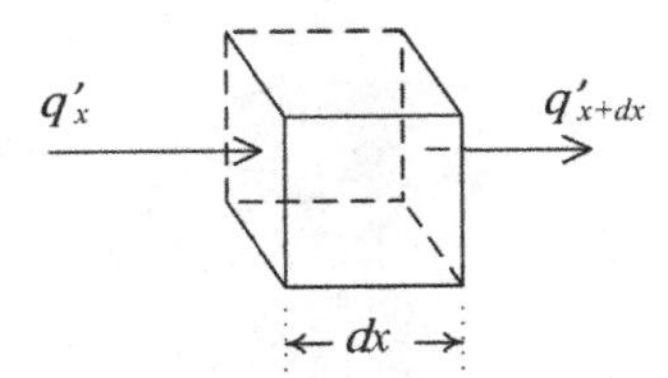

FIG. 7c.2

Heat flux through an element of thickness *dx*.

Generalising to three dimensions for a unit volume, then

$$\underline{q}' = \frac{\partial q'_x}{\partial x}\hat{\underline{i}} + \frac{\partial q'_y}{\partial y}\hat{\underline{j}} + \frac{\partial q'_z}{\partial z}\hat{\underline{k}} = \underline{\nabla} \cdot \underline{q}'$$

and,

$$\underline{q}' = -\underline{\nabla} \cdot k\underline{\nabla} T$$

To give the thermal conduction equation in terms of the change in internal energy, heat production and heat loss,

$$\rho c_v \frac{dT}{dt} = H + \underline{\nabla} \cdot k\underline{\nabla} T$$

CASE STUDY 7.3: RADIAL TEMPERATURE DISTRIBUTION IN A NUCLEAR FUEL ELEMENT

Often we are interested in the steady-state in a nuclear reactor so that the thermal conduction equation reduces to,

$$H = -\underline{\nabla} \cdot k\underline{\nabla} T$$

If we assume radial symmetry, that is, conduction only occurs radially then in cylindrical polar coordinates,

$$H = -\frac{k}{r}\frac{d}{dr}\left(r\frac{dT}{dr}\right)$$

which can be integrated, first with respect to r and then with respect to T,

$$\int_0^r d\left(r\frac{dT}{dr}\right) = -\int_0^r \frac{Hr}{k} dr$$

$$r\frac{dT}{dr} = -\frac{Hr^2}{2k} + c$$

At $r = 0$, $dT/dt = 0$ so $c = 0$ and hence,

$$\frac{dT}{dr} = -\frac{Hr}{2k}$$

Integrating between a radial point, r, in the fuel element and the radius of the element, a,

$$\int_r^a dT = \frac{H}{2k}\int_a^r r\, dr$$

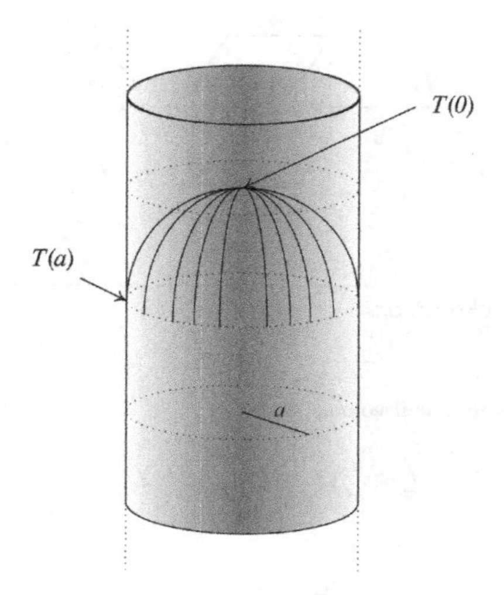

FIG. 7c.3

Heat flux through a section of a cylindrical fuel element of radius *a*.

$$T(a)-T(r)=\frac{H}{2k}(r^2-a^2)$$

$$T(r)=T(a)+\frac{H}{2k}(a^2-r^2)$$

Therefore, we have shown that the temperature, T, throughout a cylindrical fuel volume along which there is no conduction in the axial direction has a parabolic dependence with radius, r. It is offset by the temperature at the surface, $T(a)$. This is a simple approach to a complex problem as we have assumed that no conduction occurs in the axial direction and that the thermal conductivity is constant when, actually, it will have a significant dependence on temperature. However, it illustrates that if the rate at which heat produced in the fuel element, the thermal conductivity, temperature at the surface and radius of the fuel are known, then a variety of related parameters such as the temperature at the centre of the fuel and the maximum heating rate per unit radius (Fig. 7c.3) can be determined.

REVISION GUIDE

On completion of this chapter you should:

- understand that the function of nuclear power reactors is constrained fundamentally by the *laws of thermodynamics* and is based on the elementary principles of *heat engines*
- appreciate the theoretical limits and operation of the *Carnot* and *Rankine cycles*
- be able to summarise the ideal requirements of the *working substance* in a heat engine
- be aware of the relevant practical options that are open for use as working substances in the primary, secondary and where relevant tertiary cycles in nuclear power reactors
- appreciate the requirements for *exotic coolants*, be aware of relevant examples and the background to their suitability

- be aware of the conceptual operation of *steam turbines*
- understand the fundamentals of *thermal hydraulics*, particularly the background to the requirement for *thermal hydraulic analysis, boiling heat transfer, flow* and their relevance to accident prevention and safety analysis
- appreciate the relevance of *thermal conductivity* and the *heat transfer coefficient* in the design of reactor systems

PROBLEMS

1. For a system comprising a sink (outlet) temperature consistent with the ambient, ~293 K, plot a graph of the Carnot efficiency for source temperatures from 293 K through to 893 K. Hence or otherwise, assuming the working substance in this case is water, determine the maximum Carnot efficiency possible for nonsupercritical systems.
2. If the ratio of expansion in a heat engine is 3.90 and the working fluid in this case is a perfect gas, what is the pressure before expansion if the final pressure is 2.00×10^5 Pa? For the case that the working fluid is not a perfect fluid, but follows $PV^{1.15} =$ constant, what is the effect on the pressure before the expansion stage of the cycle?
3. Estimate the void fraction, α, for a homogeneous two-phase (gas-liquid) flow path in which the area occupied by vapour is 95% of the total. Hence or otherwise, assuming a density of liquid water of 1.00 g cm^{-3} and the density of the vapour is 6.00×10^{-4} g cm^{-3}, estimate the static quality χ_{st}. How would you describe this state in terms of the % mass of vapour?
4. From what you have read and performing your own research, describe the phase state in terms of flow patterns, static and flow quality of the coolant in a reactor core at full power that is subject to a rapid and significant loss of coolant, but that is within design basis.
5. Given that the specific heat capacity of uranium dioxide at 600 K is 0.30 J g^{-1} K^{-1}, and the heat transfer coefficient decreases from 0.975 to 0.600 kW m^{-2} K^{-1} as a result of dryout, calculate the temperature increase in a fuel pellet of diameter and length 1.00 cm in 1.00 s. Explain any assumptions that you make, given the following data:
   Inlet temperature: 286°C, clad temperature: 345°C, density of UO_2 = 10.97 g cm^{-3}.

REFERENCES

[1] S. Untermyer, Steam forming neutronic reactor and method of operating it, filed: 28 June 1955, US 2936273 A.

[2] P.Z. Rosta, Specific Heat and Enthalpy of Liquid Heavy Water, AECL-3689, 1971.

[3] See for example:Yu. Plevachuk, et al., Some physical data of the near eutectic liquid lead-bismuth, J. Nucl. Mater. 373 (2008) 335–342.

[4] See for examples: W.J. O'Donnell et al., The Thermophysical and Transport Properties of Eutectic NaK Near Room Temperature, ANL/FPP/TM-237, 1989.

[5] G. Meyer, E. Stokke, Data for Sizewell B, in: Description of Sizewell B Nuclear Power Plant, NKS/RAK-2 (97)TR-C4, 1997, http://www.nks.org/en/nks_reports/view_document.htm?id=111010111119605.

[6] E. Nonbøl, Data for Torness, in: Description of the Advanced Gas-Cooled Type of Reactor (AGR), NKS/RAK-2(96)TR-C2, 1996, http://www.nks.org/en/nks_reports/view_document.htm?id=111010111119559.

[7] G.A. Filippov, R.G. Bogoyavlenskii, N.N. Ponomarev-Stepnoi, A.O. Gol'tsev, Modular high-temperature helium-cooled nuclear reactor with spherical fuel elements for electricity and hydrogen production, At. Energy 96 (3) (2004) 152–158.
[8] M.-S. Yim, F. Caron, Life cycle and management of carbon-14 from nuclear power generation, Prog. Nucl. Energy 48 (2006) 2–36.
[9] H.G. Lyall, A comparison of helium and CO_2 as reactor coolants, J. Nucl. Energy 26 (1972) 49–60.
[10] H. Petersen, The properties of helium: density, specific heats, viscosity and thermal conductivity at pressures from 1 to 100 bar and from room temperature to about 1800 K, Risø Report No. 224, Danish Atomic Energy Commission, 1970.
[11] Investigation of chemical characteristics of primary helium gas coolant of HTTR (high temperature engineering test reactor), Nucl. Eng. Des. 271 (2014) 487–491.
[12] F.A. Kozlov, V.N. Ivanenko, Sodium – coolant for nuclear plants with fast reactors, At. Energy 80 (5) (1996) 318–324.
[13] G.H. Balmer, Fire safety considerations related to the large scale usage of sodium, IChemE Symp. Ser. 33 (1972) 79–89.
[14] Nuclear Energy Agency, Handbook on Lead-Bismuth Eutectic Alloy and Lead Properties, Materials Compatibility, Thermal Hydraulics and Technologies, NEA6195, https://www.oecd-nea.org/science/reports/2007/nea6195-handbook.html, 2007.
[15] Y. Sakamoto, J.-C. Garnier, J. Rouault, C. Grandy, T. Fanning, R. Hill, Y. Chikazawa, S. Kotake, Selection of sodium coolent for fast reactors in the US, France and Japan, Nucl. Eng. Des. 254 (2013) 194–217.
[16] E.S. Lea, E. Meden, The De Laval steam-turbine, Proc. Inst. Mech. Eng. 67 (1) (1904) 697–714.
[17] See for example the recommended correlations given in L.S. Tong, J. Weisman, Thermal Analysis of Pressurised Water Reactors, American Nuclear Society, ISBN: 0-89448-038-3, 1996.

CHAPTER

ELEMENTARY REACTOR PRINCIPLES

8

8.1 SUMMARY OF CHAPTER AND LEARNING OBJECTIVES

The objectives of this chapter are to:

- introduce the range of time domains over which *reactor control* is required
- introduce the concept of *dynamic neutron populations* and how they influence reactor systems
- define the *multiplication factor*, both in terms of infinite and finite reactors and across the various equivalent definitions that exist, and to introduce the *neutron lifecycle*
- explain the basis for the *four-* and *six-factor formulae*
- introduce the context of *reactor kinetics* and to define *prompt-* and *delayed neutron effects*
- describe the variety of *short-term effects* and *reactor feedback mechanisms*
- define what constitutes a *reactor poison* and discuss the various effects that arise during *reactor start-up*, *steady-state operation*, *post-shutdown* and *restart*

8.2 HISTORICAL CONTEXT: HYMAN GEORGE RICKOVER, 1900–86

Hyman George Rickover led the development of the compact pressurised water reactor that was used, initially, in nuclear-powered ships and submarines in the United States. He controlled the US naval nuclear propulsion programme for 30 years (including the construction of over 200 nuclear submarines) and was also instrumental in the development of the Shippingport nuclear power station; this was the first reactor in the US to generate electricity on a commercial basis. In recognition of his achievements he was made a four-star Admiral in 1973 and was the longest-serving officer in the history of the US Navy (Fig. 8.1).

Rickover was born in Poland and his family emigrated to the US in 1906 when he was still a child. He became instrumental in the development of a nuclear reactor for marine propulsion. Such a reactor needed to be compact and be able to produce steam (a first then with nuclear energy). This required components made from materials such as zirconium and hafnium via techniques that were

Nuclear Engineering. https://doi.org/10.1016/B978-0-08-100962-8.00008-1

FIG. 8.1

Admiral Hyman G. Rickover.

new at that time. Rickover was renowned for his tenacity, his capacity for hard work and his stringent requirements for the highest standards of safety, quality and design integrity. These themes remain important elements of the philosophy at the heart of nuclear engineering today.

8.3 INTRODUCTION

In Chapter 2 the development of the world's first nuclear reactor (CP-1) was described with which the first self-sustaining nuclear chain reaction was achieved. To sustain a nuclear chain reaction, control is essential because, while the energy available in a nuclear reactor is potentially enormous, it is only useful for the production of energy if derived gradually, at as constant a rate as possible and with respect for the materials and infrastructure around it. The objectives of reactor control are straightforward in principle: we wish to prevent the chain reaction from escalating beyond the performance limitations of the reactor but, conversely, also prevent it from dying away. Safety is the principal requirement of reactor control to ensure that the radioactive inventory of an operating reactor remains isolated from the biosphere [1]. Secondly, but significantly in an operational context, sustained operation as a result of careful control returns benefits in terms of capacity factor, the planning of maintenance outages and it also minimises the possibility of dynamic, unanticipated changes in plant operation that might otherwise exert undue strain on reactor components.

Since the first chain reaction was achieved at CP-1, a great deal has been learnt about the various response phenomena that arise as a result of external stimuli of a nuclear reactor. Some traits of the behaviour of reactors in this context can be counter-intuitive and interdependent. Meanwhile, the focus of control has developed significantly from merely sustaining the chain reaction to ensuring that a constant level of power output is achieved to bring about efficient and uniform burnup of the fuel. A discussion of these issues is the focus of this chapter.

8.4 THE DOMAINS OF CONTROL IN NUCLEAR REACTORS

In Chapter 7, a nuclear reactor was considered in simple terms as being a like-for-like substitute for a conventional source of heat in an external heat engine. In that comparison it was suggested that the primary coolant in a reactor might correspond loosely to the flue gas in a conventional, fossil-fuelled power plant. Similar priorities exist across the control requirements of both types of heat source in terms of steam quality, feedwater supply (the supply of water to the secondary circuit) and that of maintaining the integrity of the coolant circuit as a whole. However, there are some very different and generally more complex control issues associated with the operation of the nuclear reactor, and some nonlinearity of control response that is specific to this type of power plant. These can be considered across the domains of *time* and *space*, as follows:

1. *Time*
   a. *Long-term control*: A critical distinction of the operation of nuclear reactors over most other thermal sources of energy is that the fuel is resident for prolonged periods of operation lasting months, years and sometimes decades. Frequent refuelling is not desirable (as this introduces significant interruptions in electricity generation), especially since relatively few reactors in the world now have or exploit the facility to refuel while operating. Consequently, a means for the *long-term control* of a reactor, i.e. weeks and months, is essential to ensure a safe, predictable, uninterrupted and unwavering production of power along with efficient fuel consumption. Similarly, all of the waste for a given period of operation (that comprises fission products and minor actinides) is contained within the fuel elements while operating, and is not accessed until the fuel is removed. This contrasts with the continuous fuel-in/waste-out basis on which most other conventional power plants operate. Some fission products and minor actinides contribute positively to the performance of the reactor, while others contribute negatively; in particular, some introduce neutrons to the system while others can be strong neutron absorbers, respectively. The abundance of these isotopes varies with time and power, and can thus be dependent on the recent operational profile of the reactor fuel. Consequently, the influence of these isotopes on the control of the reactor needs to be understood, managed and accounted for. Some isotopes are formed that are fissile and make a contribution in-situ while others continue to make a contribution to the power output of the core long after the chain reaction is shut down (of the order of days and weeks) and potentially when restarting commences. This also has to be catered for, both under normal operation and in the event of emergencies.
   b. *Short-term control*: A nuclear reactor operating in steady-state is a system in dynamic equilibrium; neutrons are introduced and absorbed on a continuous basis, notwithstanding the long-term influences summarised above. The control objective during steady-state operation is to maintain a constant overall population to achieve a constant level of power production and to avoid unforeseen, short-term *transient*[1] changes in reactivity. While these changes do not implicate the integrity of the reactor as a whole (a reactor is designed so that their effects are catered for within anticipated ranges of performance termed the *design basis*), there is an important requirement for short-term control mechanisms, of the order of seconds and minutes, that exploit negative feedback mechanisms to alleviate the influence of transients where

[1] More detail concerning the definition of reactor transients follows in Chapter 14.

possible. This removes the requirement, for example, for excessive operator intervention and the unnecessary actuation of installed systems. From the perspective of safety, a variety of *protective* control measures is also needed with which to shut down the reactor quickly if needed and to maintain it in a safe state following shut down. As per the discussion in Chapter 4, neutron cross sections can be highly dependent on neutron energy. In this chapter the neutron population is explored further in terms of its influence on the behaviour of neutrons and their lifecycle in a nuclear core. In a reactor system where there can be a wide spectrum of neutron energies, together with what is often complex, heterogeneous structure, a wide range of symbiotic feedback mechanisms can be at work. Some of these aid the control of a reactor while others impede it.

2. *Space*: The conceptual operation of a reactor system is most easily understood in the context of there being no boundaries to the volume in which neutrons are spawned and consumed; this is termed an *infinite* reactor. However, a real reactor has to have stringent physical limitations, not only to contain it but also to integrate the necessary structural infrastructure and instrumentation essential for its operation. Consequently, to ensure a safe, efficient and balanced use of fuel across a reactor core, it is necessary to control the neutron flux in a *spatial context* as well in time. Usually the objective is to optimise the neutron flux within the confines of the core and to avoid unnecessary peaking that might lead to localised heating and stress on reactor components. All commercial power reactors operated to date use nuclear fuel in the solid phase comprising a heterogeneous spatial arrangement that remains stationary relative to its surrounding infrastructure. This is pervaded by either liquid or gaseous coolant either in a vessel or via individual tubes (liquid-fuelled, molten salt reactors are an exception summarised in Chapter 11). This places significant emphasis on the nature of the static, spatial distribution of fuel in the core and the optimum neutron flux distribution. Some effects, most notably the influence of significant neutron absorbers in the fuel, can bring about a time dependence of the spatial distribution of neutron flux; this also needs to be controlled and managed.

8.5 POPULATION DYNAMICS AND CHANGING NEUTRON POPULATIONS

8.5.1 THE MULTIPLICATION FACTOR

A population of neutrons in a reactor is not unlike any system comprising a large number of entities that have the facility to reproduce or decline; the number in the current generation being influenced by events in the preceding generations. For example, an analogy might be considered with the human population on Earth where the only *source* of the population is by birth and the only *sink* is by death. In this example the rate of change of the population with time is related to the difference between the rate of *production* (the birth rate) and the rate of *removal* (the death rate),

$$\text{rate of change of population} = \text{birth rate} - \text{death rate} \tag{8.1}$$

Clearly, if the birth rate is greater than the death rate, then the population will grow and thus the rate of growth is positive, and *vice versa*. For the case of the steady-state scenario the population does not change, by definition. Thus the rate of change is zero and

$$\text{birth rate} = \text{death rate} \tag{8.2}$$

If we compare the size of the population in the current generation (by which we infer a specific time) with that in the previous generation, then we can characterise the rate of change of the population by a *multiplication factor* such that

$$\text{Multiplication factor} = \frac{\text{Population in current generation}}{\text{Population in previous generation}} \tag{8.3}$$

By way of example, consider the population of the United Kingdom; this was 63.2 million[2] in 2011 and 58.8 million[3] in 2001. Hence, the multiplication factor in this case is 1.08. It is greater than unity reflecting the fact that the population in the UK in 2011 increased relative to 2001. In nuclear engineering the conventional notation for the multiplication factor is k. Typically, there are three different scenarios to consider

$$k = 1 \tag{8.4}$$

describes the situation in which the neutron population is not changing. This is analogous to the situation in which the birth rate and death rate are equal, as per Eq. (8.2). In this case the state of the neutron population is termed *critical*. Alternatively, if,

$$k > 1 \tag{8.5}$$

then the neutron population is growing and the state of the reactor is termed *supercritical*, whereas for

$$k < 1 \tag{8.6}$$

the neutron population is falling and the state is termed *subcritical*.

In a nuclear reactor neutrons can be introduced from a variety of sources. These will be explored later in this chapter. However, once a chain reaction is established the prominent source of neutrons is the stimulated fission of a fissile isotope; most often this being ^{235}U. Neutrons are lost from a reactor via absorption (assuming the reactor in question is infinite so that conceptually neutrons cannot be lost by leakage). This process is depicted schematically in Fig. 8.2. For the case of neutrons in a reactor, Eq. (8.7) can be written by analogy to Eq. (8.1),

$$\text{rate of change of neutron population} = \text{neutron production rate} - \text{neutron absorption rate} \tag{8.7}$$

If the neutron population does not change with time, then as for Eq. (8.2),

$$\text{neutron production rate} = \text{neutron absorption rate} \tag{8.8}$$

For this case, and in reference to the discussion in Chapter 4, the fission rate per unit volume in the current generation is equal to the product of the neutron flux ϕ and the macroscopic fission cross section Σ_f. The number of neutrons in the current generation is proportional to the neutron production rate per unit volume. This is the product of the number of neutrons produced per fission ν and the fission rate,

$$\text{Neutron production rate} = \nu \Sigma_f \phi \tag{8.9}$$

In an infinite medium where, by definition, leakage cannot occur, the number of neutrons in the previous generation must be equal to the number of neutrons that have been absorbed. The absorption rate is thus,

[2] https://www.ons.gov.uk/peoplepopulationandcommunity/populationandmigration/populationestimates/.

[3] http://www.ons.gov.uk/ons/rel/census/census-2001-first-results-on-population-in-england-and-wales/.

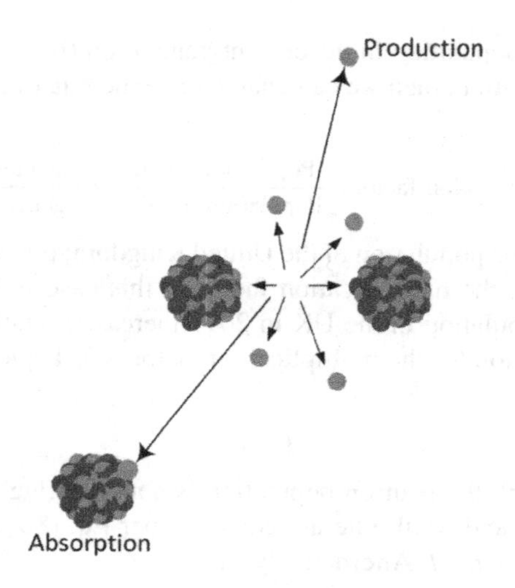

FIG. 8.2

A schematic diagram of a fissioning nucleus leaving behind two fission fragments, six neutrons and showing the destiny of two: one neutron being absorbed on a neighbouring nucleus and one resulting in production via subsequent fission. Other radiations emitted in fission have been neglected for clarity.

$$\text{Neutron absorption rate} = \Sigma_a \phi \tag{8.10}$$

By defining the multiplication factor for an infinite reactor, k_∞, as the ratio of 8.9 to 8.10 yields,

$$k_\infty = \frac{\nu \Sigma_f}{\Sigma_a} \tag{8.11}$$

There are two important and yet a little unrealistic assumptions implicit to the above analysis. First, a uniform composition has been assumed in the sense that the macroscopic cross sections and the neutron flux do not change with position, in what is an infinite, hypothetical reactor medium. Second, these parameters also vary with energy but it is assumed at this stage that they are invariant in this respect also, such that $\Sigma(E, r) = \Sigma$ and $\phi(E, r) = \phi$.

As discussed in Chapter 6, the majority of neutrons in an operating reactor arise from stimulated fission. As a result, they are introduced 'fast', that is, with a distribution of energies that is elevated relative to that of the surroundings. They are absorbed with the greatest probability when slow, notwithstanding the influence of intermediate resonances in neutron absorption cross sections, particularly of ^{238}U, that were described in Chapter 4. However, it is important to remember that the slowing-down process is almost always comprised of many individual interactions. Consequently, absorption can take place at any energy with a probability varying as reflected by the corresponding cross section dependence with energy. There are thus a variety of potential pathways that an individual neutron can take which can be considered in terms of the *neutron cycle* within a reactor.

8.5.2 THE NEUTRON CYCLE

An ideal approach to determine the reproduction state of a given neutron population might be to follow every neutron from its birth in fission, through to its absorption at a lower energy state at some time later, folding in the relevant neutron cross sections at the corresponding energies throughout. It would be necessary to do this many times to account for the probabilistic nature of neutron interactions and the variety of isotopes that they might interact with. Their *transport* in doing this is dependent on the geometry and composition of the environment in which they interact.

Something analogous to this is done in the Monte Carlo method. This approach has the benefit of associating the energy-dependent influence on the trajectory of a neutron, in this context from fission through to absorption, that is indicative of the changes that occur with changes in energy. However, to illustrate the concept of the neutron cycle a widespread approach is to first consider just *three* domains in neutron energy: *fast*, *resonance* and *thermal*.

In Fig. 8.3 a schematic representation of the life cycle of a neutron, from fission through to absorption, is provided. As a cycle we can consider a neutron within it from any point but it is conventional to start with the fission event. The cycle is also summarised in Table 8.1.

Taken together, the definition of the multiplication factor k_∞ given in Eq. (8.11) can be revised to account for these effects explicitly to give what is termed the *four-factor formula*, such that

$$k_\infty = \varepsilon p f \eta \tag{8.12}$$

where it is emphasised, for the avoidance of doubt, that the scenario in which neutrons are lost from the reactor by leakage has been ignored thus far. Hence the definition here of k_∞ ('k infinity') reflecting its association with the infinite reactor concept.

It is instructive to consider the factors defined above in more detail in terms of their equivalent cross section relationships:

- The macroscopic cross section for *absorption* occurring somewhere in the reactor core (e.g. coolant, moderator, infrastructure, fuel, etc.) can be defined as Σ_a. For the fuel alone it is usually distinguished as Σ_{aF}. The thermal utilisation factor, f, is equal to the ratio Σ_{aF}/Σ_a.
- Similarly, defining the macroscopic fission cross section for *fission* and *absorption* in the fuel as Σ_{fF} and Σ_{aF}, respectively, η follows as the ratio of the product of the average number of neutrons produced in fission ν and the macroscopic fission cross section, to that of absorption: $\upsilon\Sigma_{fF}/\Sigma_{aF}$. Note

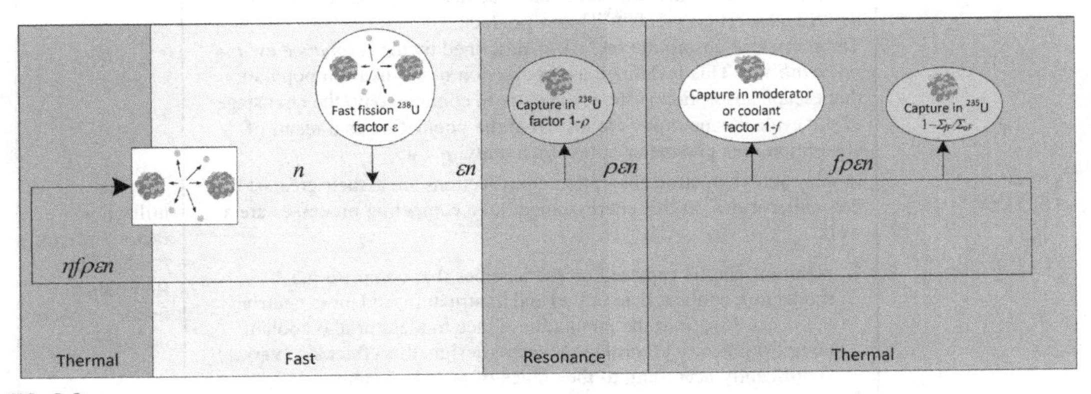

FIG. 8.3

A schematic representation of the neutron life cycle in a reactor across fast, resonance and thermal energy ranges. The symbols used in this figure are defined in Table 8.1.

Table 8.1 A Summary of the Various Effects on Neutron Population for an Infinite Reactor System in Terms of Neutron Energy Used to Define the Factors of Fermi's Four-Factor Formula (Eq. 8.12)

Neutron Energy Domain	Description of the Effect on Neutron Population	Associated Factors in the Multiplication Factor k_∞
Fast $E_n > 0.5\text{MeV}$	Neutrons arise from fission with the majority of them having fast energies, as per the Watt distribution depicted in Fig. 4.11. Relative to the interaction possibilities at lower energies, the interaction with the highest probability when a neutron is fast is scattering, notwithstanding the influence of absorption resonances for a small number of specific isotopes. This results in moderation and progression further along in terms of energy in the cycle as the neutrons from fission are slowed. Scattering interactions neither introduce more neutrons to the system nor cause them to be lost and thus do not affect the neutron population. However, *above-threshold* fission (i.e. for $E_n > 1$ MeV) in ^{238}U is possible in this energy domain. This introduces 2–3 neutrons per fission (increasing with energy [2]). The positive contribution that this makes to the multiplication factor is described by the *fast fission factor* ε; this is the ratio of the total number of neutrons arising from fission in the cycle to the number arising from thermal fission events. Thus, given n neutrons are introduced by the primary fission event(s), at the end of the fast region of the cycle the number is εn. $\varepsilon > 1$ reflecting the positive effect of this contribution.	Fast fission factor ε
Resonance E_n: $1\text{eV} \rightarrow 300\text{eV}$	In Chapter 4 some isotopes, especially the actinides, were described that have a number of resonances in their neutron absorption cross sections. As neutron energy is reduced by moderation, absorption associated with these becomes significant. This constitutes a sink by which neutrons are removed from the total population that is available to cause fission and is especially relevant for ^{238}U. However, since the neutron energy in this range is now below the threshold for fission, this mechanism does not result in fission (which would increase the neutron population). Rather, neutron absorption yields ^{239}U as a product. The resonance absorption on ^{238}U is described by the *resonance escape probability* ρ. This is defined as the fraction of the neutron population that escape absorption in the resonances to continue onto the next stage of the cycle. As neutrons are lost from the population as a result of absorption, this parameter is less than unity, $\rho < 1$.	Resonance escape probability ρ
Thermal ~0.0253 eV	In this context, neutron absorption cross sections are at their greatest for thermal energies. In this energy range three competing processes are at work: 1. Neutrons can be captured on the isotopes that comprise the moderator, coolant, reactor fuel and its structure, and these neutrons are hence lost from the population. Since moderator and coolant composition vary according to reactor design, this effect also varies significantly according to the choice of reactor system. 2. Neutrons are absorbed by ^{235}U but this does not result in fission.	Thermal utilisation factor f where, $f = \Sigma_{aF}/\Sigma_a$ Reproduction factor[a] η where, $\eta = \nu\left(\Sigma_{fF}/\Sigma_{aF}\right)$

Table 8.1 A Summary of the Various Effects on Neutron Population for an Infinite Reactor System in Terms of Neutron Energy Used to Define the Factors of Fermi's Four-Factor Formula (Eq. 8.12)—cont'd

Neutron Energy Domain	Description of the Effect on Neutron Population	Associated Factors in the Multiplication Factor k_∞
	3. Neutrons are absorbed and cause fission in ^{235}U. The effects of these phenomena on the neutron population are reflected by two factors: • The *thermal utilisation factor f* is defined as the ratio of the number of neutrons absorbed by ^{235}U to those absorbed by the reactor system in total; $f < 1$. • The number resulting in fission is described by the *reproduction factor* η, which is the ratio of the number of neutrons yielded by fission to the total number of neutrons absorbed by ^{235}U.	

For the purposes of clarity it has been assumed that the fuel isotope is ^{235}U.
[a]η *is referred to universally as 'eta' in both verbal and written exchanges. M. Ragheb refers to it as the* regeneration factor *whilst the Department of Energy Nuclear Physics and Reactor Handbook (DoE, 1993) refers to it as the* reproduction factor.

that macroscopic cross sections are used to reflect the reality of fission being possible in a variety of fissile isotopes in a given fuel mixture.

Specific values for each of the factors introduced above vary dependent on reactor design. In general, η has the most significant influence being of the order of 60%–70%. This is because it describes the injection of neutrons from the thermal fission event and this is usually the strongest influence on neutron population in steady-state operation. The thermal utilisation factor f often has the next most significant influence, of the order of 30%. It combines the effects of two main influences: first, because power reactors tend to be associated with low-enriched fuels, there is a large proportion of ^{238}U present in unavoidable proximity with the diffusing neutrons that constitutes a sink for neutrons. Second, coolant, moderator and control infrastructure is essential to the function of the plant but it is not immune to the absorption of neutrons. The fast fission factor, ε, is generally small (of the order of 2%) but, by contrast, the abundance of ^{238}U and the size of the absorption resonances can yield a resonance escape probability ρ of the order of 85%–90%.

As emphasised above, thus far we have considered the hypothetical case of an infinite reactor because this enables neutron loss by leakage from the system to be ignored. However, real reactor systems are not infinite and moreover there are practical engineering requirements that favour some core shapes over others. This means we do not have complete design flexibility with which to minimise neutron leakage by choice of reactor geometry. For example, the best choice in terms of space utilisation would be a sphere (having a minimum ratio of surface area to volume) but this is rarely a practical option because of the need for infrastructure to assist in fuel loading and retrieval, and control rod insertion favours axial symmetry. As a result, most reactor cores are cylindrical because it is easier to arrange for the penetration of control/shutdown rods and associated machinery. In

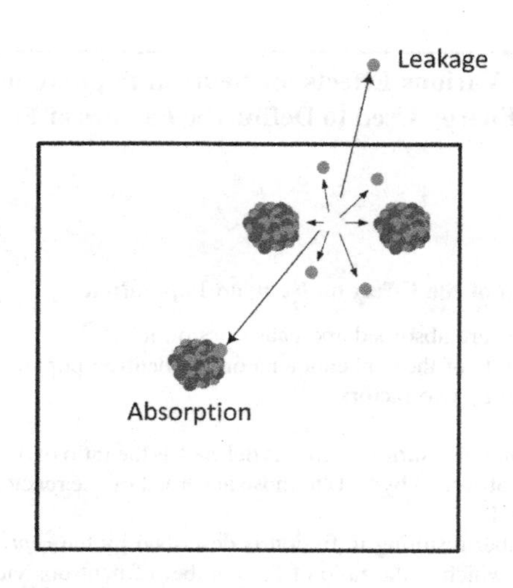

FIG. 8.4

A schematic diagram of a fissioning nucleus leaving behind two fission fragments and six neutrons. One neutron is shown being absorbed on a neighbouring nucleus and another that is leaking from the system.

considering leakage, Fig. 8.2 can be modified to illustrate the concept as shown in Fig. 8.4 where, in addition to absorption, the additional sink for neutrons leaving the finite boundaries of the reactor is included.

Reactors are designed to minimise leakage through the use of materials that reflect neutrons back into the core, and also by configuring the core to include fuel elements of greater enrichment towards its extremities to encourage greater absorption in the fissile isotopes in these regions. Nonetheless some neutrons do leave the core and do not find their way back; they are termed to have *leaked*. Also, it is plausible for neutrons to leak either while they are fast (i.e. prior to moderation) or afterwards when they are slow. The former is dependent on the time taken for a neutron to be thermalised (τ_{th}, a parameter known as the *age to thermal*) whereas the latter is dependent on the distance the neutron diffuses before it is absorbed (L_{th}, the *diffusion length*). The greater τ_{th} or L_{th}, the greater the leakage of fast or thermalised neutrons, respectively.

To fold these two scenarios into our description for the multiplication factor (Eq. 8.12), two further terms are introduced: the *fast nonleakage probability*, P_{NLf}, which describes the likelihood that a fast neutron does not leak. P_{NLth} is the corresponding factor for thermalised neutrons. As for the other factors, the nonleakage factors vary with reactor design each being of the order of a few percent. Taken together it is possible to define the *effective multiplication factor*,[4] k_{eff}, by combining what is known as Fermi's four-factor formula with the leakage probabilities, to give the *six-factor formula* as per,

$$k_{eff} = \varepsilon p f \eta P_{NLf} P_{NLth} \tag{8.13}$$

or

$$k_{eff} = k_{\infty} P_{NLf} P_{NLth}.$$

[4]Widely referred to as 'k effective'.

8.5.3 THE SIGNIFICANCE OF DELAYED NEUTRONS

In Chapter 4 the qualitative dynamics of nuclear fission were introduced and, in particular, the classification by which neutrons are emitted in fission being termed *prompt* was defined. The majority of these neutrons are emitted extremely rapidly by the highly-excited fission fragments as they accelerate away from the point of fission, with the rest suspected to arise at the point of scission. The prompt neutrons arise so rapidly that the delay between the time at which scission occurs and their emission can barely be discerned, ranging from 10^{-18} to 10^{-13} s.

While the majority of neutrons emitted in fission are prompt (usually accounting for > 99% of those emitted), the prospect exists for neutrons to be emitted from the secondary fission products, following the first β^- decay of the fission fragments. This intervening class of decay processes are not long-lived, relative to the half-lives of some fission fragments, but on the temporal scale of fission they do introduce an important source of hysteresis. Consequently, if a reactor is operated on the basis that the neutron population necessary to sustain criticality comprises a combination of both the prompt proportion *and* the small fraction that is delayed, a degree of control is achieved that is not possible if we rely on prompt neutron population alone. This important component of the neutron population in a reactor is comprised of what are termed universally as *delayed neutrons*.

By way of conceptual illustration we can consider a reactor that sustains a chain reaction on the basis of prompt neutrons alone—it is *prompt critical*—and compare it with one that is reliant on the delayed neutron component with which to perfect the population necessary to reach criticality. The former is almost a theoretical scenario because, as we will discuss shortly, it constitutes an uncontrollable case only conceived accidentally in small number of serious criticality excursions (e.g. Chernobyl in 1986 and SL-1 in 1961) and in early criticality experiments.

To illustrate the *qualitative* benefit of delayed neutrons in controlling a reactor system, consider the two scenarios shown in Fig. 8.5 in terms of the k_{eff} scale shown on the left. In these diagrams the dotted, horizontal line corresponds to the critical state where the neutron population is in equilibrium; above it corresponds to supercriticality and below it to subcriticality. The case on the left depicts how the entire neutron population is prompt and sufficient to reach the critical state corresponding to $k_{eff} = 1$, whereas on the right the delayed neutron fraction is necessary to make up the necessary population. It is the

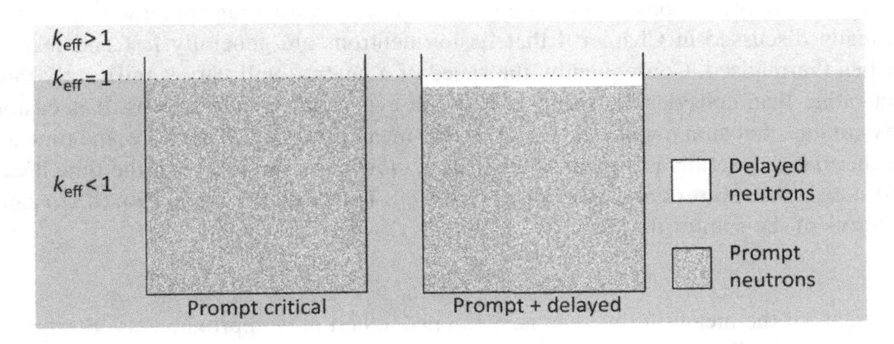

FIG. 8.5

A schematic representation illustrating a comparison of the critical reactor state comprising only prompt neutrons (the prompt-critical state) and that comprising both prompt and delayed neutrons.

reliance on this minor component that renders the system on the right controllable in comparison with that on the left. The delayed fraction is only able to change slowly, relative to the prompt majority, because of the intervening β^- decays. This constrains the reactor response in terms of the neutron population to changes in the fission rate to this limiting factor in terms of response time. For simplicity, the delayed fraction on the left is not indicated but there will always be a contribution whether a core is prompt critical or not. The important feature of the prompt critical case on the left is that the prompt component is sufficient to reach $k_{eff}=1$ independent of the delayed fraction, whereas that on the right is not.

To illustrate the role of delayed neutrons in a *quantitative* context, it is necessary to consider the temporal characteristics of neutrons from production (fission) through to consumption (absorption). Since we are interested in a relationship with time, we consider first the rate of change of the neutron population N with time t,

$$\frac{dN}{dt}=\frac{dP}{dt}-\frac{dL}{dt} \tag{8.14}$$

where dP/dt and dL/dt are the rate of production and loss, respectively. A dimensionless parameter describing the difference between the multiplication factor k and the critical state is often introduced that is known as the *reactivity, ρ,* where

$$\rho=\frac{k-1}{k} \tag{8.15}$$

the numerator $\Delta k=k-1$ is defined as the *excess multiplication* and ρ is measured typically in terms of one thousandths of a percent or *per cent mille* (pcm). If we think about the time a neutron exists, l, between being produced in a fission event and being absorbed by another nucleus, this can be related to the mean free path of the neutron λ (introduced in Chapter 4) and the average speed of the neutron v, thus,

$$l=\frac{\lambda}{v} \tag{8.16}$$

where l is known as the *prompt neutron lifetime*. This assumes implicitly that the medium is infinite such that absorption is the only sink for neutrons. Of course, some neutrons might be absorbed very quickly while others may take longer to succumb; the lifetime l is an average of these extremes. Further, we have already discussed in Chapter 4 that fission neutrons are generally fast and that most are absorbed when thermalised. Consequently, the speed of a neutron will vary significantly through its lifetime but rather than consider the speed in between every pair of interactions, it is customary to define two extremes that sum together to comprise the mean lifetime, l. These are: the time a neutron takes to be thermalised, t_s (this is known as the drift or *slowing-down time*) and the time it exists in a thermalised state before absorption (the *diffusion time, t_d*). The mean lifetime of Eq. (8.16) can then be defined in terms of the sum of these,

$$l=t_s+t_d. \tag{8.17}$$

Both components of the mean lifetime can be estimated, albeit only approximately. Starting with the slowing down time, in Chapter 6 we learned that on average in graphite (carbon) ~120 scattering interactions are necessary to slow a neutron down from 1 MeV to 0.0253 eV (thermal), with an average energy loss of ~14% at each interaction. We also know that the mean free path for graphite is of the

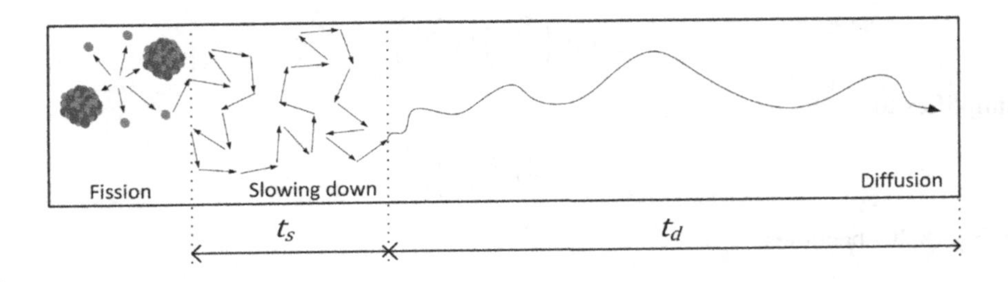

FIG. 8.6

A schematic representation of a neutron arising from fission, slowing down and diffusing until its point of absorption.

order of 5 cm, so we can estimate t_s to be of the order[5] of 0.01 ms (0.001 ms in light water). By contrast, following a similar calculation for absorption on ^{12}C, the diffusion time t_d can be estimated at approximately 1 ms (0.1 ms in light water).[6] Hence it can be concluded that neutrons diffuse for much longer than it takes for them to be slowed down, so $t_d \gg t_s$ and thus,

$$l \sim t_d \tag{8.18}$$

This is illustrated schematically in Fig. 8.6 where the lifetime of a neutron is depicted from its production by fission through slowing down and finally to the period of diffusion prior to absorption. In this illustration, which is not to scale, the horizontal range has been associated loosely with time.

The influence of delayed neutrons can be appreciated in terms of a simplified version of the one-energy group neutron balance equation, ignoring leakage. This is developed fully in Chapter 9 and is included here in terms of the thermal neutron flux, $\phi = Nv$, the macroscopic absorption cross section for the combination of moderator and fuel, Σ_a, and S which is the rate at which neutrons are supplied per unit volume. It represents the change of neutron population with time in terms of the rate neutrons are produced (S) and that at which they are absorbed, $\Sigma_a \phi$, thus,

$$\frac{1}{v}\frac{d\phi}{dt} = S - \Sigma_a \phi \tag{8.19}$$

and hence,

$$\frac{1}{v}\frac{d\phi}{dt} = \rho \varepsilon \nu \Sigma_f \phi - \Sigma_a \phi \tag{8.20}$$

where, as before, ε is the fast fission factor, v the average number of neutrons produced per thermal neutron-induced fission and ρ is the resonance escape probability; the product of all three factors corresponds to the number of neutrons per fission (both fast and thermal) that escape capture in ^{238}U resonances and are thermalised. This, multiplied by the fission rate $\Sigma_f \phi$, is equal to S. Referring to the definition of k_∞, as per Section 8.5.2, written in terms of the definitions of f and η (see Table 8.1) then

[5]This simple assessment is based on the assumption that the elastic scattering microscopic cross section for neutrons is constant between the energies of 1 MeV and 0.0253 eV as per Fig. 6.2.
[6]By way of comparison t_s for thermalisation by hydrogen in light water is ~0.01 ms.

$$k_\infty = p\varepsilon\left(\frac{\Sigma_{aF}}{\Sigma_a}\right)\left(\nu\frac{\Sigma_{fF}}{\Sigma_{aF}}\right) \tag{8.21}$$

which simplifies to

$$k_\infty = \varepsilon p \nu\left(\frac{\Sigma_{fF}}{\Sigma_a}\right) \tag{8.22}$$

and thus Eq. (8.20) becomes

$$\frac{1}{v}\frac{d\phi}{dt} = k_\infty \Sigma_a \phi - \Sigma_a \phi \tag{8.23}$$

Factorising and substituting for $v = \lambda/l$ and $\Sigma_a = 1/\lambda$ gives

$$\frac{d\phi}{dt} = \frac{(k_\infty - 1)}{l}\phi \tag{8.24}$$

Eq. (8.24) is prominent in nuclear engineering because it provides a conceptually simple route to illustrate the response sensitivity of nuclear reactors. Solving it gives the neutron flux at a time t in terms of the prompt neutron lifetime, l,

$$\phi(t) = \phi(0)e^{\frac{(k_\infty - 1)}{l}t} \tag{8.25}$$

where $T = l/(k_\infty - 1)$ is the time in which the neutron population increases by the factor e, often referred to as the *reactor period*. If the corresponding estimates for l for thermal neutron reactor designs are substituted into Eq. (8.25), and scenarios of very modest increases in reactivity are considered that neglect the delayed neutron contribution, uncontrollable increases in neutron flux and thus reactor power are evident.

The contribution made by the delayed neutrons acts to lengthen the doubling time (the period) of the neutron population. Delayed neutrons are contributed by a number of different isotopes; these are known as *delayed neutron precursors*. To consider their collective influence on the period it is necessary to consider the average time at which the neutrons are contributed in terms of the precursors grouped according to half-life.

A full analysis of all possible precursors is an involved exercise. Fortunately, a summary treatment of the most prominent contributors grouped in terms of their half-life provides a satisfactory conceptual assessment. It is worth re-emphasising that these isotopes are formally classified as secondary fission products having usually undergone prompt neutron emission, a first stage of β^- decay and are subsequently susceptible to further neutron emission; it is the intervening β^- decay that is the most significant source of the 'delay' between prompt and delayed neutron emission. The influence of the delayed neutrons is catered for by revising our definition of the neutron lifetime so that it is no longer limited to prompt neutrons only, but also includes that of those that are delayed. If the fraction that are delayed is β, then the neutron lifetime that includes these, $\bar{l}$, can be written,

$$\bar{l} = \underbrace{(1-\beta)l}_{\text{prompt component}} + \underbrace{\sum_{i=1}^{6}\beta_i(\tau_i + l)}_{\text{delayed component}} \tag{8.26}$$

Table 8.2 Delayed Neutron Precursor Groups for ^{235}U in Terms of Their Half-Life and Relative Yield

Group	Half-Life, $t_{1/2}$/s	Life Time τ_i/s	Relative Yield/β	Absolute Yield/Neutrons per Fission	Fraction of Fission Neutron Yield β_i
1	54.5	78.6	0.04	0.00068	0.00028
2	21.8	31.5	0.21	0.00357	0.00147
3	6.0	8.7	0.19	0.00323	0.00134
4	2.2	3.2	0.41	0.00697	0.00289
5	0.5	0.7	0.13	0.00221	0.00092
6	0.2	0.3	0.02	0.00034	0.00014
Totals	–	–	1.00	0.0170	0.00704

where the prompt component is equal to the total number of neutrons adjusted for the delayed fraction and the reactor period is $T=\bar{l}/(k_\infty-1)$. Eq. (8.26) depicts how each delayed neutron has a lifetime that is the sum of the prompt lifetime l plus the average lifetime of the corresponding group of precursors. Conventionally, six groups of precursors are assigned where, by definition,

$$\sum_{i=1}^{6}\beta_i=\beta \tag{8.27}$$

Unsurprisingly there is a different set of precursor groups associated with each different fissile isotope [3] but ^{235}U (and to a lesser extent ^{239}Pu due to its formation and consumption in thermal reactors) remains the most relevant to power reactors currently in operation. Example data for the precursor groups for ^{235}U are given in Table 8.2 where the total absolute delayed neutron yield is 0.017 neutrons per fission, or approximately 0.7%.

The delayed component in Eq. (8.26), adjusted as per Table 8.2, results in a mean neutron lifetime of $\bar{l}=0.09$ s. This results in a very significant change in reactor period relative to the prompt scenario, of a factor of $\sim$1000, rendering the system much more controllable as a result. Eq. (8.26) can be simplified further since the prompt component is very small in comparison with the delayed component and $\tau_i \gg l$, thus,

$$\bar{l}\sim\sum_{i=1}^{6}\beta_i\tau_i \tag{8.28}$$

8.6 SHORT-TERM EFFECTS AND REACTOR FEEDBACK MECHANISMS

8.6.1 ILLUSTRATIONS OF NEGATIVE AND POSITIVE FEEDBACK

In a nuclear power reactor the aim is always to ensure a safe, consistent and long-lived conversion of nuclear energy to heat. However, as a large system operated in a state of dynamic equilibrium and yielding significant quantities of heat and ionising radiation, a reactor undergoes both expected and

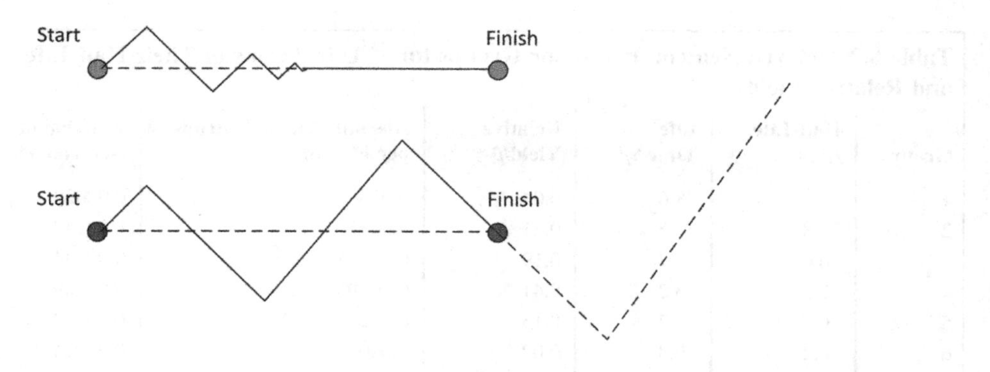

FIG. 8.7

Contrasting feedback mechanisms illustrated by the example of a person trying to follow a straight line. In the negative feedback example *(top)* the person subtracts 50% from the distance they walk every time the cross the line and change direction. The variance either side of the line falls to ~3% within five iterations. In the positive feedback example *(bottom)*, the person adds 50% each time. As a result the extent of the variance in this case increases rapidly.

occasionally unexpected physical changes throughout the time that it operates. Such changes include temperature, isotopic composition and geometry. Further, the phase of materials used in the reactor (most often coolants and moderators) can be affected on both a microscopic scale (e.g. in terms of the liquid-gas phase change associated with nucleation as discussed in Chapter 7) and also a macroscopic scale (i.e. boiling).[7] All reactors are designed and operated to ensure that such changes are small. However, controlling a reactor in response to such changes is simplified if the system's natural response is one of negative feedback, such that the action causing the change is reduced by the change. There are numerous factors associated with the behaviour of reactivity in a nuclear reactor and the feedback mechanisms associated with these are not always intuitive.

To illustrate the importance of feedback in control, a simple example is provided in Fig. 8.7 associated with a person trying to follow a straight line (an action usually accomplished subconsciously as part of the act of walking). In the negative feedback example, depicting what might be considered normal behaviour, the person changes direction every time they realise they are moving away from the line. To accomplish this, by way of illustration for the purposes of this example, they *subtract* 50% from the distance they walk when heading in a direction away from the line they are following. Consequently, the variance their path exhibits either side of the line falls rapidly to follow the line efficiently and accurately. By contrast in the positive feedback example, the opposite is illustrated in which the person *adds* 50% each time and the variance escalates rapidly. There are other useful illustrations of positive feedback in the physical world such as a resonant structure in an electronic system (an oscillator) or a structural system under resonant excitation (such as in the case of the Tacoma Narrows Bridge collapse in 1940).

[7]Fuels and cladding can be subject to changes in shape and materials characteristics but these are generally irreversible and much longer term than the feedback-related changes at the focus of this section.

8.6.2 SOME IDIOSYNCRASIES OF CRITICALITY CONTROL

Most fission power reactors comprise a heterogeneous array of nuclear fuel elements perfused with moderator, the latter being either a solid or a liquid. This enables a critical chain reaction to be achieved in fuel of relatively low enrichment. The ratio of the number of moderator atoms to the number of fuel atoms (the *moderator-to-fuel ratio*) is an important parameter in understanding of the response the reactor to physical changes in the moderator. Examples of this might include: if the density of the moderator is reduced gradually with time due to radiation damage, when the moderator expands with temperature or if the moderator changes phase. A common scenario associated with the last example is the liquid-to-vapour transition for a liquid-phase moderator. The general assumption for this section is that such changes are small. To illustrate this, the dependence of critical mass of a fissile isotope with dilution with water is often considered [4], recognising that the presence of more than one fissile isotope (i.e. ^{239}Pu in a uranium-fuelled core) renders the situation more complicated than will be considered here. This dependence, in terms of critical mass versus critical radius, is shown in Fig. 8.8 for both bare and reflected (by light water) arrangements. It is worthy of note that this is an extreme case in the context of nuclear energy, spanning the range in dilution from homogeneous metal through to relatively low dilutions, but it illustrates the scientific basis for some of the trends that are observed.

Implicit to our discussion of reactivity thus far has been that neither the phase nor the density of the various components of a reactor (such as the fuel, moderator and coolant) change quickly with time. In commercial power reactors currently in operation and under construction the fuel is used in the solid phase. However, the widespread use of liquid coolants and combined coolant/moderators, particularly in LWRs, requires that scenarios that might cause changes in density and influence the moderator-to-fuel ratio are considered.

8.6.2.1 Void feedback

Arguably the most significant changes in density of relevance to reactor moderators occur as a result of the formation of voids (bubbles) in liquids. These usually arise on the solid surfaces that constitute the interfaces between the fuel and the liquid moderator and/or coolant. This phenomenon is commonly known as *voiding* and encompasses the several specialised sub-classifications of boiling that were introduced in Chapter 7. Positive changes in reactivity can bring about increases in fuel temperature that can, in turn, be manifest by the rapid formation of voids.

To simplify the manual response needed to accommodate such developments the aim is to exploit reactor designs that respond to such a change with negative feedback, such that the voiding response acts to reduce the change in reactivity that caused it. However, the response to such changes is often complex, it can be counter-intuitive and often comprises several simultaneous, conflicting physical effects. This can be appreciated by considering the dependence of critical mass when diluted with light water, especially since light water is currently the most common medium for both coolant and moderator in use in nuclear reactors throughout the world.

In its undiluted state, the mass of fissile material at the smallest radius corresponds to the *undiluted critical mass*. Dilution influences the critical mass from this point. The reflector, acting to reflect leaked neutrons back into the mass and thus increasing the neutron population, results in a *reflected critical mass*. This is also always less than the bare equivalent (as depicted by the lower curve in Fig. 8.8). For masses smaller than the undiluted critical mass (smaller radii) neutron leakage is too great for fission to be self-sustaining.

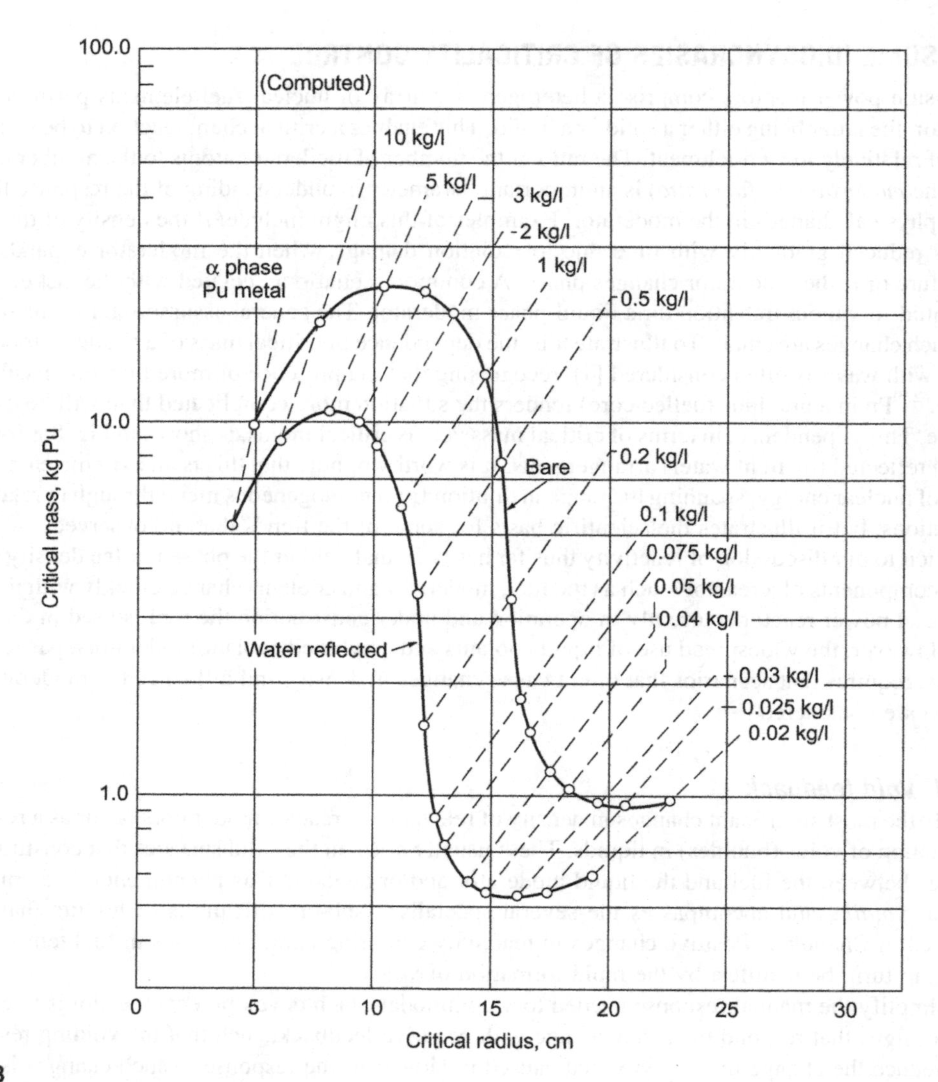

FIG. 8.8

Critical mass versus critical radius for a spheres of water-dispersed fissile material (estimated). The upper curve is for bare spheres and the lower curve for spheres immersed in light water.

Reproduced with permission of Pacific Northwest National Laboratory. E.D. Clayton, Anomalies of Nuclear Criticality, Pacific Northwest National Laboratory, PNNL-19176, revision 6, 2010.

Dilution has two effects:

1. It increases the distance between fissile atoms making them less easy for neutrons to find.
2. It increases moderation and this changes the neutron interaction probabilities.

At low levels of dilution (small critical radius), the effect of (1) is to increase neutron leakage and of (2) to soften the neutron spectrum; to begin with, given the absence of moderation, the latter increases resonance absorption. Since both mechanisms (leakage and absorption) act to increase neutron loss, the mass necessary to sustain the critical reaction increases. Continued dilution (and thus increasing moderation) results in a reduction in resonance absorption, because neutron energies are now too low for this to remain significant, and conversely since they begin to approach more favourable energy regimes of the fission cross section. This causes the critical mass to reach a maximum and then decline rapidly to a minimum. For further dilution beyond this, neutron capture on hydrogen takes over and the critical mass increases.

The majority of thermal-spectrum power reactors use nuclear fuel of low enrichment under a relatively large degree of dilution due to there being a dispersed moderator. Consequently, it is the area in Fig. 8.8 beyond the peak critical mass at higher levels of dilution that is most relevant in terms of control, particularly for critical masses of less than 1 kg.

The region in Fig. 8.8 on the left of the minimum (critical radius $<\sim 15$ cm) can be used to illustrate the *under-moderated* reactor state while the region on the right (critical radius $>\sim 15$ cm) can be used for the *over-moderated* case. Consider the case of a reactor core that is just critical consistent with the unreflected, upper line in Fig. 8.8 with a critical mass of ~1 kg and dilution of 0.04 kg l^{-1}. A transient increase in reactivity[8] results in a temperature increase that is manifest as voiding. Voiding introduces a reduction in dilution, since steam formation constitutes a reduction in the number density of moderator atoms, analogous in this context to a reduction in critical radius. As a result, the neutron spectrum is not thermalised as significantly relative to the solid (un-voided) state, which reduces the probability of fission. This renders the mass subcritical since the core moves to the left into the subcritical region (given there is less moderation but the same amount of fissile material). Thus reactivity falls; the overall consequence of voiding in under-moderated reactor systems is one of negative feedback with such a moderator-fuel arrangement referred to as having a *negative void coefficient*. This is a desirable response to voiding that might arise from short-term increases in reactivity.

By contrast, for the over-moderated case to the right of the minimum at a dilution of ~0.02 kg l^{-1}, voiding brings about the same reduction in moderation as for the under-moderated case. However, the scope for hardening of the neutron spectrum from this position, in terms of the ratio of the number of moderator atoms to fuel atoms (the *moderator-fuel ratio*), is more limited since it is, by definition, over moderated. However, voiding *does* reduce the extent by which neutrons are absorbed by hydrogen thus leaving more in the system than there were before voiding. Consequently the core moves to the left, passing above the line and becomes supercritical. If unattended, this effect would be manifest as a further increase in reactivity leading to more voiding and so on, resulting in a state of positive feedback. This is highly undesirable due to it being difficult to control. Thus over-moderated reactor systems tend to be associated with a *positive void coefficient*.

---

[8]In the context of this discussion we are not interested in the origin of the insertion since as a thermodynamic system in dynamic equilibrium small fluctuations in reactivity are occurring all the time.

While voiding due to a reactivity insertion is often the main exponent of void feedback in reactor systems, any changes in moderator-fuel ratio can present important control issues. For example, voiding can be brought about by: the rapid movement of shutdown/control rods in liquid moderators, agitation (particularly in process chemical applications) and the use of low reactivity worth components of control systems when inserted into liquid moderators (such as components of control rod infrastructure). Light-water reactors are usually designed to exhibit a negative void feedback coefficient for these reasons. However, the concentration of soluble poison ('shim') used as a long-term control agent can require consideration in this context since voiding in a concentrated poison could inadvertently undermine the desired negative feedback state. Schematic illustrations of the under-moderated and over-moderated cases associated with void feedback are given in Fig. 8.9A and B, respectively.

8.6.2.2 Thermal feedback

While void feedback is perhaps the most easily visualised reactor response to changes in reactivity, particularly because it received widespread attention following the Chernobyl accident, feedback due to changes in temperature is also of interest. These can be associated with four components in general: fuel, moderator, coolant and to a lesser extent the reactor structure. For this reason, the thermal feedback reactivity state of nuclear reactors is often distinguished on the basis of whether the reactor is *cold*, that is it has not recently been at power, or *hot* due to being in a recently operated state. The temperature coefficient α of reactivity ρ is defined as the change in reactivity per unit temperature where,

$$\alpha = \frac{d\rho}{dT} = \frac{1}{k}\frac{dk}{dT} \tag{8.29}$$

which is generally measured in units of pcm K^{-1}, recalling that $\rho = \Delta k/k$. A positive temperature coefficient indicates that reactivity increases with increasing temperature while a negative coefficient infers the inverse, by definition. Since an increase in reactivity is usually associated with an increase in power, and thus an increase in temperature, a passive safety response corresponding to a negative temperature coefficient is an important goal of reactor design. Each component, for example, the fuel, moderator, and so on, is assigned a corresponding temperature coefficient.

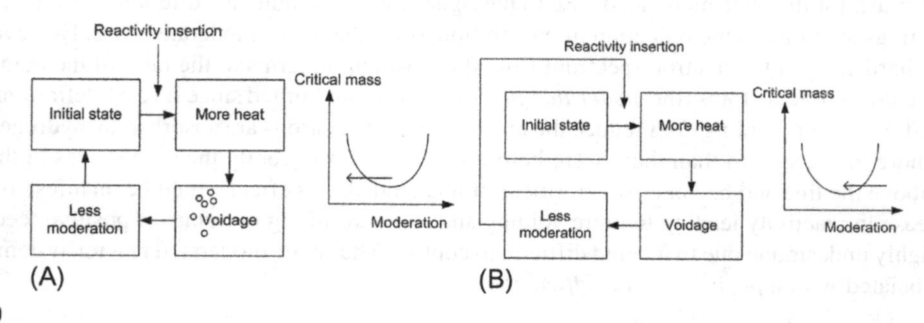

FIG. 8.9

Schematic illustrations of void feedback behaviour in terms of critical mass versus amount of moderation (such as moderator-to-fuel ratio) for (A) the under-moderated case and (B) the over-moderated case.

There are three main phenomena of interest in relation to temperature, which can be considered to a lesser or greater extent, for each of the four components of a reactor system:

- *Doppler broadening* of neutron capture resonance cross sections. This is most often associated with fuel and the resonances in ^{238}U and ^{239}Pu.
- Changes in the energy distribution of the neutron population with temperature, often referred to as *spectral hardening* or *spectral softening*.
- Temperature changes leading to changes in density.

These phenomena are important because, as discussed in Chapter 4, the reaction rate is proportional to the product of the microscopic cross section and atomic number density (Eq. 4.10); both are open to variance given the potential changes listed above.

Doppler broadening

When the concept of microscopic cross section was introduced in Chapter 4, an implicit assumption was made that the atom undergoing an interaction with a neutron is stationary, since only the velocity of the neutron incident on the substance in question was considered. However, the atoms in the substance that the neutron interacts with will be subject to thermal motion according to the Maxwellian distribution of velocities. This is manifest macroscopically as the temperature of the substance and causes the discrete lines observed in the absorption spectra associated with the substance to broaden.

Thus the probability of absorption at a particular energy is reduced while the width (the range to which it extends to both lower and higher energies) is increased, relative to what would be expected theoretically in the absence of thermal motion. In the context of neutron absorption this arises as a broadening of the line structure in the resonance region of the energy dependence of the microscopic absorption cross section. This is referred to widely in nuclear engineering as *Doppler broadening*. An example of the effect of broadening based on a simulation of a neutron capture resonance in ^{58}Ni is shown in Fig. 8.10.

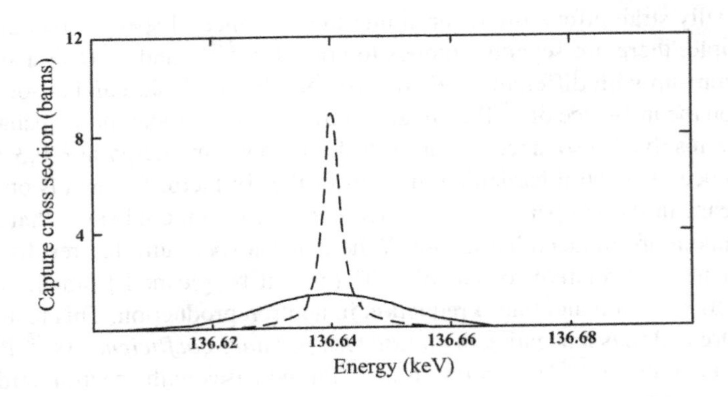

FIG. 8.10

The neutron capture cross section as a function of energy for ^{58}Ni modelled without Doppler broadening *(dashed curve)* and with Doppler broadening based on a Maxwellian distribution of velocities *(solid curve)*.

Reproduced with permission from N.M. Larson, Introduction to the theory and analysis of resolved (and unresolved) neutron resonances via SAMMY, Oak Ridge National Laboratory ORNL/M-6576, US, 1998.

The influence of Doppler broadening is not that the effect of temperature introduces a departure from the theoretical optimum, because cross sections measured experimentally will incorporate broadening implicitly at the temperature at which the measurement is made. Rather, it is the change that occurs with a change in temperature and, in turn, the effect that changes in reactivity exert on the extent of broadening. For example, an increase in reactivity will generally result in an increase in temperature and hence an increase in broadening. As a result, the range of neutron energies falling within the resonance will also increase and thus the number of neutrons that are absorbed will increase too.

This is particularly beneficial in terms of passive safety because the resonant structure of the cross section for ^{238}U renders the fuel in a reactor particularly susceptible to Doppler broadening. The associated coefficient is often referred to as the *fuel temperature coefficient* or *prompt temperature coefficient*. An insertion of reactivity will result in an almost instantaneous increase in temperature of the fuel. The immediate Doppler broadening of ^{238}U resonances acts to increase the consumption of neutrons, thus reducing reactivity. This is a very important reactor design feature because the fuel heats up more quickly as a result of an increase in reactivity than either the moderator or coolant.

Spectral hardening

Thus far it has been assumed that the neutrons in a thermal spectrum reactor are thermalised by the moderator, that is, constituting a distribution with an average energy corresponding to the mean energy of the Maxwellian distribution of energies, kT, for a temperature of 293 K. Clearly for a *cold* reactor some variance around this temperature is likely, dependent on the ambient temperature of the surroundings and because, in moving to steady-state power generation, the temperature of the reactor will increase. Since the neutron population can only be moderated to a level in equilibrium with the temperature of the moderator, the transition from cold to hot results in a hardening of the neutron spectrum. This is observed as an upwards shift in the distribution of neutron energies after moderation from that in the cold state.

While conceptually straightforward, determining the influence of spectral hardening can be complicated. For example, there are several isotopes to consider: ^{235}U and ^{238}U in a standard uranium-fuelled reactor at start-up with different levels of susceptibility to fission and absorption. Then, after a period of operation the influence of ^{239}Pu will also start to become apparent. Its behaviour is different to ^{238}U because the resolved-resonance region of ^{239}Pu is lower in energy at $\sim$0.3 eV.

The main influence of spectral hardening in terms of the six-factor formula is on the reproduction factor, η. The increase in the temperature of a reactor from a clean, cold state, that is, with no ^{239}Pu present, causes temperature-induced hardening of the neutron spectrum that results in a reduction in neutron absorption in ^{235}U relative to that of ^{238}U (as can be gleaned from the cross sections in Fig. 4.8), a reduction in fission and thus a reduction in neutron production. This results in a reduction in η with temperature and thus a negative *moderator temperature coefficient*. As ^{239}Pu is bred into the reactor system from capture in ^{238}U, absorption in ^{239}Pu increases with spectral hardening due to the resonance structure referred to above but fission of ^{239}Pu becomes less likely with lower-energy neutrons. The former is usually greater than the latter resulting in an overall increase in η and a positive moderator temperature coefficient for reactors in steady-state operation. However, the speed of response to temperature is much slower than the temperature effect in the fuel. It is worthy of note that the temperature coefficient also depends on the type of moderator in use.

Density effects of temperature

The expansion and contraction of materials used in reactor systems can cause changes in the neutron energy spectrum and hence changes in cross sections, most often those associated with the moderator. The transition from cold to hot state results in a reduction in density of the moderator due to expansion. The macroscopic cross section falls due to a reduction in the atomic number density, which can be manifest as an increase in neutron leakage from the reactor system. The corresponding reduction in neutron number in the reactor leads to a fall in fission rate; hence density effects in thermal power reactors are usually associated with negative feedback and have a negative reactivity coefficient.

8.7 LONG-TERM EFFECTS AND REACTOR POISONING

8.7.1 THE PRINCIPLE OF POISONING

Most fission products have short half-lives and thus do not exist long in a reactor after they are formed. Of those that endure, a few have the special property of an extremely large neutron absorption cross section relative to the vast majority of fission products. This is of particular interest in terms of long-term reactor control because, in comparison with the state of a reactor when loaded with fresh fuel that has not been brought to a critical state (henceforward referred to as 'clean'), these isotopes constitute a new and dynamic sink for neutrons. This arises symbiotically with the operation of the reactor such that the more power, the higher the fission rate and hence the higher production rate of fission products, including those with anomalously large neutron absorption cross sections. Since these isotopes can be evolved to an extent sufficient to change the neutron population in a reactor core, they are usually termed *reactor poisons*.

Reactor poisons tend to be proton-rich. Once they have absorbed a neutron, they are transformed into the slightly heavier isotope with a dramatically reduced neutron absorption cross section. An important consequence of this is that neutron absorption causes the poison to be consumed; this introduces sympathetic effects in terms of neutron population and reactivity. Some poisons are themselves radioactive too; thus in the absence of a neutron flux, that is, if the reactor is shut down, this variant will decay away with time. Such complexity in neutron poison composition is reflected by them exhibiting a nonlinear dependence with time, especially following reactor shutdown. The dynamic abundance of these isotopes can cause the reactivity that is required to restart a reactor shortly after shutdown to be significantly different to what is required some time afterwards, since in the latter scenario the poisons have had sufficient time to decay away. We shall consider the cases of ^{135}Xe and ^{149}Sm as cases in point as these are by far the most prominent isotopes in terms of fission product yield and absorption cross section. A summary of the relevant physical properties of these isotopes is given in Table 8.3.

8.7.2 XENON-135

Xenon-135 is the most prominent example of a nuclear reactor poison because it has an enormous neutron absorption cross section. Both ^{135}Xe and its parent tellurium-135 (^{135}Te) fall within the $130 < A < 150$ high-mass fraction of the fission fragment mass distribution and ^{135}Te is produced with

Table 8.3 The Salient Physical Properties of the Reactor Poisons ^{135}Xe and ^{149}Sm

Isotope	Origin	Production Pathway	Thermal Neutron Absorption Cross Section [5,6]/b	Radioactive Decay Pathway
^{135}Xe	Primary fission	$\gamma_I = 6.1\%$ $\gamma_{Xe} = 0.3\%$	2.65×10^6	β^- decay of ^{135}Xe, $t_{1/2} = 9.14h$
	^{135}Te decay chain	β^- decay of ^{135}I, $t_{1/2} = 6.58h$		
^{149}Sm	Fission ^{149}Pm (^{149}Nd) decay chain	$\gamma_{Pm} = 1.4\%$ β^- decay of ^{149}Pm, $t_{1/2} = 2.21d$	4.0×10^4	Stable

a significant yield. Further, ^{135}Xe is radioactive with a half-life long enough for it to build up during operation and also for it to remain present in a core for some time after a reactor is shutdown.

A relatively small amount of ^{135}Xe (a yield of 0.3%) is produced as a primary fission product. However, approximately 20 times as much ^{135}Te is produced and thus this constitutes the main source of ^{135}Xe. Like many fission products, ^{135}Te is neutron-rich and decays rapidly via β^- emission, in this case to iodine-135 (^{135}I). This then decays with a half-life of 6.58 hours via β^- emission to ^{135}Xe. Xenon-135 has a short half-life (β^- decay, $t_{1/2}$=9.14 hours) and hence none of it exists naturally, save for the fleeting, trace amounts that might arise from spontaneous fission in naturally occurring ^{238}U. Geological concentrations of ^{238}U are too dilute to result in a steady-state concentration of ^{135}Xe and thus none exists in nuclear fuel before exposure in a reactor. The ^{135}Te decay chain can be summarised as

$$^{135}_{52}Te \xrightarrow{\beta^-} {}^{135}_{53}I \xrightarrow{\beta^-} {}^{135}_{54}Xe \xrightarrow{\beta^-} {}^{135}_{55}Cs \xrightarrow{\beta^-} {}^{135}_{56}Ba \quad (8.30)$$

A reactor loaded with fresh fuel that is devoid of any poison content is termed *clean*. This distinguishes it from one that has recently been in operation and is poisoned as a result and hence referred to as *dirty*. There are two sources of ^{135}Xe (fission and the β^- decay of ^{135}I) and two sinks (radioactive decay and neutron absorption). It is important to emphasise that the neutron absorption cross section of ^{135}Xe (2.75×10^6 b) is so significant, relative to those of its parent and daughter isotopes (that are related by decay or neutron capture) that when these processes occur by which a ^{135}Xe is destroyed the neutron absorption influence effectively disappears too. We can thus visualise the factors governing ^{135}Xe concentration as shown below in Fig. 8.11.

The presence or otherwise of poisons like ^{135}Xe (xenon is the most significant example) influences the medium-term control requirements of a nuclear reactor and there are several competing effects. For example, the formation of ^{135}Xe is dependent on the fission rate, that is, reactor power, and time. Its depletion is dependent on time, the amount present and the neutron absorption rate (which is also dependent in turn on reactor power).

The formation of ^{135}Xe by fission and its consumption by neutron absorption occurs only while a reactor is operating whereas its radioactive decay occurs irrespective of the reactor state. The influence of reactor poisons thus extends throughout and beyond the operational envelope of a nuclear reactor. To illustrate this further, we can consider four scenarios of reactor operation: *start-up, steady-state operation, shutdown* and *restart*.

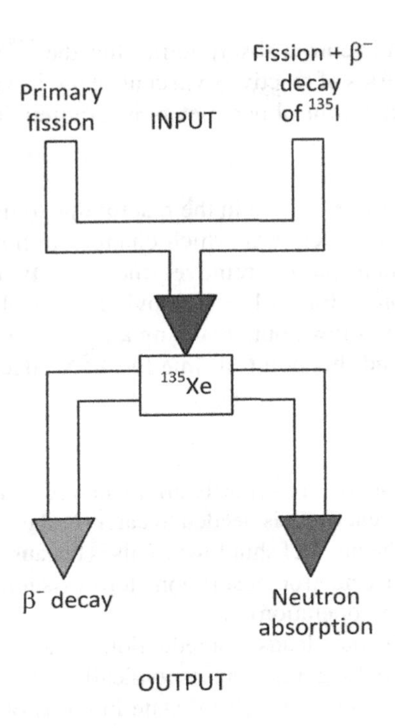

FIG. 8.11

A schematic of the dynamic equilibrium governing ^{135}Xe concentration in an operating reactor.

Start-up:

- At this stage a reactor core is devoid of poisons and comprises only fresh nuclear fuel (generally uranium dioxide comprising a low-enriched mixture of ^{238}U and ^{235}U). Thus no account for neutron absorption by poisons needs to be made since the sink associated with absorption in poisons is absent at this time.
- During the first couple of days following start-up the concentration of ^{135}I increases (the half-life of ^{135}Te, at 19.2 s, is short enough for us to consider the iodine isotope as the head of the decay chain). Via the decay of ^{135}I, the concentration of ^{135}Xe typically reaches a state of equilibrium after 30 hours. In terms of Fig. 8.11, at this point the growth in ^{135}Xe occurs until there is enough to result in a rate of decay and destruction by neutron capture equal to its rate of formation. During this dynamic period the number of neutrons absorbed by the in-growth of ^{135}Xe increases dynamically too and thus the control of the reactor has to be adapted to account for this. This can be achieved, for example, by gradually withdrawing control rods to account for the absorption by the ^{135}Xe.

Steady-state operation:

- At this point (typically $t > 30$ hours) with a constant power level and neutron flux, the rate of production of ^{135}Xe (via primary fission but mostly due to the decay of ^{135}I) is equal to the rate of consumption (via β^- decay and neutron absorption). Thus the ^{135}Xe abundance remains constant. The allowance made for it in terms of reactivity via control rod insertion is also constant and thus no further adjustment of the reactor control parameters is required, in principle.

Shutdown:

- At shutdown ^{135}Xe continues to be formed in the reactor due to the decay of ^{135}I (with a half-life that is long relative to the timescales over which changes in power might be orchestrated). However, the absence of the neutron flux removes one of the two main routes by which ^{135}Xe is consumed (neutron absorption in Fig. 8.11). Removing this sink causes the concentration of ^{135}Xe to grow over a period of a few hours reaching a maximum when its production via ^{135}I is equal to its own decay. Beyond this point in time the ^{135}Xe decays away according to its own half-life.

Restart:

- If a rapid restart of the reactor is required shortly after shutdown, along timescales of $t < 10$ hours, then significantly more excess reactivity is needed to cater for the effect of xenon than was required to keep the reactor running at the point of shutdown. This is because of the build-up that occurs after shutdown when consumption by neutron absorption stops (assuming there has been a significant, preceding period of steady-state operation).
- Typically, an interval of $t > 40$ hours leads to a reduction in the reactivity necessary relative to the point of shutdown. If a relatively long time elapses (typically 4–5 days), then all the ^{135}Xe will have decayed away returning the reactor to its 'clean' state in terms of ^{135}Xe.

8.7.3 SAMARIUM-149

The influence of samarium-149 on reactor control is conceptually similar to that of ^{135}Xe but differs in that ^{149}Sm, while having a very large thermal neutron capture cross section, is stable. Thus it is not consumed by radioactive decay in the same way that ^{135}Xe is. A further difference is that the absorption cross section of ^{149}Sm is significantly less than that of ^{135}Xe.

Samarium-149 has a negligible direct fission yield and thus is formed in reactors entirely from the β^- decay of the fission product neodymium-149 (^{149}Nd) as follows,

$$^{149}_{60}\mathrm{Nd} \xrightarrow{\beta^-} {}^{149}_{61}\mathrm{Pm} \xrightarrow{\beta^-} {}^{149}_{62}\mathrm{Sm} \tag{8.31}$$

The decay of ^{149}Nd to promethium-149 (^{149}Pm) is 100% and rapid ($t_{1/2} = 1.73$ h) in comparison with the subsequent decay of ^{149}Pm to ^{149}Sm ($t_{1/2} = 2.21$ d). Hence it is usually the promethium isotope that is taken as the start of the chain in terms of the equilibrium state needed to predict the concentration dependence of ^{149}Sm with time. A schematic diagram depicting the formation and destruction of ^{149}Sm is provided in Fig. 8.12.

The change of concentration of ^{149}Sm with time is dependent on the rate of production and the rate of consumption, as follows,

$$^{149}\mathrm{Sm\ concentration\ rate} = {}^{149}\mathrm{Sm\ production\ rate\ via}\,{}^{149}\mathrm{Pm\ decay} - {}^{149}\mathrm{Sm\ neutron\ absorption\ rate} \tag{8.32}$$

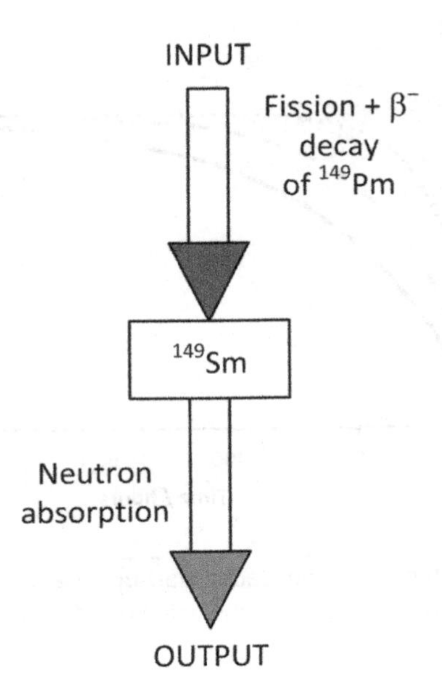

FIG. 8.12
A schematic of the dynamic equilibrium governing ^{149}Sm concentration in an operating reactor.

The production rate of ^{149}Pm comprises the rate at which it arises from fission minus the rate at which it decays via a development of Eq. (3.3) to include the yield arising from fission,

$$\frac{dN_{Pm}}{dt} = \gamma_{Pm}\Sigma_f\phi_{th} - \lambda_{Pm}N_{Pm} \tag{8.33}$$

where N_{Pm} is the number of ^{149}Pm nuclei, γ_{Pm} is the fission yield of ^{149}Pm (0.014 atoms per fission) and λ_{Pm} is the decay constant for ^{149}Pm. Integrating from time zero when $N_{Pm}(0)=0$ and t, gives,

$$N_{Pm}(t) = \frac{\gamma_{Pm}\Sigma_f\phi_{th}}{\lambda_{Pm}}\left[1 - e^{-\lambda_{Pm}t}\right] \tag{8.34}$$

Similarly, the rate of change of ^{149}Sm comprises,

$$\frac{dN_{Sm}}{dt} = \lambda_{Pm}N_{Pm} - \sigma_c N_{Sm}\phi_{th} \tag{8.35}$$

where N_{Sm} is the number of ^{149}Sm nuclei and σ_c is its microscopic thermal neutron capture cross section. Substitution of Eq. (8.34) in Eq. (8.35) and solving for N_{Sm} gives,

$$N_{Sm}(t) = \gamma_{Pm}\Sigma_f\phi_{th}\left[\frac{1}{\sigma_c\phi_{th}}\left(1 - e^{-\sigma_c\phi_{th}t}\right) - \frac{1}{(\sigma_c\phi_{th} - \lambda_{Pm})}\left(e^{-\lambda_{Pm}t} - e^{-\sigma_c\phi_{th}t}\right)\right] \tag{8.36}$$

The dependence of N_{Sm} with time for three different levels of flux for a clean reactor from start-up is shown in Fig. 8.13.

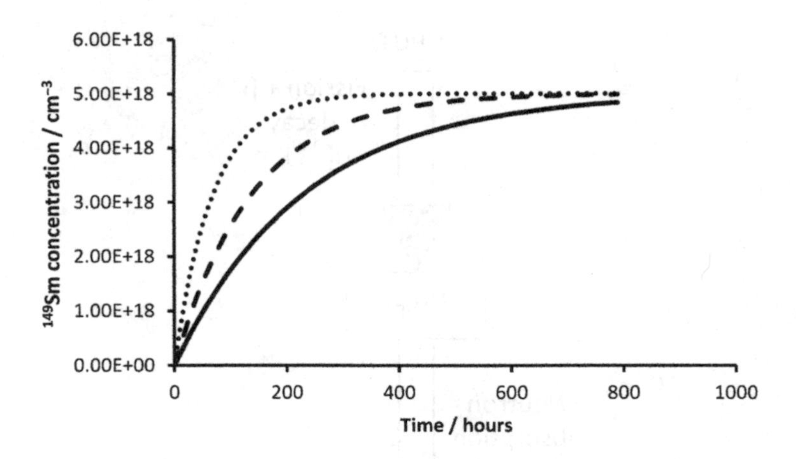

FIG. 8.13

A plot of ^{149}Sm concentration versus time following reactor start-up for a clean reactor for three different levels of flux (power): *solid line*=lower flux, *dashed line*=intermediate flux, *dotted line*=higher flux.

In Fig. 8.13 it is clear that the equilibrium level of ^{149}Sm is reached at a rate dependent on neutron flux (the greater the flux the more rapid the equilibrium level is reached). The equilibrium level N_{Sm}^{eq} is independent of neutron flux and is given for the case that $t \rightarrow \infty$,

$$N_{Sm}^{eq} = \frac{\gamma_{Pm}\Sigma_f}{\sigma_c} \tag{8.37}$$

Revising the narrative provided above for the case of ^{149}Sm we find the following:
Start-up:

- Assuming a clean reactor devoid of poisons, the reactor is brought up to power without any allowance necessary for the samarium isotope.
- ^{149}Pm starts to be generated from the decay of ^{149}Nd immediately but its half-life is much longer (relatively speaking) than for the decay chains involved in ^{135}Xe. Hence, it takes much longer for the ^{149}Sm to reach equilibrium concentration of the order of 800 hours (~1 month), with control rod insertion levels adjusted accordingly to accommodate the in-growth of the poison.

Steady-state operation:

- At this point (typically $t > 800$ hours) the rate of production of ^{149}Sm and its consumption are equal. Thus the reactor can continue to be operated without further control adjustment.
- Since the yield of ^{149}Pm and the neutron capture cross section in ^{149}Sm are smaller than in the case of ^{135}Xe, the allowance in terms of reactivity is less than for the case of xenon which remains the dominant poison.

Shutdown:

- At shutdown, consumption via neutron capture ceases but production via the decay of ^{149}Pm continues; hence the concentration of ^{149}Sm increases. Since ^{149}Sm is stable its concentration stabilises with the decline of the concentration of ^{149}Pm.
- Due to the relatively long half-life of ^{149}Pm, its decline is a slower process than in the case of ^{135}Xe and in contrast the stabilisation level remains constant after shutdown because of the stability of ^{149}Sm.

Restart:

- If a rapid restart of the reactor is required shortly after shutdown, that is $t < 24$ hours, then the effect of ^{149}Sm will be similar to what it was before shutdown. Waiting longer than this, say $t > 50$ hours, leads to an excess reactivity being necessary, due to the in-growth of ^{149}Sm from the decay of ^{149}Pm, which will need to be accounted for. It is worth emphasis, however, that this effect, while long-lived is much less than the poisoning effect of ^{135}Xe, for the reasons described above.
- Since a neutron flux in a reactor by which ^{149}Sm might be consumed is unlikely in the absence of the production of ^{149}Pm, once poisoned with ^{149}Sm a reactor remains in that state until the entire fuel load is refreshed.

The variation of ^{149}Sm concentration with time during operation and post-shutdown is shown in Fig. 8.14. The dependence of N_{Sm} with time post-shutdown follows as Eq. (8.39),

$$N_{Sm}(t) = N_{Sm}(t_s) + N_{Pm}(t_s)\left(1 - e^{-\lambda_{Pm}(t-t_s)}\right) \tag{8.39}$$

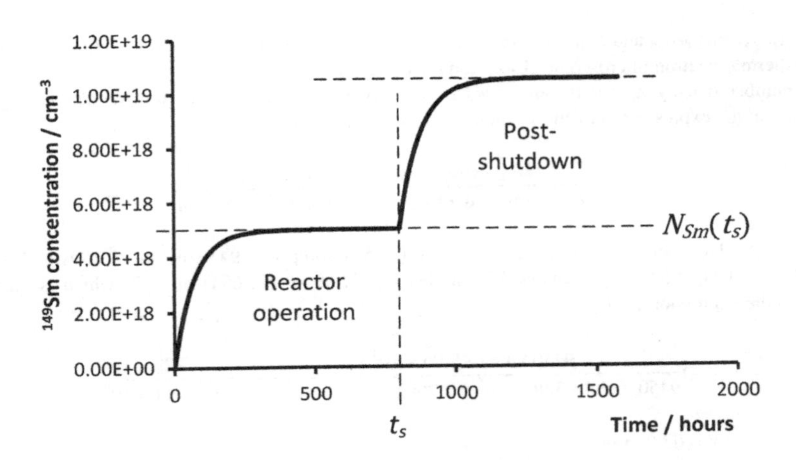

FIG. 8.14

A plot of ^{149}Sm concentration versus time following reactor start-up for a clean reactor and post-shutdown where shutdown occurs at $t = t_s$.

CASE STUDIES

CASE STUDY 8.1: ESTIMATING THE THERMAL UTILISATION FACTOR, *f*

Given the definition for the thermal utilisation factor *f*,

$$f = \Sigma_{aF} / \Sigma_a$$

to obtain an estimate for *f* for a reactor moderated with light water it is necessary to consider the composition of both the fuel and the coolant/moderator. Given the following information:

- The reactor contains 85 t fuel in the form of UO_2 with an average ^{235}U enrichment of 3% wt. and a combined, light-water moderator and coolant.
- The control/shutdown rod infrastructure and instrumentation, and so on, can be ignored for the purposes of this illustration.
- The core comprises fuel assemblies perfused by the water, which can be simplified by assuming the core and coolant are a mixture of UO_2 and H_2O of uniform composition with negligible absorption in the cladding.
- The core dimensions are 3.0 m in diameter, 4.3 m in height and it thus has a volume of 30.4 m^3.
- The molar masses of ^{235}U, ^{238}U, UO_2 and H_2O are 235.04 g mol^{-1}, 238.05 g mol^{-1}, 270.03 g mol^{-1} and 18.02 g mol^{-1}, respectively.
- The microscopic absorption cross sections for ^{235}U, ^{238}U, hydrogen and oxygen are $\sigma_{235} = 684$ b, $\sigma_{238} = 2.7$ b, $\sigma_H = 0.329$ b and $\sigma_O = 0.19$ mb (~0), respectively.
- The density of UO_2 is 10.97 g cm^{-3} and the density of water is 1.00 g cm^{-3}.

The expression above can be developed thus since $\Sigma_a = \Sigma_{aF} + \Sigma_{aH_2O}$,

$$f = \frac{\Sigma_{aF}}{\Sigma_{aF} + \Sigma_{aH_2O}}$$

where $\Sigma_{aF} = n_{235}\sigma_{235} + n_{238}\sigma_{238}$ and $\Sigma_{aH_2O} = n_H\sigma_H$ and n is the number density of the corresponding element and σ is the corresponding thermal neutron microscopic absorption cross section.

Since the number density n_Z for the mass M_Z of a hypothetic substance Z of molar mass m_Z can be written $n_Z = n_A M_Z / V m_Z$, in the expression for *f* the volume V and N_A cancel to give,

$$f = \frac{(M_{235}/m_{235})\sigma_{235} + (M_{238}/m_{238})\sigma_{238}}{(M_{235}/m_{235})\sigma_{235} + (M_{238}/m_{238})\sigma_{238} + (M_H/m_H)\sigma_H}$$

Hence, since 85 t UO_2 comprises 3.15×10^5 mol, 3% (2.55 t) comprises 9450 mol of $^{235}UO_2$ and 97% (82.45 t) 3.06×10^5 mol of $^{238}UO_2$. 85 t UO_2 occupies 7.75 m^3 leaving 22.65 m^3 (22.65 t) of H_2O, which is 1.25×10^6 mol). Substituting into the expression for *f*,

$$f = \frac{9450 \times 684 + 3.06 \times 10^5 \times 2.7}{9450 \times 684 + 3.06 \times 10^5 \times 2.7 + 2 \times 1.25 \times 10^6 \times 0.329} = \frac{7.29 \times 10^6}{8.11 \times 10^6}$$

$$= 0.90 \text{ or } 90\%$$

CASE STUDY 8.2: DERIVING THE DEPENDENCE OF ^{135}Xe WITH TIME FROM CLEAN REACTOR START-UP

The rate of change of ^{135}Xe concentration with time can be written,

^{135}Xe concentration rate = ^{135}Xe production rate via ^{135}I + ^{135}Xe production rate via fission − ^{135}Xe decay rate − ^{135}Xe neutron absorption rate,

Hence, recalling the decay equation of Eq. (3.3) and that the production rate is the product of the yield γ_{Xe}, macroscopic fission cross section Σ_f and the thermal neutron flux, ϕ_{th}, and that the macroscopic capture cross section is $\sigma_{cXe} N_{Xe}$,

$$\frac{dN_{Xe}}{dt} = \lambda_I N_I + \gamma_{Xe} \Sigma_f \phi_{th} - \lambda_{Xe} N_{Xe} - \sigma_{cXe} N_{Xe} \phi_{th}$$

Collecting terms,

$$N_{Xe}(\lambda_{Xe} + \sigma_{cXe}\phi_{th}) + \frac{dN_{Xe}}{dt} = \lambda_I N_I + \gamma_{Xe} \Sigma_f \phi_{th}$$

and since $N_I(t) = N_I(0)e^{-\lambda_I t}$,

$$N_{Xe}(\lambda_{Xe} + \sigma_{cXe}\phi_{th}) + \frac{dN_{Xe}}{dt} = \lambda_I N_I(0)e^{-\lambda_I t} + \gamma_{Xe} \Sigma_f \phi_{th}$$

This is slightly more involved than a simple case of two isotopes but is simplified by defining terms such as a, b and c where $a = \lambda_{Xe} + \sigma_{cXe}\phi_{th}$, $b = \lambda_I N_I(0)$ and $c = \gamma_{Xe}\Sigma_f \phi_{th}$ and hence,

$$aN_{Xe} + \frac{dN_{Xe}}{dt} = be^{-\lambda_I t} + c$$

which is solved as for Case study 3.2 to give,

$$N_{Xe}(t) = \frac{b}{(a - \lambda_I)}\left[e^{-\lambda_I t} - e^{-at}\right] + \frac{c}{a}\left[1 - e^{-at}\right]$$

and,

$$N_{Xe}(t) = \frac{\lambda_I N_I(0)}{(\lambda_{Xe} + \sigma_{cXe}\phi_{th} - \lambda_I)}\left[e^{-\lambda_I t} - e^{-(\lambda_{Xe} + \sigma_{cXe}\phi_{th})t}\right] + \frac{\lambda_{Xe}\Sigma_f \phi_{th}}{(\lambda_{Xe} + \sigma_{cXe}\phi_{th})}\left[1 - e^{-(\lambda_{Xe} + \sigma_{cXe}\phi_{th})t}\right]$$

This describes all start-up scenarios, both clean and poisoned. For the 'clean' start-up case, there are no fission products and hence $N_I(0) = 0$, so,

$$N_{Xe}(t) = \frac{\gamma_{Xe}\Sigma_f \phi_{th}}{(\lambda_{Xe} + \sigma_{cXe}\phi_{th})}\left[1 - e^{-(\lambda_{Xe} + \sigma_{cXe}\phi_{th})t}\right]$$

A plot of $N_{Xe}(t)$ versus time is shown in Fig. 8c.1 for three different levels of neutron flux. This shows that as the flux is increased, ^{135}Xe concentration reaches equilibrium more rapidly and the equilibrium level increases.

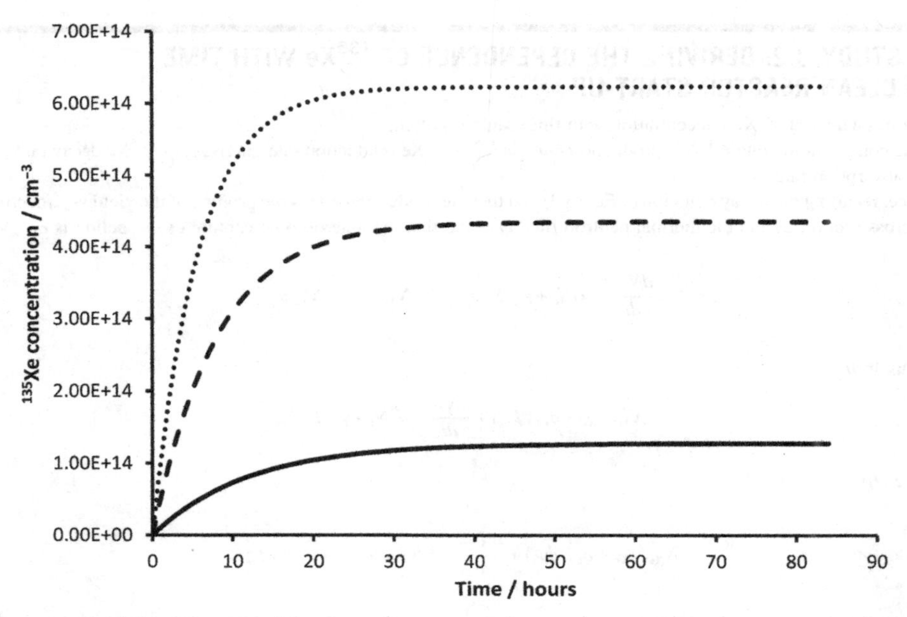

FIG. 8C.1

A plot of ^{135}Xe concentration, $N_{Xe}(t)$, versus time for three different neutron fluxes. *Solid line*—$\phi_{th}=10^{12}$ cm^{-2} s^{-1}, *dashed line*—$\phi_{th}=5\times10^{12}$ cm^{-2} s^{-1} and *dotted line*—$\phi_{th}=10^{13}$ cm^{-2} s^{-1}.

CASE STUDY 8.3: DERIVING THE DEPENDENCE OF ^{135}Xe FOLLOWING SHUTDOWN

Using the rate of change of ^{135}Xe concentration with time once again with the same notation as for Case study 8.2,

$$\frac{dN_{Xe}}{dt}=\lambda_I N_I+\gamma_{Xe}\Sigma_f\phi_{th}-\lambda_{Xe}N_{Xe}-\sigma_{cXe}N_{Xe}\phi_{th}$$

Prior to shutdown at $t=t_s$, assuming equilibrium has been reached, $\frac{dN_{Xe}}{dt}=\frac{dN_I}{dt}=0$ so given the rate of change of the concentration of ^{135}I, N_I, formed as a result of fission and lost via radioactive decay is,

$$\frac{dN_I}{dt}=\gamma_I\Sigma_f\phi_f-\lambda_I N_I$$

at the point of shutdown the equilibrium ^{135}I concentration is,

$$N_I(t_s)=\frac{\gamma_I\Sigma_f\phi_{th}}{\lambda_I}$$

and similarly for ^{135}Xe,

$$N_{Xe}(t_s)=\frac{(\gamma_I+\gamma_{Xe})\Sigma_f\phi_{th}}{\lambda_{Xe}+\sigma_{cXe}\phi_{th}}$$

After shutdown, the production of ^{135}Xe by fission and its destruction by neutron absorption ceases since $\phi_{th}=0$, so the integrand of the ^{135}Xe concentration rate becomes,

$$dN_{Xe}=\lambda_I N_I dt-\lambda_{Xe}N_{Xe}dt$$

where $N_I(t)=N_I(t_s)e^{-\lambda_I t}$ since this is a problem similar to Case Study 3.2,

$$dN_{Xe}=\lambda_I N_I(t_s)e^{-\lambda_I t}dt-\lambda_{Xe}N_{Xe}dt$$

$$N_{Xe}(t)=\frac{\lambda_I N_I(t_s)}{(\lambda_{Xe}-\lambda_I)}\left(e^{-\lambda_I t}-e^{-\lambda_{Xe}t}\right)+N_{Xe}(t_s)e^{-\lambda_{Xe}t}$$

When the production of ^{135}Xe ceases sometime after shutdown when all the ^{135}I has decayed away, the ^{135}Xe concentration reverts to the radioactive decay law $N_{Xe}(t_s)=N_{Xe}(t_s)e^{-\lambda_{Xe}t}$. The dependence of ^{135}Xe concentration following shutdown is shown in Fig. 8c.2 for three different neutron flux levels.

$$N_{Xe}e^{\lambda_{Xe}t}\Big|_0^t=\frac{\lambda_I N_I}{\lambda_{Xe}}e^{\lambda_{Xe}t}+\frac{\gamma_{Xe}\Sigma_f\phi_{th}}{\lambda_{Xe}}e^{\lambda_{Xe}t}\Bigg|_0^t$$

$$N_{Xe}(t)=\frac{\lambda_I N_I}{\lambda_{Xe}}\left(e^{\lambda_{Xe}t}-1\right)+\frac{\gamma_{Xe}\Sigma_f\phi_{th}}{\lambda_{Xe}}\left(e^{\lambda_{Xe}t}-1\right)$$

$$N_{Xe}(t)=\left(e^{\lambda_{Xe}t}-1\right)\left(\frac{\lambda_I N_I}{\lambda_{Xe}}+\frac{\gamma_{Xe}\Sigma_f\phi_{th}}{\lambda_{Xe}}\right)$$

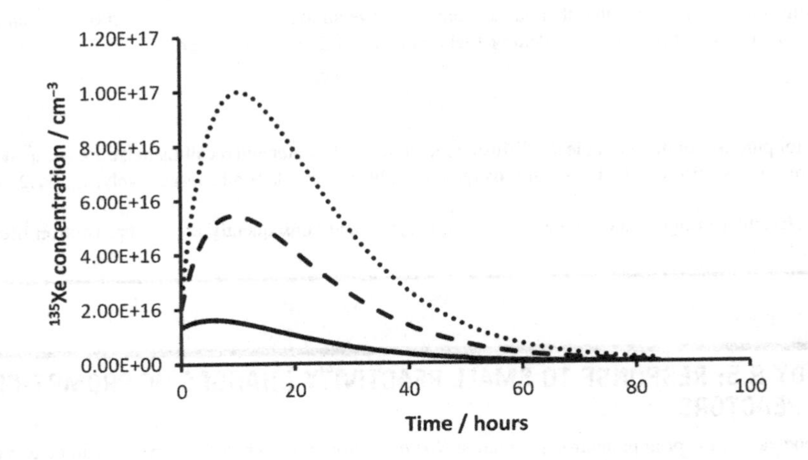

FIG. 8C.2

A plot of ^{135}Xe concentration, $N_{Xe}(t)$, versus time after shutdown. The data for three different neutron fluxes prior to shutdown assuming both ^{135}Xe and ^{135}I had reached equilibrium before shutdown. *Solid line*—$\phi_{th}=10^{13}$ cm^{-2} s^{-1}, *dashed line*—$\phi_{th}=5\times10^{13}$ cm^{-2} s^{-1} and *dotted line*—$\phi_{th}=10^{14}$ cm^{-2} s^{-1}.

CASE STUDY 8.4: ESTIMATES FOR t_S AND t_d FOR LIGHT WATER AND AN ILLUSTRATION OF $t_d \gg t_S$

In Chapter 6 the concept of neutrons being slowed down through a succession of scattering interactions was introduced whereas in this chapter the prompt neutron lifetime has been defined: $l = t_s + t_d$. For the case of hydrogen it was shown that, on average, this results in the transfer of half the neutron energy at each interaction. The time taken, t_s, for a neutron to be thermalised can be calculated provided the initial energy (e.g. 1 MeV), the final energy (0.0253 eV) and the average macroscopic scattering cross section over the slowing down range, $\overline{\Sigma_s}$, are known. Recalling the logarithmic decrement in energy lost per collision ξ (Chapter 6) and the rate $\overline{\Sigma_s} v$ at which scattering interactions occur where v is the speed of a thermal neutron, Glasstone and Edlund [7] describe that, for a loss of energy dE in a period dt, the normalised energy loss (hence the negative sign) is equal to the product of the collision rate, the period and ξ,

$$-\frac{dE}{E} = \xi \Sigma_s v dt$$

Re-arranging to include the dependence of v with E and integrating,

$$\int_0^t dt = -\frac{1}{\xi \Sigma_s}\sqrt{\frac{m}{2}}\int_E^{E'} \frac{1}{E^{3/2}} dE$$

$$t_s = \frac{\sqrt{2m}}{\xi \Sigma_s}\left[\frac{1}{\sqrt{E'}} - \frac{1}{\sqrt{E}}\right]$$

For the case of light water, $\overline{\alpha} = 0.03$ and thus $\xi = 0.89$, $\overline{\Sigma_s} \sim 1.2$ cm giving $t_s \sim 0.01$ ms.

For the diffusion time, t_d, given that thermalised neutrons have an average speed v of ~2200 m s^{-1} and if the macroscopic scattering cross section for the moderator-fuel combination Σ_a is known, then,

$$t_d = {}^1\!/_{\Sigma_a v}$$

Assuming for purposes of this example the diffusion medium is light water and recalling that the corresponding thermal neutron capture cross sections of hydrogen and oxygen are 329 mb and 0.19 mb, respectively, $\Sigma_a = 2.28$ m^{-1} to give $t_d \sim 0.2$ ms.

Thus it is clear that in light water as per this example $t_d \gg t_s$ and consequently the prompt neutron lifetime can be approximated: $l \sim t_d$.

CASE STUDY 8.5: RESPONSE TO SMALL REACTIVITY CHANGES OF PROMPT-CRITICAL THERMAL REACTORS

The time dependence of the prompt neutron population $N(t)$ for a prompt-critical infinite system can be expressed as per Eq. (8.25),

$$N(t) = N(0)e^{\frac{(k_\infty - 1)}{l}t}$$

where l is the prompt neutron lifetime and the period is $T = l/(k_\infty - 1)$.

Let us consider the effect on the neutron population of a very small reactivity insertion of 1 pcm after 5 seconds in a light-water reactor, a graphite-moderated reactor and a heavy-water reactor, each with $f = 0.95$ for the purposes of this illustration. As per Case study 8.4 where $v = 2200$ m s^{-1},

$$l \sim t_d = \frac{(1-f)}{\Sigma_{aM} v}$$

Given the density (ρ), microscopic absorption cross sections for thermal neutrons (σ_a) and molecular mass (M), the parameters in italics can be calculated (neglecting the relatively small absorption cross sections for oxygen).

Moderator	ρ/g cm^{-3}	σ_a/b	M/g mol^{-1}	Σ_{aM}/m^{-1}	l/ms	T/s
Light water	1.00	0.329	18.02	*2.20*	*~0.01*	*~1*
Graphite	1.70	3.86×10^{-3}	12.00	*0.033*	*~0.69*	*~69*
Heavy water	1.11	5.50×10^{-4}	20.03	*0.002*	*~11.36*	*~1140*

Hence, after 5 seconds for light water, returning to the expression for $n(t)$, the ratio of the neutron population at this time relative to that at time zero is,

$$\left(N(5)/N(0)\right)_{LWR} = e^5 \sim 150,$$

that is a 150-fold increase in 5 seconds.

For the graphite and heavy water options, the reduced energy loss per collision results in a longer prompt neutron lifetime and therefore a longer reactor period.

CASE STUDY 8.6: THE INFLUENCE OF THE DELAYED NEUTRON FRACTION ON REACTOR RESPONSE

What is the effect on the neutron population of a very small reactivity insertion of 1 pcm after 5 seconds in the case for light water, as per Case study 8.5, with a delayed component of 0.09 s?

As per earlier discussion,

$$N(t) = N(0)e^{\frac{(k_\infty - 1)}{\bar{l}}t}$$

where $\bar{l}$ is the mean neutron lifetime incorporating the delayed neutron component.

The reactor period $T = \bar{l}/(k_\infty - 1)$ becomes 2.5 hours rather than 1 s in the prompt neutron scenario for light water and thus the ratio, as per that considered in Case Study 8.5, is,

$$\left(N(5)/N(0)\right)_{LWR} \sim 1.0006$$

that is, a 0.06% increase in neutron population.

REVISION GUIDE

On completion of this chapter you should:

- understand that controlling a nuclear reactor spans a number of time domains, from the very short associated with the timescale of the decay of *primary fission products*, through to the build-up and consumption of *reactor poisons*

- appreciate that reactor control is closely associated with the *neutron population* and that this is characterised by the variety of ways in which neutrons can be introduced into a reactor and those by which they can be removed from it
- understand that, as a system during operation in a state of *dynamic equilibrium*, the time dependence of the neutron flux can be described in terms of the rates of *neutron production* and *absorption*, at a *single (thermal) energy* in an *infinite environment* of *uniform composition*
- know that the *prompt neutron lifetime* can be subdivided in terms of the time taken for them to be thermalised and the time that they exist afterwards until they are absorbed, and that the latter is usually much greater than the former
- be able to explain qualitatively that the contribution of *delayed neutrons* to the neutron population is essential in controlling nuclear reactor systems
- understand the various *feedback mechanisms* that exist associated with nuclear reactor systems, and the concepts of *negative* and *positive* feedback under which all such mechanisms are considered
- know of the two principal reactor poisons associated with uranium-fuelled nuclear fission reactors, understand the origin of their influence on reactor operation and appreciate the typical dynamics in terms of reactivity that would be observed as a result

PROBLEMS

1. The time taken for ^{135}Xe to reach equilibrium is dependent in part on the time taken for ^{135}I to reach equilibrium in a reactor fuelled with uranium dioxide of natural enrichment. Using your knowledge of secular equilibrium, based on the concepts covered in Chapter 3, write down the first-order differential equation describing the rate of change of the concentration of ^{135}I and solve it to obtain an expression for the concentration $N_I(t)$. Finally, determine the time taken to reach 99% of the concentration at equilibrium, using the data given below.
2. Derive the expression for the concentration of ^{149}Sm as a function of time given below.

$$N_{Sm}(t) = \gamma_{Pm}\Sigma_f\phi_{th}\left[\frac{1}{\sigma_c\phi_{th}}\left(1-e^{-\sigma_c\phi_{th}t}\right) - \frac{1}{(\sigma_c\phi_{th}-\lambda_{Pm})}\left(e^{-\lambda_{Pm}t}-e^{-\sigma_c\phi_{th}t}\right)\right]$$

Show that for start-up of a clean reactor $N_{Sm}(0)=0$ and $N_{Sm}^{eq} = \dfrac{\gamma_{Pm}\Sigma_f}{\sigma_c}$.

3. Derive the diffusion time for a graphite-moderated reactor given that the density of nuclear graphite is 1.7 g cm^{-3}, the thermal utilisation factor is 0.9, the atomic mass of carbon is 12 u and the microscopic absorption cross section averaged over the thermal neutron spectrum is 2.99 mb. Explain any differences you observe with regard to the same estimate for a light-water reactor and what implications they might have for the reactor design.
4. Given the delayed neutron fractions for ^{233}U given below, calculate the corresponding reactor period for a 1% increase in reactivity and comment on the origin of the difference in result you obtain in comparison with the period for ^{235}U. The absolute delayed neutron fraction in ^{233}U is 0.0074 neutrons per fission or 0.3%.

Group	Half-life, $t_{1/2}$/s	Life time τ_i/s	Relative yield/β	Absolute yield/neutrons per fission	Fraction of fission neutron yield β_i
1	55.1	79.5	0.09	0.00067	0.00027
2	20.7	29.9	0.27	0.00200	0.00080
3	5.3	7.7	0.23	0.00170	0.00068
4	2.3	3.3	0.32	0.00237	0.00095
5	0.6	0.9	0.07	0.00052	0.00021
6	0.2	0.3	0.02	0.00014	0.00006
Totals	–	–	1.00	0.00741	0.00296

5. Explain why the combination of a negative fuel temperature feedback coefficient might be tolerated with a positive moderator coefficient, but the converse is not desirable.

REFERENCES

[1] H. van Dam, Physics of nuclear reactor safety, Rep. Prog. Phys. 11 (1992) 2025–2077.

[2] K.H. Bockhoff (Ed.), Nuclear Data for Science and Technology: Proceedings of the International Conference, 6–10 September 1982, Antwerp, 1982, p. 18.

[3] R.J. Tuttle, Delayed neutron data for reactor-physics analysis, Nucl. Sci. Eng. 56 (1) (1975) 37–71.

[4] E.D. Clayton, Anomalies of nuclear criticality, Pacific Northwest National Laboratory, 2010. PNNL-19176, revision 6.

[5] G. Leinweber, et al., Neutron capture and transmission measurements and resonance parameter analysis of samarium, Nucl. Sci. Eng. 142 (2002) 1–21.

[6] D. Santry, R.D. Werner, Neutron capture cross-section of ^{135}Xe, J. Nucl. Energy 27 (1973) 409–413.

[7] S. Glasstone, M.C. Edlund, The Elements of Nuclear Reactor Theory, MacMillan, London, 1952.

CHAPTER

THE REACTOR EQUATION AND INTRODUCTORY TRANSPORT CONCEPTS

9

9.1 SUMMARY OF CHAPTER AND LEARNING OBJECTIVES

The focus of this chapter is on the development of the reactor equation and related aspects of elementary neutron transport that support our understanding of the requirements of reactor design and composition necessary to achieve a self-sustaining chain reaction. Current capacity of more than 400 nuclear power reactors worldwide is testament to there being very successful configurations of fissile material and ancillary systems by which nuclear energy can be harnessed for steady-state power supply needs. However, this has required a relatively sophisticated degree of human intervention and conceptual ingenuity.

In this chapter we shall learn that the transport behaviour of neutrons can rarely be categorised as 'simple' although a conceptual solution can be obtained by considering what influences the balance of the neutron population. This results in a relatively uncomplicated equality between the two overarching requirements of *geometry* and *materials composition*. Conversely, while a solution across these domains is tractable for uniform, homogeneous compositions of simple geometrical arrangements, real power reactors are necessarily heterogeneous and significantly more sophisticated. We revisit the fundamental concepts of *neutron production*, *absorption* and *leakage* from a necessarily mathematical perspective but not losing sight of the principal underlying concepts. Subsequently, a short introduction to the neutron transport processes at work in a mono-energetic, steady-state scenario is provided.

The objectives of this chapter are to:

- introduce the concept of neutron balance via a word-based equation that relates the processes of production, leakage and absorption
- introduce the concepts of *neutron current density*, *partial neutron current*, *diffusion coefficient* and the relationship between neutron leakage and neutron flux
- develop the word-based equation referred to above, in terms of its corresponding arithmetic quantities, assuming a neutron flux that is only weakly dependent on angle
- define the neutron diffusion equation
- explain how the neutron diffusion equation can be rationalised for steady-state scenarios to give the *reactor equation*, in terms of material quantities
- illustrate the solution of the reactor equation for several geometries, especially the *infinite slab*, and hence to highlight the materials and geometrical requirements that arise from this development

Nuclear Engineering. https://doi.org/10.1016/B978-0-08-100962-8.00009-3

- explain how the anisotropy in neutron scattering on light isotopes might be taken into account
- introduce elementary concepts of neutron transport and the derivation of the one-velocity neutron transport equation for a steady-state, mono-energetic scenario

9.2 HISTORICAL CONTEXT: JOHN VON NEUMANN, 1903–57

John von Neumann's legacy serves to epitomise the intellect that came together in the late 1930s associated with the dawn of the nuclear age and specifically the Manhattan project; he was regarded by his contemporaries at the time as a genius's genius (Fig. 9.1).

Von Neumann was educated in Hungary attending what was reputed to be one of the best schools in Budapest: the *Fasori Evangelikus Gimnázium*. He studied both Chemical Engineering and Mathematics. He left Europe at the time of the Nazi persecution of the Jews prior to the Second World War, working subsequently at Princeton University in the United States. He was renowned for his prolific output, publishing typically one paper per month on a wide range of mathematics and related theoretical physics problems. These included quantum theory, game theory and areas stimulated at the time by the research of David Hilbert. He was also known for extraordinary feats of memory, such as being able to recount entire volumes at parties and had a preference for there to be very loud music whilst he worked.

Most significantly in the context of nuclear energy Von Neumann pioneered the Monte Carlo method of probabilistic simulations along with Ulam, Szilard and Teller. He also pioneered the computing architecture in which data and the corresponding programme are stored in the same memory space; this approach still bears his name and is the basis for modern computer systems.

FIG. 9.1

John Von Neumann.

9.3 INTRODUCTION

Nuclear power reactors are large, sophisticated, precision-engineered systems comprising a variety of materials, liquids and gases. Their compositions often comprise exotic substances selected primarily for their nuclear properties that are used alongside other materials selected for their thermo-physical or structural attributes. The structure of a reactor results in an unavoidable collection of interfaces at which the transport properties of neutrons can change dramatically. Construing such changes, to ensure safe operation whilst optimising the performance of the reactor, is often the design intent.

However, *real* reactors differ significantly from the rather idealised hypothetical environments we might contrive to improve our understanding of reactor operation. Given the complexity that the possible range in neutron energy, the dependence of neutron cross sections on energy and the variable that time adds to this context, it is perhaps not surprising that a complete understanding of neutron behaviour in heterogeneous nuclear reactors remains a significant mathematical challenge. The theory of neutron transport in nuclear reactor systems can be challenging to the reader only seeking to gain a conceptual appreciation of its significance. Fortunately, a small number of important simplifications can be made whilst retaining the insight that derives from such an analysis. This chapter provides a summary of the concept of neutron balance, which leads to the development of the neutron diffusion equation, the reactor equation and then some concepts on elementary neutron transport are introduced.

9.4 RELATING THE NEEDS OF COMPOSITION AND GEOMETRY IN REACTORS

Before embarking on a conceptual description as to how the neutron population in a given environment ebbs and flows—for it is this that dictates how fission reactors work—it is important to define in an engineering context what we seek to achieve in gaining such a mathematical description. From a qualitative perspective we already know that the properties of the materials involved, either by design or as a necessity of the structure and operational requirements of the reactor, have a strong influence on whether a self-sustaining chain reaction is possible. Consider, for example the influence of *enrichment*, *neutron scattering* (moderation) and *neutron absorption*. Similarly, the size and shape of a reactor system influences the extent of *moderation* and *leakage*; these also have a bearing on whether a self-sustaining chain reaction is possible. Hence from our analysis we would like to know how *materials* and *geometry* relate to each other to enable steady-state, time-independent neutron production to be achieved consistent with the production of power.

When considering a problem mathematically, in which one or more variables are subject to change, it is often preferable to consider an infinitesimally small component over which the variation can be approximated as linear. A large number of these components can then be summed together, as is the basis of integral calculus. In this context, we seek a relationship that will equate the various contributions to the neutron population, defined in a small component, of each of the range of parameters that has an influence on the population. With such a relationship a specific solution can be derived numerically or, alternatively, a general solution might be obtained. Out of necessity this is often based on a number of assumptions associated with composition, isotropy and so forth.

9.4.1 NEUTRON BALANCE AND THE DIFFUSION EQUATION

In a given volume the rate of change of the neutron population, $\partial N/\partial t$ where $N=\phi/v$, is equal to the number of neutrons produced in the volume minus the number that leak and minus the number that are absorbed,

$$\frac{1}{v}\frac{\partial\phi}{\partial t}=\text{rate of production}-\text{rate of leakage}-\text{rate of absorption} \tag{9.1}$$

The conventional notation for the terms in Eq. (9.1) is: S for the rate of production per unit volume unit time, the absorption rate is the product of the macroscopic neutron absorption cross section and the neutron flux, $\Sigma_a\phi$, and the leakage rate is equal to $-D\nabla^2\phi$ where D is the *diffusion coefficient*. Thus the *neutron diffusion equation* can be written as

$$\frac{1}{v}\frac{\partial\phi}{\partial t}-D\nabla^2\phi+\Sigma_a\phi=S \tag{9.2}$$

For the time-independent case, $\partial\phi/\partial t=0$, and in a reactor where the main source of neutrons is fission then $S=v\Sigma_f\phi$ such that,

$$D\nabla^2\phi+(v\Sigma_f-\Sigma_a)\phi=0 \tag{9.3}$$

and dividing through by D we arrive at the *reactor equation* (or Helmholtz equation),

$$\nabla^2\phi+B^2\phi=0 \tag{9.4}$$

where B is a function of neutron production, absorption and diffusion as per,

$$B^2=\frac{(v\Sigma_f-\Sigma_a)}{D} \tag{9.5}$$

and B^2 is termed the buckling.

9.4.2 A CONCEPTUAL SOLUTION OF THE REACTOR EQUATION: THE INFINITE SLAB

Whilst a wide range of reactor geometries are feasible [1], conceptually one of the simplest and most illustrative approaches of a solution to the reactor equation is to consider the infinite, fissile 'slab'. This is a plane of thickness d in the x dimension, which is limitless in the y and z dimensions. Conventionally the material composition is assumed to be uniform such that the neutron flux is constant as a function of y and z and is symmetrical either side of the centre line of the slab due to the identical boundaries on each side. These are usually referred to as the *free surfaces* and for the purposes of this example are defined at $x=\pm d/2$. Hence the slab is positioned such that its centre lies on $x=0$ so that it extends from $-d/2$ through to $d/2$ on either side. The reference to 'free' infers that it is assumed that neutrons leaving across such a boundary do not scatter back into the slab. The slab is depicted in Fig. 9.2.

Since the dependencies in the y and z plane have been removed due the infinite extent in these dimensions, Eq. (9.2) in this case reduces to a function of x and time, t. To obtain a solution for flux ϕ as a function of distance x and time t the usual approach is to separate these variables in terms of the product of two separate functions. Here these are defined as X and Y thus,

$$\frac{1}{v}\frac{\partial\phi}{\partial t}-D\frac{\partial^2\phi}{\partial x^2}+\Sigma_a\phi(x,t)=S(x,t) \tag{9.6}$$

where, given we define $\phi(x,t)=X(x)Y(t)$ and substitute into Eq. (9.6),

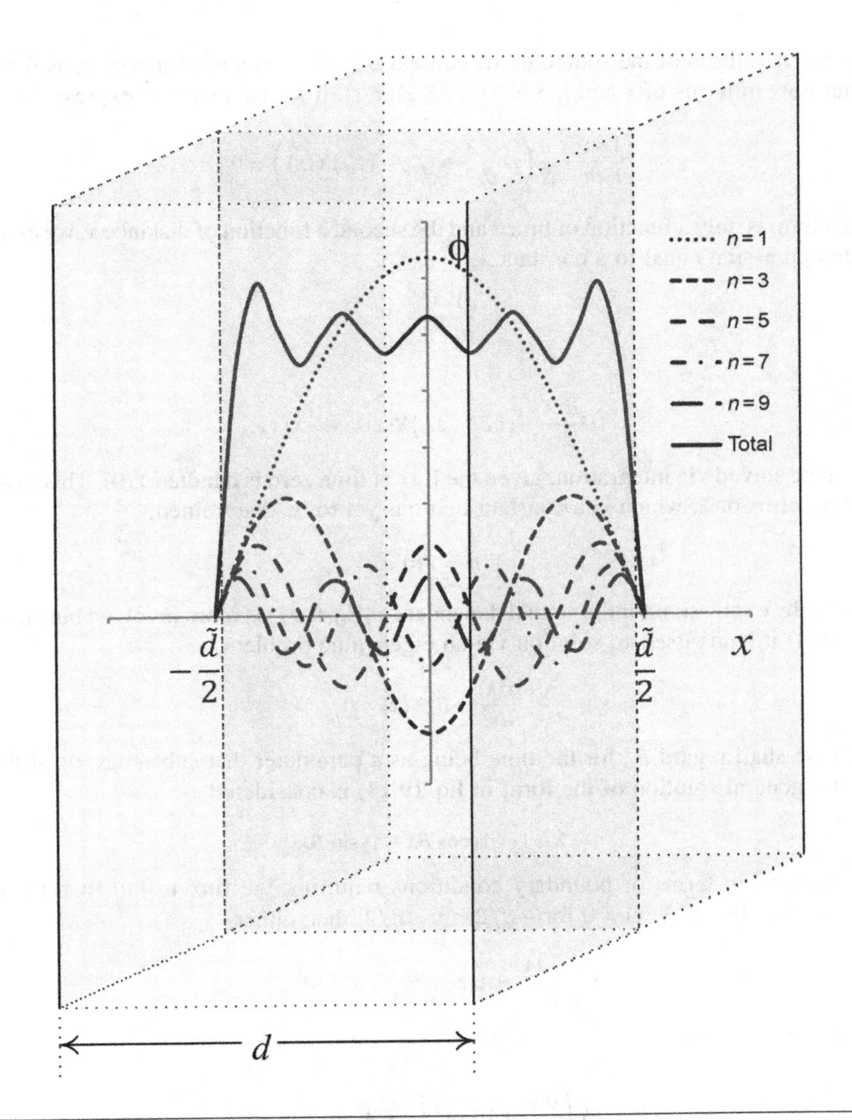

FIG. 9.2

A plot of distance x versus neutron flux ϕ for a uniform slab of fissile material of width d depicting the five modes from $n=1$ through $n=9$ with the total assuming $\phi_0(x)=$ constant for a fixed point in time, a negligible period after $t=0$.

$$\frac{X(x)}{v}\frac{\partial Y}{\partial t}-Y(t)D\frac{\partial^2 X}{\partial x^2}+\Sigma_a X(x)Y(t)=S(x,t) \quad (9.7)$$

and dividing through by $X(x)Y(t)$ we obtain

$$\frac{1}{vY}\frac{\partial Y}{\partial t}-\frac{D}{X}\frac{\partial^2 X}{\partial x^2}+\Sigma_a-S(x,t)=0 \quad (9.8)$$

Referring to the definition of the source in the context of fission in a reactor where, as for the case of Eq. (9.3), but now in terms of x and t, $S(x,t) = v\Sigma_f \phi(x,t)$, it is customary to express this as

$$\frac{1}{Y}\frac{\partial Y}{\partial t} - \frac{v}{X}\left(D\frac{\partial^2 X}{\partial x^2} + (\Sigma_a - v\Sigma_f)X(x)\right) = 0 \tag{9.9}$$

Since the first term is only a function of time t and the second a function of distance x, we can write each as a separate expression equal to a constant λ,

$$\frac{dY}{dt} = -\lambda Y \tag{9.10}$$

and

$$D\frac{d^2 X}{dx^2} + (v\Sigma_f - \Sigma_a)X(x) = -\frac{\lambda}{v}X(x) \tag{9.11}$$

Eq. (9.10) can be solved via integration, given the flux at time zero is denoted $Y(0)$. This gives the time dependence in terms of λ, which is a constant that has yet to be determined,

$$Y(t) = Y(0)e^{-\lambda t} \tag{9.12}$$

The solution of the expression for the spatial dependence (Eq. 9.11) is more involved but since it has the form of Eq. (9.4) it lends itself to solution via an eigenvalue problem,

$$\frac{d^2 X_n}{dx^2} + B_n^2 X(x) = 0 \tag{9.13}$$

In Eq. (9.13) we shall regard B_n for the time being as a parameter that subsumes all of the terms in Eq. (9.11). If a general solution of the form in Eq. (9.14) is considered,

$$X(x) = A_1 \cos Bx + A_2 \sin Bx \tag{9.14}$$

and this is analysed in terms of boundary conditions requiring the flux to fall to zero beyond the confines of the slab, that is $X(x) \neq 0$ for $-\tilde{d}/2 < x < \tilde{d}/2$, then either,

$$X\left(\frac{-\tilde{d}}{2}\right) = A_1 \cos {}^{B\tilde{d}}/_2 - A_2 \sin {}^{B\tilde{d}}/_2 \tag{9.15}$$

or

$$X\left(\frac{\tilde{d}}{2}\right) = A_1 \cos {}^{B\tilde{d}}/_2 + A_2 \sin {}^{B\tilde{d}}/_2 \tag{9.16}$$

Here the tilde indicates that the estimate for where the flux falls to zero is approximate because it will extend marginally into the vacuum surrounding the slab; $\left(\tilde{d} - d\right)/2$ corresponds to what is known as the *extrapolation distance* associated with this phenomenon.

Combining Eqs (9.15), (9.16) requires that for the flux to fall to zero either side of the slab simultaneously,

$$A_1 \cos {}^{B\tilde{d}}/_2 = 0 \text{ and } A_2 \sin {}^{B\tilde{d}}/_2 = 0 \tag{9.17}$$

Clearly if both A_1 and A_2 were zero this would yield the mathematically viable and yet trivial solution of $X(x)=0$ for all x. Reference to the cosine and sine functions, respectively, highlights the periodicities at which these expressions are satisfied, or the case of either $A_1=0$ or $A_2=0$, such that the eigenfunctions are,

$$X(x)=A_n\cos\frac{n\pi x}{\tilde{d}} \quad \text{for } n=1,3,5\ldots \quad \text{or} \quad X(x)=A_n\sin\frac{n\pi x}{\tilde{d}} \quad \text{for } n=2,4,6\ldots \tag{9.18}$$

and the eigenvalues are defined as

$$B_n^2=\left(\frac{n\pi}{\tilde{d}}\right)^2 \tag{9.19}$$

In a reactor we are interested in symmetric solutions. These are only given for the cosine case so we arrive at a series of modes (effectively a Fourier series) for contributions to the flux in the form of $\phi(x,t)=X(x)Y(t)$,

$$\phi(x,t)=\sum_{n\ odd}A_n e^{-\lambda_n t}\cos\frac{n\pi x}{\tilde{d}} \tag{9.20}$$

To illustrate this important and yet perhaps a little obscure concept, the dependence of the neutron flux ϕ with x is given in Fig. 9.2, for the first five modes and also the sum of these. This 'snapshot' is meant to represent the state shortly after $t=0$ before any indication of decay with time is evident. Note that the coefficients are derived as per such a series,

$$A_n=\frac{2}{\tilde{d}}\int_{-\tilde{d}/2}^{\tilde{d}/2}\phi_0(x)\cos B_n x dx \tag{9.21}$$

and the corresponding terms are given as a function of n in Table 9.1.

Returning to Eq. (9.10) and using Eq. (9.11) it is possible to define λ, the *time eigenvalue*,

$$\lambda_n=vDB_n^2-\left(v\nu\Sigma_f-v\Sigma_a\right) \tag{9.22}$$

Since, with increasing n, the coefficients B_n also increase so then do the time eigenvalues. This implies, as per the negative exponent in Eq. (9.12), that higher-order modes decay more quickly than lower-order modes. Thus we tend to be concerned primarily with the fundamental mode that remains, associated with B_1^2. This is referred to specifically as the *geometric buckling* B_g^2,

$$B_g^2=\left(\frac{\pi}{\tilde{d}}\right)^2 \tag{9.23}$$

9.4.3 ACCOUNTING FOR ANISOTROPIC SCATTERING ON LIGHT ISOTOPES

Interestingly, our consideration of leakage thus far via the reactor equation, and particularly in terms of the definition of the diffusion coefficient D, assumes that neutrons are scattered isotropically in the laboratory frame of reference. This is a satisfactory approximation for heavy scatterers but neutrons exhibit an anisotropy as a result of their interaction on light isotopes, and of course light isotopes are abundant in a reactor as part of the moderator and coolant. Scattering on light isotopes has the potential to result in greater energy transfer, target recoil and moderation (as discussed in Chapter 6) than on heavier isotopes. The level of maximum energy transfer for heavier isotopes is limited in comparison with light isotopes with greater isotropy observed as a result, relative to light isotopes where is a greater tendency for neutrons

Table 9.1 Representative Eigenvalues B_n, the Corresponding Coefficients A_n and the Spatial Variables $\cos n\pi x/\widetilde{d}$

n	$B_n = \dfrac{n\pi}{\widetilde{d}}$	$A_n = \dfrac{2}{\widetilde{d}} \displaystyle\int_{-\widetilde{d}/2}^{\widetilde{d}/2} \phi_0 \cos B_n x\,dx$	$\cos \dfrac{n\pi x}{\widetilde{d}}$
1	$B_1 = \pi/\widetilde{d}$	$4\phi_0/\pi$	$\cos \dfrac{\pi x}{\widetilde{d}}$
3	$B_3 = 3\pi/\widetilde{d}$	$-4\phi_0/3\pi$	$\cos \dfrac{3\pi x}{\widetilde{d}}$
5	$B_5 = 5\pi/\widetilde{d}$	$4\phi_0/5\pi$	$\cos \dfrac{5\pi x}{\widetilde{d}}$
7	$B_7 = 7\pi/\widetilde{d}$	$-4\phi_0/7\pi$	$\cos \dfrac{7\pi x}{\widetilde{d}}$
9	$B_9 = 9\pi/\widetilde{d}$	$4\phi_0/9\pi$	$\cos \dfrac{9\pi x}{\widetilde{d}}$

to be scattered forwards. Of course, such scattering processes are stochastic and it is necessary to consider the *average* scattering angle. This is usually parameterised by the corresponding cosine, $\overline{\mu}_0$. This yields, as demonstrated in Case Study 9.4 and depicted by the schematic diagram in Fig. 9.3, the following expression,

$$\overline{\mu}_0 = \frac{2}{3A} \tag{9.24}$$

With this convention we can now return to the anisotropy in neutron scattering on light isotopes. The bias towards forward scattering in the laboratory frame of reference, that is not accounted for in the isotropic case, will result in a larger mean free path in the forward direction than is inferred by the reciprocal of Σ_t. This is manifest mathematically by a modulation of macroscopic scattering cross

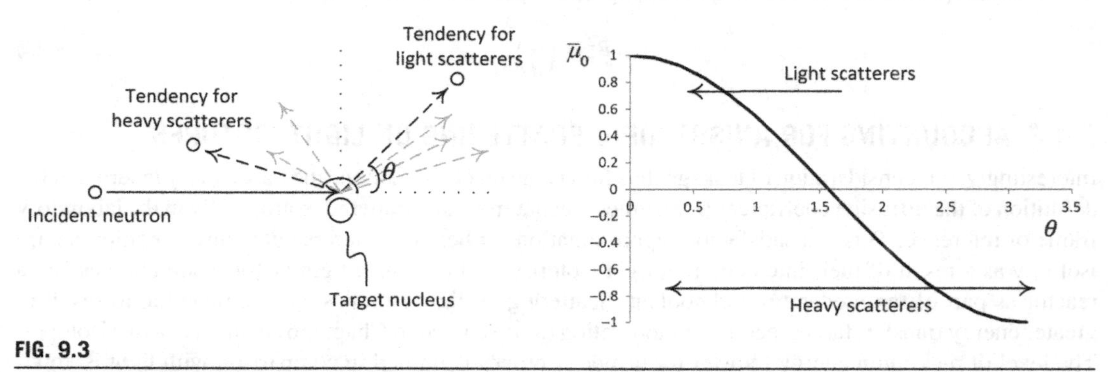

FIG. 9.3

A schematic diagram of a neutron scattering off a nucleus (left) depicting the scattering angle θ and the scattering angle cosine dependence with θ, right.

section Σ_s by the average scattering angle cosine. Hence, a modification to the macroscopic scattering cross section is introduced which is usually referred to as the *transport cross section*, Σ_{tr}, such that

$$\Sigma_{tr} = \Sigma_S(1 - \bar{\mu}_0) \quad (9.25)$$

and the diffusion coefficient is thus revised, assuming $\Sigma_s >> \Sigma_a$,

$$D = \frac{1}{3\Sigma_{tr}} \quad (9.26)$$

Clearly, given Eq. (9.24) for the anisotropic extreme, the forward polarisation of neutron scattering on light isotopes acts to increase diffusion and therefore leakage whereas for greater isotropy $A \rightarrow \infty$ and $D \rightarrow 1/3\,\Sigma_s$. This highlights an important materials-dependent factor in reactor design. Note that, for compounds, the corresponding average scattering angle cosine for each constituent element is usually included for each of the individual contributions to the total macroscopic scattering cross section. For hydrogenous substances, experimental values are often used [2].

9.4.4 THE CONDITION FOR CRITICALITY

The expression for geometric buckling expressed in Eq. (9.23) describes the size and shape of the neutron flux where fission is the sole source of neutrons. For time-dependent scenarios contributions to the flux die away with time as a result of absorption and leakage, in the absence of additional sources. The fundamental mode exhibits the longest time dependence reflected by the time eigenvalue λ_1 being smallest for this mode.

When Fermi and his team achieved the first criticality by gradually withdrawing 'the' control rod they observed short-lived increases in neutron flux on neutron monitors at each stage of withdrawal. These then died away in a manner that was qualitatively consistent with the time-eigenvalue dependence explored earlier in this chapter, albeit not quantitatively exact since the CP-1 geometry was not the idealised fissile slab used as the example in Section 9.4.2. Eventually, a position of rod withdrawal was reached where the flux no longer changed with time but remained constant due to the emission of neutrons from ^{235}U fission being sufficient to sustain further fission. This, as we know from earlier discussions in Chapter 4, corresponds to the critical state characterised by $k_{\text{eff}} = 1$. However, this historical recollection reminds us of the other perhaps more tangible indication of a controlled nuclear criticality: since the reaction is self-sustaining the neutron flux at this point is *time-independent*.

This perspective applied to the expression for the first mode, since this is the mode that sustains longest, requires for time independence that the time eigenvalue (expression (9.20)) for the case of $n = 1$ is zero, thus $\lambda_1 = 0$. This yields the corresponding expression for B_1 which is associated with the material properties and is thus termed the *material buckling* B_m,

$$B_m^2 = \frac{\nu\Sigma_f - \Sigma_a}{D} \quad (9.26)$$

A greater *material* buckling (inferring a larger $\nu\Sigma_f$, a smaller absorption cross section and/or smaller diffusion constant) infers a more favourable propensity for criticality. Conversely, a small *geometric* buckling is favoured with respect to criticality (shape notwithstanding) as this infers a larger d and thus a larger core. However, the material composition of a core and its size and shape are clearly sympathetic factors; a criticality is not possible without the combination of the two being satisfied. This condition is met when the buckling equality is satisfied,

$$B_m^2 = B_g^2 \quad (9.27)$$

which corresponds to the state at which $k = 1$. Therefore for the case that $B_m^2 > B_g^2$ supercriticality is inferred, that is, $k > 1$ and conversely $B_m^2 < B_g^2$ infers the subcritical condition.

Returning to the flux time dependence of Eq. (9.12), this can be expressed in more familiar terms now that the separation of variables notation can be dispensed with. The time eigenvalue for the fundamental flux mode is the inverse of the reactor period, $\lambda = 1/T$, and this can be written as

$$\phi(t) = \phi(0)e^{-t/T} \tag{9.28}$$

9.5 NEUTRON TRANSPORT MECHANISMS AND CONCEPTS

Whilst the scope of neutron transport is beyond the anticipated level of this text, it would be remiss to leave the reader with the impression that diffusion accounted for all of the likely neutron population scenarios that might arise. Fundamentally, diffusion assumes that the neutrons comprising a population have velocities where the angular distribution of the corresponding vectors is isotropic. A more complete assessment of the neutron distribution is made if the individual velocity vectors are taken into account; that is, if the direction of these vectors is included as a variable in our analysis.

As was our approach in the context of diffusion and the definition of the neutron current density, we shall at first confine our thinking to a limited volume. The various quantities of interest, neutron number, flux, etc. will be defined as a function of a number of dependent variables. Some of these are associated with space but there are others that can require more thought; these are introduced below. Interestingly, in adopting a more comprehensive set of dependent variables, the conventional definition of space being one limited to geometry has to be replaced by one of a *phase space*. This enables account of both position and velocity-related variables to be made. This is necessary because in an infinite reactor with, by definition, no boundaries or interfaces it is plausible to assume that the neutron flux has a weak dependence on angle; neutron transport in this scenario can be approximated as being due to diffusion alone. However, in a system where the neutron transport is frequently anisotropic then this more comprehensive approach is desirable.

To visualise this concept, Fig. 9.4 provides an illustration of an elemental volume and the mechanisms by which the number of neutrons in the volume element is influenced. Note that since the aim is now to take account of directionality, a more significant use of vector notation is made.

9.5.1 MECHANISMS THAT INFLUENCE THE NEUTRON POPULATION IN A SMALL VOLUME

Firstly, we might consider the mechanisms by which neutrons might *join* and those by which they might *leave* the elemental volume described above:

- *Production*: Perhaps the most obvious contribution; neutrons are produced *in* the volume from induced fission and, to a much lesser extent, spontaneous fission and α, n reactions. This contribution is conventionally assigned the notation S (Eq. 9.2), and as per Chapter 8. It is related to the product of the number of neutrons emitted per fission, ν, and the macroscopic fission cross section, Σ_f.
- *Diffusion*: Neutrons pass into the volume and pass out of it. This is also referred to as *streaming* or *flow*.

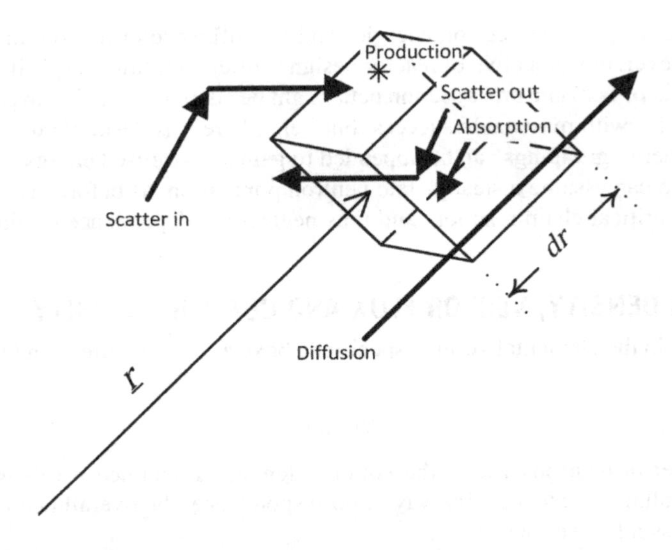

FIG. 9.4

An elemental volume of dimension $d\underline{r}$ at $\underline{r}$ depicting the various mechanisms by which neutron number can change within it.

- *Scattering* (*in*): Neutrons can also have their trajectory changed as a result of a scattering interaction outside of the volume. This interaction causes them to enter the volume, thus leading to an increase in the number of neutrons in the volume.
- *Scattering* (*out*) *and absorption*: Similarly, neutrons can be scattered whilst within the volume causing them to scatter out. This is related to the macroscopic scattering cross section, Σ_s. Neutrons can also be lost due to absorption whilst in the volume, which is dependent on the macroscopic absorption cross section, Σ_a. Both of these mechanisms result in a net loss of neutrons in the specified volume.

9.5.2 DEPENDENT VARIABLES AND THE CONCEPT OF PHASE SPACE

Having summarised the fundamental influences on neutron balance, it is necessary to consider the variables on which they depend, firstly in the absence of any assumptions. In particular, because isotropy is no longer assumed, we should include variables relating to velocity (vectors), in addition to those relating to position (scalars). Put simply, we need to account for the direction in which the neutrons are moving. Hence we have:

- *Position*: the element of volume is defined as being of dimension $d\underline{r}$ at position $\underline{r}$
- *Direction*: in addition to physical space, the element defines an element of solid angle $d\underline{\Omega}$ about a direction $\hat{\underline{\Omega}}$. The latter is a unit vector specified by the neutron velocity, viz. $\hat{\underline{\Omega}} = \underline{v}/|\underline{v}|$
- *Energy E* and *time t*

There are two significant assumptions that can be made at this stage. First, the energy dependence E can be removed if it is assumed that neutrons in a given environment all have the same energy. This is known as the *one-speed approximation*. It might appear a little drastic given earlier discussions concerning the range of possible neutron energies and the influence of this on, in particular, neutron cross sections. However, one objective of reactor design is often to fashion a specific neutron spectrum (for example, thermal or fast) and cross section data might be adapted by taking averages across a specified group of energies with reasonable success. Further, where it is desirable to extend this simplification, additional energy groupings can be appended to gain a discretised energy-dependent picture if needed.[1] Second, we can assume a steady-state neutron population, as before, as would be characteristic of a sustained, critical chain reaction, and thus neglect the dependence on time.

9.5.3 NEUTRON DENSITY, VECTOR FLUX AND CURRENT DENSITY

The neutron density in the elemental volume specified above at $\underline{r}$ and in direction $\hat{\underline{\Omega}}$ in phase space can be defined,

$$N(\underline{r},\hat{\underline{\Omega}}) \tag{9.29}$$

such that the number of neutrons within the volume element is obtained if this is multiplied by the dimensions of the volume $d\underline{r}$ and $d\hat{\underline{\Omega}}$. By way of correspondence, the overall neutron density is given by the integral across all directions,

$$N(\underline{r}) = \int_0^{4\pi} N(\underline{r},\hat{\underline{\Omega}})\,d\hat{\underline{\Omega}} \tag{9.30}$$

and the *neutron flux* $\phi(\underline{r})$ is as per the discussion in Chapter 8,

$$\phi(\underline{r}) = vN(\underline{r}) \tag{9.31}$$

It is also necessary to define the number of neutrons travelling in a particular direction, per unit area, per unit time. This is different to the *neutron* flux $\phi(\underline{r})$ because the direction is specified.[2] Hence, it is often referred to as the *angular neutron flux*, $\varphi(\underline{r},\hat{\underline{\Omega}})$. Of course the neutrons involved are the 'same' and are thus related, as per,

$$\phi(\underline{r}) = \int_0^{4\pi} \varphi(\underline{r},\hat{\underline{\Omega}})\,d\hat{\underline{\Omega}} \tag{9.32}$$

where the integration is performed across all directions. Further to Eq. (9.3) it follows that

$$\varphi(\underline{r},\hat{\underline{\Omega}}) = N(\underline{r},\hat{\underline{\Omega}})\,v\hat{\underline{\Omega}} \tag{9.33}$$

where $v = |\underline{v}|$.

It is also necessary to define the *neutron current density*, $\underline{J}(\underline{r})$, which is the sum of all the individual angular neutron currents $j(r,\hat{\underline{\Omega}})d\hat{\underline{\Omega}}$ acting in the direction $\hat{\underline{\Omega}}$ within $d\hat{\underline{\Omega}}$ over 4π. $\underline{J}(\underline{r})$ is the net number of neutrons flowing in a specific direction and has units of neutron flux. As per the earlier discussion, the neutron current density is related to the *gradient* of the neutron flux via,

[1]Computational methods quickly become attractive as geometrical and/or materials complexities escalate.
[2]The direction is specified as orthogonal to the plane of the area through which the neutrons are envisaged passing through.

$$\underline{J}(\underline{r}) = -D\underline{\nabla}\phi(\underline{r}) \tag{9.34}$$

as per Fick's law where D is the diffusion coefficient, $\underline{\nabla} = \frac{\partial}{\partial x}\hat{\underline{i}} + \frac{\partial}{\partial y}\hat{\underline{j}} + \frac{\partial}{\partial z}\hat{\underline{k}}$ and $D = 1/3\Sigma_{tr}$.

9.6 DEVELOPMENT OF THE ONE-GROUP TRANSPORT EQUATION

Having defined the fundamental variables, the dependent variables and explained the assumptions associated with energy and time, we can return to the summary of the constituent mechanisms described earlier in this section and associate some quantitative meaning to them.

9.6.1 PRODUCTION

Beginning with neutron production that occurs in the elemental volume of phase space, it is reasonable to assume that this contribution is isotropic, that is, fission yields neutrons uniformly throughout 4π; hence s (the density of those produced) is a function of position only. The neutrons are produced throughout 4π, such that $S(\underline{r})$ is the total, then

$$s(\underline{r}) = \frac{S(\underline{r})}{4\pi} \tag{9.35}$$

where $S(\underline{r}) = \nu\Sigma_f\phi(\underline{r})$.

9.6.2 DIFFUSION

The diffusion mechanism is related to the divergence of the angular neutron flux, as per earlier discussions and derivations, but given the direction. Hence this corresponds to,

$$\hat{\underline{\Omega}}.\underline{\nabla}\varphi(\underline{r},\hat{\underline{\Omega}}) \tag{9.36}$$

9.6.3 SCATTERING IN

Scattering results in a change in the trajectory of a neutron causing it to enter $d\hat{\underline{\Omega}}$. This contribution is dependent on the direction of the neutron *before* the interaction, which is denoted by $\hat{\underline{\Omega}}'$, as opposed to $\hat{\underline{\Omega}}$ afterwards. Thus the neutron density prior to scattering is $N(\underline{r}, \hat{\underline{\Omega}}')$. The interaction rate is the product of the neutron velocity v, the macroscopic scattering cross section Σ_s and the vector flux, and to take account of all possible prior trajectories it is necessary to integrate over $\hat{\underline{\Omega}}'$, as per,

$$d\underline{r}\, d\hat{\underline{\Omega}} \int \Sigma_s(\hat{\underline{\Omega}}' \to \hat{\underline{\Omega}})\, \varphi(\underline{r},\hat{\underline{\Omega}}')\, d\hat{\underline{\Omega}}' \tag{9.37}$$

Importantly, from a conceptual perspective, $\Sigma_s(\hat{\underline{\Omega}}' \to \hat{\underline{\Omega}})$ is the differential neutron scattering cross section as was introduced in Chapter 6, in this case for scattering from $\hat{\underline{\Omega}}$ to $\hat{\underline{\Omega}}'$. If the assumption that scattering is isotropic in the laboratory frame of reference is made[3] then

[3]This is sound in the centre-of-mass frame but scattering on light isotopes (consider moderation by hydrogen) is not isotropic in the laboratory frame as discussed earlier in this chapter.

$$\Sigma_s(\hat{\underline{\Omega}}' \to \hat{\underline{\Omega}}) = \frac{\Sigma_s}{4\pi} \tag{9.38}$$

9.6.4 SCATTERING OUT AND ABSORPTION LOSSES

The interaction rate governing scattering and absorption losses is the product of the neutron velocity v and the total macroscopic cross section, Σ_t, where $\Sigma_t = \Sigma_s + \Sigma_a$. Thus the number lost from the volume per unit time is

$$v\Sigma_t\varphi(\underline{r}, \hat{\underline{\Omega}})d\underline{r}\, d\hat{\underline{\Omega}} \tag{9.39}$$

Now, summing each of the terms defined above, gives the steady-state, one-speed neutron transport equation,

$$-\hat{\underline{\Omega}}.\bar{\nabla}\varphi(r, \hat{\underline{\Omega}}) + \int \frac{\Sigma_s}{4\pi}\varphi(\underline{r}, \hat{\underline{\Omega}}')\, d\hat{\underline{\Omega}}' - v\Sigma_t\varphi(\underline{r}, \hat{\underline{\Omega}}) + \nu\Sigma_f \int \varphi(\underline{r}, \hat{\underline{\Omega}})\, d\hat{\underline{\Omega}} = 0 \tag{9.40}$$

where the total macroscopic cross section Σ_t for this case is independent of position[4] and energy.

CASE STUDIES

CASE STUDY 9.1: DERIVING A RELATIONSHIP FOR A PARTIAL NEUTRON CURRENT IN A HOMOGENEOUS MEDIUM

If the neutron flux varies only slowly (i.e. is limited to a linear dependence in $\underline{r}$) it is possible to derive an approximation between neutron current and neutron flux which is important in determining the relationship between neutron leakage and neutron flux.

Neutrons scatter in and out of an element of volume in all three dimensions, with the difference between the input and the output per unit time per unit area being the current density. This can be determined for one dimension and then generalised to all space. The geometry central to this case study is illustrated in Fig. 9c.1.

The scattering interaction rate per unit volume at $\underline{r}$ is the product of the corresponding macroscopic cross section Σ_s and the flux at $\underline{r}$. Thus the rate in an elemental volume dV is

$$\Sigma_s\phi(\underline{r})dV$$

If we define an element of area dA in the $x-y$ plane centred at the origin relative to $\underline{r}$, the proportion of the total neutron flux that scatters from dV at $\underline{r}$ towards dA is the ratio of the area subtended by the angle of $\underline{r}$ relative to the vertical,

$$dA\cos\theta \Big/ 4\pi r^2$$

It is still possible for a neutron to undergo a scattering interaction in between leaving dV and reaching dA, *and* so it is necessary to fold in the corresponding form of the Lambert-Beer law. Thus the fraction becomes

$$e^{-\Sigma_s r} dA\cos\theta \Big/ 4\pi r^2$$

and the element of volume $dV \approx r^2 \sin\theta\, dr\, d\theta\, d\psi$, further to Fig. 9c.1, noting that the small angle approximation is assumed for $d\theta$ and $d\psi$ such that $\sin d\theta \approx d\theta$ and $\sin d\psi \approx d\psi$. Hence, the rate of neutrons passing through dA in the negative z direction from dV is

[4]Homogeneity has been assumed.

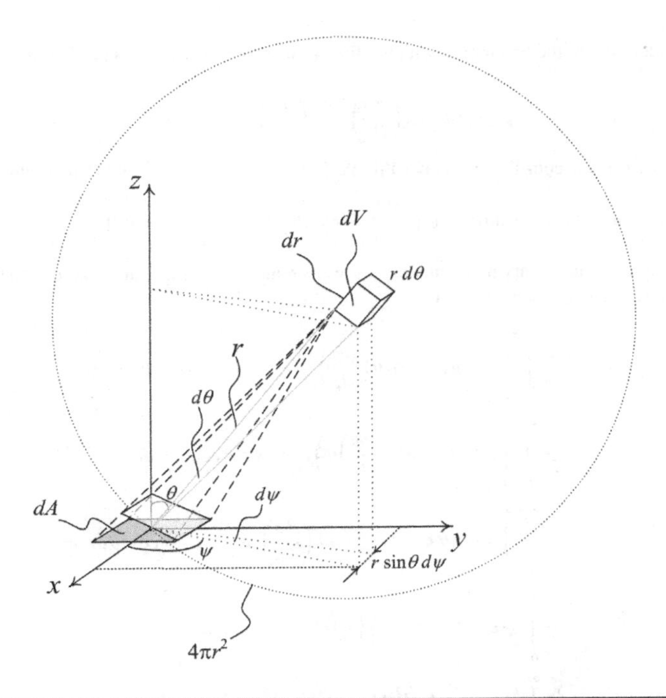

FIG. 9C.1

A schematic diagram depicting an elemental volume dV and an element of surface area dA subtending an angle θ from the vertical and ψ from the x axis.

$$\Sigma_s e^{-\Sigma_s r}\phi(\underline{r})\, dA\, \sin\theta\, \cos\theta\, dr\, d\theta\, d\psi \Big/ 4\pi$$

To obtain the total rate of neutron scatters passing through the z plane from top to bottom (thus in the negative direction), it is necessary to integrate through $0 \le \theta \le \pi/2$ (from the vertical through to the $x-y$ plane) and then throughout the whole of $x-y$ (i.e. $0 \le \psi \le 2\pi$) over $0 \le r \le \infty$, dividing through by the area dA (to yield the corresponding partial neutron current) J_{z-},

$$J_{z-} = \frac{\Sigma_s}{4\pi}\int_0^{2\pi}\int_0^{\pi/2}\int_0^{\infty} e^{-\Sigma_s r}\phi(\underline{r})\, dA\, \sin\theta\, \cos\theta\, dr\, d\theta\, d\psi$$

CASE STUDY 9.2: DERIVING AN APPROXIMATE RELATIONSHIP BETWEEN NEUTRON CURRENT AND NEUTRON FLUX

The net neutron current through a given plane is equal to the difference in the input and output partial currents but we do not have a generic form for $\phi(r)$. If we assume the flux varies only slowly with r, then over small distances we might assume ϕ is linearly related with r. Using a Maclaurin series, a function $f(x)$ can be approximated at $x=0$ as follows:

$$f(x) \approx f(0) + \frac{x}{1!}\left(\frac{\partial f}{\partial x}\right)_0 + \frac{x^2}{2!}\left(\frac{\partial^2 f}{\partial x^2}\right)_0 + \frac{x^3}{3!}\left(\frac{\partial^3 f}{\partial x^3}\right)_0 + \cdots$$

Limiting our consideration to the linear case with the flux at the origin being ϕ_0, as per Fig. 9c.1, we can approximate $\phi(\underline{r})$ as

$$\phi(\underline{r}) \approx \phi_0 + x\left(\frac{\partial \phi}{\partial x}\right)_0 + y\left(\frac{\partial \phi}{\partial y}\right)_0 + z\left(\frac{\partial \phi}{\partial z}\right)_0$$

Re-writing in spherical polar coordinates, as per Fig. 9c.1, $x = r\sin\theta\cos\psi$, $y = r\sin\theta\sin\psi$ and $z = r\cos\theta$,

$$\phi(r) \approx \phi_0 + r\sin\theta\cos\psi\left(\frac{\partial \phi}{\partial x}\right)_0 + r\sin\theta\sin\psi\left(\frac{\partial \phi}{\partial y}\right)_0 + r\cos\theta\left(\frac{\partial \phi}{\partial z}\right)_0$$

Fortunately, in terms of the necessary mathematics, the terms relating to ψ integrate to zero with the limits $0 \le \psi \le 2\pi$, which simplifies the partial current density J_{z-} to,

$$J_{z-} = \frac{\Sigma_s}{4\pi}\int_0^{2\pi}\int_0^{\pi/2}\int_0^{\infty}\left(\phi_0 + r\cos\theta\left(\frac{\partial \phi}{\partial z}\right)_0\right)e^{-\Sigma_s r}\sin\theta\cos\theta\, dr\, d\theta\, d\psi$$

$$= \frac{\Sigma_s}{2}\int_0^{\pi/2}\int_0^{\infty}\left(\phi_0 + r\cos\theta\left(\frac{\partial \phi}{\partial z}\right)_0\right)e^{-\Sigma_s r}\sin\theta\cos\theta\, dr\, d\theta$$

$$= \frac{\Sigma_s}{2}\int_0^{\pi/2}\int_0^{\infty}\phi_0\sin\theta\cos\theta\, e^{-\Sigma_s r} + r\left(\frac{\partial \phi}{\partial z}\right)_0 e^{-\Sigma_s r}\sin\theta\cos^2\theta\, dr\, d\theta$$

$$= \frac{\Sigma_s}{2}\int_0^{\infty}\frac{\phi_0}{2}\sin^2\theta\, e^{-\Sigma_s r} - \frac{r}{3}\left(\frac{\partial \phi}{\partial z}\right)_0 e^{-\Sigma_s r}\cos^3 dr\Bigg|_0^{\pi/2}$$

$$= \frac{\Sigma_s}{2}\int_0^{\infty}\frac{\phi_0}{2}e^{-\Sigma_s r} + \frac{r}{3}\left(\frac{\partial \phi}{\partial z}\right)_0 e^{-\Sigma_s r}\, dr$$

$$= \frac{\Sigma_s}{2}\left(-\frac{\phi_0}{2\Sigma_s}e^{-\Sigma_s r} + \frac{1}{3}\left(\frac{\partial \phi}{\partial z}\right)_0\left(-\frac{re^{-\Sigma_s r}}{\Sigma_s} - \frac{e^{-_s r}}{\Sigma_s^2}\right)\right)\Bigg|_0^{\infty}$$

$$J_{z-} = \frac{\phi_0}{4} + \frac{1}{6\Sigma_s}\left(\frac{\partial \phi}{\partial z}\right)_0$$

Similarly,

$$J_{z+} = \frac{\phi_0}{4} - \frac{1}{6\Sigma_s}\left(\frac{\partial \phi}{\partial z}\right)_0$$

Hence the net current loss for all three dimensions as per $J_{z+} - J_{z-}$, and so on, yields the diffusion coefficient $D = 1/3\Sigma_s$, notwithstanding anisotropy in the laboratory frame discussed in Section 9.4.3, and thus,

$$\underline{J}(\underline{r}) = -D\underline{\nabla}\phi(\underline{r})$$

CASE STUDY 9.3: NEUTRON CURRENT DENSITY AND DIFFUSION.

If we think of neutrons diffusing through a cubic volume of dimension $dx \times dy \times dz$, then there are three faces that they can diffuse into, i.e. those of areas $dx\,dy$, $dy\,dz$ and $dx\,dz$, and three that they can diffuse out of (of the same area but at dz, dx or dy beyond the face they entered through, respectively). To determine the net number of neutrons that diffuse through the volume, it is usual to consider the component of the neutron current density $\underline{J}$ travelling in the x direction. This travels through the surface at x that is of area $dydz$ and the number is related to the component J_x of $\underline{J}$ via,

$$J_x dy\, dz$$

whereas that at $x+dx$ that passes out of the volume is

$$\left(J_x + \frac{\partial J_x}{\partial x}dx\right)dy\,dz$$

Hence, the net number passing through *dydz* stems from the difference between the two expressions above,

$$\left(J_x + \frac{\partial J_x}{\partial x}dx\right)dy\,dz - J_x dy\,dz = \frac{\partial J_x}{\partial x}dx\,dy\,dz$$

Generalising to three dimensions by repeating for J_y and J_z gives the net number per unit volume (since the volume is $dx\,dy\,dz$) per unit time as

$$\frac{\partial J_x}{\partial x}\underline{\hat{i}} + \frac{\partial J_y}{\partial x}\underline{\hat{j}} + \frac{\partial J_z}{\partial x}\underline{\hat{k}} = \underline{\nabla} \cdot \underline{J}(\underline{r})$$

Since $\underline{J}(\underline{r}) = -D\,\underline{\nabla}\phi\,(\underline{r})$ then

$$\underline{\nabla} \cdot \underline{J}(\underline{r}) = -D\,\nabla^2\phi(\underline{r})$$

where the Laplacian:

$$\nabla^2 = \frac{\partial^2}{\partial x^2} + \frac{\partial^2}{\partial y^2} + \frac{\partial^2}{\partial z^2}$$

CASE STUDY 9.4: THE RELATIONSHIP BETWEEN THE AVERAGE SCATTERING ANGLE COSINE AND ATOMIC MASS

The relationship between the mass of the isotope on which a neutron scatters A and the angle θ between the direction of the neutron before $\underline{\hat{\Omega}}'$ and afterwards $\underline{\hat{\Omega}}$ in the laboratory frame of reference is described by the cosine of that angle. From elementary vector algebra,

$$\underline{\hat{\Omega}}'.\underline{\hat{\Omega}} = \cos\theta$$

since $\underline{\hat{\Omega}}'$ and $\underline{\hat{\Omega}}$ are unit vectors. The average of the cosine is used conventionally to account for influence of the anisotropy of the scattering on the mean free path of the neutron, with the notation $\overline{\mu}_0$. This allows the macroscopic neutron scattering cross section to be adjusted for the effect of anisotropic scattering (predominant for light isotopes), to reflect the fact that forward scattering is more likely in this case leading to a longer mean free path in the forward scattering direction than in other directions.

To determine $\overline{\mu}_0$ the macroscopic scattering cross section Σ_s is defined as a function of the cosine thus $\Sigma_s(\mu_0)$ where the isotropic case is $\Sigma_s/4\pi$ since scattering is symmetrical either side of the trajectory of the incident neutron. Integrating across the product of the cosine and $\Sigma_s(\mu_0)$ from +1 to −1 $(0 \le \theta \le \pi)$, and normalising to the isotropic case gives the average cosine,

$$\overline{\mu}_0 = \frac{4\pi}{\Sigma_s}\int_{-1}^{1} \mu_0 \Sigma_s(\mu_0)d\mu_0$$

For the case of isotropy when $\Sigma_s(\mu_0) = \Sigma_s/4\pi$,

$$\overline{\mu}_0 = \int_{-1}^{1} \mu_0 d\mu_0 = 0$$

and consequently $\Sigma_{tr} = \Sigma_s(1-\overline{\mu}_0) = \Sigma_s$ as expected.

In the centre-of-mass frame of reference the assumption that scattering of a neutron from an atomic nucleus is isotropic is reasonable whilst it is in the laboratory frame anisotropy is evident, especially for light isotopes. The scattering angle relative to the direction of incidence in the laboratory frame can be defined as θ_L and in the centre-of-mass frame as θ_C. It is useful to consider two planes in the kinematics following a neutron scattering interaction: that which is at right angles to the incident direction (related by the sine of the velocity) and that parallel (related to the cosine). As for the scattering angles, the magnitude of the velocities are defined v_L' and v_C' for the laboratory and centre-of-mass frame, respectively, such that for the plane orthogonal to the incident direction of the neutron is

$$v_L' \sin\theta_L = v_C' \sin\theta_C$$

whereas the distinction in the direction parallel to the direction of incidence is the offset due to the velocity of the centre-of-mass V_c,

$$v_L' \cos\theta_L = V_C + v_C' \cos\theta_C$$

where $V_C = m_n v_L/(m_n + M)$ and m_n is the mass of the neutron, M is the mass of the scattering nucleus and v_L is the magnitude of the incident velocity of the neutron. For the case of $m_n = 1$ and $M = A$ and combining the expressions above,

$$\tan\theta_L = \frac{A\sin\theta_C}{1 + A\cos\theta_C}$$

Returning to the expression for $\overline{\mu}_0$ above, because scattering is assumed isotropic in the centre-of-mass frame, the cross section as a function of θ_C is $\Sigma_s/4\pi$. The differential cross sections in the centre-of-mass frame and laboratory frame ($\sigma_L(\theta_L)$ and $\sigma_C(\theta_C)$, respectively) are related as per,

$$\sigma_C(\theta_C)\sin\theta_C d\theta_C = \sigma_L(\theta_L)\sin\theta_L d\theta_L$$

so that

$$\overline{\mu}_0 = \frac{1}{2}\int_0^{\pi} \sin\theta_C \cos\theta_L d\theta_C$$

From this expression, relating θ_L to θ_C via tan θ_L and using the identity $\sin^2\theta + \cos^2\theta = 1$, we obtain

$$\cos\theta_L = \frac{1 + A\cos\theta_C}{\sqrt{A^2 + 1 + 2A\cos\theta_C}}$$

Substituting into the expression for $\overline{\mu}_0$,

$$\overline{\mu}_0 = \frac{1}{2}\int_0^{\pi} \frac{\sin\theta_C(1 + A\cos\theta_C)}{\sqrt{A^2 + 1 + 2A\cos\theta_C}} d\theta_C$$

This can be solved via the substitution $u = \cos\theta_C$ and then by parts to yield

$$\overline{\mu}_0 = {}^2/_{3A}$$

Sometimes the following approximation is used instead: $\overline{\mu}_0 \sim {}^1/_A$.

CASE STUDY 9.5: REAL REACTOR GEOMETRIES

The fissile slab is perhaps the simplest geometry with which to illustrate solutions of the reactor equation. However, it does not present the most efficient use of either space or fissile material in practise because of the significant leakage on either of the free surfaces defined by $-d/2 \leq x \leq d/2$. More practical geometries are, for example, the cylinder and the cube. The sphere offers the most efficient geometry but is less practical in terms of structural requirements of the fuel, shutdown rod, control rod and coolant infrastructure.

A right circular cylinder of radius *g* height *h*

Including the Laplacian in terms of cylindrical polar coordinates (radius r and height z) into the reactor equation of Eq. (9.4) gives the following expression:

$$\frac{\partial\phi^2}{\partial r^2} + \frac{1}{r}\frac{\partial\phi}{\partial r} + \frac{\partial\phi^2}{\partial z^2} + B^2\phi = 0$$

As for the case of the infinite slab, ϕ is separated into variables, this time of R and Z,

$$\phi(r,z)=R(r)Z(z)$$

with each resulting in a separate equation for R and Z equal to constants α^2 and β^2, respectively, such that $\alpha^2+\beta^2=B^2$,

$$\frac{\partial^2 R}{\partial r^2}+\frac{1}{r}\frac{\partial R}{\partial r}=-\alpha^2 R(r) \tag{9c.1}$$

$$\frac{\partial^2 Z}{\partial z^2}=-\beta^2 \tag{9c.2}$$

Similar to the case for the infinite slab, a solution to Eq. (9c.2) for Z is of the form

$$Z(z)=A_1\sin\beta z+A_2\cos\beta z$$

We require the flux to be non-negative (a negative flux would be unphysical) and symmetric in z; a sinusoid is not and thus $A_1=0$. Also, the flux needs to conform with the boundary condition that it is zero just beyond the defined height h of the cylinder, i.e. $\phi(z)=0$ for $z=\pm\tilde{h}/2$, giving $Z_n(z)=\cos n\pi/\tilde{h}$ with eigenvalue $\lambda_n^2=\left(n\pi/\tilde{h}\right)^2$ and thus $Z_1(z)=\cos\pi/\tilde{h}$ and $\lambda_1^2=\left(\pi/\tilde{h}\right)^2$ due to the constraint for non-negativity.

The solution for R is simplified if we substitute for $x=\alpha r$ such that

$$\frac{dF}{dr}=\frac{dF}{dx}\cdot\frac{dx}{dr}=\alpha\frac{dF}{dx}$$

and remembering that a combination of the chain rule *and* the product rule is needed to derive second derivatives thus,

$$\frac{d^2F}{dr^2}=\frac{d}{dr}\left(\frac{dF}{dx}\frac{dx}{dr}\right)=\frac{d}{dr}\left(\frac{dF}{dx}\right)\frac{dx}{dr}+\frac{dF}{dx}\frac{d^2x}{dr^2}$$

$$u=\frac{dF}{dx},\quad \frac{du}{dr}=\frac{du}{dx}\cdot\frac{dx}{dr}$$

$$\frac{d^2F}{dr^2}=\frac{d^2F}{dx^2}\left(\frac{dx}{dr}\right)^2+\frac{dF}{dx}\frac{d^2x}{dr^2}=\alpha^2\frac{d^2F}{dx^2}$$

Thus,

$$\alpha^2\frac{d^2R}{dx^2}+\frac{\alpha}{r}\frac{dR}{dx}+\alpha^2R(x)=0$$

and multiplying through by r and factorising in *terms of* $x=\alpha r$,

$$x^2\frac{d^2R}{dx^2}+x\frac{dR}{dx}+x^2R(x)=0 \tag{9c.3}$$

which is a form of Bessel's differential equation with the general solution,

$$R(x)=a_1J_0(x)+a_2Y_0(x)$$

where, since Eq. (9c.3) is a 2nd-order differential equation, there are two linearly-independent solutions characterised by J_0 and Y_0 for the case when the integer $v=0$. J_0 and Y_0 are Bessel functions of zeroth order of the first and second kind, respectively, where

$$J_v(x)=x^v\sum_{m=0}^{\infty}\frac{(-1)^m x^{2m}}{2^{2m+v}m!\Gamma(v+m+1)}$$

and

$$Y_v(x)=\frac{J_v(x)\cos v\pi-J_{-v}(x)}{\sin v\pi}$$

and the gamma function Γ is defined,

$$\Gamma(\alpha+1)=\int_0^\infty e^{-t}t^{\alpha-1}dt=\alpha!$$

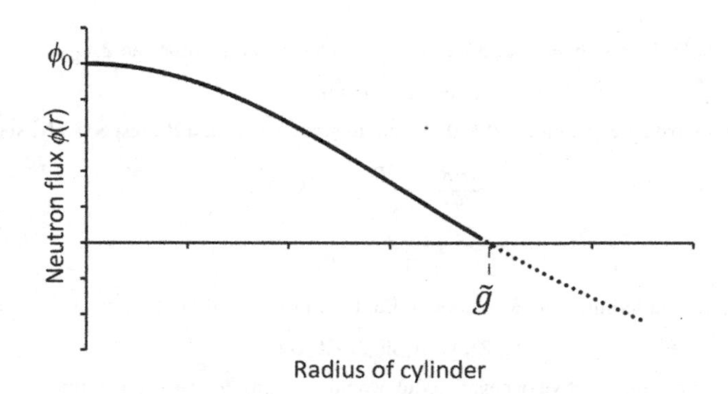

FIG. 9C.2

The dependence of neutron flux $\phi(r, z)$ versus radius r for a right, circular cylindrical reactor of radius g at the central position in height, $z=0$ with maximum flux ϕ_0 and extrapolation radius $r=\tilde{g}$ [3].

So,

$$J_0(x) = 1 - \frac{x^2}{4} + \frac{x^4}{64} - \frac{x^6}{2304}\ldots, \quad Y_0(x) = \frac{J_0(x)(\cos v\pi - 1)}{\sin v\pi}$$

For $x = \alpha r$ and when $r \to 0$, $Y_0(\alpha r) \to -\infty$ which is unphysical since the flux must be finite at the centre of the cylinder, so $a_2 = 0$ and hence,

$$R(r) = a_1 J_0(\alpha r)$$

At the extrapolation distance $\tilde{g}$ the flux should fall to zero, $R(\tilde{g}) = 0$, which corresponds to the position of the first zero in J_0. We know that the first zero of $J_0(\alpha r)$ is at $\alpha r = 2.405$ (from tables of Bessel function zeroes) so $\alpha = 2.405/\tilde{g}$ and thus combining the expressions for the separated variables R and Z,

$$\phi(r, z) = \phi_0 \cos\frac{\pi z}{\tilde{h}} J_0\left(\frac{2.405}{\tilde{g}} r\right)$$

where ϕ_0 is the flux at the centre of the cylinder, for which $z = 0$, $r = 0$, and $B_g^2 = \left(\pi/\tilde{h}\right)^2 + (2.405/\tilde{g})^2$. The dependence of flux with radius described by $J_0(2.405r/\tilde{g})$ is depicted in Fig. 9c.2.

CASE STUDY 9.6: THE EQUIVALENCE OF THE BUCKLING EQUALITY AND THE REPRODUCTION CONSTANT

It is illustrative to consider the equivalence and therefore the practical significance of the time eigenvalue arising from the solution of the reactor equation, λ_1,

$$\lambda_1 = v\Sigma_a - v\nu\Sigma_f + vDB_g^2$$

where for clarity v is the neutron velocity, ν is the number of neutrons emitted per fission, Σ_f is the macroscopic fission cross section, Σ_a is the macroscopic absorption cross section, D is the diffusion constant and B_g is the geometric buckling. Each term of the time eigenvalue represents a rate of either neutron absorption, production or leakage, respectively, and thus it is clear that it is related to the reciprocal of time. The expression can be rearranged by factorising in terms of Σ_a,

$$\lambda_1 = v\Sigma_a\left(1 - \frac{\nu\Sigma_f}{\Sigma_a} + \frac{D}{\Sigma_a}B_g^2\right)$$

where the diffusion constant is defined in terms of the *neutron diffusion length*, L, where $L = \sqrt{D/\Sigma_a}$ such that,

$$\lambda_1 = v\Sigma_a\left(-\frac{\nu\Sigma_f}{\Sigma_a} + 1 + L^2B_g^2\right)$$

The probability that neutrons are absorbed without leaking, per unit volume, is the ratio of the absorption rate to the sum of the absorption rate and the leakage rate,

$$P_{NL} = \frac{\Sigma_a\phi}{\Sigma_a\phi + DB_g^2\phi}$$

which, when divided through by $\Sigma_a\phi$ leads to the significant result for the nonleakage probability,

$$P_{NL} = \frac{1}{1 + L^2B_g^2}$$

Hence,

$$\lambda_1 = \frac{v\Sigma_a}{P_{NL}}\left(1 - P_{NL}\frac{\nu\Sigma_f}{\Sigma_a}\right)$$

Two things are significant here: since $k_\infty = \nu\Sigma_f/\Sigma_a$ then $P_{NL}\nu\Sigma_f/\Sigma_a$ is the multiplication factor, k. Secondly, $v\Sigma_a$ is the interaction rate for absorption assuming no leakage and thus the reciprocal is the time elapsed until a neutron is absorbed; multiplying this by P_{NL} gives the neutron lifetime in a finite reactor since we have then accounted for leakage, that is, $l = P_{NL}/v\Sigma_a$, thus

$$\lambda_1 = -\frac{(k-1)}{l}$$

which, returning to Eq. (9.28), it is clear that the time eigenvalue is the reciprocal of the reactor period, $\lambda_1 = 1/T$ and hence,

$$\phi(t) = \phi(0)e^{-\lambda_1 t}$$

REVISION GUIDE

On completion of this chapter you should:

- understand the concept of neutron balance
- be able to write down a word equation relating neutron balance and hence the neutron diffusion equation
- understand the origin of the terms in the neutron diffusion equation, and particularly the relationship of these terms to materials characteristics (via for example the macroscopic cross section)
- be able to define the *neutron flux*, *neutron current density*, *partial neutron current* and so forth
- be able to develop the reactor equation from the diffusion equation
- be able to solve the reactor equation for a simple geometry and to identify what is meant by *material buckling* and *geometrical buckling*
- understanding the processes and terminology associated with the development of the one-velocity, neutron transport equation

PROBLEMS

1. For a geometry of a cube (a rectangular parallelepiped) of dimensions $a \times b \times c$ solve the reactor equation to obtain expressions for the reactor flux (x, y, z), the geometric buckling B_g^2 and the maximum flux ϕ_0. The flux should be symmetric, non-negative and fall to zero at the extrapolation distance in all dimensions, i.e. $x = \pm\tilde{a}/2$, $y = \pm\tilde{b}/2$ and $z = \pm\tilde{c}/2$.

2. Explain what is meant by the terms *neutron flux* and *angular neutron flux*. What is the difference between them and, hence how are they related in mathematical terms?
3. For a geometry of a sphere, of radius p, solve the reactor equation to obtain expressions for the reactor flux $F(r)$, the geometric buckling B_g^2 and the maximum flux F_0, and sketch the dependence of F with r. The flux should be symmetric, non-negative and fall to zero at the extrapolation distance $\tilde{p}$.
4. Plot the dependence of the average cosine $\overline{\mu}_0$ for $A = 1$ through to $A = 238$. Hence determine the corresponding values of the diffusion coefficient D for the cases of light water, graphite, iron, tungsten, lead and uranium metal (the latter of sufficiently low enrichment to constitute being approximated as 100% ^{238}U) at thermal neutron energies, assuming $\Sigma_s >> \Sigma_a$. Plot the dependence of D versus A and hence calculate the difference between D that accommodates scattering anisotropy at each point in A with that that assumes isotropy. What do you conclude can be assumed about D as $A \rightarrow \infty$?
5. For a homogeneous, thermal reactor, with a right circular cylindrical geometry comprising light water perfused with UO_2 of low enrichment, describe the various contributing factors to the geometric buckling and the materials buckling. Explain what might comprise the major influence(s) preventing such a design from functioning.

REFERENCES

[1] N.G. Sjöstrand, Calculation of the geometric buckling for reactors of various shapes, http://www.iaea.org/inis/collection/NCLCollectionStore/_Public/38/088/38088353.pdf, 1958.
[2] K. Drozdowicz, Thermal neutron diffusion parameters dependent on the flux energy distribution in finite hydrogenous media, Henryk Niewodniczanski Institute of Nuclear Physics Report no. 1838/PN, 1999.
[3] E. Kreyszig, Advanced Engineering Mathematics, Wiley & Sons, London, 1972. ISBN: 0-471-88941-5.

FURTHER READING

[1] J.J. Duderstadt, L.J. Hamilton, Nuclear Reactor Analysis, Wiley, Delhi, 1976. ISBN: 978-81-265-4121-8.
[2] D.J. Bennett, J.R. Thomson, The Elements of Nuclear Power, Longman, New York, 1989. ISBN: 0-582-02224-X.
[3] R.A. Kneif, Nuclear Engineering, Taylor & Francis, 1992. ISBN: 1-56032-089-3.

CHAPTER

MAINSTREAM POWER REACTOR SYSTEMS 10

10.1 SUMMARY OF CHAPTER AND LEARNING OBJECTIVES

The conceptual requirements for the materials composition and geometry of nuclear fission reactors have been considered in the analysis provided in the previous chapter including the highly simplified case of the infinite slab; now we turn to the designs of *real* reactors generating electricity around the world at present. These are distinct from conceptual designs—as Hyman Rickover is reputed to have remarked, typically forthright: 'A practical reactor plant can be distinguished by the following characteristics: (1) it is being built now, (2) it is behind schedule, (3) it is requiring an immense amount of development on apparently trivial items, (4) it is very expensive, (5) it takes a long time to build because of the engineering development problems, (6) it is large, (7) it is heavy and (8) it is complicated.' Rickover's summary highlights the general distinction that real reactors are much more sophisticated than their conceptual counterparts.

There are a variety of potential options by which a nuclear reactor might be configured in terms of its composition, size and shape. However, as a result of the significant investment during and following the Second World War, many of these alternatives have been explored at some point in the past. Some of these developments have been taken no further while others have been adopted as mainstream designs replicated across many units. An interesting facet of the nuclear engineering discipline is that the salient features of the mainstream designs often reflect historical and geographical influences as well as engineering benefits. Some of these traits stem from the dawn of the industry whereas the modern era is typified by influences from lessons learnt from major accidents and a desire for reactor designs to be more adaptable to forecasted macroeconomic priorities.

In this chapter mainstream power reactor designs have been divided into five categories: pressurised water reactors (PWRs), boiling water reactors (BWRs), gas-cooled reactors (GCRs), heavy-water reactors (HWRs) and light-water, graphite reactors (LWGRs). The discussion is not exhaustive in terms of the specific details of all of the designs in operation today. This classification has been adopted on a somewhat arbitrary basis in terms of, for example, the material used for the reactor moderator because this is, in the author's experience, an effective way by which the salient concepts can be appreciated. Some of the designs that might not be classified as mainstream, together with advanced reactors and fast reactor systems are considered in Chapter 11.

Nuclear Engineering. https://doi.org/10.1016/B978-0-08-100962-8.00010-X

The objectives of this chapter are to

- provide a comprehensive and yet succinct summary of the five mainstream reactor designs currently in operation throughout the world
- provide background to the reasons a particular reactor design was selected and has been used
- summarise the reactor core design, including the salient dimensions and power ratings of the reactor system for each of the categories that are discussed
- explain how the reactivity in each of the corresponding reactor systems is controlled

10.2 HISTORICAL CONTEXT: OTTO HAHN 1879–1968

Otto Hahn was a radiochemist who spent many years of his early career studying heavy radioactive isotopes and discovered, for example, ^{228}Ra (Fig. 10.1). Fermi had used neutrons to irradiate many of the elements up to uranium and had produced what were believed to be short-lived transuranium isotopes corresponding to neighbouring, heavier elements with atomic numbers 93, perhaps the heavier 94 and a variety of isotopes with short half-lives that decayed via β decay. Otto Hahn, Lise Meitner and latterly Fritz Strassman repeated Fermi's experiments since they were very familiar with the chemical properties of protactinium that was believed to be one of the short-lived mystery elements. They proved it could not be protactinium, uranium, actinium, thorium and thus anticipated it was a transuranium isotope as yet unknown, that is one of the heavier isotopes beyond uranium. Until that time only isotopic neighbours of the irradiated isotope had been identified when irradiated with neutrons, usually those heavier by one neutron. The possibility of light isotopes being produced had been excluded as physically implausible despite an earlier, largely ignored hypothesis that this might result by Ida Noddack in 1934.

FIG. 10.1

Otto Hahn.

Marie Curie and co-workers had observed the production of a substance from the irradiation of uranium with neutrons with a 3.5-hour half-life and the chemical properties of a rare earth element. The options were that it could be either barium or radium on the basis of these properties. Hahn and Strassman precipitated it with barium sulphate and barium chloride but observed no radioactivity associated with radium in the precipitate. Next they attempted to separate the mystery substance via fractional crystallisation but again without success, confirming the absence of radium. They thought they could separate known radium isotopes from barium but they could not extract the mystery rare earth isotopes in the same way, in order that they were distinct from the barium carrier. Finally, they checked the products of the β decay of the rare earth: the decay of radium would leave behind actinium whereas they observed lanthanum instead. This could only be consistent with the element barium, which being drastically lighter than uranium, was hypothesised as arising from a 'bursting' or splitting of the nucleus. Hahn and Strassman were able to confirm this in correspondence with Meitner and Frisch, who were in exile. Hahn was awarded the Nobel Prize for Chemistry for this discovery in 1944. Interestingly, the heavier elements that Hahn et al. had sought originally in their experiments are likely to have been present too but in too weak abundance to be identified from Hahn's preparations at the time.

10.3 INTRODUCTION

Nuclear power reactors use very little fuel and operate for a long time by the standards of all other base load electricity generation options: typically 30%, ~30 t of a core being replaced annually. One tangible result of this engineering trait is that while energy strategies may change, the world's stock of power reactors spans a variety of designs operating with timescales that overlap significantly. As time has passed, some designs have become the precursors of plant currently under construction, while others have been superseded but continue to make a valid and important contribution to national electricity needs, and to influence future developments. Increasing the power output of existing plant in preference to building new plant (*uprating*) and extending the life of existing plant (referred to as *long-term operation* or *life extension*) are important influences in this regard.

Given our focus on commercial energy production, it is important to appreciate that there is a variety of reactor systems operating throughout the world at present for research, materials irradiation and propulsion. These are based on similar engineering principles but we do not consider them further here. The first generation of power production reactors that followed the military developments after World War II (Generation I) are now shutdown and at various stages of decommissioning. A significant number of the subsequent Generation II reactor systems continue to contribute to the world's electricity needs but most are forecast to end generation by ~2025 with the remainder being Generation III plant built in the 1990s. The mainstream reactor designs that are options for current and future build (Generation III+) comprise the PWR, the BWR and the pressurised heavy-water reactor (PHWR) designs. The PWR and BWR are often referred to as light-water reactor (LWR) designs whereas the heritage of the PHWR design draws on the CANada Deuterium Uranium reactor design known as the CANDU. A variant of the PWR principle in use in Eastern Europe and Russia is the *water–water energetic reactor* (VVER). There are also reactor systems currently in operation that are unlikely to be pursued further commercially when they come to the end of their operating life, either due to economic reasons or as a result of operational experience. These comprise the GCR and the LWGR. The former is associated almost entirely with the advanced gas-cooled reactors (AGR) used in Britain while the latter

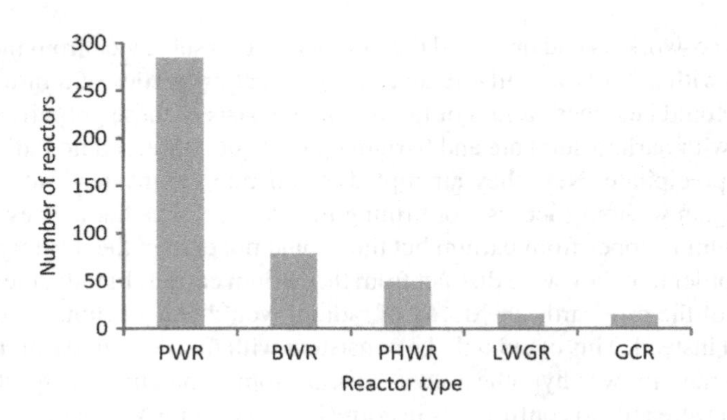

FIG. 10.2

The number of operable thermal-spectrum fission reactors as a function of type in the world at the time of writing. PWR: pressurised water reactor; BWR: boiling water reactor; PHWR: pressurised heavy-water reactor; LWGR: light-water graphite reactor; GCR: gas-cooled reactor.

with the Reaktor Bolshoy Moshchnosty Kanalny (RBMK) that are in use in Russia. A histogram of the number of operable nuclear power reactors in the world by type is provided in Fig. 10.2 with a summary of the main technical attributes provided in Table 10.1.

10.4 PRESSURISED WATER REACTORS

10.4.1 INTRODUCTION

The prominent features that distinguish PWRs from other reactor designs are that the same volume of light water is used as both moderator and coolant *and* that this is maintained in the liquid state under pressure. This has significant benefits in terms of, for example, the compact nature of the core, a negative void feedback coefficient and the relative ease of disposal of the moderator when the plant is decommissioned. However, it also introduces constraints in terms of neutron absorption (high relative to other choices of moderator) and the requirements to have a high-integrity reactor pressure vessel and pressuriser as principal components in the primary reactor coolant system. Perhaps most significantly in comparison with the BWR that constitutes the other mainstream LWR, is that a means by which the heat from the core is exchanged with a secondary coolant is necessary to accomplish the liquid-to-vapour phase transition; this is usually accomplished via a steam generator system. Thus PWRs exploit an *indirect* coolant cycle with the primary coolant cycle being in contact with the reactor core, while the secondary is in contact with the electricity generation system, with the heat exchanged between the two via a steam generator.

The PWR is the most widespread reactor design currently in use across the world, accounting for 287 of the world's 446 operating reactors (~65%),[1,2] at present. It also accounts for the majority of the power reactors currently under construction or in plan throughout the world. While PWRs cannot be refuelled on-load due to the requirement to maintain the primary coolant under pressure in a high-

[1]www.world-nuclear.org.
[2]https://www.iaea.org/PRIS.

Table 10.1 A Summary of the Main Technical Attributes of the World's Current, Mainstream Operating Reactors

Design Attribute	PWR[a]	BWR[b]	GCR[c]	PHWR[d]	LWGR[e]
Number of fuel channels/assemblies	121–193	748	332	380	1661
Number of control rods/assemblies	53	177	89	25[f] and 28[g]	221
Core diameter/m	3.4	4.9	11.0	7.0	11.8
Core height/m	3.7	4.3	9.8	6.0	7.0
Inlet coolant temperature/°C	288	278	339	266	270
Outlet coolant temperature/°C	326	288	639	310	284
Gas flow rate/kg s^{-1}	–	–	4067	–	–
Operating pressure/bar	160	70	41	100	70
Peak coolant flow rate/kg s^{-1}	17,438	1820	14	24 per ch.	13,000
Rated thermal power/MW_{th}	3400	3320	1623	2064	3200
Electrical power/MW_e	1150	1100	600	~675	925
^{235}U enrichment/%	2.1–3.1	0.71–3.05	2.2–2.7	0.7–2.1	2.0–2.6
Pellet diameter/mm	8	10	14.5	12	11.5
Pellet length/mm	10	10	–	16	15
Cladding	Zircaloy-4	Zircaloy-2	Stainless steel	Zircaloy-4	Zircaloy
Fuel element diameter/mm	9.1	11.2	14.5	13.1	13.6
Fuel element length/mm	3658	3708	1036	495.3	3.64
No. elements per fuel assembly	264	55	36	37	2 × 18

[a]*4-Loop Westinghouse design example.*
[b]*General Electric BWR/6 design example.*
[c]*Advanced gas-cooled reactor (AGR) example.*
[d]*Canada deuterium uranium (CANDU6) example.*
[e]*RBMK-1000 example.*
[f]*Reactor-regulating system (RRS).*
[g]*Shutdown system (SDS#1).*

integrity, steel pressure vessel, any disadvantages this might suggest in terms of operation are often overshadowed by very high capacity factors[3] achieved by consistent operation at high output and short outages. A summary of operable PWRs by nation is provided in Table 10.2.

10.4.2 BACKGROUND TO THE USE OF LIGHT WATER

The merits of light water for moderation and cooling in thermal fission reactors were discussed in Chapters 6 and 7, respectively. To recapitulate: it is *cheap*, *abundant*, has a *high specific heat capacity* and has *well-understood flow characteristics*. However, its critical point sets the upper limit on the thermodynamic efficiency if it is to be retained in the subcritical,[4] liquid phase under pressure.

[3]The capacity factor is the ratio of the actual output of a power plant to the potential output if it were able to operate at full power continuously for the same period of time.
[4]In terms of its thermodynamic phase state.

Table 10.2 The World's Operable Pressurised Water Reactors by Country[1]

Country	Number of PWRs	Notes[a]
Armenia	1	Armenian-2, VVER designs
Belgium	7	Two 2-loop, five 3-loop PWR designs
Brazil	2	Two PWRs plus one further under construction
Bulgaria	2	VVER designs
China	31	Plus 20 under construction
Czech Republic	6	
Finland	2	Plus one (Olkiuoto-3 under construction)
France	58	Plus Flamanville-3 under construction
Germany	6	
Hungary	4	
India	1	VVER plus Kudankulam-2 under construction
Iran	1	
Japan	21	Operation halted since Fukushima incident in 2011
Netherlands	1	
Pakistan	2	Plus 3 under construction
Russia	18	VVER plus 8 under construction
Slovakia	4	VVER plus two under construction
Slovenia	1	
South Africa	2	
Republic of Korea	21	Plus three under construction
Spain	6	
Sweden	3	
Switzerland	3	
Taiwan	2	
Ukraine	15	Plus two under construction
United Kingdom	1	Sizewell B
United States of America	66	Plus four under construction

[a]*https://www.iaea.org/PRIS.*

Recalling our focus on neutron economy adopted in Chapter 8, the cost-benefit of light water used in PWRs is balanced on the one hand by the excellent elastic neutron scattering properties of its constituent hydrogen with the relatively high neutron absorption cross section of hydrogen on the other. The latter constitutes a sink for neutrons evident in, for example, a greater Σ_a in terms of materials buckling than for non-LWR reactor designs; this is usually offset in part by using low-enrichment fuel that is now widespread for almost all operating reactors in any case. A schematic diagram of the side elevation of a PWR reactor core is shown in Fig. 10.3.

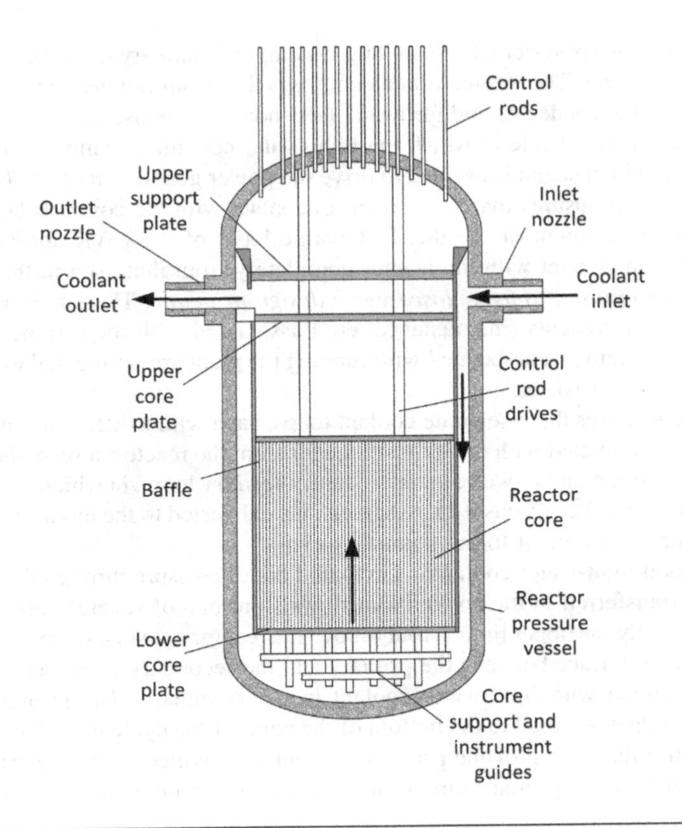

FIG. 10.3

A schematic diagram of the side elevation of a PWR reactor pressure vessel and core showing the main features (not to scale).

10.4.3 DESIGN OVERVIEW

The Westinghouse PWR design basis has been assumed for the purposes of this chapter with recent developments in the related designs such as the evolutionary PWR (EPR), AP1000, APR1400, etc. discussed in Chapter 11. The Westinghouse design has been utilised in a variety of power capacities across the range from 600 to 1200 MWe. This has been achieved by adjusting the yield in coolant output through design advances, including: increasing the number of coolant loops (each comprising a dedicated outlet, coolant pump and steam generator to transfer heat to the generation stage), extending the length of fuel assemblies, increasing the number of fuel assemblies comprising the core or via a combination of all of these approaches.

In common with most other reactor designs, a PWR plant tends to be delineated in terms of the nuclear systems—usually referred to as the *nuclear island*—and the power generation systems that are collectively known as the *turbine island*. This constitutes not only a physical separation in terms of the boundaries on-site and the reactor containment civil infrastructure but also in terms of organisational controls separated by, for example, permits to operate and radiation dosimetry requirements. The nuclear island comprises the reactor coolant system, control system, fuel handling systems and auxiliary fluid systems (purification and emergency coolant supplies). The turbine island comprises the steam turbine, the generator from which electricity is produced, the condenser and feedwater (secondary) systems, etc.

PWRs exploit an *indirect* cycle in which the light-water coolant flowing through the core is separated from that from which steam is derived to drive the power generation plant. This has two important advantages: firstly, it ensures that the coolant in contact with the core can be maintained in the liquid phase (a key requirement of the thermodynamic basis of the PWR approach) and secondly it isolates the coolant in contact with the reactor completely from that used in the power generation stage; a concept often referred to as a *closed-cycle design principle*. The latter arrangement simplifies the operational requirements (maintenance, etc.) associated with the turbine apparatus because the nuclear-related requirements associated with running the plant are segregated to the set of facilities confined within the nuclear island.

A PWR design comprises three separate coolant loops, each with a different function. These are: a *primary* coolant loop associated with direct heat transfer from the reactor; a *secondary* associated with the liquid–vapour transition and power generation, and a *tertiary* loop via which waste heat (as per the second law of thermodynamics discussed in Chapter 7) is exhausted to the environment and feedwater is condensed to return to the input to the secondary cycle.

In the primary loop light-water coolant is circulated under pressure through the reactor core. Heat from the reactor is transferred to the coolant that flows from one of several outlets from the reactor pressure vessel (generally the upper limit is four) above the reactor core to a steam generator. The steam generator provides an interface between the primary and the secondary loops across which heat from the primary is exchanged with light-water coolant in the secondary. The primary coolant, having exchanged its heat, is then returned to the bottom of the core for the cycle to be repeated. The pressure necessary to prevent bulk boiling of the primary coolant is provided by an electrically-heated pressuriser system located in the primary circuit and the coolant is circulated by electrically-powered coolant pumps.

In the secondary loop heat from the steam generator causes some of the coolant in this part of the cycle to turn to steam. This is then dried and fed to the turbine system. It is important that the steam is dry because the presence of water droplets would reduce the efficiency of the conversion from thermal energy to mechanical energy, and would also risk damage as a result of impact and corrosion of the precisely machined components of the turbine. The steam is exhausted from the turbine to a condenser that returns it to its liquid state.

The tertiary loop constitutes a separate coolant system that feeds a separate stream of light-water coolant to a condenser system, in order that the waste heat from the secondary coolant exhausted from the turbines can be transported to the ambient. A variety of arrangements for this are used dependent on the location of the plant; for example, the heat can be transferred to a river or the ocean if these are close by or alternatively via a cooling tower system; some interest also exists in using this energy directly for district heating, etc.

10.4.4 REACTOR DESIGN

The epicentre of the PWR is the reactor pressure vessel (RPV); this is a cylindrical, low-alloy carbon steel container of typical internal diameter 4 m which houses the reactor core. The top of the RPV is removable to allow access to the core during outage for refuel and maintenance. It is connected to the base of the RPV via a flange and gasket-lined seal. The upper part of the RPV also comprises the coolant inlets and outlets;

these are located around its radius and have penetrations in its top for the control rod drive mechanisms (CRDM). Flux monitoring instrumentation systems usually enter the RPV from the bottom. The surfaces of the RPV in contact with the reactor coolant are lined, for example with a 3-mm austenitic stainless steel, for protection against corrosion. Although it is the central feature of the reactor system and the component from which the reactor's thermal energy emanates, the RPV can appear rather diminutive in scale relative to some of the other components positioned around it in the reactor containment building, particularly in comparison with the enormous steam generators, for example.

The PWR reactor core is cylindrical in design with its axis of symmetry in the vertical plane, of typical diameter 3 m and height 3.7 m, and comprising up to ~200 fuel assemblies (dependent on the specific design and specified power output). Each assembly typically comprises 17×17 fuel elements in a square array with each element housing a number uranium dioxide fuel pellets of low ^{235}U-enrichment (a distinction of the VVER design is that VVER fuel assemblies have a hexagonal arrangement but the fuel and cladding composition and design are otherwise similar). A typical fuel weight in the core, again dependent on the specific design, would be between 50 and 85 tonnes. A typical PWR fuel assembly under inspection following manufacture is shown in Fig. 10.4.

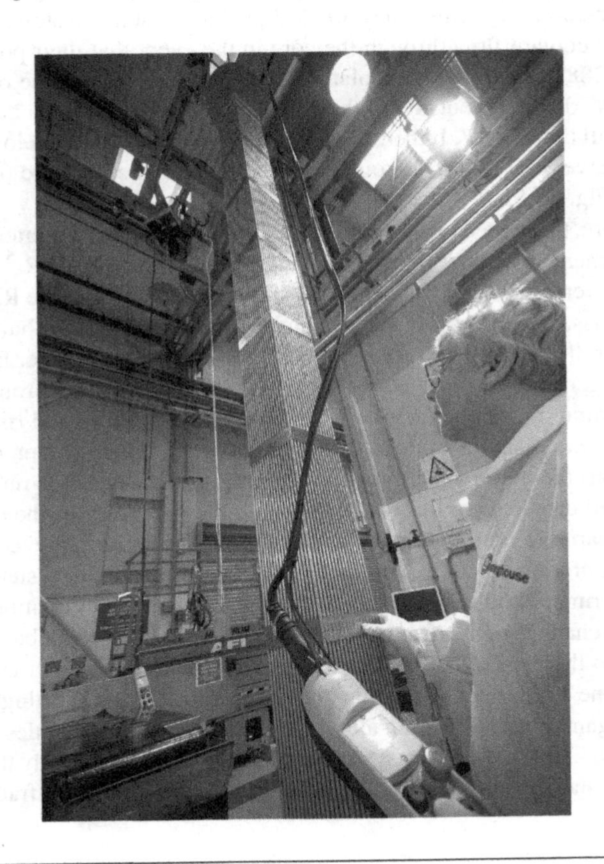

FIG. 10.4

A PWR fuel assembly under inspection following manufacture.

With permission from the Westinghouse Electric Company.

Fuel is arranged in the core in such a way to yield efficient consumption of ^{235}U while ensuring a uniform energy yield across the core; this tends to result in higher-enrichment fuel assemblies being positioned at the periphery of the core (where neutron leakage competes most significantly with fission in ^{235}U) and lower enrichments towards the centre. The core is designed to provide power within regulated margins and to withstand the worst perceivable deviations from normal operation; the latter are often referred to as *design-basis accidents* or *transients*. Since the reactor core is initially in a low-activity state (operation results in the production of minor actinide isotopes that emit neutrons), start-up neutron sources are installed: for example one of ^{252}Cf and another that is activated as a result of early operational irradiation.

10.4.5 THERMO-HYDRAULIC SYSTEMS

The coolant in the primary loop is circulated by electrically-driven pumps with usually one for each of the parallel coolant loops that constitute the primary circuit. These pumps are generally vertical, three-phase, single speed, air-cooled systems. They are designed to coast down slowly in the event of loss of power to maximise the coolant flow through the core in the event that their power supplies are interrupted. The cold (i.e. 288°C) light-water coolant/moderator flows in via one of the inlets around the upper waist of the RPV, down the outside of the core in the space between a baffle that surrounds the reactor core and the wall of the RPV. It then flows up through the core via the lower core plate. The hot coolant (i.e. 326°C) leaves via one of the outlets in the vessel. A schematic of a typical PWR plant layout is provided in Fig. 10.5.

From the reactor core, the heated primary coolant passes to the steam generator(s). Although subject to continuous improvement since they were used in the first commercial PWR,[5] the design objective of PWR steam generators remains largely unchanged; primary coolant from the RPV flows into the generator at its base and passes up through a set of U-shaped tubes in the lower half of the system. Hence these systems are generally referred to as *U-tube steam generators*. Feedwater from the secondary circuit enters the steam generator at just above its midpoint, flows down around the outside of the U-shaped tubes containing the primary coolant. The feedwater heats up and rises through the middle of the steam generator, vaporising as it does so. On entering the upper section of the steam generator (the *steam drum* section) the majority of moisture remaining in the steam is removed by spinning the vapour so that entrained droplets are thrown outwards to the periphery of the drum. Together with a subsequent stage of steam driers, steam quality above 99.75% can be achieved.

The *pressuriser* is a prominent component in the PWR reactor coolant system because it maintains pressure levels in the primary coolant loop and also acts to remove the variations in pressure that can arise from temperature changes in the reactor. The pressuriser in reactor number 2 at Three Mile Island featured prominently in the events that lead to the accident in 1979, as is discussed in Chapter 14.

There is normally one pressuriser per primary coolant circuit. In the Westinghouse design the pressuriser is made of manganese–molybdenum alloy steel, clad inside with stainless steel. It comprises an electrical heater (~1000–2000 kW), spray systems and relief valves. Typically the pressuriser contains a combination of water and steam with a volume ratio of ~60:40. The steam fraction (often referred to

[5]Yankee Rowe, Rowe, Massachusetts, United States, commissioned 1960.

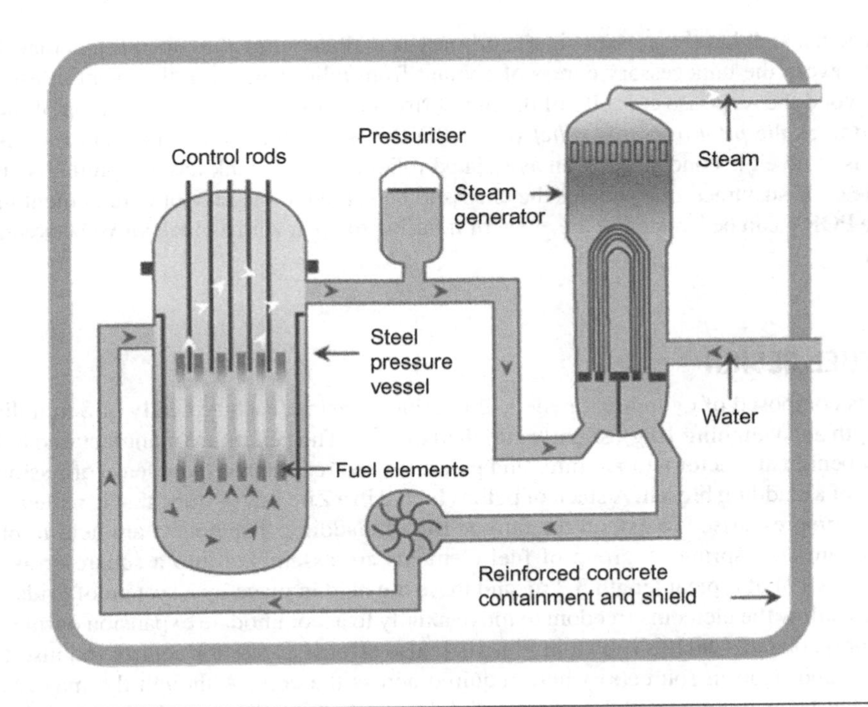

FIG. 10.5

A schematic of a PWR plant layout.

With permission from the World Nuclear Association.

as the *steam bubble*) is maintained by the heater that can be switched in or out of operation or controlled on a proportionate basis. The latter function is necessary, for example, when the reactor is being brought on or off load to counteract the increase or decrease in pressure due to the expansion or contraction of the primary coolant, respectively.

Two scenarios can be anticipated that illustrate the function of the pressuriser:

1. When the electrical load on the reactor is reduced as a result of, for example, a fall in demand, a reduced amount of energy is withdrawn from the primary coolant. Since the surplus energy remains in the primary coolant, this raises the temperature as a consequence. The coolant expands, compressing the steam bubble in the pressuriser. This pressure increase actuates spray lines to inject cold coolant into the pressuriser that condenses a proportion of the steam bubble thus absorbing the increase in volume and pressure.
2. Alternatively, an increase in demand reduces the primary coolant temperature and thus the coolant contracts. This depressurises the steam bubble causing water in the pressuriser to vaporise while the pressure drop actuates the heater to generate more steam.

Thus pressure changes arising within the bounds of normal reactor operation are stabilised by the pressuriser. This ensures that reactor operation is adjusted effectively in response to changes in demand and minor transients, and that the pressure in the primary is maintained within operational margins. It also functions to avoid the unnecessary egress of coolant from relief valves. In the event of escalations in pressure beyond the response capacity of the pressuriser control system, a valve on top of the pressuriser referred to as the *pilot operated relief valve* (PORV) is designed to open to relieve pressure. The steam that is evolved is condensed in an associated relief tank; this tank has a rupture disc in case the coolant release is sustained that enables the excess to be vented to the reactor containment in an emergency. The PORV can be isolated in the event of it failing to open with a block valve between it and the pressuriser.

10.4.6 FUEL DESIGN

PWR fuel is composed of cylindrical pellets of uranium dioxide (each typically of 8 mm diameter by 10 mm length and weighing 10 g), slightly enriched in ^{235}U. The pellets are manufactured to be largely chemically benign at reactor temperatures and pressures and resistant to the release of fission products in the event of a cladding breach. A stack of pellets is clad in a Zircaloy-4[6] tube that is sealed by welding having been prepressurised to extend the fatigue life of cladding. The pellets are held in place by an internal compression spring. A group of fuel elements are assembled into a square array (typically 16×16 or 17×17) of typical length 3.7 m, and these are held in place by a system of grids and spring straps. These allow the elements freedom to move axially to accommodate expansion during operation. Each assembly, one of which is shown in Fig. 10.4, also affords access for control rod insertion, burnable poisons and neutron source(s) where required across the core. Although the majority of PWR experience has been obtained with fuel that is slightly enriched in ^{235}U, mixed oxide fuels[7] comprising a combination of ^{235}U and ^{239}Pu as the fissile component have also been used, particularly in France and Japan. An image of the typical PWR fuel assembly is provided in Fig. 10.6.

10.4.7 OPERATION

There are generally three modes of power reactor operation: *start-up*, *normal operation* and *shutdown*. Start-up procedures differ depending on whether the reactor is *cold* (as a result of having been left to cool for maintenance and refuel) or *hot* such as when restarting the reactor without significant delay following a turbine trip. In either case, prior to withdrawing control rods to bring the reactor to criticality, it is necessary to bring the reactor to a state of *hot standby*. In a typical PWR this is characterised by a primary circuit temperature and pressure of the order 285°C and 153 atm., respectively. Starting up a power reactor is more sophisticated than in the case when Fermi first brought the CP-1 reactor to criticality because not only is a self-sustaining chain reaction required but it is also necessary to establish a uniform, extensive fission reaction throughout the core that is capable of a high-power, sustained level of output. This is a state that has to be approached gradually. Further, there are many ancillary systems associated with power generation, including the secondary circuit and turbine systems, that must be synchronised with the reactor's operation.

[6]Typical composition 98.23% zirconium, 1.45% tin, 0.21% iron, 0.10% chromium and 0.01% hafnium.
[7]Known as MOX fuel.

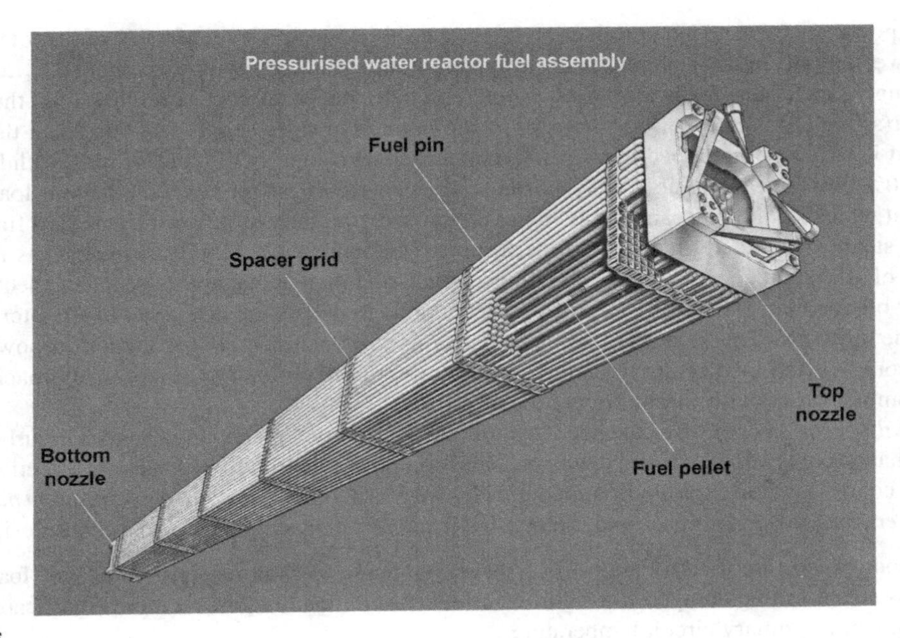

FIG. 10.6

A photograph and schematic diagram of a PWR fuel assembly.

With permission from the Westinghouse Electric Company.

Prior to start-up, extensive checks of instrumentation, the safety injection system and the boron concentration level of the primary coolant are made. The heat and associated pressure at this stage is provided by the heaters in the pressuriser and steam is generated on the secondary side drawing on the auxiliary feedwater supply, that is not the main feedwater supply and at significantly reduced levels compared with mainstream power operation since the core is not in operation yet. Since the plant is not connected to the turbo generator at this stage, the steam is dumped to the main condenser. This state is characterised by the reactor coolant being at the average temperature (T_{avg}) necessary to maintain a saturation pressure under no load; this is the *hot-standby state*.

To approach criticality the boron concentration in the primary coolant is adjusted to ensure that a criticality is feasible when the control rods are withdrawn to a specified degree of insertion. The control of reactor systems is regulated by strict limits on the extent to which control rods are inserted and a criticality achieved. These are known as *power-dependent insertion limits* (PDILs) beyond which it is not possible to insert the control rods further while a criticality established. This constraint prevents a scenario developing in which the primary boron concentration is diluted to such an extent that it is not possible to offset the critical state by insertion of the control rods.

With the control rods inserted the reactor shutdown rods are withdrawn. As part of the staged approach to criticality the boron concentration is adjusted further and the appropriate combination of control rods are withdrawn to an extent necessary to facilitate a critical state but producing negligible heat output at this stage. Further adjustments to a wider selection of control rods are made to raise the power further and heat production begins, at this stage to a level of ~1% of the total design basis power output. Further increases in power are made gradually with increasing T_{avg} and, correspondingly, auxiliary feedwater flow is increased to maintain the water level in the steam generator(s) until it is necessary to transfer to the main feedwater supply. At this stage the steam generated is still being dumped to the main condenser. However, with control rods withdrawn to yield ~10% power, it is possible to synchronise the generators and connect the reactor load to the turbine, at which point the steam dump valves are closed gradually as power is increased further. In this way steam is directed gradually to the turbine cycle and steam output increased. As the temperature of the plant increases, the combined effect of inherent negative feedback mechanisms (Doppler broadening, spectral hardening and reductions in density as discussed in Chapter 8) start to become apparent. This is termed a *power defect* as it is manifest as a reduction in power. It is usually counteracted by diluting the boron concentration in the primary coolant to approach power levels commensurate with normal operation.

Normal operation comprises the need to sustain power output, respond to changes in turbine load (due to changes in the demand) and to ensure a uniform power distribution. There are effectively five means of control that are exploited sympathetically with each other: the *control rods*, *steam dump*, the *pressuriser*, the *steam generator* and the *passive feedback characteristics* of the PWR design:

- The control rods are used to control T_{avg} on an automatic basis in response to turbine load.
- Excess steam can be dumped if a significant reduction in load requires a more significant adjustment to primary circuit temperature.
- Pressure can be adjusted via the spray, relief valves and heater in the pressuriser.
- Water level is maintained in the steam generator(s) via feedwater flow valves.
- The distribution of power is maintained to be as uniform as possible to avoid xenon oscillations and also to optimise fuel burnup.

Shutdown is not merely a reversal of the start-up procedure, primarily because a fission reactor continues to generate heat long after the chain reaction is halted due to the radioactive decay of short-lived radioactivity in the fuel. This feature is commonly referred to as *decay heat* and is amongst the most important features of reactor operation because, as per all forms of nuclear decay, it is not possible to accelerate the decline in decay heat production while the stored energy associated with it is significant, assuming shutdown is preceded by a reasonable period of operation. *Shutdown* as a term in power reactor operation is distinct from a *reactor trip* since the former is an anticipated action ahead of, for example, a plant outage, whereas the latter is generally not.

The first procedure towards shutdown is to reduce the load on the turbine generator, usually via an automatically programmed reduction of T_{avg} until the level is sufficiently low for reactor control and steam generator operation to be transferred to manual operation. Then station power is transferred from the turbine generator to the off-station grid, since the plant still requires power to run the *residual heat removal system* (RHRS) (see below). At this point the turbine can be unloaded totally and shut down completely. With the control rods inserted the reactor enters a state of *hot stand-by*; a minimum number of coolant pumps remain in service to maintain coolant flow and heat transfer but with the associated steam being dumped (not used). To proceed to *cold shutdown* the boron concentration in the primary circuit is increased to meet the required shutdown margin and the reactor then enters a shutdown state.

As the primary coolant cools, it also contracts and hence it is necessary to raise the pressuriser level to accommodate this change in coolant level. The pressuriser heater can then be switched off with the spray used to cool the pressuriser. The RHRS comprises typically two independent combinations of pump and steam generator that are separate from the main reactor coolant heat removal system (that is also fed from separate electrical lines to ensure redundancy). This is used to maintain the cooling rate below temperatures of approximately 177°C. The specific state sought in terms of coolant temperature and readiness for restart is dependent on the subsequent plan for the reactor; that is, as to whether a relatively quick return to operation is intended or, conversely, a longer outage involving extensive maintenance is scheduled.

10.4.8 REACTIVITY CONTROL AND REFUELLING

The three main control mechanisms in a PWR are the *control rods*, the *chemical shim* (boric acid added to the primary coolant) and *burnable poisons*:

- The chemical shim is used to offset the extent the control rods need to be inserted (as per the discussion on PDILs above), to enable long-term control via its gradual dilution as fuel enrichment is depleted and also to consolidate the core in a long-term shutdown state when required for maintenance, etc.
- The control rods enable automatic, short-term control in response to relatively small changes in reactivity; they are configured in rod cluster control assemblies extending the whole length of the core when inserted. They are coupled to a drive shaft enabling groups of control assemblies to be moved together electrically or released by electromagnetic clutches to drop into the core under gravity in the event that an emergency shutdown is required.
- Burnable poisons (typically boron carbide elements present alongside fuel elements in assemblies throughout the core) offset the need for excessive boronation of the coolant and are depleted as neutron poisons build up in the fuel and the abundance of ^{235}U is depleted with burnup.

Approximately one-third of the fuel in a PWR is renewed at a time to optimise burnup and to maintain the uniformity of power output across the core; generally fuel at the periphery of the core is of a higher enrichment than that at the centre with the central region comprising the low- and intermediate-enrichment assemblies occupying alternating positions. A typical refuel programme would involve the removal of the assemblies from the central region of the core and the transfer of the outer assemblies to the centre, as per a specified reload arrangement.

10.4.9 PROTECTION SYSTEMS

A wide range of developments have been made to the protection systems on the PWR as a result of operational experience. Consequently, specific PWR designs may differ from the generic design basis described here. The generic purpose of PWR reactor protection systems (or *safeguards systems* as they are sometimes referred not to be confused with the need to safeguard nuclear materials) is to prevent any scenario in which damage to the reactor might result in the release of radioactive material to the environment. As is described in Chapter 14, a ubiquitous philosophy in nuclear reactor safety is the concept of *defence-in-depth*, in which systems, operations and behaviours are designed to provide multiple layers of protection against scenarios that might undermine safe operation. For a reactor system, the three principal layers or barriers are the fuel cladding, the reactor vessel and the containment. The protection systems prevent these from being compromised and span three main functions: *safety injection, emergency feedwater supply* and *component cooling*.

The *safety injection system* provides a source of borated cooling water to counteract the effects of an unexpected loss of coolant. This constitutes a source of replacement coolant and negative reactivity that can be injected to counteract the effect of unexpected transients or the need for a greater shutdown margin. The primary purpose of the response to a loss of coolant is to prevent the reactor core from becoming uncovered; both emergency core cooling and long-term cooling fulfil this requirement following such an accident scenario. This response is provided by a supply from passive accumulator tanks located inside the containment and a combination of high-head and low-pressure injection apparatus located outside the containment.

The injection system has a design philosophy of two-train redundancy in terms of system and power supply. The high-head system serves no function under normal operation while the low-pressure supply is usually derived from the RHRS described in the summary of the shutdown procedure. A typical order of operation might be as follows: high flow rate refill is provided from the accumulator tanks, the high-head safety injection pumps are actuated and subsequently the RHRS is operated, deriving coolant from the refuelling water storage tank. All systems inject into the cold legs of the primary circuit loops. They are stimulated by a pressure decrease in the reactor cooling system relative to that of the emergency supply systems via pressuriser signals or steam pressure drops (the latter protecting against breaks in the secondary circuit that might constitute a loss of heatsink to the primary circuit). High containment pressure levels or manual override can also start the safety injection system which also starts emergency, on-site, diesel-fuelled generators, high-head safety injection pumps and the RHRS pumps. If coolant loss is sustained then recirculation of surplus coolant collected in the containment sump can be administered to maintain cooling indefinitely.

The *emergency feedwater supply system* caters for the event that the main system coolant supply to the steam generators in the secondary circuit fails. As for the high-head injection system, the emergency feedwater supply is often derived from two identical, redundant trains with separate electrical supply systems and is not used in normal operations. It is set into operation by the indication of a low level in the steam generators or by actuation of the safety injection system; the emergency feedwater system can supply water to a tripped reactor at the rate needed to maintain cooling in the hot-standby state until the temperature for the RHRS to be activated is reached.

The *component cooling water system* provides cooling to all the components in a PWR that have the potential to carry radioactive fluids. It provides cooling during normal operation, during cooldown following shutdown and in the event of an accident. It comprises two subsystems: one associated with component cooling requirements in the event of an accident—referred to as the *safeguards subsystem* - and an *auxiliary subsystem* for all component cooling requirements during normal operation. The safeguards system comprises two independent system trains with separate heat exchangers, surge tanks and pumps powered from independent supplies. Examples of the components cooled by this system are the reactor coolant pumps and components in the RHRS.

10.5 BOILING WATER REACTORS

10.5.1 INTRODUCTION

Initial research [1] on LWR designs in the late 1940s and early 1950s focussed on what would become known as the pressurised water concept. At that time this was referred to as a 'nonboiling reactor'. The emphasis was to ensure that the light-water coolant/moderator remained in the liquid phase due to concerns that boiling would result in a loss of flow that could result subsequently in overheating of the fuel and hence core damage. Also, sustaining a controlled nuclear chain reaction amidst the highly disruptive and unpredictable boiling state of the moderator was considered unfeasible, or was at least poorly understood, because it was anticipated that the sudden formation and collapse of bubbles would result in unpredictable fluctuations in reactivity and thus power. Further, it was unknown at that time as to whether significant issues might arise from radioactivity that might be transferred to the vaporised coolant. Conversely, it was known that heat transfer was significantly more favourable between fuel and light-water coolant when boiling than in the liquid state, and also that direct-cycle designs enabled by the boiling approach would require significantly less apparatus such as steam generators and higher-pressure components, than the corresponding PWR. Thus, boiling offered the potential for cheaper and easier reactor construction.

With light-water moderated and cooled reactors exhibiting negative fuel coefficients of reactivity, due to Doppler broadening of ^{238}U resonances, and negative void coefficients, heating of the fuel as a result of an increase in reactivity results in a rapid reduction in reactivity (as discussed in Chapter 8) while the removal of the coolant/moderator as a result of voiding leads to a further (albeit slower) quenching of the chain reaction. Thus the key thermodynamic question as to the feasibility of the boiling water concept at the time of its inception was which process was faster following a positive transient: the escalation of the chain reaction and thus the temperature of the fuel, or the voiding of the water around it. Pioneering experiments at the time indicated that the water was ejected more rapidly than the prompt neutron lifetime and thus there was no possibility of the chain reaction outrunning the phase change of the coolant, and hence leading to an escalation. Furthermore, it was observed in related experiments that the significant reduction in density of the vaporised coolant resulted in much reduced levels of radiation being emitted from it, relative to that emitted from the liquid phase, and that far from the boiling phenomenon being a source of instability it presented a means by which the nuclear reaction could be controlled by coolant flow. The boiling concept is now exploited widely for the production of power: a summary of the world's BWRs is provided in Table 10.3.

10.5.2 DESIGN OVERVIEW

The BWR design[8] development has comprised six stages beginning with the BWR/1 (embodied by the example of the Dresden plant in the United States, 180 MWe commissioned in 1955) through to the BWR/6 at 1100 MWe pioneered in the early 1970s. A BWR plant layout comprises the *reactor vessel* at its centre that is surrounded by the *shield wall* of reinforced concrete. This is contained within the *drywell* within a steel containment which is itself contained within the outer, reinforced concrete *shield building*. A schematic of the elevation of a BWR/1 is provided in Fig. 10.7.

[8]We shall assume the General Electric design basis for this discussion of the BWR highlighting any variances associated with other designs where they arise: 'BWR/6 General description of a boiling water reactor', General Electric, 2012.

Table 10.3 The World's Operable Boiling Water Reactors by Country[1]

Country	Number of BWRs	Notes
Finland	2	
Germany	2	
India	2	
Japan	22	Operation interrupted following Fukushima incident in 2011
Mexico	2	
Spain	1	
Sweden	7	
Switzerland	2	
Taiwan	4	Plus two under construction
United States of America	33	

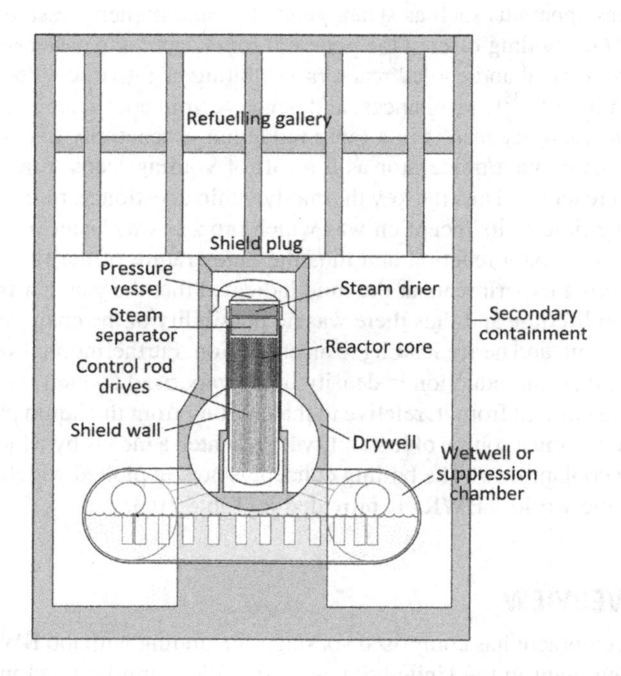

FIG. 10.7

A schematic diagram of the side elevation of a BWR reactor core (BWR/1). Note that some details have been omitted in the interests of clarity including the spent fuel pool and the refuelling gantry.

The drywell is a cylindrical reinforced concrete structure that carries out several structural functions and shields the environment in the containment beyond it to radiation levels sufficiently low to allow personnel access during normal reactor operations. Steam outlet/inlet connections to the generator side of the installation pass through the *steam tunnel* to the auxiliary building, while on the other side of the containment there is the *fuel building* that deals with fuel receipt and transfer. The reactor vessel contains (moving from the bottom of the vessel through to its top): the vessel support structure, the control rod drive system, the lower plenum, reactor core, steam separators, steam driers and the upper plenum. The reactor core within the reactor vessel is surrounded by a stainless steel shroud; this separates the downward flow of the cold coolant to its outside from the upward flow of coolant in the central regions.

Light water in a BWR is circulated through a heterogeneous core of nuclear fuel rods arranged vertically. The upright, cylindrical core sustains a critical chain reaction causing the water circulating in the fuel channels to boil and to be converted to a two-phase, steam–water mixture with the wet steam processed by a variety of systems in the plenum directly above the core. The liquid and gaseous phases of the coolant are not separated between primary and secondary loops as in a PWR but instead steam evolved from the primary coolant is used directly for power generation, as per the boiling water concept introduced earlier. However, the wet steam generated in a BWR has to be separated from any entrained water and then it is dried before being sent to the generator stage; this avoids damage and corrosion of the turbine that might result otherwise. The *steam separators* remove the entrained water centrifugally by setting the steam–water mixture into a vortex to cause the water to be cast away horizontally from the vertically rising flow; subsequently the *steam driers* force the flow in many different directions through a series of drier panels with the finer, particular moisture thrown aside.

Perhaps the two most distinctive design features of the BWR are that (i) the control rod drive system is positioned *below* the reactor core entering from beneath rather than from the top (since the steam separator and drier assemblies are positioned above the core) and (ii) it is possible to control the reactor via the rate of coolant circulation in addition to the use of control rods. This latter feature aids the flexibility of this reactor design to follow the electricity demand (load) placed upon it. Further beneficial characteristics of the BWR design are a large, negative moderator void coefficient and also that the operating pressure is typically half of that of a PWR producing the same flow of steam. Thus BWRs are regarded cheaper to install in part due to the reduced pressure performance requirement on nuclear island components.

An important feature of the BWR design is the *jet pump*, of which there is typically an array of between 16 and 24 (dependent on the specific reactor boiler arrangement of a given design) located between the core shroud and the vessel housing. The jet pumps allow coolant flow through the core to be controlled without the need for moving parts within the vessel that would otherwise be prone to failure and complex maintenance requirements. Their principle of operation exploits the Venturi effect: the reduction in fluid pressure that results when a fluid flows through a constricted section of a pipe. In a jet pump water is pumped through a narrow nozzle in the jet pump housing by a recirculation pump. In passing through the nozzle the velocity of the flow increases and generates suction that acts on a second flow of coolant entering the jet pump housing from the vessel lower plenum; both streams mix beyond the nozzle aperture resulting in flow to the base of the core where it is directed to each of the fuel channels. The flow rate and thus its control is achieved by adjustment of the flow rate through the jet pump nozzles via control of the recirculation pump speed; the pumps are located outside of the vessel within the containment building.

10.5.3 OPERATION

The operation of BWRs can be interpreted in terms of the type of boiling heat transfer in the core. This spans *nucleate boiling*, *transition boiling* and *film boiling* and is classified in terms of the extent of the nonlinear dependence of cladding temperature with power in order of increasing temperature. Nucleate boiling is the mode in which BWRs are anticipated to operate, yielding efficient heat transfer at relatively modest cladding temperatures; transition boiling and film boiling occur at higher powers and result in temperatures at which cladding could be weakened and its corrosion enhanced. These modes are thus avoided by maintaining a significant margin between extreme design-basis operating scenarios and the conditions associated with the transition boiling state.

The operation and performance of the BWR benefits from a prompt, negative fuel Doppler coefficient and a large, negative, moderator density reactivity coefficient. Thus the delay between a power increase, the associated increase in temperature of the fuel (and the natural response acting to oppose reactor transients through neutron absorption in broadened resonances in ^{238}U) is minimised. The moderator density reactivity coefficient (due to temperature-induced expansion and voiding) has an influence that is generally more significant than the effect of neutron absorption in xenon-135, thus overriding the effect of poisoning. Related to this, and described in more detail later in this section, load changing by flow control exploits the voiding component of the moderator density reactivity coefficient.

In terms of maintenance and refuelling, a new reactor is normally first refuelled 1–2 years after start-up. Refuel outages then follow at 18-month intervals with typically one-third of the fuel replaced at a time. Fresh fuel charges tend to have a higher gadolinia composition to offset the effect of burnup, described in more detail later in this section. A schematic of a typical BWR plant layout is provided in Fig. 10.8.

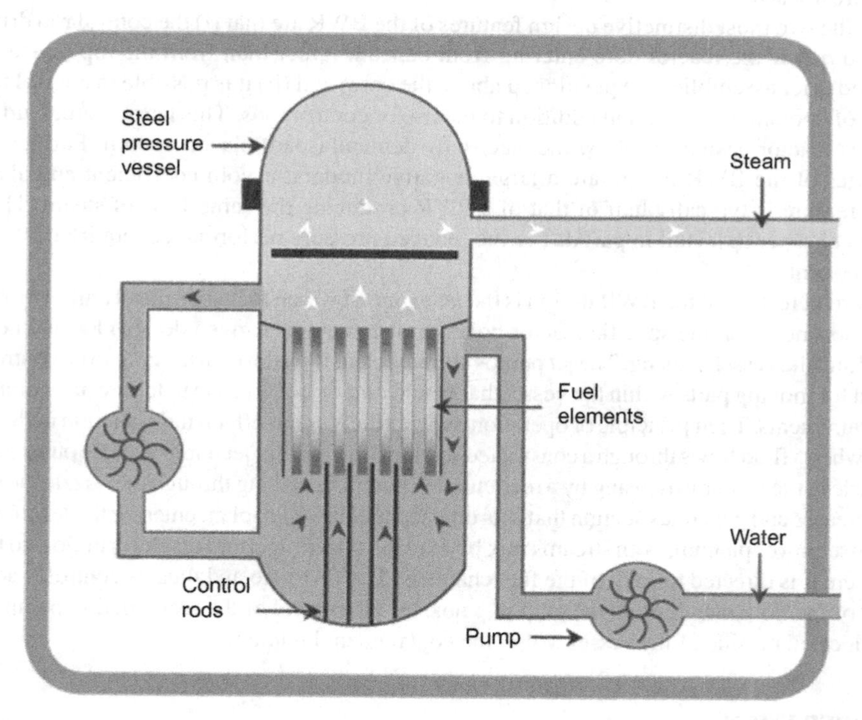

FIG. 10.8

A schematic of the BWR plant layout.

With permission from the World Nuclear Association.

10.5.4 FUEL DESIGN

A BWR core is composed of *fuel modules* each of which is a square arrangement of four square *fuel bundles*. These are separated by a cruciform-shaped control blade in the centre. Each bundle comprises 64 rods in a square array (8 × 8 in the case of BWR/6) held in place by brackets top and bottom (usually referred to as *tie plates*) made of 304 stainless steel. Not all of the rods contain fuel: there are *water rods* (typically one Zircaloy-2 tube devoid of fuel at the centre of the bundle), that ensure moderator is present in the centre of the fuel bundles, and there are also *tie rods* (8 per bundle in BWR/6) that support the weight of the bundle when it is lifted and moved during refuel; these are positioned around the perimeter of each bundle.

Each of the *fuel rods* that comprise the rest of the bundle is a stack of uranium dioxide pellets lightly enriched in ^{235}U, contained in a Zircaloy-2 tube of diameter 11.2 mm and 4 m in length. A space of approximately 240 mm is left in each rod to accommodate fission gas that is evolved during use with the pellets held in place with a spring. The fuel rod is filled with helium and closed with Inconel plugs welded in each end. The bundle design enables individual rods to be removed if required and ensures the flexibility for the rods to expand during use. The combination of the four bundles constituting a fuel module is placed in a fuel channel (a square tube made of Zircaloy-2) that constitutes a *fuel assembly* [2].

10.5.5 REACTIVITY CONTROL

The primary control mechanisms in a BWR are the insertion or withdrawal of the control rods and the adjustment of the coolant flow rate via the jet pumps. While up to ~25% of rated power control can be achieved via flow adjustment without requiring control rod movement, the control rods (blades) are the principal means of shutting down the reactor. In a BWR the control blades typically comprise boron carbide (B_4C) and hafnium. They are activated hydraulically by drives located beneath the reactor core using condensate water as the operating fluid. From this position the control rod system does not interfere with refuelling apparatus in operation above the reactor vessel. The bottom-entry arrangement provides good axial flux shaping capabilities and offers high operational reliability because no special working substance is necessary; the use of water assures that there is very little contamination and no special seals are needed.

The control blades are inserted with locking piston drives incorporating a latch and spring. They are used to control power distribution in the core (particularly to cater for the formation of steam voids in the top of the core), for power distribution shaping across the core, to achieve large changes in power (such as weekly changes in load of more than 25% rated power) and also to trip the core in emergencies. Each blade comprises a cruciform-shaped, stainless steel sheath containing cylindrical tubes of diameter 4.8 mm with boron carbide powder sealed inside them; each blade of a rod is typically 250 mm across with an active length of ~3.7 m and overall length 4.4 m.

Control via coolant flow exploits the negative void coefficient in the moderator/coolant. By way of illustration, if a power increase is required then circulation flow is increased. This temporarily reduces the proportion of voids in the moderator, moderation increases and so does the fission rate and thus power. Steam generation increases as a result and the void distribution is re-established at the higher power level. Conversely, to reduce power: flow is reduced, steam generation and voiding increases as the coolant is resident for longer in the core, moderation is reduced and along with it so is power generation; the void fraction balance is then regained at the lower power level. Necessarily, the void effect needs to be larger than the fuel Doppler effect for this phenomenon to be exploited.

A solid burnable poison, in the form of digadolinia trioxide (Gd_2O_3) mixed with the UO_2 in some of the fuel pellets, is also used to offset the long-term effects of burnup and the depletion of ^{235}U enrichment.

10.5.6 PROTECTIVE SYSTEMS

There are a variety of engineered safeguard systems in the BWR design. Most are designed to counteract the thermo-physical effects in the event of a loss of coolant from the system from a line break or loss of flow. For example, to ensure that a significant increase in pressure in the core resulting from such an event does not cause the control rods to be ejected from it, the base of the reactor is protected by a *control rod blowout restraint*. A further engineering safeguard specific to BWRs is that the jet pump arrangement allows for the core to be flooded to two-thirds of its height because there is no intermediate recirculation line leading to a drain. There is also a flow restriction nozzle in each steam line to staunch the flow of steam in the event of a break of the main steam line that could otherwise result in the core being uncovered. To counteract a scenario in which it is not possible to insert the control blades, the *standby liquid control system* enables a solution of sodium pentaborate to be injected into the core to bring the reactor to cold shutdown from hot. The *reactor core isolation cooling system* provides sufficient water to cool the core in the event it becomes isolated from the main condenser following a trip (due to, e.g. there being a break in this flow path).

From the perspective of in-core engineered safeguards, the *emergency core cooling system* (ECCS) comprises the *low-pressure injection*, *high-* and *low-pressure spray systems*. The ECCS is designed to prevent cladding fragmentation in the event of incidents in which the coolant is lost. The high-pressure spray system introduces water to the reactor vessel at a high rate to depressurise it and to cool the fuel assemblies. The low-pressure system comes into operation when the core pressure has been reduced but supplies are still needed to maintain the water level. ECCS systems are driven from diesel standby generators in the event of a loss of power to the site. The system is designed so that water flowing from the vessel collects in the drywell basement and flows to the suppression chamber forming a closed loop to continue to feed the spray system indefinitely.

10.6 HEAVY-WATER REACTORS

10.6.1 INTRODUCTION

Heavy water (D_2O) has thermo-physical properties very similar to light water but contrasts in two important respects relevant to its use in thermal spectrum nuclear reactors: (i) deuterium has a much smaller neutron absorption cross section than hydrogen and (ii) D_2O is much more expensive to source. As a result, rather than being like-for-like design analogues of LWRs, HWRs are very different largely because of these two features and also due to the heritage that surrounds the development of these reactor designs. A summary of the world's operable heavy water reactors is provided in Table 10.4.

The smaller neutron absorption cross section of heavy water results in heavy water reactors being compatible with a greater variety of fuel types and enrichments; in this respect they offer greater *fuel*

Table 10.4 The World's Operable Heavy-Water Moderated Reactors by Country[1]

Country	Number of PWRs	Notes
Argentina	3	Pressure vessel-based HWR designs by Siemens-KWU, pressurised coolant and moderator, vertical fuel assemblies
Canada	19	Various
China	2	2 × CANDU 6 (Qinshan phase III)
India	18	Various
Pakistan	1	CANDU
Romania	2	CANDU 6 (Cernavoda)
South Korea	4	CANDU 6 (Wolsong)

flexibility than other designs. Conversely, unnecessary heavy-water coolant loss has to be guarded against wherever it might arise primarily because of its cost. A further distinction is that neutron activation of the deuterium component in heavy water can lead to tritium arisings that are more significant than for other mainstream reactor designs. In general, the prominent advantage of the use of heavy water (as both moderator and primary coolant) is that the neutron economy of the reactor is high relative to other designs as fewer neutrons are lost via absorption. This enables, for example, fuels of natural enrichment to be used to a high degree of ^{235}U utilisation and there is also consideration being given to the potential for fuels spent from other reactors to be used in the future, if desired.

The natural abundance of heavy water is just one molecule in 44 million water molecules and thus it has to be extracted and concentrated. Various techniques for this exist including electrolysis, distillation and the hydrogen sulphide gas–liquid exchange process. The majority of the world's heavy water is produced in Canada. An efficient separation process should be able to handle abundant quantities of feedstock, be simple, be efficient in terms of energy consumption and return as high a separation factor as possible. Heavy water can be produced at the site of a nuclear power reactor but is generally received from larger-scale, off-site production facilities offering greater production efficiencies and lower supply cost.

10.6.2 DESIGN OVERVIEW

The development of the heavy water reactor is characterised by several design variants of the use of heavy water but the CANDU design based on pressure-tubes is the most successful. The development of the CANDU arose out of early Canadian experience with related research reactor designs. It has allowed the use of indigenous Canadian uranium reserves without the requirement to use military enrichment facilities in foreign countries. Related heavy water reactor designs have been built outside of Canada in India, Pakistan, Argentina, South Korea, Romania and China.

Although heavy water flows and transports heat to an extent that is very similar to light water, this is effectively where parallels between fission reactor designs exploiting these substances end. The CANDU reactor[9] is a design in which heat is generated from a critical reaction in uranium in a cylindrical vessel on its side. The fuel assemblies lie horizontally in tubes through which heavy-water

[9]Specifically the CANDU 6 for the purposes of this discussion, see also [3].

coolant flows under pressure. Several hundred of these pressure tubes (cf. 380 in the CANDU 6 design) pass through a cylindrical austenitic stainless steel tank of heavy water; it is this that constitutes the moderator, and which is known universally as the *calandria*. The primary coolant circuit comprises two separate loops passing in both directions through the calandria (i.e. each supply and return path to/from a steam generator constitutes a loop), typically with two pumps per loop and a pressuriser in each. The pressuriser maintains the primary coolant loop pressure with a combination of heaters and steam relief valve systems.

An important distinction between the two heavy-water components is that during operation the *coolant* in the pressure tubes is hot and pressurised (the latter to ensure that it remains in the liquid phase) whereas in the calandria the *moderator* is maintained at a relatively cold temperature (~70°C) and at low pressure by comparison. Each Zircaloy-4 calandria tube contains a zirconium–niobium pressure tube, typically of 6 m length, in which the fuel bundles are placed. Each tube contains 12 or 13 fuel bundles of 0.5 m in length (depending on the specific design variant).

The calandria is perhaps the most prominent distinguishing design feature of the CANDU reactor in comparison with light-water cooled/moderated reactor designs. The heavy-water moderator has two water-based shield tanks on the face of each end of the calandria and it receives ~4.5% of the rated power of the reactor as heat. This derives mostly from γ radiation, some from fast neutrons and a little heat directly from the tubes themselves. Gadolinium or boron is added to the moderator for long-term reactivity control. The calandria has its own separate coolant circuit which comprises two pumps and two heat exchangers (cooled by light water). The moderator is top-filled with helium as a cover gas to inert it and to aid the recovery of gaseous products formed as a result of radiolysis (deuterium and hydrogen). The secondary and tertiary coolant stages are based on light water. Schematic diagrams of the front and side elevations of a CANDU reactor design are given in Fig. 10.9A and B, respectively.

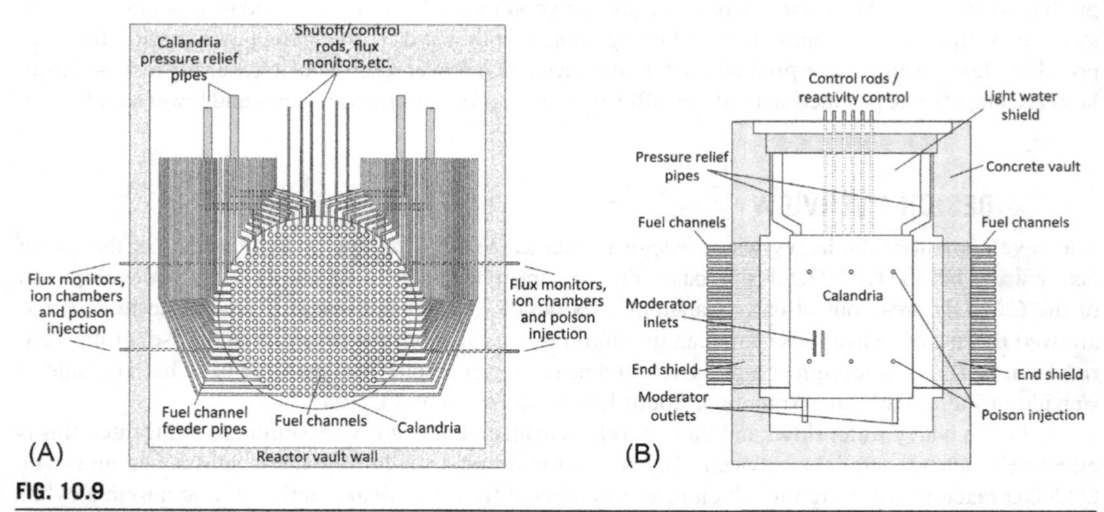

FIG. 10.9

(A) A schematic diagram of the front elevation of a CANDU reactor core. (B) A schematic diagram of the side elevation of a CANDU reactor core. Note that some details have been omitted in the interests of clarity, particularly the detail of the control rod drives, ionisation chambers and flux monitors, and that the figure is not to scale.

10.6.3 OPERATION

Controlled fission in the several hundred fuel channels of a CANDU reactor generates heat. Heavy-water under pressure is circulated in a closed primary circuit from the manifold of the pressure tubes and is fed to steam generators to produce light-water steam and subsequently electricity via the turbine generator stage.

An important operational feature of the CANDU reactor design is that it can be refuelled while on-load (either partially or through to 100% if necessary). This requires a relatively simple shuttling of several fuel bundles out of one end of a pressure tube and the replacement of them with fresh bundles at the other. A refuelling machine operates at each end of the calandria and between 4 and 8 of the 12 bundles per tube can be exchanged at a time. This enables the fuel loading to be adjusted on a basis that is more routine than for the case of most other reactor designs leading to less variation in the reactivity state of the core through life. It is also not as necessary to trim the reactivity to compensate for significant changes in reactivity due to burnup and thus there is no need to add boron to the primary coolant. However, as mentioned earlier, soluble agents for trim (boron) and shutdown (gadolinium nitrate) are added to the moderator to quench the reactivity of this component of the system.

Tritium[10] typically builds up to equilibrium in ~40 years in CANDU systems. It is necessary to manage the environmental and occupational dose hazards of this in terms of operation. Tritium constitutes a relatively low radiological hazard as discussed in Chapter 5. The management philosophy in general terms is to: minimise heavy-water loss, to isolate areas where the risk of water loss is high, recover or discharge escaped water, to protect workers via controls and personal protective equipment (PPE), to assess their exposure (via bioassay) and to employ de-tritiation processes. A schematic diagram of a typical CANDU plant is given in Fig. 10.10.

10.6.4 FUEL DESIGN

The fuel design for the CANDU comprises sintered, uranium dioxide pellets naturally enriched in ^{235}U placed in a Zircaloy-4 cladding tube to constitute a fuel element; 37 fuel elements are arranged in concentric rings to form a cylinder with two plates welded at each end to constitute a fuel assembly or bundle. Bearing pads are welded to the outer ring of fuel elements to prevent the fuel cladding coming into contact with the inner surface of the pressure tube. A fuel bundle constitutes the module that is inserted into a given pressure tube and that is retrieved at refuel.

10.6.5 CONTROL AND PROTECTION SYSTEMS

CANDU reactors offer a variety of passive and active protection features that are specific to this particular design. For example, the relatively long neutron lifetime (~1 ms) that is characteristic of the CANDU design (due to the low neutron absorption cross section of the deuterium in heavy water) and low ^{235}U enrichment is manifest in that increases in power that result from positive transients in reactivity are slow to materialise. Reactor periods are relatively long and are compatible with the response time of the reactor system shutdown rods. Further, the control rods in a CANDU

[10]Arising from the neutron exposure of deuterium and from neutron reactions on boron and lithium where used as additives in the water.

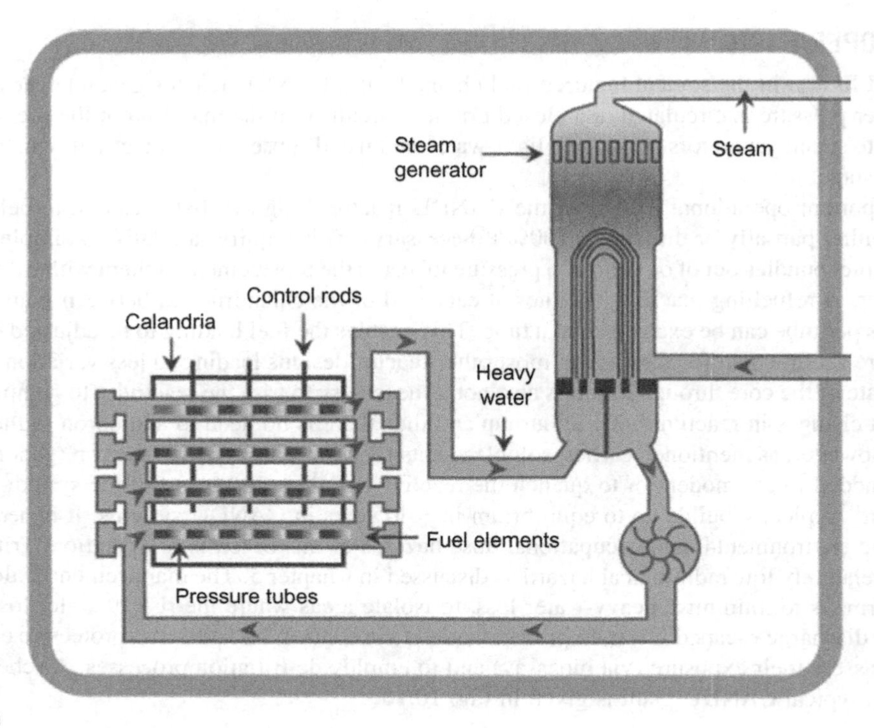

FIG. 10.10

A schematic of the CANDU plant layout.

With permission from the World Nuclear Association.

design penetrate the calandria (moderator) as opposed to quenching the neutron number in the pressurised, coolant side of the system or in the individual fuel bundles directly. This results in a relatively low risk of fuel rods being ejected in the event of a very significant positive increase in reactivity.

The combination of the relatively significant quantity of moderator and coolant together with fuel of natural enrichment results in a relatively high moderator-to-fuel ratio and the possibility of a positive void coefficient. However, in an emergency this is offset by the injection of poison into calandria (note not the coolant) and power-assisted shut down rods.

The ability to use low-enrichment fuel due to the high neutron economy that is characteristic of the low neutron absorption of deuterium results in the probability of an inadvertent criticality following a severe accident being low. The use of low-enriched fuel also results in simplified storage and handling of the fuel prior to use because the risk of criticality is low in the absence of heavy water.

The pressure tubes in a CANDU are replaceable and tube failures can usually be anticipated via the evidence of precursor, minor leaks before they happen. This aids the planning of shutdown and repair procedures rather than it being necessary to react to significant, unexpected damage. Tube failure is also usually limited to a single pressure tube as the calandria arrangement provides extensive multi-interdependency.

The moderator also serves as an emergency heatsink in the event of a loss of coolant or the loss of emergency core cooling. Consequently, the implications of severe accident scenarios are limited by design with the moderator and its shield tank assembly providing sufficient capacity to cater for the management of emergency decay heat if necessary. In the event that this capacity was undermined and excess heat generation due to loss of coolant were to result in steam generation in the moderator, this would cause the reactor to shut down due to the loss of moderation and hence the loss of thermal-spectrum neutrons necessary to sustain the critical reaction.

In the context of station blackout scenarios in which all power is lost to the plant, the rundown of the reactor coolant pumps is designed to maintain coolant circulation in the short term with natural circulation providing the necessary cooling capacity afterwards.

The ECCS in the CANDU reactor design refills the pressure tubes with light water in the event of a loss of heavy-water coolant. The heightened neutron absorption in the light water replacement causes the critical reaction to cease and thus there is no need for boron to be added to the coolant to hold down the reactor in a shutdown state. CANDU reactors have a large containment building to ensure low residual pressures in the event of an accident causing vapour under pressure to be vented to the containment. Further, the reactor systems employ two independent shutdown systems comprising shutdown rods and the injection of a neutron poison. There are also two independent trains of diesel generators as back-up electricity emergency supplies should grid-derived power sources fail.

10.7 GRAPHITE-MODERATED, GAS-COOLED REACTORS

10.7.1 INTRODUCTION

Mainstream GCRs are almost always based on a combination of graphite as moderator and carbon dioxide as coolant. While there have been several GCRs built across the world, Britain is unique in having adopted this design for the majority of its nuclear power reactor fleet and is the only country that has operating GCRs today. There are 14 reactors of the second-generation design (Generation II) known as the *advanced gas-cooled reactor* (AGR) in operation at the time of writing; they are amongst the most thermodynamically efficient reactor designs (Table 10.5).

The first generation of reactors based on the combination of graphite and carbon dioxide that produced power on a commercial basis were known as *Magnox* on account of the magnesium–aluminium alloy used as the fuel cladding that resisted oxidation: 'magnesium no oxidation'. These Generation I reactors were generally smaller than the AGRs mentioned above that followed, but the Magnox programme did culminate in larger-capacity variants. These include the Wylfa plant on Anglesey in Wales that was commissioned in 1971 and was the last to cease power production in 2015.

Table 10.5 The World's Operable Gas-Cooled Reactors

Country	Number of LWGRs	Notes
Great Britain	14	Three design classes of 660-MWe reactors constructed in pairs

10.7.2 BACKGROUND TO THE USE OF GRAPHITE AND CARBON DIOXIDE

In Chapter 2 early reactor designs were described that were derived from the Manhattan project to produce nuclear materials for the construction of nuclear weapons. In Britain the prominent exponents of these were the Windscale piles; graphite-moderated systems in which the nuclear fuel was arranged horizontally in cartridges and cooled by air blown through the core by large fans. As discussed in Chapter 6, graphite was chosen to provide both the physical matrix to contain the fuel, favourable moderating properties and also enabling access to the fuel after relatively short burnup periods. The very small neutron absorption cross section of graphite is also compatible with the use of fuel of natural enrichment which was desirable at that time. However, graphite undergoes radiation damage which reduces its density (and thus moderation effectiveness) and it also becomes contaminated which presents significant challenges in terms of decontamination and disposal. The Windscale reactors did not produce power as the emphasis for their use was on the production of plutonium.

The fire in pile 1 at Windscale in 1957 (discussed in Chapter 14), together with the desire to generate electricity from the reactor systems that followed, caused reactor designers to turn to alternative coolants. They selected carbon dioxide due to its favourable thermo-physical properties at high temperature and pressure; so began a long association between British power reactor systems and carbon dioxide coolant cycles. Unlimited by the critical point ceiling that constrains upper operating temperatures of reactors reliant on pressurised water-based coolants,[11] GCR systems are a prominent candidate for the high-efficiency, high-temperature reactor systems of the future. A schematic diagram of a typical GCR is provided in Fig. 10.11.

10.7.3 DESIGN OVERVIEW

The AGR design comprises a cylindrical, prestressed concrete pressure vessel containing the reactor core (fuel and moderator), 12 boilers and gas circulators (pumps). The vessel annulus is subdivided into four quadrants—each containing three boiler units and a gas circulator—with the reactor core at its centre in a steel, domed vessel known as the *gas baffle*. This is a steel vessel of approximately 13.7 m diameter that contains the reactor core. The principal role of this is to direct coolant flow through the core; it comprises the dome (upper), the cylinder (middle) and the skirt (lower section). The gas baffle has a typical external diameter and height of ~30 m, wall thickness of ~5 m and is lined with stainless steel. The reactor core comprises a 16-sided stack of interconnected graphite bricks within a steel restraint tank that is supported by a system of steel support plates known as the *diagrid*. The core comprises a central region of cylindrical fuel channels interspersed by interstitial channels that are used for control and protection systems. Around this central region there is extensive radiation shielding comprised of a combination of graphite (above the core) and steel (to the periphery and the base). This enables the boilers to be accessed safely for maintenance purposes when the reactor is shutdown. The core is typically 11.0 m in diameter, 9.8 m height and mass 1300 t.

While GCRs such as the AGR design are capable of higher-temperature operation relative to LWR alternatives, as the coolant is already in the gas phase, thermal oxidation of the graphite moderator has to be inhibited by cooling it in addition to the vertical fuel channels that pass through. This also minimises undesirable temperature gradients in the core. This is achieved by encouraging re-entrant flow of the

[11]Notwithstanding the potential use of supercritical water as a coolant.

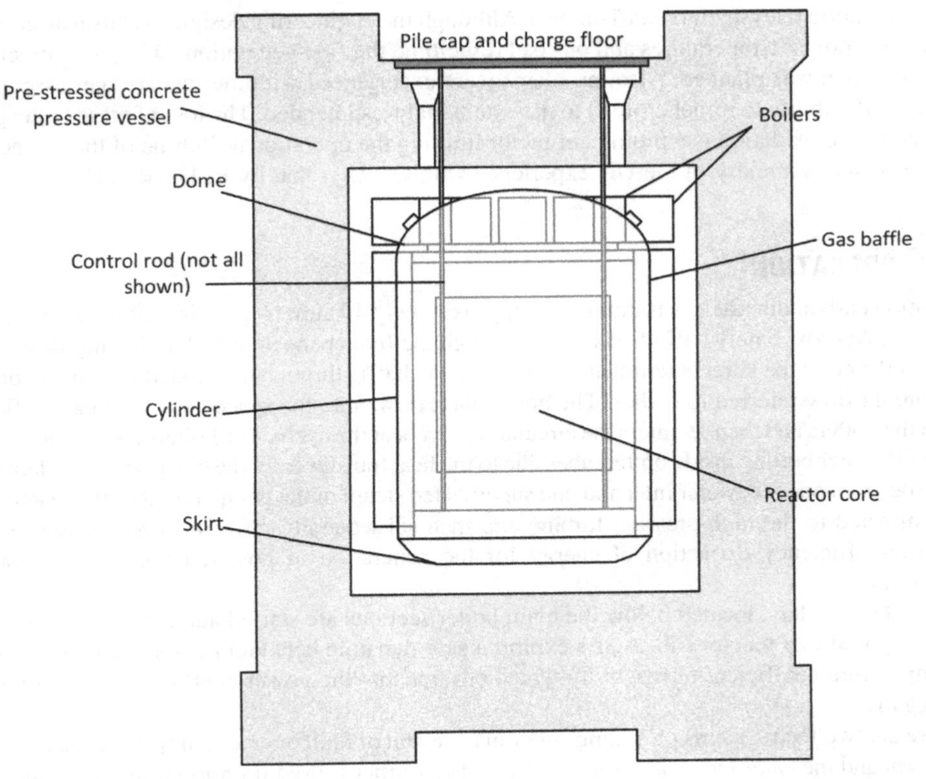

FIG. 10.11

A schematic diagram of the side elevation of a GCR reactor core, fuel channels and coolant routing have been omitted for clarity.

carbon dioxide coolant through the moderator with the gas baffle referred to above; when the coolant reaches the top of the baffle it is redirected back down through the moderator. The baffle also separates the hot gas from cold to feed to the boilers and has penetrations for the fuel assemblies and control rod guide tubes.

The concrete pressure vessel design has a separate opening for each fuel channel above the core which permits refuelling while the reactor is operating at pressure or alternatively when off-load at atmospheric pressure. AGRs were built in pairs adjacent to one another in a single station housing. This enables one refuelling machine (essentially a hoist with a housing to enable a pressurising seal to be established with each channel) to serve both reactor cores from above through port plugs in the top of the pressure vessel, with each port plug corresponding to each fuel channel.

Britain's fleet of 14 AGRs comprise seven stations: Dungeness, Hartlepool, Heysham I and II, Hinkley B, Hunterston B and Torness. The AGR design generation is subdivided into three chronological phases due to modifications to the first generation (Dungeness B, Hinkley Point B and Hunterston B) that resulted in the modified first generation (Hartlepool and Heysham I) and led subsequently to the second generation (Heysham II and Torness). Although the engineering design has intrinsic advantages such as high coolant temperatures and on-load refuelling, the first generation AGRs took much longer to construct than was planned. Also, problems were experienced with the fuel stringer assembly that reduced the flexibility to refuel on-load to the extent that was intended. The loss of mass of the graphite due to radiolytic oxidation is a prominent factor limiting the operational lifetime of these reactor systems. A succinct summary of the UK experience with GCRs is that by R. Davies [4].

10.7.4 OPERATION

In operation carbon dioxide gas is pumped at approximately 40 atm. by gas circulators from the bottom of the core. Approximately half of this flows through the fuel channels with half having flowed up the outside of the core; the latter is returned at the top by the baffle through the moderator which constitutes the re-entrant flow referred to earlier. The hot coolant flows into the space above the gas baffle, down through the boilers and then re-enters the circulators to repeat the cycle. The boilers are of a once-through design with water boiling inside boiler tubes due to the heat transfer from the hot gaseous coolant flow on the outside, with one feedwater inlet and one superheated steam outlet per quadrant of the reactor vessel. This is directed to the high-pressure turbine and then subsequently through lower-pressure stages to ensure high-efficiency extraction of energy for the generation of power, typically via a 660-MW generator unit.

Decay heat boilers, located below the main boiler sections are started automatically, as part of the system response to a reactor trip. AGRs exhibit a safe dynamic behaviour associated with a negative fuel temperature coefficient of reactivity which ensures that the power coefficient of reactivity is always negative.

There are two feed systems providing water in the event of fault or emergency: the *decay heat boiler feed system* and the *emergency boiler feed system*. The former is used for normal shutdown and reactor trips where the core remains pressurised; the decay heat feedwater is provided from a tank that is condensed and returned to become the feed input again. The system is designed to afford redundancy so that 2 out of 4 pumps can provide sufficient feedwater flow to the boiler system. The decay heat boiler coolers are served by air-fed cooling towers, with a 2-from-4 redundancy philosophy, and the main and decay heat boilers provide diversity in the event of a boiler failure. A single quadrant is designed to service the reactor in the event of a trip. Natural circulation, that is coolant supplied to the decay heat boilers without the requirement for pumping, is possible via two quadrants given feedwater flow is maintained and the core remains pressurised. The emergency boiler feed system takes water from the station reserve feed tanks in the event of a trip; a scenario in which all circulators are lost is offset as long as at least two boilers are fed by natural circulation when the core is pressurised.

10.7.5 FUEL DESIGN

To distinguish the individual tubes of fuel from the packs used in an AGR, the former are known as *fuel pins* and the latter are known as *fuel assemblies*. Each pin comprises a stack of uranium dioxide pellets of relatively low enrichment (of between 2 and 3% wt. ^{235}U) encased in a stainless steel clad. Each cladding tube is approximately 900-mm long and is welded shut after being filled with helium to inert the atmosphere in the pin. Each AGR fuel element comprises 36 fuel pins set in a graphite sleeve which is held in place by a support brace at each end; a *fuel stringer assembly* comprises 8 of these on top of one another inserted into a fuel channel in the graphite assembly.

10.7.6 REACTIVITY CONTROL

The control rod system occupies a proportion of the interstitial positions between fuel channels with the other positions between elements being dedicated to the other protection measures, described below. The control rods comprise a chromoly sheath containing 4.4% boron stainless steel inserts with graphite spacers and end sections, the latter forming part of the upper neutron reflector. There is a combination of both *grey* and *black* control rods corresponding to their relative opacity to neutrons; grey rods have less boron due to there being none in the lower sections, and are thus less absorbent than black. The grey rods are the prominent agent in the day-to-day control of the reactor whereas the more absorbent black rods are the *primary shutdown device* or PSD. The control rods are raised and lowered by motor-operated systems with electromagnetic clutches designed to enable them to be dropped in under gravity in an emergency.

10.7.7 PROTECTIVE SYSTEMS

There are three protection systems in the AGR design constituting three layers of protection in terms of defence in depth. The PSD as described above are the control rods that can be inserted under gravity in an emergency by de-energising their clutches. To cater for the possibility that this is not possible, a system by which nitrogen can be injected rapidly from on-site reserves to the interstitial positions between the fuel assemblies is also provided. Nitrogen absorbs neutrons more readily than carbon dioxide and this system is designed to place the reactor in a shutdown in an emergency, in place of the control rods assuming the reactor remains pressurised. In the event that reactor pressure cannot be maintained (preventing the use of nitrogen injection) together with a control rod problem preventing PSD insertion, the third layer of defence is a pneumatic system by which 3-mm diameter boron glass beads can be introduced from nearby hoppers to a sufficient number of the interstitial channels in the reactor. This is designed to render it shutdown indefinitely without the requirement for pressurisation.

10.8 LIGHT-WATER GRAPHITE REACTORS

10.8.1 INTRODUCTION

The sole exponent of the light-water graphite reactor (LWGR) is the Soviet RBMK (high-power channel BWR) design [5] (Table 10.6). The RBMK series comprises two generations: the RBMK-1000 and RBMK-1500, although the corresponding reactor designs differ significantly only in terms of capacity. The series was commissioned in the 1970s and 1980s and the reactors were built in pairs; examples of the RBMK-1000 include those reactors operating at the complexes at Kursk, Smolensk and Leningrad in Russia while the reactors at Ignalina in Lithuania are RBMK-1500s (the latter now shut down). The RBMK design will be perhaps forever synonymous with the Chernobyl accident in 1986. Significant changes to the design were made after this as discussed in more detail below and in Chapter 14. A schematic elevation of the RBMK is provided in Fig. 10.12.

Table 10.6 The World's Operable Light-Water, Graphite-Moderated Reactors by Country[1]

Country	Number of LWGRs	Notes
Russia	15	11 RBMK-1000 and 4 of the smaller EGP-6 variant, 12-MWe power

RBMK-1500 designs in Ignalina, Lithuania are shut down and thus not listed.

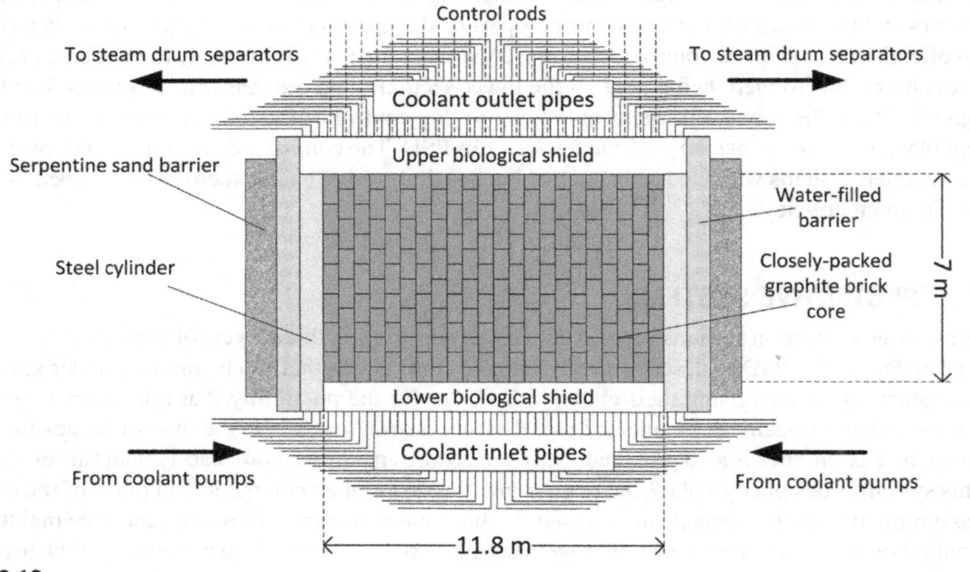

FIG. 10.12

A schematic diagram of the side elevation of a RBMK reactor core. Note: fuel assemblies are not shown as they would be contained within the pressure tubes passing through the graphite stack and the coolant pipe inlets and outlets are elevated in the diagram for clarity rather than their real, more horizontal perspective. Coolant pumps and steam driers have also been omitted for clarity.

10.8.2 DESIGN OVERVIEW

The RBMK uses boiling light water as its coolant in conjunction with a graphite moderator and uranium fuel that is lightly enriched in ^{235}U. The reactors are large in comparison with, for example, LWRs with a core geometry of a vertical cylinder, typically of 11.8 m in diameter and 7 m in height. The core is comprised of a stack of a large number of closely packed graphite blocks of a total typical mass of 1700 tonnes. This is placed on a steel baseplate that is supported by a steel cylinder of 14.5 m in diameter and 2 m in height. The graphite is contained within a steel cylinder also of 14.5-m diameter. This is perfused with a mixture of nitrogen and helium to inert it, to improve heat transfer and to facilitate sampling of the core environment. A large number (~1660) of fuel channels are configured to run vertically throughout the graphite along with ~200 separate absorber rod channels. There are also a number of special channels to accommodate instrumentation.

Each fuel channel comprises a pressure tube in which two fuel assemblies are positioned. The light-water coolant flows through each channel from the bottom of the core to the top. Each pressure tube comprises a central region of zirconium–niobium alloy with upper and lower steel regions. The pressure tube design was selected to enable online refuelling to combine abilities to yield both plutonium and power from the reactors; individual fuel channels can be isolated from coolant flow and the fuel extracted when desired without shutting the reactor down.

The RBMK coolant system comprises two parallel coolant loops, one for each half of the core. Each individual loop has two steam generators and four coolant pumps; three pumps are used for normal operation and one is on standby for emergencies. Each fuel channel has its own coolant supply and individual flow rate that results in a relatively complicated arrangement to control during operation. The steam evolved from boiling in the pressure tubes is separated from the saturated component in steam drum separators before being supplied to the steam turbines. After this it is condensed. RMBK reactors are characterised by having no containment as such but are housed within a lined, reinforced concrete cavity.

10.8.3 FUEL DESIGN

The fuel in RBMK reactors is comprised of uranium dioxide pellets encased in Zircaloy cladding 3.65-m long and 13.6-mm diameter with each such rod constituting a *fuel element*. A *fuel bundle* comprises 18 elements fixed in a cylindrical arrangement of two concentric rings. Two bundles are placed end-to-end per pressure tube to constitute a *fuel assembly*; an assembly extends to just over 7 m total length. Each individual bundle contains approximately 147.5 kg uranium.

10.8.4 REACTIVITY CONTROL AND PROTECTION

Reactivity control in RBMK reactors is largely associated with the control rod infrastructure. This is organised into four groups comprising: a set of shortened absorber rods for axial control (×24, length ~3 m), full-length (~5 m) absorbing rods for radial control (×24), absorbing rods for the automatic control of reactor power (×139) and emergency rods (×24). There are 211 in total and all are made of boron carbide. RBMK control rods are distinguished in one way in particular by the graphite displacers (or 'followers') that form part of their design. These are designed to remain in the core when the absorbers are withdrawn and act to displace the light-water coolant from the space that would otherwise be left behind. Without these, undesirable neutron absorption on the hydrogen in the water that would fill the space left by the rod would occur and undermine neutron economy.

There are a number of control issues that are specific to the RBMK design. Firstly, the reactor when first fuelled harbours a reactivity surplus that is large relative to that of other mainstream reactor designs. This requires a greater number of absorbers to be in the core as a counterbalance. Hence, the reactivity of a fresh fuel load is offset by the incorporation of an additional number of absorbers in fuel channel positions until sufficient burnup of ^{235}U has occurred to offset the surplus. The fuel composition of the core can be increased as burnup develops from a typical level of 165 tonnes at the start to approximately 192 tonnes during steady-state operation, with the difference made up with additional absorber rods at the time of fresh-fuel core load. When control rods are withdrawn some zones in the core can be relatively devoid of absorbers (especially peripheral areas) and localised criticalities can result.

The relatively low enrichment of the fuel together with extensive moderating properties of the graphite and the light-water coolant results in a high moderator-to-fuel ratio that can precipitate a positive reactivity void coefficient. When control rods are withdrawn and replaced by graphite followers the moderator-to-fuel ratio increases further, potentially exacerbating the void coefficient, as it does with fuel burnup too. The effect on the control behaviour of the reactor reduces the reactor period further complicating the ease of reactivity control. This is offset by the permanent presence of absorber rods in place of fuel which, in turn, requires the enrichment of the fuel to be increased (typically from 1.8 to 2.4% wt.).

A further feature of the reactivity control in RBMKs is that when the control rods are in their fully withdrawn position, a column of coolant is present below the graphite displacer at the end of the absorber rod because the rods are shorter than the channel in which they move. When the rod is inserted from this position the water (of relatively high neutron absorption) is displaced by the graphite (of relatively low neutron absorption). This constitutes an insertion of reactivity despite the intent being an insertion of absorber and hence a reduction in reactivity. This is known as the 'end-rods effect'.

10.8.5 DESIGN MODIFICATIONS AFTER CHERNOBYL

Following the accident at Unit 4 of the Chernobyl nuclear power plant in 1986, a variety of modifications have been made to improve the safety of the RBMK reactor design. For example, to reduce the extent of the positive void coefficient, additional absorbers were added to the reactor cores and the fuel enrichment was increased from an average of 2.0–2.4% wt. to reduce the moderator-to-fuel ratio. The effectiveness of the emergency protection system during the first few seconds of control rod insertion has been improved by revising the manual control rod design to extend the proportion of absorber at the bottom end. This change increases the proportion of the rod that is absorbing neutrons as opposed to that of the previous combination of water column and graphite displacer. The rod drives were also improved so that the full insertion time was reduced from 19 to 12 s reducing the insertion time for the 24 emergency control rods to 2.5 s. The control and monitoring system was improved to enable manual trips in response to low power (below 700 MWth) or too low a reactivity margin. Further modifications have been made to the ECCS including, for example, increasing the number of feedwater pumps and the number of emergency core cooling lines.

REVISION GUIDE

On completion of this chapter you should:

- be able to name and describe the *five mainstream nuclear power reactor systems* in operation in the world today
- be able to describe the design of each of the five examples, with particular reference to the *core*, choice of *moderator*, *coolant*, *control* and *fuel* materials
- understand the background to the *choice of the materials* and *layout* of each of the designs
- where applicable, understand the *protection systems* and *protocols* used to ensure the *emergency management* of each reactor design
- have the knowledge to be able to recall *approximate numbers* and *distribution of each reactor system* in use in each case throughout the world

PROBLEMS

1. For each of the five nuclear reactor designs summarised in this chapter, describe qualitatively what would happen to the reactivity as a result of reducing the rate of primary coolant flow.
2. Explain why LWRs are generally smaller in dimension for the same power rating than graphite-moderated reactors.
3. Explain why it might be possible for PHWRs (assume CANDU for the purposes of this problem) to utilise spent fuel from LWRs for the production of power but the converse is not possible.
4. Explain the national context and its influence on the choice of reactor designs in operation producing power today in the United States, Canada, the United Kingdom and Russia.
5. All of the reactor systems described in this section for the large-scale production of power are heterogeneous designs, that is they exploit a distribution of separate fuel elements dispersed throughout a combination of substances necessary for moderation and cooling. Explain why this is preferable in most cases to a homogeneous arrangement, particularly for the production of power.

REFERENCES

[1] A.W. Kramer, Boiling Water Reactors, Addison-Wesley, Reading, MA, US, 1958.

[2] W.B. Weihermiller, G.S. Allison, LWR Nuclear Fuel Bundle Data for Use in Fuel Bundle Handling, Topical Report, PNL-2575, 1979.

[3] Heavy Water Reactors: Status and Projected Development, Technical Reports Series 407, International Atomic Energy Agency, Vienna, 2012.

[4] R. Davies, The sizewell B nuclear enquiry: an analysis of public participation in decison making about nuclear power, Sci. Technol. Hum. Values 9 (3) (1984) 21–32.

[5] M.V. Malko, The chernobyl reactor: design features and reasons for accident, in: T. Imanaka (Ed.), Recent Research Activities About the Chernobyl NPP Accident in Belarus, Ukraine and Russia, Research Reactor Institute, Kyoto University, Japan, 2002, p. 11.

CHAPTER

11 ADVANCED REACTORS AND FUTURE CONCEPTS

11.1 SUMMARY OF CHAPTER AND LEARNING OBJECTIVES

In this chapter the future prospects for nuclear energy production are summarised in terms of advanced reactor system designs, breeding and related concepts. Some of these areas of the field have been appreciated as possibilities for the future almost since fission itself was first harnessed for the commercial production of power; these include fast breeder reactors, fusion and thorium. Other areas are new by comparison and are receiving attention at present; these include Generation III+ reactors, small modular reactors and Generation IV systems. There is a wide variety of potential schemes by which nuclear energy might benefit the world in the future. This chapter provides a summary of the most prominent examples.

The objectives of this chapter are to:

- introduce the range of current developments associated with fission reactor systems currently under development, construction or commissioning
- summarise four prominent exponents of the small modular reactor concept
- introduce the scientific basis for breeding and consider how fertile materials might be converted to fissile materials compatible with use in future reactors
- review the six Generation IV designs under consideration for long-term nuclear energy generation
- discuss the potential for a thorium-based fuel cycle, its advantages and disadvantages
- present the scientific basis for fusion and the options for this source of nuclear energy in the future

11.2 HISTORICAL CONTEXT: HOMI JEHANGIR BHABHA 1909–66

Homi J. Bhabha progressed to theoretical physics specialising in nuclear physics following a first degree in Mechanical Sciences. His nuclear physics studies started with his doctorate that was associated with cosmic rays, their absorption features and studies of the associated electron shower. He was first to calculate the cross section of electron–positron scattering that was to become known as *Bhabha scattering* in his honour. Much of his research was done at the Cavendish laboratory, Cambridge before he returned to India in Sep. 1939 (Fig. 11.1).

Alongside his scientific achievements, Bhabha is regarded as the *father of Indian nuclear power* because of his prominent role in the development of several dedicated institutes, their research focus

Nuclear Engineering. https://doi.org/10.1016/B978-0-08-100962-8.00011-1

FIG. 11.1

Homi Jehangir Bhabha.

and India's national, long-term nuclear strategy. On returning to India he took up a position at the Indian Institute of Science from where he convinced the Indian government to embrace an ambitious nuclear programme; he became the first chairman of the Indian Atomic Energy Commission and was also a strong advocate of nuclear weapons development in India. Bhabha's vision was that without this programme India would be reliant on nuclear expertise from overseas. His strategy was to bring together the competency and skills of existing researchers from across India in one place so as to develop this talent to support the future of nuclear energy he anticipated in India. In particular he established what is now known as the Bhabha Atomic Research Centre in Mumbai and proposed India's three-stage nuclear power programme for the extraction of energy from India's significant reserves of thorium.

11.3 INTRODUCTION

The energy made available from changes in the nuclear composition of matter is generic; how it is exploited in a sustained way as a source of heat and subsequently as electricity is dependent on the engineering approach that is adopted. Thus far in this book both existing and legacy reactor designs have been discussed. However, these reactors only represent the second or in some cases the third generation of nuclear power systems. Since they were commissioned a great deal has been learnt about their design and optimisation that has resulted in new ideas by which future designs might be made safer, more efficient and that produce less waste. The lessons learnt from nuclear accidents have resulted in revised designs, safety and regulatory practices while the perceived disadvantages of earlier designs that caused cost and schedule overruns have been identified and are influencing the design of future stations to make these easier to build.

Looking to the future, a dynamic set of priorities is influencing the way in which nuclear power is generated and exploited. Not only is the way that nuclear plant are built being revised but also the way in which they are deployed and used is attracting greater attention. Perhaps most importantly, reactor systems are now required to be more efficient, operate for longer and to complement a diversity of other low-carbon energy options that are essential if we are to combat climate change. This is a major influence on current research activities, particularly concerning the design of fuels and thermohydraulic cycles to operate at higher temperatures and for longer durations. Also, while some of the most promising nuclear generation concepts have been known almost since the discovery [1] of nuclear energy itself, taking these ideas to commercial levels of exploitation poses some of very significant engineering challenges. These will take time to resolve but offer significant potential benefits in return. Looking further ahead, breeding of new fissile fuel offers the potential to significantly extend fuel utilisation exploiting ^{238}U and ^{232}Th, and fusion offers a route to utilise nuclear energy from tritium and deuterium. A discussion of these developments is the aim of this chapter.

11.4 CURRENT DEVELOPMENTS: GEN III+ DESIGNS

In recent years, mainly due to strategies to decarbonise electricity supplies to avert climate change, many countries have been considering nuclear energy as a source of low-carbon electricity alongside other options. A significant number of active projects are under construction with a number of these taking place in countries that have not embraced nuclear energy before. For those countries with existing nuclear power stations, there can be a degree of urgency since the existing reactors will eventually reach the end of their operational life and replacements take several years to plan and build. Nuclear energy is an effective means of providing the baseload proportion of electricity supply needs that can complement renewables generation at a time when many countries are not seeking to renew their fossil fuel generation infrastructure, especially coal.

A number of issues have influenced many of the reactor designs that are currently under development on which this chapter is focussed. The first are the lessons learnt from prominent incidents, particularly Windscale, Three Mile Island, Chernobyl and Fukushima. These have brought with them an increased requirement for integrated safety systems (both active and passive) and improved containment. The second has been a desire to increase power output in order to maximise revenues from the plant once they are operational. The third development is also economic in origin: the emphasis given to matters that ensure that nuclear plant can be built to time and cost, particularly in terms of modularity and ease of licencing. Simplified, modular plant designs are being considered to reduce the risk of escalation in the upfront investment cost (this is discussed further in Chapter 16) as this has been regarded as a key risk in nuclear power infrastructure projects in the past.

At the time of writing, most designs subject to these developments are under construction and not yet completed. Specific notation has been adopted for stages of plant following such a development path: they are either first-of-a-kind ($FOAK_1$) or first of a particular design in a particular country ($FOAK_2$) where, as discussed in Chapter 16, a degree of contingency is often anticipated by vendors because of the experience that is gained in completing the first plants of a given design. The expectation is that n^{th}*-of-a-kind* (NOAK) projects are cheaper and quicker to complete as a result.

At the same time, the international electricity generation markets have changed more profoundly than has been the case perhaps since widespread electrification began in the early 20th century. Recently, significant investment in intermittent renewables production and the rapid build and start-up of combined cycle gas turbine systems has begun to offset the extent of the baseload requirement. This has caused some reactor vendors to recast their designs in terms of power-output-per-unit to offer the flexibility to meet demand via a combination of several modules of the same, small modular design. The nature of the energy supply options is also being challenged in terms of whether it is, for example, power or process heat; the latter destined to provide direct energy to process plant and hydrogen production systems. None of these small-to-medium-sized reactors have yet been constructed or operated on a commercial scale of late, but these developments reflect the adjustment of the nuclear industry to the changing energy market.

Meanwhile, in many countries the way in which electricity is supplied has changed from regulated markets to liberalised and competitive systems that tend to favour smaller upfront investments and quicker returns to investors, thus not always favouring the very large capacity reactor designs of the Gen III+ variant as strongly as smaller capacity, quick start-up alternatives. The situation is a little different in Asia and the Middle East, where governments have made significant investments in new-build programmes of the order of >$30 Bn.

11.4.1 THE EPR (EUROPEAN PRESSURISED REACTOR OR EVOLUTIONARY POWER REACTOR)

The EPR design is derived from earlier designs developed in France (associated with the N4 reactors) and those developed in Germany (associated with the Konvoi series of rectors of the 1980s), arising from a collaboration between Framatome and Siemens. At the time of writing EPRs are being constructed in France (Flamanville), Finland (Olkiluoto) and China (Taishan). EPRs are planned in the United Kingdom for sites at Hinkley in Somerset and Sizewell in Suffolk. A cutaway schematic diagram of the EPR is given in Fig. 11.2A and a photograph of the EPR nearing completion at Olkiluoto in Finland is given in Fig. 11.2B.

Conceptually the EPR is one of the largest capacity nuclear power reactors ever designed and having progressed to construction, rated at 1600 MWe. The major design developments since the N4/Konvoi era have been influenced heavily by the imperatives to introduce greater active redundancy of systems into the design and to be able to resist external hazards. These include, for example, being robust against aircraft strike after the September 2001 terrorist attacks on New York and against flooding after the Great East Japan earthquake in 2011. Advances to enhance containment in the event of a major core accident have also been made.

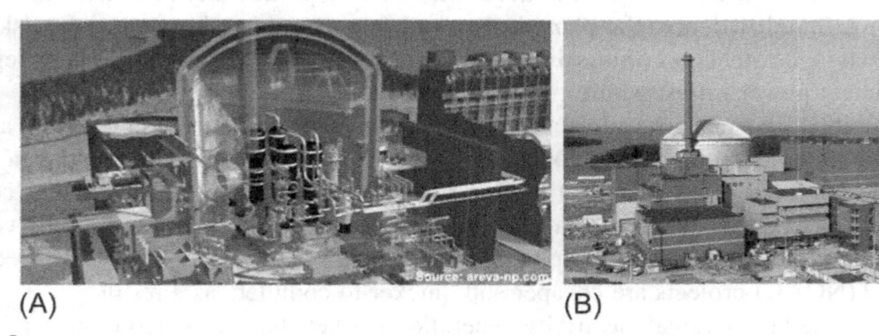

FIG. 11.2

A design-cutaway of the EPR (A) (reproduced with permission, copyright: AREVA/Image and Process) and the EPR approaching completion at Olkiluoto, Finland (B) (reproduced with permission, copyright: AREVA).

The EPR is an indirect-cycle PWR design with four-train redundancy of its coolant circuitry comprising four independent steam generators. It also has a large, concrete 'core catcher' at the base of its containment providing greater substantiation to the prevention contamination from reaching the environment in the event that the reactor pressure vessel is compromised. Aircraft strike is resisted by designing the reactor buildings to withstand anticipated crash loadings. The safety-classified systems, such as diesel generator buildings and service water pump buildings, are separated geographically to reduce the possibility of all the systems being compromised at the same time by such an event.

11.4.2 THE AP1000

The AP1000 reactor design is a development of the earlier Westinghouse AP600 PWR reactor with a nominal electrical power output of 1110 MWe (Fig. 11.3). The design emphasis with the AP1000 is quite different from the EPR. It is focussed on the integration of greater plant simplification to make it cheaper to build, operate and maintain, reflecting the context associated with reducing investment risk introduced earlier in this chapter. With a focus on the use of passive safety systems, such as coolant circulation under natural convection, these simplifications result in fewer components, less cabling and a smaller seismic volume than previous designs.

In addition, there is a focus on a *modular* design in which as much of the reactor as possible is manufactured off-site in a factory and brought to the site for assembly. The main reason behind adopting this approach is to ensure a more consistent degree of compliance with legislative certification, in the same manufacturing environment by the same people and organisations, rather than risk the variance that can occur between one site and another. Once again, this strategy aims to limit the risk of unforeseen cost escalations and project overruns. At the time of writing, four AP1000 reactors are under construction in China (Haiyang and Sanmen) and three have been proposed at Moorside, near Sellafield in the United Kingdom.

FIG. 11.3

Two AP1000 reactors approaching completion at the Sanmen site in China.

Courtesy of the Westinghouse Electric Company.

11.4.3 THE ADVANCED BOILING WATER REACTOR

The ABWR is a design offered by GE Hitachi and Toshiba and is rated at 1350-MWe output. As a direct-cycle boiling water reactor, its main areas of development include: fully digital reactor control and protection systems, and 10 additional coolant pumps at the base of the reactor pressure vessel. The latter eliminate the need for large pumps in the containment and remove the need for the associated large-bore piping and complex routing. At the time of writing, four operational ABWR plant are located in Japan with another under construction and two are being built in Taiwan. The ABWR is being proposed for construction at Anglesey, Wales in the United Kingdom.

11.4.4 THE APR1400

The APR1400 reactor is a PWR system that is a development of the South Korean OPR1000 design. There are 12 OPR1000 units operating in South Korea. The APR1400 is rated at 1455 MWe and comprises an indirect circuit (as is usual for PWRs), two primary coolant loops and four coolant pumps with one steam generator per loop. The APR1400 is offered by the KEPCO Engineering & Construction Co. Inc. with four units currently under construction in South Korea and four in the United Arab Emirates (Fig. 11.4).

11.4.5 THE PRESSURISED HEAVY WATER REACTOR

At the time of writing there are five pressurised heavy water reactors under construction in the world: four in India and one in Romania. The capacities of Indian PHWRs are a little smaller in terms of power generation than the Gen III+ examples described thus far, of approximate rating 700 MWe. The choice

FIG. 11.4

The New Kori APR1400 plant, a Korean-design Generation III reactor, under construction in Kyungnam, South Korea.

Reproduced with permission and copyright of Korea Hydro and Nuclear Power Company Ltd.

of heavy water coolant and moderator, and the associated benefits in terms of neutron economy, originate from a desire in India to use uranium of natural enrichment obviating the need for enrichment facilities and, since there is not a pressure vessel, this removes the need for the heavy casting capabilities needed for PWRs. The improved neutron economy over other water-based reactor designs is compatible with India's staged programme associated with thorium exploitation [2] (see discussion that follows later in this chapter).

11.5 SMALL MODULAR REACTORS

Although not a new concept, SMRs have received significant attention in the last 10 years as the appetite for significant upfront investment in large plant has declined, coupled with a strong desire for a faster, more risk-tolerant return on investment. Technically, an SMR is defined as a reactor with a power output rating of less than 300 MWe, i.e. between a third and a fifth of the capacity of those reactors of traditional scale described above and in the preceding chapter.

There are several drivers behind the current commercial interest in SMRs. Firstly, amidst established electricity markets that have a significant and growing contribution from intermittent renewables and thus a reduced baseload requirement, SMRs may provide a source of backup supply for periods of intermittency and for periods of downtime of ageing fossil fuel plants. SMRs are also compatible with the production of what are referred to as *multiple energy end products*. This infers that, in addition to electricity, they can provide process heat directly to plant in the bulk chemicals sector or to neighbouring hydrogen production facilities. This approach removes the need for electricity conversion infrastructure between the power plant and the process plant, significantly increasing production efficiencies and reducing conversion costs. Perhaps most importantly, SMRs are designed to be built on a modular basis in factories and transported in a state of near completion to site. At a given site several of them could be installed to constitute a battery, providing flexibility in terms of the overall operating capacity required of a given plant. This property is referred to as being *scalable*, offering flexibility in terms of growth and contraction in an uncertain scenario of demand.

SMRs are designed to be cheaper than large-scale plant unit-for-unit (recognising that this might not be the case on a per unit electricity production basis) in order to reduce the scale of upfront investment prior to a return materialising in terms of generated power. The factory-build and the consistency of design and quality between units (the assumption is by definition that successive units are identical) promises more consistent licencing and thus a reduced risk of project overruns with a quicker route to production revenues. In many cases each SMR is designed to be self-contained with an operating pressure vessel, reduced requirements for large-scale pipework and an emphasis on passive safety and co-location with spent fuel storage facilities. Although there are many design concepts under discussion [3], there are four reactors that are nearest to deployment, as discussed below.

11.5.1 THE URANIUM BATTERY (U-BATTERY)

The *U-Battery* is classed as a *micro*-SMR due to its power output being smaller by the standards of other SMRs at approximately 8 MWe per plant; each plant comprises two units (Fig. 11.5). The design originated from a partnership between the Universities of Delft and Manchester and the initiative is being driven currently by a partnership between URENCO, Atkins and AMEC-Foster-Wheeler in

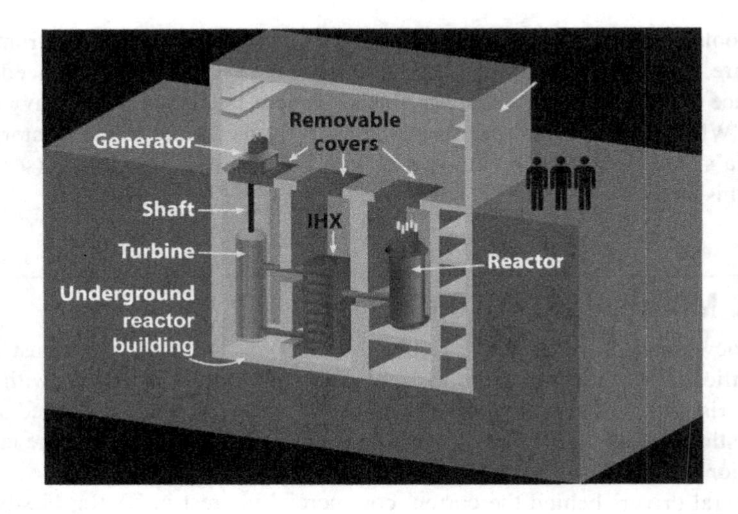

FIG. 11.5

A conceptual cutaway diagram of the U-Battery primary circuit design.

With permission and copyright URENCO and U-Battery.

the United Kingdom. U-Battery is designed to be intrinsically safe based on the use of the TRISO-type uranium fuel that is immune to melting beyond the design-basis accident temperatures of the core, were it to be devoid of coolant.

U-Battery is designed to provide high-temperature process heat (20 MWth at 800°C) and electrical power. The units are constructed as modules in a factory and the intent is that they will be transported by conventional road transport to their site of operation. The footprint of the site needed for the plant is very compact and the thermal cycle is based on the use of a helium primary coolant with a nitrogen-based secondary coolant to drive the turbines. The anticipated operational life of the reactor is 60 years, with refuelling anticipated to be needed every 5 years.

The target applications of the U-Battery are as dedicated power plant on existing nuclear-licenced sites, as back-up power supplies for large nuclear plant, for power and heat on heavy industrial sites, for independent power requirements (such as isolated datacentres and requirements in remote locations), for desalination applications & for transport power requirements.

11.5.2 mPOWER

The *mPower SMR* was a Babcock & Wilcox design[1] originally offered by Generation mPower LLC (a partnership of Babcock and Wilcox Nuclear Energy Inc. and Bechtel Power Corporation). It is rated at 180 MWe and was launched in 2009. The reactor is an evolution of the maritime-type reactor used in the Otto Hahn ship in the 1960s, being an integrated PWR in which the steam is generated within the reactor's pressure vessel.

[1]http://www.generationmpower.com/.

The conceptual approach for the mPower plant was that it would be operated in a secure underground containment where the spent fuel from would be stored in a pond located between each pair of units. The reactor design has no electrically driven pumps and its relatively low-power operation means that in the event of a malfunction all coolant operation can be accommodated under natural circulation, with significant stores of water installed above the plant, in the event more cooling is necessary. The plant is designed to be transported in modules by rail and constructed in 3 years for a 4-year operating cycle using standard, low-enriched ($^{235}U < 5\%$ wt.) PWR fuel.

11.5.3 THE WESTINGHOUSE SMR

The Westinghouse SMR is an evolution of the AP1000, de-rated to 225 MWe and effectively 1/25th of the size of the AP1000 design.[2] This gives it the advantage of sharing many of its components with the larger plant but the entire plant contained in a dedicated pressure vessel, 4 m in diameter. In this design, the confinement of the steam supply system within the containment removes the need for large-bore piping and it has a passive cooling emergency supply system based on evaporation, condensation and gravity. A schematic of the pressure vessel and internals is provided in Fig. 11.6A and a computer-based visualisation of an SMR installation is provided in Fig. 11.6B.

11.5.4 NUSCALE™

The *NuScale Power™ SMR* is rated at 50 MWe and is the evolution of a PWR design established in 2000 by the Idaho National Environment and Engineering laboratory with support from Oregon State University. This design is an integrated PWR (iPWR) and is cooled entirely by light water under natural circulation. The project has significant investment from the Fluor Corporation. A schematic of the NuScale™ iPWR is shown in Fig. 11.7.

11.6 BREEDER REACTORS

Perhaps the most fascinating concept in the nuclear energy field is that of *breeding*. Breeding is the process in which transuranic isotopes that are not fissile can be transmuted to isotopes that are, usually via neutron capture and subsequent β^- decay. Isotopes susceptible in this way are termed *fertile*. While the production of plutonium (^{239}Pu) from uranium (^{238}U) in uranium-fuelled thermal-spectrum reactors is unavoidable, as was discussed in Chapter 5, this does not constitute a positive yield because in this case more fissile nuclei are consumed than are produced (in this example more ^{235}U nuclei are consumed than those of ^{239}Pu are produced). By contrast, the term *breeding* in a nuclear energy sense refers to the production of a positive yield of fissile material relative to the quantity of fissile material that is consumed in doing so. A different reactor configuration is required to achieve a net positive yield, as is discussed in this section.

Breeding is a very attractive, long-term nuclear energy strategy as it enables access to the nuclear energy in isotopes of much greater abundance than is possible via ^{235}U. Uranium-235 is generally regarded as being too limited to sustain power production much beyond 100 years, particularly if it

[2] http://www.westinghousenuclear.com/New-Plants/Small-Modular-Reactor.

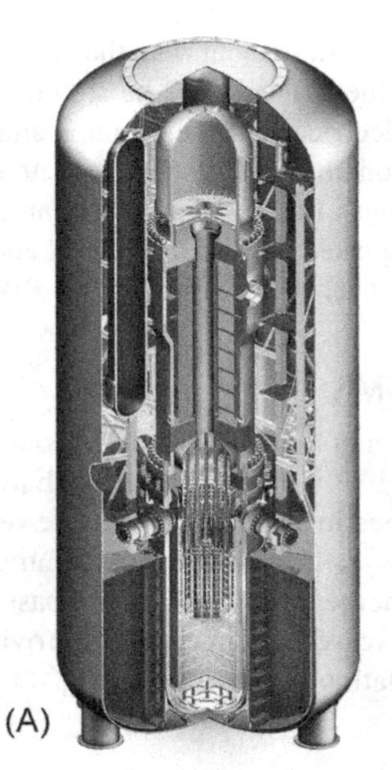

FIG. 11.6

(A) Cutaway schematic of the Westinghouse SMR[4]. (B) Visualisation of a Westinghouse SMR installation.

Courtesy of Westinghouse Electric Company.

is used on a once-through basis in which the fuel is not recycled to recover the ^{235}U that is unused. Estimates of this vary depending on the forecast of uranium reserves, current rates of uranium consumption and the economic viability of the extraction of uranium ores of lower yield. However, the variation is not too significant; the possibility of recycling uranium from spent fuel in which the unused ^{235}U is extracted and used in the manufacture of new fuel only extends such estimates to approximately 250 years.

The extent to which recycling in this way can continue is limited by the build-up of some specific isotopes, principally ^{236}U; while ^{236}U is neither significantly fissile nor fertile, it does make a

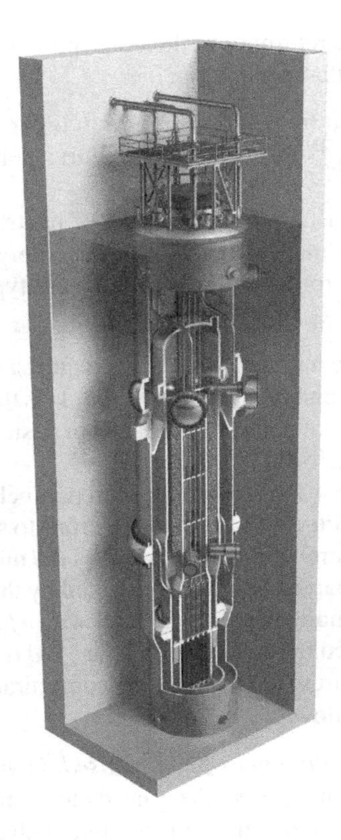

FIG. 11.7

Cutaway schematic of the NuScale™ iPWR.

Reproduced with permission, copyright http://www.nuscalepower.com/.

substantial contribution to the radioactivity of the material and is difficult to separate isotopically from ^{235}U. Breeding, by contrast, provides a route to the use of ^{238}U, which constitutes the vast majority of naturally occurring uranium. Uranium-238 is converted to the fissile isotope ^{239}Pu via double β^- decay following neutron capture (as described in Chapter 5 and Eq. 5.2). The latter is reproduced below for clarity,

$$^{238}\text{U} + n \rightarrow {}^{239}\text{U} \underset{23.5\text{m}}{\overset{\beta^-}{\rightarrow}} {}^{239}\text{Np} \underset{2.4\text{d}}{\overset{\beta^-}{\rightarrow}} {}^{239}\text{Pu} \tag{11.1}$$

The definition of breeding more fissile material than is consumed is described mathematically by the *breeding ratio, B*. This is the ratio of the number of new fissile atoms created to the number that are consumed in doing so. To achieve $B > 1$ requires a neutron economy that is sufficient to (i) sustain the nuclear chain reaction that is responsible for the production of neutrons, (ii) convert ^{238}U to ^{239}Pu and (iii) cater for the unavoidable losses via leakage and capture. This requirement has some implications for the design of breeder reactors when compared with standard, thermal-spectrum power reactor systems. By way of illustration, and returning to the conceptual framework of the six factor formula introduced in Chapter 8, it is clear that per fission:

- At least one neutron is consumed by neutron capture in order to convert one fertile nucleus (say ^{238}U) to a fissile isotope (i.e. ^{239}Pu). This number corresponds to the *breeding ratio*, B, defined above.
- A second neutron is needed to sustain the production of neutrons via the chain reaction, that is to stimulate fission in a fissile nucleus resident in the core, usually in ^{235}U or ^{239}Pu.
- There are losses in terms of absorption and leakage that are typically of the order of 0.2 neutrons per fission.

Considering the above reasoning, the number of neutrons required for at least as many fissile nuclei to be produced as are consumed (a conversion rate of unity, $B = 1$) is the sum of these contributions, that is ~2.2. This corresponds to the number of neutrons produced in fission per nuclei destroyed, η, necessary to yield a net positive conversion factor of $\eta > 2.2$.

Returning to the data for η (Table 4.1 and reproduced in part below in Table 11.1), it is clear that at thermal energies neither ^{235}U nor ^{239}Pu yield sufficient neutrons to sustain breeding. However, increasing the energy of the stimulating neutron generally increases the number of neutrons produced per fission, $\overline{\nu}$ (as was also considered in Chapter 4). This is reflected by the fast spectrum values for ^{235}U and particularly for ^{239}Pu in Table 11.1, indicating that a reactor with a *fast* neutron spectrum is necessary to yield breeding with neutrons produced from fission of either ^{235}U or ^{239}Pu; hence the term *fast* breeder. On this basis the salient design features of a *fast* reactor configuration that might yield a net positive yield of ^{239}Pu start to emerge, as follows:

- *Moderation of the neutron population must be minimised.* Given that the second law of thermodynamics must endure, that is, waste heat needs to be removed, it is necessary that the coolants used for this must not moderate the neutron flux inadvertently. This is a major distinction

Table 11.1 Relevant Actinide Isotopes With Their σ_f and σ_γ (Microscopic Fission and Neutron Capture Cross Sections, Respectively), $\overline{\nu}$, The Mean Number of Neutrons Per Fission[a] and, η, Mean Number Of Neutrons Produced Per Fission Per Nuclei Destroyed[b] at Thermal and Fast (1 MeV) Energies

	Thermal (0.0253 eV)				Fast (1 MeV)			
Nuclide	σ_f/b	σ_γ/b	$\overline{\nu}$	η	σ_f/b	σ_γ/b	$\overline{\nu}$	η
^{232}Th	–	7.40	–	–	–	0.14	–	–
^{233}U	**529.1**	**45.5**	**2.50**	**2.30**	**1.95**	**0.08**	**2.60**	**2.50**
^{235}U	582.6	98.8	2.43	2.08	**1.24**	**0.11**	**2.58**	**2.37**
^{238}U	–	2.68	–	–	0.02	0.13	–	–
^{239}Pu	748.1	269.3	2.88	2.12	**1.80**	**0.05**	**3.09**	**3.01**
^{240}Pu	0.05	289.5	–	–	1.50	0.09	–	–
^{241}Pu	1011.1	362.0	3.00	2.21	1.56	0.11	3.15	2.94

Isotopes and data that are particularly relevant to the conversion of fertile isotopes are shown in bold.
[a] *http://www.oecd-nea.org/janisweb/book/neutrons.*
[b] *http://www.kayelaby.npl.co.uk/.*

between thermal spectrum reactors and fast reactors since, in the case of the former, the moderator and coolant are often the same substance. For fast breeders coolants with favourable properties of heat transfer and heat capacity but minimal moderation are often liquid metals. To ensure that flow is possible, the use of metals with relatively low melting points is favoured such as sodium, lead or bismuth. Often eutectic alloys are used.

- *Significant enrichment is necessary.* This is required to offset the fall in fission cross section as a result of raising the incident neutron energy necessary to yield a higher η.
- *A heterogeneous driver-plus-blanket arrangement is often used.* The driver sustains the chain reaction and provides the neutron flux for breeding. This is usually highly enriched ^{235}U or $^{235}U/^{239}Pu$ mixture; note in Table 11.1 the advantageous increase in η achieved at fast neutron energies. Breeding occurs in the blanket that surrounds the driver and, in the context of ^{239}Pu breeding, is made of ^{238}U.

It is also clear from the data in Table 11.1 that the exception to the fast, $^{235}U/^{239}Pu$ route is via ^{233}U, since this yields sufficient neutrons at thermal energies to sustain breeding and highlights the potential of thermal breeder reactors based on this driver substance. Uranium-233 is derived by breeding from ^{232}Th and hence this approach is discussed in detail later in this chapter when we consider thorium.

It is worthy of note that breeding is by no means a new concept as the potential it might provide for many thousands of years of nuclear energy derived from ^{238}U was appreciated early in the development of nuclear energy. However, amongst a number of influences that are considered in more detail in Chapter 16, cheap gas caused uranium consumption to migrate away from the earlier forecasts in the mid-1990s. This relaxed pressure on ^{235}U supplies that undermined the economic viability of some conceptual commercial breeder systems, several of which had been built and operated for several years at that point.

11.7 GENERATION IV DESIGNS

Beyond Gen III+ and SMR designs, reactor technologies expected to be deployed commercially on a 30- to 40-year horizon are usually referred to as *Generation IV systems* (Gen IV),[3] reflecting the significant advancements over current Generation III and Generation III+ approaches that they represent. A panel of 100 international experts constituting the *Generation IV international forum* (GIF) first met in 2000 and considered over 130 reactor concepts before they selected six reactor technologies for further research and development.

The emphasis with Gen IV systems is that such a design uses fuel more efficiently, produces less waste, is economically competitive and compliant with high standards of safety and proliferation resistance. Many Gen IV reactors have been designed to aid the closure of the fuel cycle, that is they facilitate the recycle of fuel. The aim is that they also achieve greater operating efficiencies by utilising higher outlet temperatures; some are fast-spectrum systems relevant to the discussion in the previous section.

[3]See for example: Prog. Nucl. Energy 77 (2014) 1–420, 'Technology roadmap update for Generation IV nuclear energy systems', Jan. 2014, OECD Nuclear Energy Agency, https://www.gen-4.org/gif/jcms/c_60729/technology-roadmap-update-2013 and 'A technology roadmap for Generation IV nuclear energy systems', Dec. 2002, US DoE Nuclear Energy Advisory Committee/Generation IV International Forum, https://www.gen-4.org/gif/jcms/c_40481/technology-roadmap.

11.7.1 VERY HIGH-TEMPERATURE REACTOR

As the name suggests, the *very high-temperature reactor* (VHTR) Gen IV design is focused on achieving a primary outlet temperature of 1000°C on the basis that this affords favourable increases in thermal efficiency over the typical temperature ranges of Gen III and Gen III+ reactor designs. Given the high-performance criteria of materials in these temperature ranges, the VHTR design is a thermal spectrum reactor based on the use of a gaseous coolant (helium) and a graphite moderator structure; two configurations are anticipated: either a system of TRISO-coated fuel particles or a prismatic block arrangement. The design is anticipated to be compatible with a variety of fuel cycles comprising of uranium, plutonium, thorium and mixtures of these, such as mixed oxides of uranium and thorium where desired.

The application focus of the VHTR, given its high-temperature specification, is on the cogeneration of both electricity and the direct supply of heat, as opposed to the intermediate production of electricity. The latter is targeted at process heat applications, particularly, for example, the generation of hydrogen via the sulphur–iodine process. This approach is compatible with the longer-term, anticipated needs of the chemical, oil and iron industries. It is anticipated that the electricity side of the generation system would be supplied via a direct-cycle helium-driven turbine operating in the primary coolant loop while the thermal energy supply would be derived from a heat exchanger in the secondary loop (as per an indirect cycle). An intermediate version is under consideration with slightly lower temperatures in the range 700–850°C in order to reduce the technical risk anticipated with subsequent projects designed to operate at higher temperatures.

11.7.2 MOLTEN SALT REACTOR

The reactor design concept in which a liquid mixture of fuel and coolant is used instead of the more usual heterogeneous combination of solid fuel and moderator is known as the *molten salt reactor* (MSR). This is an established concept that was subject to much development at the dawn of the nuclear era but never escalated to commercial scale. The original designs were based on a fluoride salt mixture that included the fissile content necessary to support the chain reaction. These were thermal spectrum reactors and thus used a solid graphite moderator. The Gen IV approach in this area is to apply the MSR concept to a fast spectrum system thus affording the additional benefits of breeding of ^{239}Pu or ^{233}U (the latter from thorium).

Molten salt reactors are of particular interest with regard to the overarching Gen IV goals of fuel efficiency, reduced waste and greater proliferation resistance referred to above. In an MSR the significantly greater intimacy between fuel and coolant, that is possible in comparison with heterogeneous reactor systems, reduces the extent of the fissile excess that is necessary to sustain a chain reaction over a long period of time. The absence of a solid fuel matrix removes the possibility of radiation damage (i.e. cracking, deformation and dimensional change) and enables burn-up to be optimised leading to better fuel consumption. The requirement to manufacture fuel as solid components, cladding and structural components is also removed and MSR systems are also more compatible with the reactor-based consumption of long-lived actinides which would benefit waste management. MSR systems are also anticipated to benefit in terms of safety from having large, negative coefficients of voiding and temperature reactivity feedback.

However, while the waste arising from MSRs is anticipated to be reduced relative to their net yield per unit consumption of fuel, in order to extract material for reuse (i.e. plutonium and uranium), the associated flowsheet usually includes an online reprocessing facility. This would be a complex arrangement, primarily due to its proximity to the reactor system. Other key challenges include: the requirement to limit corrosion to structural materials in the reactor, and the control of the MSR-based process in which the location of the fuel is effectively a time-dependent variable.

11.7.3 SODIUM-COOLED FAST REACTOR

The *sodium-cooled fast reactor* (SFR) Gen IV system is considered to be the design with the greatest maturity for the burning of minor actinides (MA) because fast spectrum neutrons generally have a greater probability for the fission of MA than thermal. As a fast reactor, the SFR appeals to the Gen IV requirements of greater fuel efficiency and the reduction in waste production. The SFR design benefits from several decades of related experience with sodium-cooled reactor designs; the relevant precursor examples are the Phénix in France, Monju in Japan and the BN-600 in Russia. Recent developments based on this concept include the Chinese experimental fast reactor (CEFR) and the prototype fast breeder reactor (PFBR) in India.

Three conceptual designs are considered plausible for the SFR: a modular 50- to 150-MWe system; a 300- to 1500-MWe pool-type design and a 600- to 1500-MWe loop-type reactor. The outlet temperature is specified in the range 500–550°C, in common with earlier sodium-cooled fast reactor systems, to benefit from the earlier experience gained from these designs. Further, the design is based on a closed-fuel cycle to achieve greater levels of fuel utilisation and minor actinide consumption. The SFR exploits three coolant circuits: a *primary* circuit using sodium that is in contact with the core; a *secondary* intermediate circuit also based on sodium to separate the core from the power generation loop and a *third* circuit based on either light water/steam, supercritical carbon dioxide or nitrogen to drive the turbines.

11.7.4 SUPERCRITICAL WATER-COOLED REACTOR

The key distinction of the *supercritical water-cooled reactor* (SCWR) is that in this design the water coolant is used above the thermodynamic critical point (recall: 374°C temperature and ~220 atm. pressure). The reactor design is similar to a boiling water reactor in that it operates on a direct cycle with the superheated steam evolved from the coolant fed directly to the turbines. The concept spans the possibilities of thermal or fast spectrum operation and the use of a heavy- or light-water coolant.

The higher operating temperatures are desirable because they result in significant increases thermodynamic efficiencies of some 10% over and above subcritical water moderated/cooled reactor designs. Further advantages follow since, for example, no coolant pumps are required and, by definition, the steam that is generated is already superheated and therefore ancillary heat exchangers and steam generators are not required. Further, the significantly increased enthalpy of the steam requires a smaller turbine system. This could result in significant cost savings on reactor system components.

The challenges associated with this design include: obtaining a better knowledge of the heat transfer behaviour in moving from the supercritical state to the subcritical state (as a result of sudden depressurisation for example); understanding the effect of the supercritical coolant on the reactor materials and understanding the operation of passive safety systems in response to changes in the supercritical state.

11.7.5 GAS-COOLED FAST REACTOR

While the SFR described earlier constitutes the shortest route to an operating fast reactor technology due to the extensive prior experience that exists from earlier operational activities, there are technical challenges associated with the reactivity of the sodium-based coolants with air and water, the potential for a positive void coefficient of reactivity, the opacity of the coolant and the limit that boiling exerts on core outlet temperature and hence thermodynamic efficiency. The *gas-cooled fast reactor* (GFR) utilises helium as its primary coolant to avoid these issues while addressing the same key requirements of improved utilisation of fuel, the potential for actinide burning and generating less waste.

The current design concept for the Gen IV GFR is an indirect, combined-cycle arrangement. The primary circuit in contact with the reactor core would be cooled by helium at a pressure of ~90 atm. with inlet and outlet temperatures 490°C and 850°C, respectively. This is connected, via a main heat exchanger, to an intermediate loop comprising a mixed, helium–nitrogen coolant that is connected to a heat recovery steam generator and hence to a tertiary, light-water-cooled loop to generate steam to drive a steam turbine. This tertiary loop comprises the feedwater supply, pump and condenser. Somewhat unusually in this design the secondary, helium–nitrogen loop is also integrated with a turbine by which additional electricity is produced. It is effectively the waste heat from this cycle that is used to drive the tertiary loop; hence the design is a combined-cycle system with closed secondary and open tertiary stages.

Given the high temperatures at which the GFR is required to function, the main technical challenges associated with it are ensuring the resilience of the fuel, cladding and reactor structure at high temperatures. Current intentions include the use of carbide, nitride and oxide fuel materials clad in a ceramic compound. While the use of a gas avoids the potential for what is often referred to as a *disruptive* core accident scenario due to a water/air reaction in the case of the SFR, the helium-cooled GFR has virtually no thermal inertia since there is no moderator in contrast with the case of thermal-spectrum, high-temperature, gas-cooled reactor systems. Consequently, temperature escalations as a result of unexpected transients during a period of depressurization might present a serious risk to the integrity of the core. Hence convective flow is required at all times because the gas density is too low to yield sufficient natural convection. This places great emphasis on the need for effective heat removal at shutdown and for the removal of decay heat after shutdown.

11.7.6 LEAD-COOLED FAST REACTOR

The *lead-cooled fast reactor* (LFR) is, as the name suggests, a reactor exploiting a fast-neutron spectrum for the conversion of fertile materials and the burning of actinides, the latter to yield wastes of smaller volumes and lower radiotoxicities. The coolant envisaged is either molten lead or a lead–bismuth eutectic (LBE). The LFR is anticipated to be compatible with the production of electricity, hydrogen and process heat.

Lead and its related compounds developed for the role of fast reactor coolants have the advantages of a relatively high natural abundance, they are relatively inert, have very good thermodynamic properties and can be used at low pressures. Further, there is a large temperature margin between their liquid and boiling phases that reduces concerns associated with voiding, etc. Consequently the potential for disruptive problems that were envisaged for the case of sodium and the requirement for pressurising the coolant (components specification, specialised pumps and pressurisers) are removed. However, there are research challenges associated with materials performance, opacity and corrosion.

11.8 THORIUM

Of late there has been widespread media coverage of the potential for nuclear energy to be derived from thorium and a significant increase in research interest from the academic community. This can give the impression that the use of thorium for nuclear fuel applications is a recent development. However, nuclear fuel cycles based on thorium were considered alongside the development of the uranium/plutonium (U/Pu) fuel cycle at the dawn of the nuclear era, some years before any commercial nuclear reactor was first operated. Indeed, several commercial reactors operated by independent companies (largely in the United States but also in Germany) deployed thorium-containing fuels in their plant for several years in the 1960s and 1970s. However, while not being a recent development, the global renaissance in nuclear power of late *has* stimulated renewed interest in the use of thorium. This has resulted in a diversity of views over its benefits and some of the significant technical obstacles that remain associated with its use for power generation.

What *is* clear is that, assuming a continued development and greater reliance on nuclear power in the future, more efficient use of mineral resources and methods for reducing waste streams will need to be considered further. In the medium term, plutonium management and disposal is also likely to feature significantly. These are areas of significant relevance to the thorium question. However, as discussed in this section, the use of thorium is not free from risk and neither is it a route that is ready to use in isolation of the U/Pu fuel cycle.

Thorium was discovered in 1828 and comprises six naturally-occurring isotopes all of which are unstable. However, ^{232}Th is by far the most abundant with a long half-life so long (1.4×10^{10} years) that it is regarded as being relatively stable. Thorium is approximately three times as abundant naturally as uranium; this is one reason cited as to why thorium is often considered a worthy alternative, although perhaps not the most prominent justification. In the early years of the nuclear era, when the nuclear industry underwent significant expansion, relatively conservative forecasts of uranium abundance and overestimates of its consumption advanced the belief that uranium would quickly become supply-constrained. This caused plutonium breeding to be considered on a commercial scale and also thorium.

Since that time, forecast reserves of uranium have increased steadily while the economic climate (particularly the price of gas) has acted to reduce the extent of uranium consumption compared with original forecasts. As is discussed in Chapter 12, the preference for the continuous operation of some variants of uranium enrichment facilities can further influence uranium markets. Latterly, the US administration halted the reprocessing of commercial nuclear fuel, to supress the possibility of proliferation of separated material. This influenced the design of reference thorium fuel cycles and, in the United States at least, limited energy-related interest in the recovery of ^{233}U from thorium-based spent fuels.

Conversely, the lack of a commercial thorium-based fuel cycle has meant that the full extent of thorium reserves has not been assessed to the same extent as uranium, while many of the known reservoirs of thorium have proven uneconomical to exploit. Hence, whilst heralded as an alternative to uranium in terms of mineral recovery, we are less certain of the extent of material that justifies economic recovery to serve as the feedstock for a Th/U cycle. Apart from its potential for power production in nuclear reactors, thorium has few other uses. The most prominent sources of thorium at the time of writing are as a by-product of mining for other bulk-use minerals such as titanium, rare earth elements and (somewhat ironically) uranium.

The utilisation of thorium relies on conversion as none of the thorium isotopes, whether occurring naturally or man-made, offer favourable fissile characteristics. The fissile isotope ^{231}Th, for example, is only present in trace amounts. While 'thorium' has been represented as an alternative to uranium for nuclear energy, it is actually the fissile *uranium* isotope ^{233}U that arises from neutron capture on ^{232}Th and the subsequent double β^--decay that constitutes the thorium route to fission, via:

$$^{232}\text{Th} + n \rightarrow {}^{233}\text{Th} \xrightarrow[22.3\text{m}]{\beta^-} {}^{233}\text{Pa} \xrightarrow[27\text{d}]{\beta^-} {}^{233}\text{U} \tag{11.2}$$

The important advantage of ^{233}U over ^{235}U in terms of fission is that it has more favourable neutronic properties at thermal and epithermal neutron energies, manifest as a greater $\overline{\nu}$ and η and, in particular, $\eta > 2.2$ for thermal neutrons (as per Table 11.1).

Thus, whereas for the breeding of ^{239}Pu with neutrons derived from fission in ^{235}U a fast spectrum is needed to yield sufficient neutrons from the ^{235}U to breed a positive yield of fissile nuclei while sustaining the critical chain reaction (as discussed in Section 11.6), with ^{233}U sufficient numbers of neutrons can be obtained without the need for a fast reactor, although a harder spectrum does improve conversion further. This suggests that breeding of ^{233}U might be achieved in conventional, thermal spectrum reactor designs. It is important to appreciate, however, that commercial-scale reprocessing and recycle of thorium is yet to be demonstrated and the high neutron economy associated with ^{233}U comes at the expense of the delayed neutron fraction. This places a greater emphasis on the control needs of the chain reaction itself.

The significant investment in the U/Pu fuel cycle in the 1940s and 1950s caused economic drivers to detract from thorium. This happened despite the neutronic attractiveness of thorium and, given its high melting point,[4] its superior behaviour as a fuel under irradiation. A significant issue associated with the use of thorium that influenced this decision at the time was the difficulty resulting from the high-energy γ-ray emissions arising from the decay products of ^{232}U. This isotope is formed via $(n,2n)$ reactions on ^{232}Th, ^{233}U and ^{233}Pa and decays (with $t_{1/2}=68.9$ y) via an extensive but short-lived decay chain culminating in ^{208}Pb. This is accompanied by the emission of a number of highly penetrating γ rays including a 2.6 MeV emission from ^{208}Tl ($t_{1/2}=3.1$ m).

While they are only present at the level of 2–6 ppb in uranium oxide fuels, ^{232}U levels are much higher at ~60 ppb in thorium/^{233}U fuels. This necessitates the use of remotely-operated apparatus and shielded facilities to extract the ^{233}U and to manufacture fuel based on it, prior to any subsequent spent-fuel management, but has the benefit that it would complicate attempts to proliferate the material. While it can be argued that the same precautions are necessary to recycle plutonium and to manage spent fuel in the U/Pu cycle, it is important to appreciate that fuels based on ^{235}U can be manufactured with relative ease without the need for remote handling or extensive shielding, because the radioactivity of fresh uranium is low and generally confined to α decay in ^{238}U that is easily shielded. Also, a fissile stimulus is needed to provide the neutrons to pump-prime the Th/U fuel cycle long before self-sufficiency in ^{233}U is reached, via either ^{235}U or ^{239}Pu. In short, it is generally easier to use uranium than thorium given that supplies of the former are not constrained significantly at present.

A significant distinction can exist for nations that have little or no indigenous uranium and significant reserves of thorium; India is perhaps the best example of such a case having extensive thorium reserves and also having been party to international restrictions regarding the transport of uranium.

[4] 3390°C as opposed to 2865°C for dioxides of thorium and uranium, respectively.

India has adopted a long-term energy strategy to draw on thorium to provide their baseload electricity needs that comprises three stages. The first stage is to obtain power from thermal spectrum heavy-water and light-water reactor designs and also to derive plutonium from these to fuel the second stage. The second stage is to operate fast reactors fuelled with plutonium to breed ^{233}U from thorium. The third stage sees the operation of an advanced heavy-water reactor providing high conversion factors as a result of being fuelled with ^{233}U to provide power and also a self-sustaining supply of ^{233}U from ^{232}Th. However, this was originally inspired in part by the transport restrictions alluded to above. In general, plutonium is needed to seed the production of ^{233}U from thorium and it is the availability of this in nations with little or no indigenous uranium that sets the timescale for the adoption of thorium.

A clear advantage of the use of ^{233}U for fission via thorium (assuming ^{232}Th as the fertile matrix and not ^{238}U) is that there is no route by which plutonium can be derived and also that the waste arisings of MA are reduced, relative to the U/Pu cycle. This renders thorium advantageous in terms of long-term waste management because, while the inventory in terms of radiotoxicity of the fission products is similar in both fuel cycles, it is the MA component that sets the ultimate lifetime of the waste. Further, thorium can be used as a matrix for burning plutonium; another desirable facility. The favourable physico-chemical properties of thorium fuels offer a route to fuel forms with greater accident tolerance in terms of lower water-solubility, higher melting point and higher thermal conductivity, etc.; some representations of thorium suggesting it cannot be weaponised or has no association with plutonium can detract from these authentic areas of potential.

In summary, the adoption of thorium requires a significant realignment of the nuclear fuel supply chain and fuel cycle against a background of a well-evolved U/Pu cycle. The difficulty is perhaps that the potential to offset uranium scarcity and reduce MA in wastes is too far off to constitute sufficiently tangible benefits today. Recycling and fuel manufacture based on ^{233}U have the same technological challenges as other advanced fuel cycles in terms of the need for shielding, waste arising, etc. The Nuclear Energy Agency [4] highlights three scenarios:

- In the short term, that is before 2030, thorium offers potential as an additive to existing U/Pu fuels at the level of 5%–10% wt. This will not offset uranium scarcity or benefit waste management significantly but the favourable neutronic properties of ^{233}U offer improved core power flattening without the need for as much burnable poison in the core and hence enable higher fuel burn-up. Clearly, however, reduced power is needed to encourage the production of ^{233}U over fission in ^{235}U. This could be perceived as revenues lost to generating utilities and thus constitute a short-term economic disincentive to thorium used in this way.
- In the medium term, a period of transition would be necessary in order for sufficient ^{233}U to be produced to sustain a fuel cycle based entirely on thorium. Uranium forecasts in this era need to be made on the basis of the global intent to decarbonise our electricity supplies, the need for low-carbon, baseload power and the potential for direct process heat supply to industrial processes. Thorium has potential benefits in this context and in the development of accident-tolerant fuels.

  However, it also needs to be borne in mind that the U/Pu fuel cycle represents only a small part of the levelised costs of nuclear power (~10%–20%) and, at present, uranium resources are increasing, albeit very gradually. The key drivers in the medium term are the intent to burn plutonium in light- and heavy-water reactors and the possibility of offsetting uranium reserves, given a scenario that the take-up of nuclear power by 2050 is so significant that pressure starts to influence uranium exploration. The deployment of thorium at this stage is anticipated to comprise dedicated fuel assemblies as opposed to homogeneous mixtures with either U or Pu, or indeed mixed fuel rods because there is not a comprehensive fuel cycle for mixed fuels as yet.

- In the long term, that is post-2050, the forecast is for a progressive introduction of thorium assuming sufficient quantities of ^{233}U are available to render this feasible. This would make reprocessing and recycling of thorium-based fuels necessary. This has many technological challenges associated with it because, for example, thorium dioxide is much more difficult to dissolve than its uranium counterpart.

The benefits of the introduction of thorium into the nuclear fuel cycle comprise the opportunity to burn plutonium, the possibility of higher conversion ratios with which to recycle fissile material from spent fuel, a reduction in the production of plutonium and MA relative to the U/Pu cycle and physico-chemical properties sympathetic with greater accident tolerance. However, the introduction of thorium is a long-term strategy, requiring plutonium to seed the production of ^{233}U from ^{232}Th and the development of a thorium extraction process that is complicated significantly by the radioactivity of ^{232}U decay products. Significant challenges to licencing may also arise.

In terms of proliferation, most characteristics of ^{233}U in this regard are similar to ^{239}Pu, such as critical mass and heat generation but ^{232}U complicates the ease with which it is handled. Further, ^{233}U has a lower spontaneous neutron emission rate than plutonium (since ^{240}Pu is absent) for the purposes of passive neutron assay. Conversely, there is the potential for identification via γ-ray spectroscopy of the products of the ^{232}U decay chain, especially where storage has allowed the decay products to grow in.

11.9 FUSION

11.9.1 BACKGROUND

Earlier in this book we explored why some nuclear processes (such as spontaneous fission, stimulated fission and α-decay) are feasible on the basis of the energy that is released, whilst others require an input of energy to occur. This was explained on the basis of the *mass defect* and more generally with reference to the nonlinear dependence of binding energy per nucleon with atomic mass.

Using neutrons to split heavy nuclei, as in the case of ^{235}U and ^{239}Pu, has the benefit that there is a net energy yield (essential if we are to generate energy) and also the rather more implicit advantage that neutrons are released with which trigger subsequent fissions. This latter feature is essential if the nuclear energy production process is to be self-sustaining and controllable. We recall that on average ~1 MeV per nucleon of the parent isotope was liberated during fission of either uranium or plutonium in the mass region $A \sim 230$, and that the products of the reaction have a greater sum of binding energy than the parent nucleus, since the fission products are more stable than the fissioning nucleus. In this section we are concerned with the other mass extreme, $A < 5$. Here an energy yield of a similar order is also feasible, again by exploiting benefits in binding energy. However, in this case rather than splitting nuclei this yield arises when light isotopes are merged, or *fused*, rather than split apart; hence the term *nuclear fusion*.

The potential for this principle to be exploited for energy production was appreciated early in the nuclear era but the physics involved differs from fission in a number of ways that have a significant influence on the ease with which fusion is achieved. By way of revision, the two implicit and fundamental engineering requirements that influence the ease with which nuclear energy is exploited are that:

i. A means for sustaining the nuclear reaction in question must exist.
ii. It must be possible to extract the energy yielded from the self-sustaining reaction.

In fission, neutrons satisfy both of these requirements. However, while a neutron and a light isotope might fuse to yield a more stable daughter and in some cases yield a net yield in energy, the production of a subsequent yield of neutrons to stimulate the necessary generations of reactions that follow to sustain the process unfortunately does not occur. Rather, the stimulating neutron is absorbed to yield the heavier, product isotope.

Setting neutron generation aside for the moment, it is possible to cause the nuclei of two light isotopes to fuse by raising the temperature of a fusion fuel mixture of these isotopes to a level at which the Coulomb barrier between the nuclei can be surpassed. The extent of the barrier requires that such temperatures exceed that of the Sun, at $>150 \times 10^6$ K, which poses significant challenges as to how the materials and systems are configured to confine the fusing medium. This requirement also places constraints as to how the energy produced might be extracted efficiently and converted to useable forms, the most immediate form being electricity. These challenges are being addressed by a number of large international collaborations across the world by complementary approaches, with the general forecast being that the widespread application of fusion power will be possible in 40–50 years' time.

Given the relatively mature technological state of thermal-spectrum fission power and the potential for the wider use of uranium (and thorium) resources via fast spectrum reactors, compounded with the relatively extreme terrestrial conditions necessary for fusion, the reader may wonder: why pursue fusion at all? The answer to this question is best stated in reference to the other sources of energy that comprise the current energy mix. For example, we have already highlighted some of the benefits and limitations of fission power: from a beneficial perspective fission reactors produce low-carbon, baseload power, they do not use much fuel and generate relatively small amounts of waste. However, the reactors can be expensive to build, once-through fuel supplies are forecast to be limited to <100 years and a small proportion of the wastes by volume are long lived and highly radioactive.

Coal, by contrast, is available in significant abundance providing the longest-lived and most widespread source of baseload electricity currently available commercially. However, it yields amongst the largest quantities of CO_2 emissions from which we must refrain if climate change is to be averted. Similarly, oil is a significant source of CO_2 (more from its use in transport and heating than necessarily in electricity production) and, whilst gas produces less CO_2, like oil it is the most supply-constrained fossil-based energy source in terms of natural reserves. Further, a significant proportion of the most abundant reserves are sited in places of political unrest that can present concerns over security of supply.

Renewable sources, such as wind and solar, can be expensive in terms of the cost-per-unit-of-energy-produced. This is because they are dispersed sources and intermittent, in that power is only available when the natural conditions allow, and it can be necessary to build many individual generating units. Renewable sources are very useful for offsetting the reliance of grid infrastructure on inflexible, baseload sources and to respond to changes in demand. However, some power requirements such as the needs of hospitals, factories, heating and cooling require uninterrupted power supplies assured only at present from baseload contributions to national grid systems. It is important to state that, as the population grows and economies develop, the demand for electricity also grows with changing demand profiles and supply methodologies; the focus is not perhaps on 'which' supply is best but rather 'how' contributions from a variety of sources are mixed to meet current and forecasted requirements.

Fusion has significant advantages in this context over current industrialised methods of energy production: it has the potential to provide baseload electricity unaffected by climate conditions while utilising isotopes of hydrogen that are virtually limitless and do not produce CO_2. The nuclear reactions *do* yield fast neutrons which, in turn, activate the structures of the fusion reactors rendering them radioactive. However, the level of radioactivity involved is much less significant than that produced by fission reactors, both in terms of level and half-life, and this is itself dependent on the materials selected for the fabric of the reactor system. It is forecasted that this will have decayed away within 100 years which is comparable with the industrial legacy of fossil fuel use and many orders of magnitude less than waste produced by fission reactors. From the perspective of safety there is not any extensive thermal mass to be managed in terms of decay heat, in contrast with fission, and an escalation analogous to a criticality in fission terms cannot occur because the fusion reaction stops when the magnetic confinement is removed.

Perhaps the most attractive attribute of fusion, alongside baseload power provision, is that it offers perhaps the most attractive long-term means by which our energy needs might be displaced from a reliance on fossil fuels. These requirements (principally transport and domestic heating/cooling) are currently served almost entirely by oil and gas yielding significant levels of CO_2. However, they are also amongst the most amenable uses of energy to transfer to zero-carbon systems. Such systems might exploit hydrogen- and/or battery-powered sources [5] that need a primary source of power: fusion could be this source. There is a variety of principles by which controlled fusion might be achieved; we shall concentrate on thermonuclear fusion to illustrate the concepts of fusion energy.

11.9.2 SCIENTIFIC BASIS

It is clear from the low-A portion of the graph of binding energy per nucleon versus atomic mass (Fig. 4.9) that there are several combinations of light isotopes that might constitute reactions by which fusion power might be derived in terms of binding energy alone. These have been researched extensively and each offers a number of advantages and disadvantages. Important criteria in selecting the best reaction include: the availability of the isotopic ingredients that will constitute the fusion fuel, the ease with which nuclear fusion is achieved and the energy evolved per nucleon from said reaction.

The availability of fuel materials for fusion parallels the constraints in natural abundance of fissile materials used in fission. For example, the prospect of fusing deuterium ($D = {}^2H$) and the light isotope of helium (${}^3He$) is flawed because of the relative scarceness of ${}^3He$. Otherwise, the D-${}^3He$ reaction

$$D + {}^3He \rightarrow {}^4He + p + 18.35\,MeV \tag{11.3}$$

has the benefits of a charged particle yield (hypothetically advantageous in terms of the potential for direct electrical power take-off) and a high-energy yield per nucleon. By contrast, the combination of deuterium with itself (conventionally referred to as the *D–D reaction*) and deuterium with tritium (${}^3H$, the *D–T reaction*) have distinct advantages in this regard. These reactions are as follows,

$$\begin{aligned} D + D &\rightarrow T + p + 4.03\,MeV \\ D + D &\rightarrow {}^3He + n + 3.27\,MeV \end{aligned} \tag{11.4}$$

and,

$$D + T \rightarrow {}^{4}He + n \ +17.60\,MeV \tag{11.5}$$

Deuterium is derived relatively easily from the enormous quantities present in the world's oceans in the form of D_2O (approximate concentration of 33 mg/l). Tritium, while not present naturally because of its relatively short half-life ($t_{1/2} = 12.3$ years), can be derived exothermally from lithium via its reaction with thermal neutrons, for example on ^{6}Li,

$$^{6}Li + n \rightarrow {}^{3}H + {}^{4}He \ +4.8\,MeV$$

$$^{7}Li + n \rightarrow {}^{3}H + {}^{4}He + n \ -2.5\,MeV \tag{11.6}$$

While ^{6}Li is the less abundant isotope (~7.6%) of naturally-occurring lithium that comprises almost entirely ^{7}Li (~92.4%), elemental lithium is available naturally in very large quantities. The ^{6}Li reaction above has a much more significant thermal neutron cross section (~1000 b) for the production of tritium than the corresponding reaction for ^{7}Li with fast neutrons. However, the ^{7}Li reaction yields neutrons with which to drive the ^{6}Li reaction when they are suitably thermalised.

The ease with which each candidate fusion reaction can be achieved, that is D–D or D–T given the natural scarcity of ^{3}He, is quantified in terms of the corresponding cross section as for all nuclear reactions. A comparison of the corresponding cross sections is given in Fig. 11.8. As we explored in earlier chapters (particularly Chapter 4 in the corresponding context of fission), there is a *classical* component to the probability of a given reaction (where, for example, the primary components of a reaction are considered kinematically as solid spheres with defined dimensions and boundaries) and a *stochastic* component. The latter incorporates quantum–mechanical phenomena manifest as *tunnelling* (the nonzero probability of reactions below the Coulomb barrier), *wave-particle effects* and *resonances* in the interaction cross sections. Both aspects are needed to understand the interaction probabilities witnessed in the laboratory, as these are important for selecting the reaction route offering the greatest feasibility.

What is concluded from this qualitative analysis of reaction probability is that the quantum–mechanical adjustments are beneficial to the ease with which fusion is achieved; experimental cross sections are generally higher than those estimated on a classical basis corresponding to a lower barrier in energy to overcome between the fusing nuclei. Such an analysis indicates a much greater cross section for the D–T reaction than the D–D alternative, with the former being the easier one to achieve in the short term in terms of the capability of current, supporting technologies.

The salient disadvantage of the D–T reaction, by contrast, is that the high-energy yield (relative to the D–D reaction) arises in part in the form of 14.1 MeV neutron radiation. This requires that the potential for the activation of materials comprising the reactor by these neutrons is catered for in the design of fusion systems. Further, while we are only benefiting in part from the world's almost infinite reservoirs of deuterium, sustained generation of power is plausible if tritium is bred from ^{6}Li and ^{7}Li in blanket structures around the fusion reactor core and reintroduced into the central fuel mixture; this is the scientific basis for fusion based on the D–T reaction. The ratio of tritium produced to that which is consumed is known as the *tritium breeding ratio* (TBR). This needs to be positive (to ensure the tritium supply is not exhausted) but perhaps by only a few per cent.

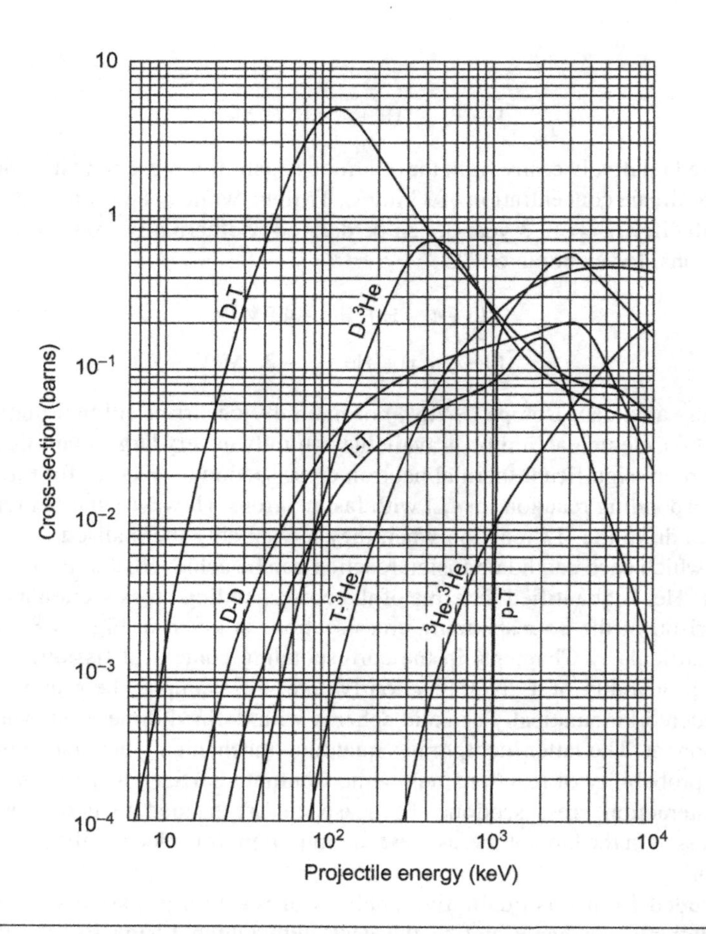

FIG. 11.8

Fusion cross sections as a function of projectile energy (http://www.kayelaby.npl.co.uk/atomic_and_nuclear_physics/4_7/4_7_4a.html).

Courtesy of the National Physical Laboratory (NPL).

11.9.3 ENGINEERING BASIS

The scale of the energy barrier to overcome to achieve fusion of a deuterium nucleus and a tritium nucleus is of the order of 20 keV. This equates to a temperature in excess of one hundred million degrees Kelvin, as stated earlier. Such temperatures are significantly in excess of the ionisation potential of the atoms comprising the D–T fuel mixture and thus at energies favourable for fusion to occur, the atoms are completely ionised and in a plasma state of a totally ionised gas. Clearly, a physical basis by which to contain such a substance at these temperatures goes beyond conventional approaches that rely on physical contact, as the plasma would destroy the associated infrastructure almost instantaneously. Therefore, two alternative methods to contain fusion plasma have been explored to date as the engineering bases for nuclear fusion: *magnetic confinement* and *inertial confinement*. By way of example, we shall concentrate on magnetic confinement.

11.9.3.1 Magnetic confinement and the tokamak

The charged state of the extremely hot fusion fuel mixture lends itself to confinement by a magnetic field which, as an action-at-distance phenomenon, provides a means of confining the plasma without the need for physical contact. However, the use of a magnetic field imposes some significant constraints on the design of a fusion reactor. Perhaps the most obvious are the requirements for the main infrastructure associated with the plasma to be confined within a vacuum vessel and the large structural components necessary to stabilise the magnet-coil systems themselves. Generically, a fusion reactor based on the principle of magnetic confinement would comprise a *nuclear island* and *external power generation infrastructure*. The former comprises a *vacuum vessel* in which the plasma is confined. This is surrounded by a *blanket* and an outer *shield*, beyond which is the *magnet and coil infrastructure* necessary to provide the magnetic field to contain the plasma.

The blanket comprises several layers: a *neutron multiplier* to increase the number of neutrons entering the blanket from the plasma; a *moderator* to thermalise the neutrons to optimise tritium production; a *breeder* layer comprising ^{6}Li for the production of tritium and a *shield* to protect people and the magnets outside of the core from the effects of the radiation emitted during operation.

The optimum geometric arrangement for the shape of the plasma in terms of the minimum surface area to volume ratio would be a sphere (or more practically a cylinder) analogous to the earlier discussion concerning fission reactors (Chapter 9). Spherical geometries are being pursued such as the Mega Amp Spherical Tokomak (MAST) at Culham in the United Kingdom. However, a toroidal (or doughnut) shape have received more attention due to the constraints of maintaining the magnetic field. One set of coils (toroidal) is used to provide a magnetic field that travels around the torus in circles. Another set of coils induces a current in the plasma and an associated poloidal field orthogonal to the toroidal field. These two fields provide the horizontal and vertical components in which the plasma is confined. Operationally, such an arrangement is known as a *tokamak* that derives from the Russian pioneers [6] of the concept in the 1960s. A small amount of gas is introduced into the vacuum vessel and heated by passing an electrical current (discharge) through it induced by a central solenoid; this solenoid constitutes the primary winding with the plasma being the secondary in what is effectively a large electrical transformer. The current induces ohmic heating of the plasma. Further heating is achieved by the interaction of a highly accelerated beam of neutral atoms (usually hydrogen) that is injected into the plasma (called *neutral beam heating*), and by electromagnetic heating at radio frequencies exploiting resonances at which the transfer of energy and thus heating of the plasma can be very high (referred to as *radio-frequency heating*).

It is, of course, necessary to supply electrical power to the electromagnets that supply the magnetic field to achieve an environment conducive to fusion. In evolving concept reactor designs this places a significant constraint on both the choice of magnet systems and their location: superconducting magnets are likely to be necessary in order to achieve power consumption levels sufficiently low to yield viable, net positive levels of power generation by future fusion reactor systems. These need to be positioned outside of the reactor core arrangement and, in particular, its shield to avoid inadvertent heating and damage by the neutron and γ radiation emitted by the fusion reaction in the plasma. In addition to the vacuum vessel, it will be necessary to contain the wider system within a cryostat to maintain the low temperatures required by the superconducting operation of the magnets. Cooling infrastructure (most probably coils that penetrate into the blanket arrangement) is necessary to extract the energy from the high-energy neutrons yielded by the D–T reaction in order for electricity to be generated.

The fusion reactors of the future are likely to be capitally expensive facilities. There are two principal reasons for this: Firstly, the structural integrity necessary to stabilise the magnet and vacuum vessel is a significant civil engineering undertaking because of the energy transfer necessary to achieve fusion. Secondly, the extraction of the energy produced *by* fusion is likely to be extracted via coolant systems that will be an integral part of the blanket and shield design. This arrangement is less compact than if it were possible to penetrate the torus to extract the heat generated but this would destabilise the plasma. Therefore, in order to achieve a positive cost-benefit in terms of the build costs relative to power generation revenues, practical fusion power generation systems will need to be of relatively large generating capacities to offset upfront construction costs.

Following on from earlier discussions, there is also the requirement to breed tritium sympathetically from lithium and this places further constraints on the design of a fusion system based on magnetic confinement. For example, in addition to the fusion cross section between deuterium and tritium, that in part defines the extent of the plasma required, the breeding requirement requires that a dedicated layer be integrated into the blanket. The extent of this layer is influenced in particular by the neutron mean free path for thermalisation and the neutron capture cross section for the tritium production reaction on lithium. Not surprisingly, this extensive and somewhat sophisticated infrastructure means that to date fusion offers a source of energy with a high capital cost, not dissimilar to fission in this respect but with greatly reduced liabilities in terms of decommissioning, waste management and safeguards needs. This influences the economic viability of fusion in terms of the cost of the electricity it generates to be within the large capacity regime, with for example each plant generating in the area of 1000 MWe and beyond. A cutaway diagram of the ITER fusion reactor design, currently under construction in France by a collaboration of 35 nations, is given in Fig. 11.9.

FIG. 11.9

A cutaway view of the ITER tokamak currently under construction in France.

Reproduced with permission of the ITER Organization (www.iter.org).

REVISION GUIDE

On completion of this chapter you should:

- understand the economic and technological bases for the *current generation* of fission power reactors in plan and construction (Generation III+), and appreciate the features of the various designs that comprise this group
- be aware the resurgent reactor design concept associated with *small, modular systems* and be familiar with the four prominent examples of this range of reactors
- appreciate that there is a long-term set of design possibilities classified as *Generation IV* and be familiar with the six examples of these
- understand the scientific basis behind *breeding*, the distinction of fast- from thermal-spectrum breeding and appreciate the main requirements of the coolants used in these reactors
- in the context of breeding and more widely in terms of a long-term energy possibility, appreciate the benefits and disadvantages of *thorium* as a fertile material for nuclear energy
- appreciate the scientific basis behind *fusion*, the challenges as a mainstream source of power and appreciate magnetic confinement as an example by which it might be realised

CASE STUDIES

CASE STUDY 11.1: THE PROTOTYPE FAST REACTOR (PFR) AT DOUNREAY IN THE UNITED KINGDOM

The fast breeder concept originated at almost the same time as did the 'conventional' thermal fission reactor systems that constitute the commercial nuclear energy supply system in use today. Immediately after the Second World War, post-war shortages of coal throughout Europe presented an attractive opportunity for countries like the United Kingdom to export coal from their indigenous reserves but the nuclear arms race rendered uranium a relatively expensive source of power. Conversely, those reactors that were built for the production of plutonium generally did not produce electricity. Those that did, from the late 1950s onwards, were restricted to very short burn-up periods in order to avoid the significant build-up of unwanted, contaminant isotopes. This resulted in very inefficient use of the uranium for the production of power with much of it left in the fuel, unused.

The McMahon Act (the Atomic Energy act of 1946)[a] restricted the ease with which nuclear information, enriched material or plutonium could be exchanged between the United States and the United Kingdom. Hence the emphasis in the United Kingdom fell to the development of indigenous reactor designs combining the capabilities of plutonium production and the extraction of energy for power as efficiently as possible (both in terms of fuel use and reactor operation). In the United Kingdom, for example, spent fuel from the weapons programmes was too depleted in ^{235}U to be used in thermal fission reactors directly, while the possibility of enriching it with ^{239}Pu was at the time considered wasteful of the material that had been produced for military uses. The United Kingdom nuclear power strategy at that time in terms of reactor designs comprised: a *uranium pile* utilising fuel of natural enrichment (which would go on to become the Magnox reactor design and would be advanced further in time to become the AGR—the mainstay of the UK thermal power reactor fleet), a *thermal breeder* and a *fast breeder*. The breeder reactors were the principal route to convert ^{238}U

[a]https://www.osti.gov/atomicenergyact.pdf.

to the neighbouring, fissile isotope ^{239}Pu (or in principle ^{232}Th to ^{233}U) by which anticipated shortages of ^{235}U-based fuels might be averted.

The operational requirements for the production of fissile material on a net positive basis in a fast reactor for the production of energy contrast significantly with the production of plutonium when $B < 1$: high power densities are a consequence of the high levels of enrichment that are needed to overcome the decline in fission cross section at high neutron energies; a fast-spectrum self-sustaining chain reaction is not possible in materials of natural enrichment. This results in higher operational temperatures than is the case for thermal reactors. The reactor systems are long-lived because the in-growth of fissile material takes extended periods of sustained operation if they are to be economically viable and they also need to be self-sustaining in terms of reactivity. At the time when they were first considered, fast breeders were deployable immediately due to the availability of ^{238}U and ^{239}Pu whereas thermal breeders were reliant on the availability of ^{233}U which in turn needed to be bred from ^{232}Th.

The *Prototype Fast Reactor* (PFR) followed on from the Dounreay Fast Reactor (DFR); the latter had been a smaller (output of ~14 MWe), largely experimental reactor developed to test materials and prototype fuel element designs for the fast reactor development in the United Kingdom [7]. The DFR was based on the use of coolant loops: three interconnected circuits with the primary and secondary using liquid sodium and the tertiary using light water, all interconnected by heat exchangers. It highlighted the four salient requirements of breeder reactor designs that remain pertinent today: the design must be safe; significant power production must be possible given the relative expense of the enriched core; high coolant outlet temperatures are necessary to ensure viable thermodynamic efficiencies; the recovery of the bred proportion of material must be straightforward and cost effective. The DFR is perhaps most often associated with its iconic ~40 m diameter steel sphere adopted from the prior example of the Knolls Atomic Power Laboratory in the United States.

The PFR by contrast was much bigger consistent with the reactor output of the thermal fission reactors of the day at ~250 MWe. Again it was sodium-cooled but was based on a pool or tank design in which the primary sodium and primary heat exchangers were immersed in a bath of liquid sodium (a eutectic alloy of sodium and potassium, NaK, which has a lower solidification temperature than either in isolation). This approach was deemed to be much more resilient to the possibility of sodium leaks. The core and primary infrastructure that routed the heated coolant from the reactor fuel to the first set of heat exchangers was made of stainless steel with the pool tank being a vessel 12.2 m in diameter and 15.2-m deep. The reactor apparatus—the core, three sodium coolant pumps and six primary heat exchangers connected to three sodium-cooled secondary loops—were suspended from the vessel lid above the pool. A total of 900 t of liquid metal coolant was inserted into the tank. This flowed up through the core to the heat exchangers at 500°C and back again into the pool at 400°C. To consolidate and again aid resilience in terms of the risk of coolant leakage, the entire vessel/reactor core apparatus was located below ground level in an excavation 15.25 m in diameter and of depth 15.25 m.

The loop-design fast breeder involves pipe routing to pumps and heat exchangers whereas the single-tank, pool design has no penetrations. A liner of reflector elements, blanket (breeder) elements (^{238}U) and a central zone filled by fuel core or driver elements (enriched uranium and ^{239}Pu) are located in the tank. Three parallel circuits (that together constitute the secondary loop) supply sodium coolant to and from six intermediate heat exchangers in the tank, providing triple redundancy in the event of a loss of sodium coolant in the secondary circuit or water in the tertiary stage. The primary loop is effectively the flow to and from the pool to the intermediate heat exchangers. An important imperative with the design of sodium-cooled fast breeders is to prevent any risk of reaction between the liquid metal coolant in the primary with water in the power generation side. Contact is prevented by the use of an intermediate loop while reaction between the primary and secondary is prevented by using the same substance that is sodium which is, by definition, chemically inert from itself. The primary sodium phase in the tank is inerted by a layer of argon gas injected into the vessel above it at a pressure of 7000 Pa (Figs 11c.1 and 11c.2).

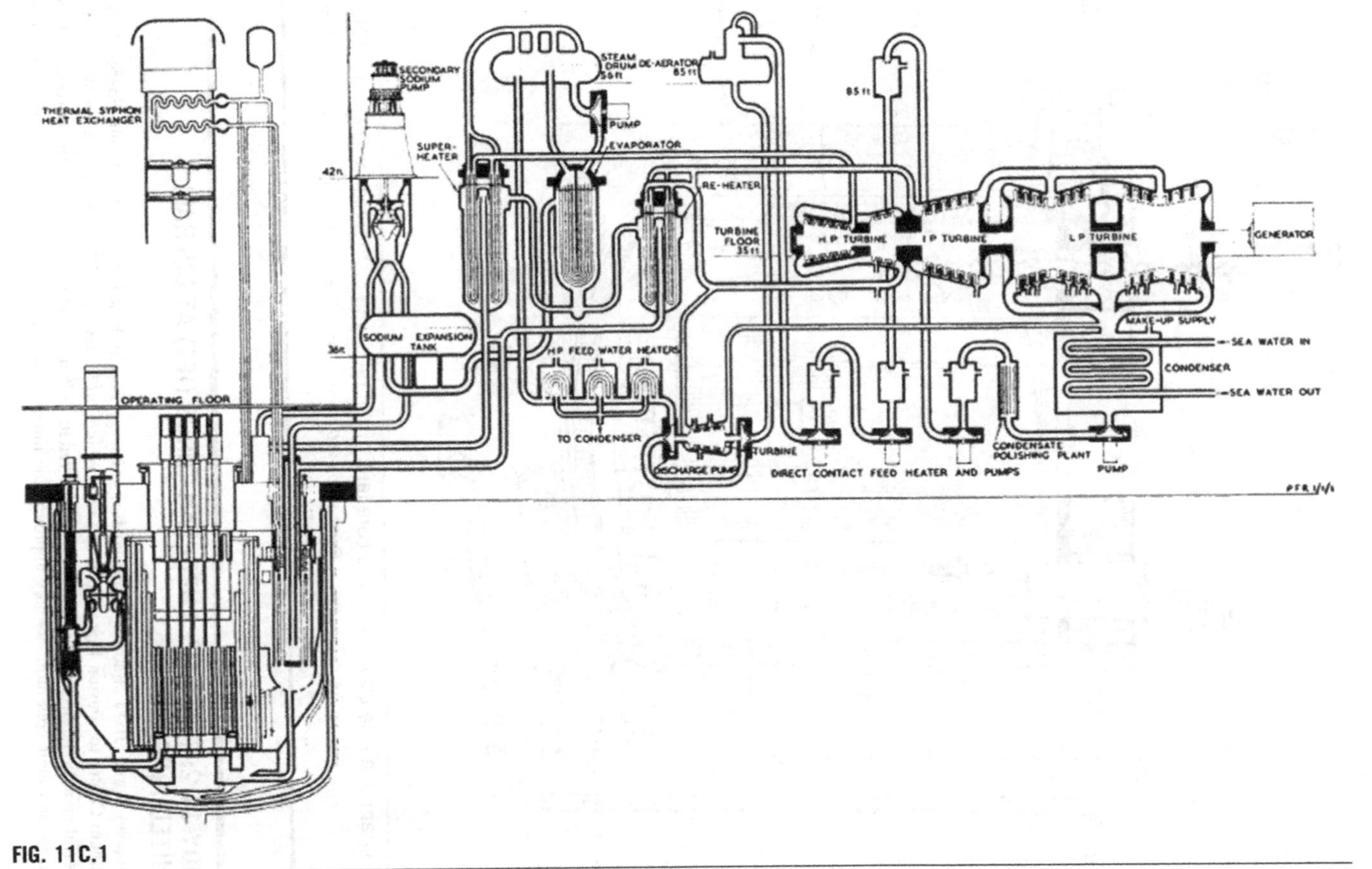

FIG. 11C.1

A schematic diagram of the pool-type prototype fast reactor designed, constructed and operated at Dounreay [8].

Reproduced with permission of the Institution of Engineering and Technology.

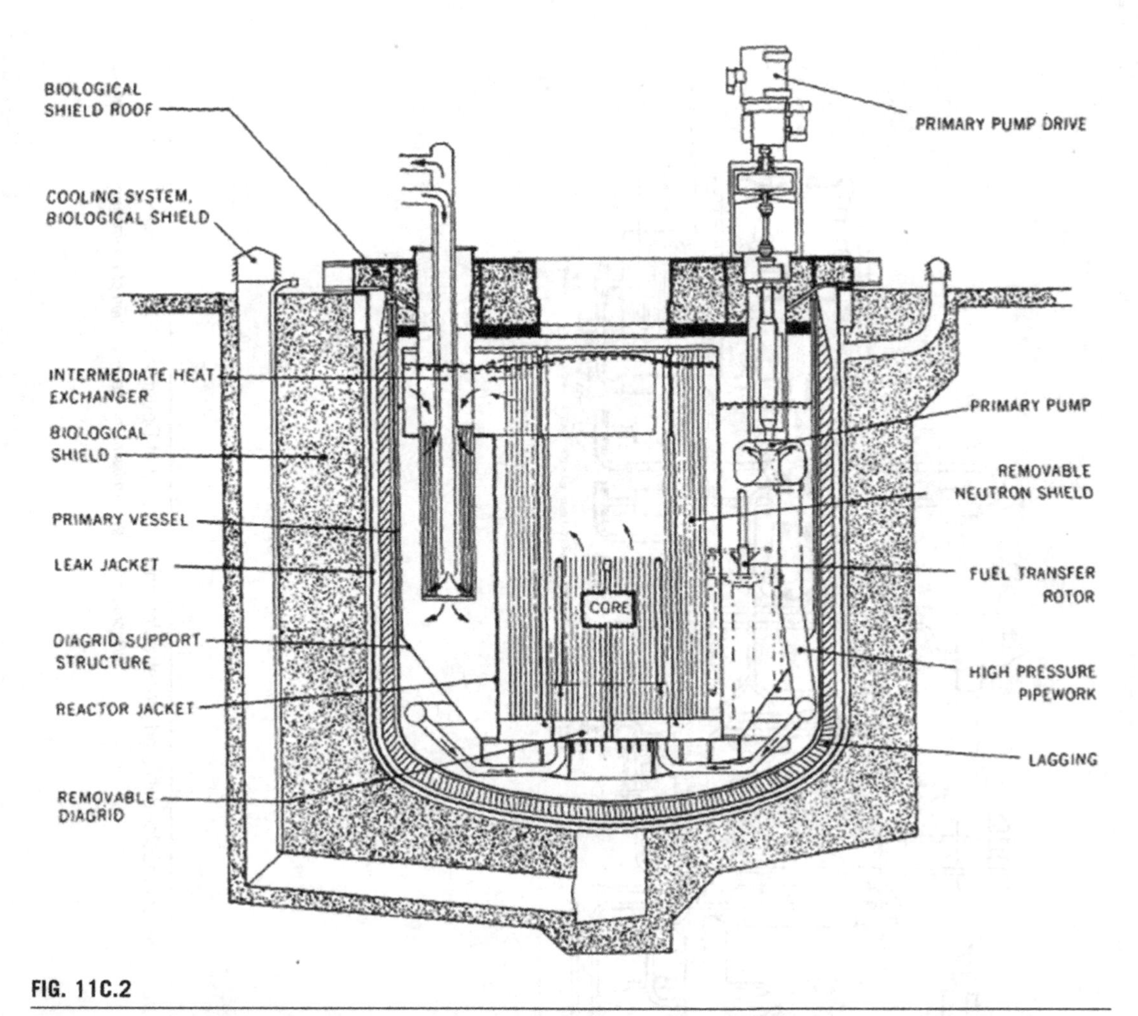

FIG. 11C.2

A schematic diagram of a side elevation of the core tank of the prototype fast reactor at Dounreay [9].

Reproduced with permission of the Institution of Engineering and Technology.

CASE STUDY 11.2: THE JOINT EUROPEAN TORUS (JET) AT CULHAM IN THE UNITED KINGDOM

The Joint European Torus (JET) is an experimental fusion energy facility based at the UK Atomic Energy Authority (UKAEA) Culham Centre for Fusion Energy (CCFE) near Abingdon in the United Kingdom; it is the World's largest and most powerful operational fusion tokamak. It was conceived in 1978 and started operation in 1983 to enable the study of fusion plasmas and their conditions as they approach the dimensions needed for commercial-scale fusion reactors (Fig. 11c.3).

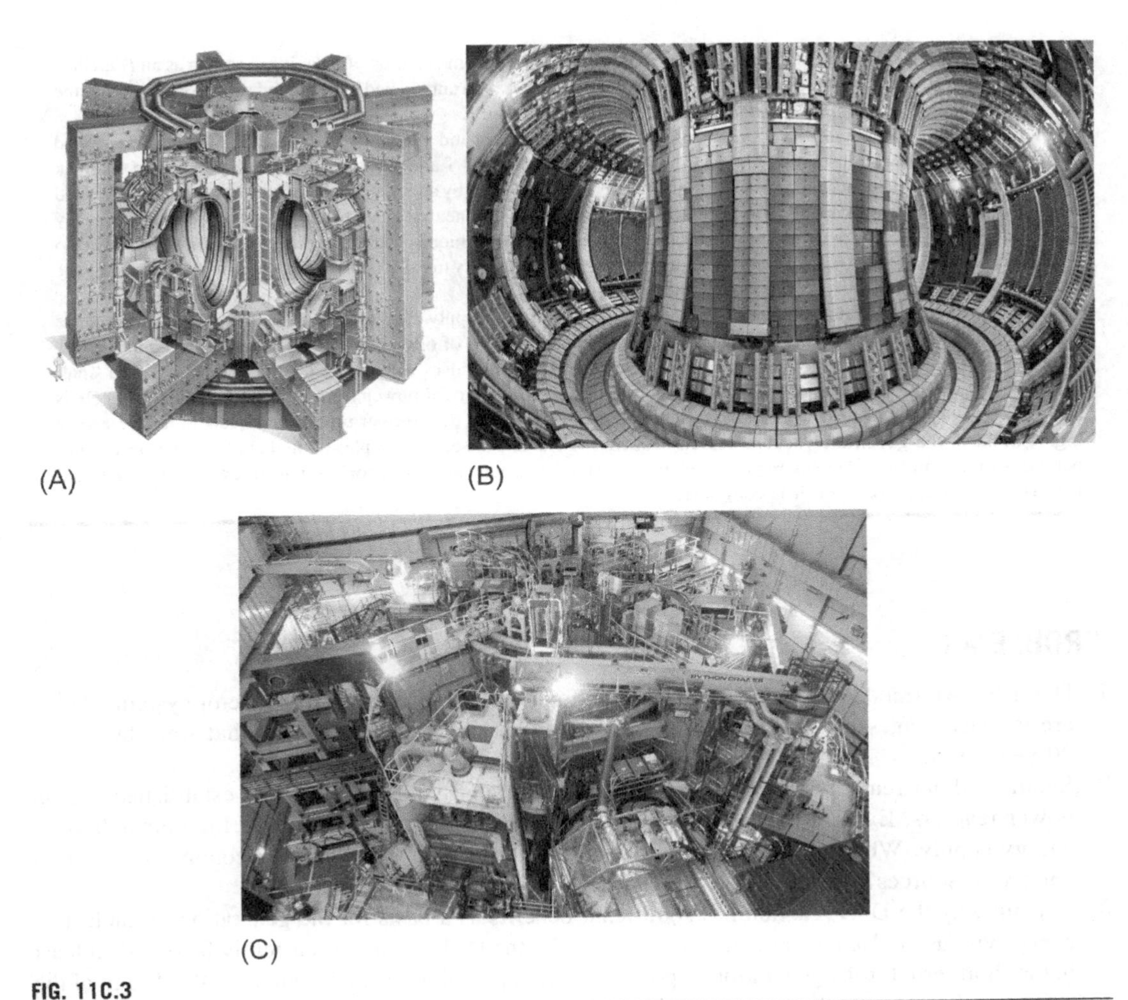

FIG. 11C.3

(A) A cutaway diagram of the Joint European Torus (with an octant removed to expose the inside of the vacuum vessel), (B) a photograph of the interior of the JET vacuum vessel and (C) an aerial photograph of the JET facility.

Reproduced by permission of EUROfusion (www.euro-fusion.com).

There are a number of challenges to harnessing fusion power for the generation of electricity which have been addressed by JET. These include: understanding the interaction of the plasma with the first wall of the device (that can result in impurities diluting the plasma and instabilities that limit the period of time the plasma can be maintained) and the choice of methods by which the plasma can be heated to achieve the temperature necessary for fusion. More recently, JET has supported the development of the ITER project[b] via activities such as: installing and testing the beryllium/tungsten

[b]ITER, the 'International Tokomak Experimental Reactor' is an international collaboration of 7 partners (Europe, USA, Japan, China, India, Russia and South Korea) set up to build the world's largest tokomak in southern France. ITER is designed to maintain fusion plasmas for hours with significant energy gain to demonstrate burning plasmas, where input heating can be largely switched off. https://www.iter.org/.

material mix required for the plasma-facing components in ITER via the installation of what is referred to as an *ITER-like wall*, to explore the regimes of plasma operation consistent with operation anticipated in ITER and to explore the use of the deuterium–tritium fuel mixture consistent with that anticipated at ITER.

JET is a large tokomak: its plasma vessel has a major radius of 3 m and a height of 5 m, enabling 90 m^3 of D-shaped plasma to be confined and sustained, with a toroidal magnetic field of 3-4 T and plasma currents of 5-7 MA. The tokomak comprises a double-walled vacuum vessel of eight octants interconnected by an arrangement of bellows and all constructed from Inconel.[c] JET has the capability to heat the plasma using a 34 MW neutral beam heating system and radiofrequency heating of 17 MW. It has a pellet injector for plasma refuelling and the unique capability among tokomaks world-wide to operate with tritium. There is also a state-of-the-art remote handling facility to carry out work in the vessel thus reducing the requirement for manned entry.

During operation, JET draws power from the national electricity supply grid to heat the plasma. However, because a great deal of electrical power is required for a relatively short period of operation (often referred to as a pulse), JET uses two large flywheels (780 tonnes each) that are spun up between pulses. When a JET pulse is due to start, a small amount of deuterium (~5–10 mg) is injected into the vessel and electrical power from the grid and the flywheels is routed to the tokamak heating system and magnetic coils. The central transformer primary coil acts to induce a very high current in the gas in the vacuum vessel, causing the gas to ionise, form a plasma and start to heat up to temperatures of 150 million degrees required for fusion. JET has a tungsten 'divertor' system to exhaust the heat and gas arising from a pulse when it is completed.

PROBLEMS

1. Describe two trends that are driving the development of Generation III+ reactor systems. What are the advantages of these design changes over previous reactor designs that were built 20 years ago?
2. Small modular reactors are often lower-rated, integrated versions of larger, established fission power reactors. Explain the justification for their consideration in terms of the demands of energy supply. What are the benefits that they offer to an energy supply economy based on a variety of sources?
3. Explain why the D–D reaction has significant benefits as a basis for the generation of nuclear energy via fusion. Hence or otherwise, explain why the D–T reaction is currently favoured, at least in the short term for the generation of power. What implications does this have for the design of the reactor and sustaining the fusion reaction?
4. The most popular eutectic coolants—bismuth–lead and sodium–potassium—are very different substances but share some common properties. List both their common features and their contrasting properties. Hence recommend specific applications for each of these materials in light of Gen IV design criteria.
5. Explain why it is necessary to use ^{235}U or ^{239}Pu in the first instance to convert ^{232}Th to yield ^{233}U that might then be used in a nuclear fuel cycle dedicated to thorium. Hence explain why thorium cycles have a reputation for being proliferation resistant. What is the origin of ^{233}U compatibility with breeding at thermal neutron energies?

[c]An alloy of nickel, chromium and iron that is robust at very high temperatures.

REFERENCES

[1] W.E. Lee, M. Gilbert, S.T. Murphy, R.W. Grimes, Opportunities for advanced ceramics and composites in the nuclear sector, J. Am. Ceram. Soc. 96 (7) (2013) 2005–2030.
[2] S.S. Bajaj, A.R. Gore, The Indian PHWR, Nucl. Eng. Des. 236 (2006) 701–722.
[3] Z. Liu, J. Fan, Technology readiness assessment of Small Modular Reactors (SMR) designs, Prog. Nucl. Energy 70 (2014) 20–28; G. Locatelli, C. Bingham, M. Mancini, Small modular reactors: a comprehensive overview of the economics and strategic aspects, Prog. Nucl. Energy 73 (2014) (2014) 75–85.
[4] Nuclear Energy Agency, Introduction of Thorium in the Nuclear Fuel Cycle., Organisation for Economic Co-operation and Development, 2015 (NEA No. 7224).
[5] J.E.T. Team, Fusion energy production from a deuterium-tritium plasma in the JET tokamak, Nucl. Fusion 32 (2) (1992) 187–203.
[6] M.A. Gashev, et al., Fundamental technical characteristics of the experimental thermonuclear system Tokamak-3, J. Nucl. Energy C7 (5) (1965) 491–499.
[7] J.W. Kendall, T.M. Fry, The Dounreay fast reactor project, Engineer (1955) 330.
[8] W. Macrae, The prototype fast reactor at Dounreay, Electron. Power 20 (14) (1974) 581–584.
[9] W.J. Outram, IEE power division: the prototype fast reactor Dounreay—some interesting aspects of its design and construction, Proc. Inst. Electr. Eng. 122 (1) (1975) 49–54.

CHAPTER

NUCLEAR FUEL MANUFACTURE 12

12.1 SUMMARY OF CHAPTER AND LEARNING OBJECTIVES

In this chapter the processes associated with the production of nuclear fuel are introduced. Unlike most other fuels that are used for electricity generation via heat engines, the production of nuclear fuel is an engineering process involving high degrees of accuracy and precision, and a variety of carefully selected materials and compounds. The process is sophisticated relative to other fuel sources drawing on a significant technology infrastructure. Once manufactured, relatively little fuel is necessary from which large amounts of energy are derived over long periods of time. The methods of uranium mining, extraction, solvent extraction, conversion, enrichment and its manufacturing into fuel pellets, elements and assemblies are described. This introduction to nuclear fuel manufacture is focussed on a single manufacturing route that is well established and in mainstream use. This does not imply that this route is the most efficient or preferred. Further, the route described may not be suitable for some reactors, especially niche designs such as reactors that are used for research.

The objectives of this chapter are to:

- describe the origin of uranium in the natural environment, its ores and extraction methods
- introduce the concepts of mining and milling, and to discuss the typical, expected relative abundances
- discuss the extraction of uranium ores via leaching to produce an aqueous solution of the raw material
- provide a description of the solvent extraction method and its application to the separation of the uranium-bearing mineral(s) from the aqueous leachate to yield a product suitable for transport
- introduce the concept of conversion by which a gaseous product comprising uranium is produced, suitable for enrichment
- describe the process of conversion by which uranium dioxide is produced and made into fuel elements and assemblies

12.2 HISTORICAL CONTEXT: FRITZ STRASSMAN 1902–80

Fritz Strassman was an analytical chemist who worked firstly on the absorption of iodine in carbonic acid and then developed the rubidium–strontium method of dating that is used widely in dating geological samples. In 1934 he joined Lise Meitner and Otto Hahn to work on the identification of the

Nuclear Engineering. https://doi.org/10.1016/B978-0-08-100962-8.00012-3

FIG. 12.1

Fritz Strassman.

products of the irradiation of uranium with neutrons. It was Strassman's significant expertise in analytical chemistry, combined with that of Hahn, that enabled the light rare-earth, barium products of neutron irradiation to be separated and distinguished from the possibility that they might be radium isotopes, and thus proved the existence of nuclear fission (Fig. 12.1).

12.3 INTRODUCTION

Unlike fuels used to generate most other forms of large-scale, baseload electricity such as coal, oil and gas, nuclear fuel requires the relatively sophisticated processes, enrichment and manufacture in addition to the common needs of extraction and refinement. This is a principal reason why the infrastructure associated with nuclear energy brings with it a requirement for advanced technical skills and the educational foundation that supports these. This can be a welcome economic stimulus where the adoption of nuclear power is under consideration for the first time.

Uranium is extracted in the form of a variety of ores. While it is present in trace quantities very widely, deposits of abundances sufficient to justify the investment to extract it on an industrial scale are located in just a few places throughout the world, such as in Canada, Namibia, Kazakhstan and Australia. There are 15 deposit types used to describe the geological environments in which uranium is found, with the major primary ore minerals being *uraninite* (principally uranium dioxide, UO_2) and *pitchblende* (principally triuranium octoxide, U_3O_8) with a range of other mineral forms present in smaller proportions. The nature of the specific mineral form can be significant as this influences the ease with which extraction might be possible and hence the efficiency with which uranium is recovered from it. Similarly, the nature of the deposit can influence the best approach to access the mineral including its depth, its association with other deposits and so forth.

Uranium cannot be used as a fuel in its raw mineral form (in the way that coal is for example) because the ^{235}U is too dispersed for a chain reaction to be self-sustaining, and the impurities with which it is combined as a natural deposit act to scatter and absorb neutrons to too significant an extent. Early developments in nuclear fuel manufacture for commercial power generation focussed on the manufacture of metal fuel. However, the response of this type of fuel to irradiation and high temperatures limits the time that it can be used in a reactor. The drive for higher capacity factors and greater thermodynamic efficiencies has demanded longer periods between refuelling and operation at higher temperatures. This caused the industry to move to the use of ceramic fuels, generally of the form of ceramic pellets of sintered uranium dioxide. While enriched fuels were used primarily in experimental systems and light-water reactor designs, several early reactor designs (for example, Magnox and CANDU) used fuel of natural enrichment, i.e. 0.71% wt. ^{235}U and 99.28% wt. ^{238}U. Nuclear fuel for commercial power reactors is now manufactured universally to a relatively low level of enrichment, that is <5% wt., and the enrichment step to these levels is a ubiquitous feature of nuclear fuel production prior to the manufacturing process.

12.4 MINING AND MILLING

Uranium ore is extracted from opencast mines, underground mines and by in situ leaching. The ore from mining is processed into a refined form of U_3O_8 known as *yellowcake* or UOC (uranium ore concentrate). This is usually done near the mines at facilities referred to as *uranium mills* prior to transport to fuel manufacturing facilities. The solid wastes from this process are known as *mill tailings* and are usually stored on the site of the mine. Significant quantities of ore have to be processed because the uranium content is relatively low, at between 0.1% and 0.2%.

The extracted ore is crushed into small pieces and the uranium is leached into solution from it, usually with sulphuric acid (alkaline leaching is also used dependent on the nature of the ore), as per the following:

$$UO_3 + 2H^+ \rightarrow UO_2^{2+} + H_2O$$

$$UO_2^{2+} + 3SO_4^{2-} \rightarrow UO_2(SO_4)_3^{4-} \quad (12.1)$$

Ammonium carbonate can also be used to yield uranium trioxide, UO_3, or uranyl tricarbonate, $UO_2(CO_3)_3^{4-}$. In situ leaching (also known as in situ recovery) involves dissolving the ore with an oxidant (such as hydrogen peroxide) while it is still in the ground and fixing the uranium-containing minerals with a complexing agent (either acidic or alkaline depending on the mineral content of the local geology). By pumping the leachate into underground rock deposits, the uranium-containing solution can be pumped out.[1] In situ leaching is regarded as environmentally preferable and cost effective, with almost half of all uranium mined in the world being leached, especially in United States, Kazakhstan and Uzbekistan.

[1]At present 42% uranium comes from mines, 51% from in situ leaching and 7% as a result of the extraction of other minerals. In addition to these primary sources, a significant secondary contribution arises from the decommissioning of nuclear warheads (from Russia and the USA) that is blended down to levels of enrichment compatible with reactor use and a little comes from the recycle of uranium from reprocessed nuclear fuel (Source: World Nuclear Association, www.world-nuclear.org).

To extract the uranium from its solution, either a *solvent extraction* or *ion exchange* process is used. Solvent extraction is an approach common to several elemental separations applications used in the nuclear fuel cycle, nuclear reprocessing and elsewhere in the process industries for metals extraction: two immiscible liquid phases—an organic solvent phase and an aqueous phase—are combined by mixing or by forcing them to move counter-currently to one another. The extraction process exploits the difference in solubility of an element or compound in the aqueous stream as opposed to the organic medium. The element or compound of choice (in this case uranium) is transferred (or *partitioned*) from one liquid to the other, in order to separate it from the other elements and compounds in the primary aqueous stream (that is in this case derived from the ore). When the transfer is complete, the mixing process is halted and the loaded immiscible media separate under gravity.

In the first stage, a tertiary amine in a kerosene diluent is used to transfer the uranium from the aqueous to the organic phase. Subsequently, the uranium is stripped from the loaded solvent by mixing with ammonium sulphate. This returns the uranium back into solution and the solid form ammonium diuranate, $(NH_4)_2U_2O_7$, is precipitated by mixing it with ammonium hydroxide. It is this form that is bright yellow in colour as per the name *yellowcake*, although this reference tends to apply to all of the general products of milling irrespective of whether they are actually yellow. The ammonium diuranate is then calcined to remove the ammonium ions that yields triuranium octoxide, U_3O_8, which is the main form of uranium produced by most milling operations.

12.5 CONVERSION TO URANIUM HEXAFLUORIDE

Yellowcake (referring in this context to U_3O_8) is only mildly radioactive, insoluble in water and thus constitutes a relatively stable compound that can be transported. However, it cannot be used to produce nuclear fuel directly because it contains impurities and neither is it in a form suitable for enrichment; the impurities need to be removed and enrichment requires a uranium compound to be in a gaseous state. Consequently, at the *conversion* stage yellowcake is converted first to uranium tetrafluoride (UF_4) and subsequently to uranium hexafluoride (UF_6). The latter is often referred to as 'hex'. UF_6 sublimes at the relatively low temperature of 56.5°C and has a triple point of 64.05°C rendering it in the gaseous state necessary for ^{235}U enrichment by *gaseous diffusion* or *gas centrifuge* (the two most prominent approaches to enrichment). UF_6 is the preferred form for enrichment as it has the important property that variations in molecular mass due to there being different constituent isotopes are due solely to the difference in uranium mass, since fluorine has only one naturally occurring stable isotope, ^{19}F. Thus enrichment results in the separation of the gaseous molecules due only to the difference in uranium mass. UF_6 is highly toxic, highly reactive with water and corrosive; hence it is the yellow cake that is transported as the uranium product from milling to the conversion plant rather than UF_6 itself.

There are two ways by which U_3O_8 can be processed into UF_6: either a dry or a wet process. The former is used in the US and involves further calcination, reduction with hydrogen to yield UO_2 and reaction with gaseous hydrogen fluoride via the endothermic reaction,

$$UO_2 + 4HF \rightarrow UF_4 + 2H_2O \tag{12.2}$$

Further reaction with fluorine produces gaseous uranium hexafluoride via,

$$UF_4 + F_2 \rightarrow UF_6 \tag{12.3}$$

and purification via fractional distillation of the UF_6 product stream.

In the wet route, used for example in France and Canada, the U_3O_8 is dissolved in nitric acid to form uranyl nitrate, this is filtered and then this is subject to a further, multi-stage solvent extraction process with tributyl phosphate dissolved in kerosene. The uranium is extracted into the organic phase and is stripped from this with nitric acid. This is then concentrated via evaporation and calcination to render UO_3. The UO_3 is reacted with gaseous hydrogen fluoride, as per the corresponding stage in the dry route, and follows the same path henceforth. UF_6 can be liquefied by reducing its temperature and stored under moderate pressure in steel shipping containers as a white crystalline solid.

12.6 URANIUM ENRICHMENT

While several of the early commercial power-generating reactors were designed to operate with uranium of natural enrichment, most operating reactors and those in plan use fuel that is enriched in ^{235}U to low levels, as mentioned earlier. Uranium enrichment, as for subsequent stages of the nuclear fuel cycle, is subject to stringent international nuclear safeguards controls because of the proliferation risk that it presents were processes adapted to yield higher enrichments for illicit purposes. These are administered by the International Atomic Energy Agency (IAEA).

All uranium enrichment techniques exploit the difference in mass between the minor isotopic constituent, ^{235}U, and the major constituent, ^{238}U. There are two mainstream enrichment processes in commercial use: *gaseous diffusion* and the *centrifuge process*. The centrifuge process is generally much more efficient in terms of energy use than gaseous diffusion (using typically 10% of the electricity required for diffusion) but is less flexible, particularly in terms of the ease of starting and stopping the procedure. Both methods require the uranium feedstock in a gaseous form at near-to-ambient temperatures; hence the requirement for UF_6 as the feedstock. The majority of commercial enrichment capacity in the world is provided by France, Germany, Netherlands, the Russian Federation, United Kingdom and the United States, with China's capability increasing significantly of late.

12.6.1 THE CENTRIFUGE PROCESS

The centrifuges used for uranium enrichment are tall, narrow cylindrical vacuum tubes of several metres in height. They comprise a rotor that spins about a shaft down the centre of the tube and are designed to run continuously for several years at a time. They are arranged in cascades, with the enriched product of one being fed to the next in the cascade while the depleted product is fed back to the previous centrifuge in the series. The geometry of the tubes is specified to yield the degree of acceleration necessary (many orders of magnitude greater than gravity) to achieve optimum separation of ^{235}U from ^{238}U. UF_6 gas is fed into the head of the cascade and, with the rotor spinning at around 50,000 revolutions per minute, the heavier ^{238}U fraction is flung to the outer radial regions of the cylinder while the lighter ^{235}U fraction is concentrated at the centre. Each fraction is drawn off axially with the UF_6 cascading, counter-currently throughout the system. A cascade might comprise typically 10–20 individual stages with the facility for a stage to be bypassed should a unit fail in use. A widespread approach to assessing the effectiveness of the enrichment process is to assess the ^{235}U content of the depleted or tail fraction, known as the tails assay, which has a complex relationship with the desired enrichment and the feed and product quantities. A typical level achieved from a centrifuge system is $\sim 0.15\%$ wt.

12.6.2 GASEOUS DIFFUSION

As a result of the demand for enriched uranium at the height of World War II and its technological maturity relative to other techniques at that time, gaseous diffusion grew to a significant extent and accounts currently for approximately 25% of world enrichment capacity. However, as the plant reach the end of their useful lives they are being replaced with the more energy efficient, centrifuge-based systems because of the benefits in terms of energy efficiency and economy. Diffusion systems comprise typically 1400 stages in a given cascade and each stage is a combination of compressor and a diffuser; a heat exchanger is necessary to remove the heat that results from the compression of the gas. Only a small amount of separation is achieved at each stage but, conversely, a large amount of gas can be handled.

12.6.3 OTHER TECHNIQUES

Two further methods of enrichment have been explored: laser processes and electromagnetic processes. In laser-based methods, laser light is used that is tuned to the energy required to ionise either ^{235}U atoms or to disassociate one of the fluorine atoms from the ^{235}U-containing molecule UF_6 to yield UF_5^+. The former enables the ^{235}U component to be isolated electromagnetically from its heavier counterpart in uranium metal vapour, whereas the latter yields the solid $^{235}UF_5$ from UF_6 feedstock. The molecular approach has attracted more research attention of late since it is compatible with existing conversion processes based on UF_6.

The electromagnetic process was established on a significant scale in the Manhattan project but has been largely abandoned since due to its inefficiency (requiring significantly more energy than gaseous diffusion). This approach utilises the principle exploited in mass spectrometers in which two atoms of different mass are deflected by different amounts dependent on their mass when accelerated by a magnetic field: the heavier isotope being deflected less than the lighter isotope.

12.7 NUCLEAR FUEL MANUFACTURE

Some of the first nuclear reactors used uranium comprising ^{235}U in natural abundance and in metal form, and a variety of cladding materials have been used such as magnesium alloy and in some cases aluminium. These reactors operated for a long time with the last of the generation associated with them only recently coming to the end of their useful lives. However, the fuel design was developed for the Gen II reactor systems that followed to a low-enriched, uranium dioxide, ceramic form and the fuel manufacturing process was adapted accordingly. This will continue to be the case for Generation III+ reactors at least and therefore this is the focus of this section.

12.7.1 CONVERSION TO URANIUM DIOXIDE

The feedstock for nuclear fuel manufacture is either uranium hexafluoride (UF_6), assuming it has been enriched, or uranium trioxide (UO_3) if it is of natural isotopic abundance. Conversion of the feedstock (either UF_6 or UO_3) is necessary because neither has a sufficiently high-melting point to be utilised directly as fuel in a reactor. By contrast, the preferred fuel compound, uranium dioxide (UO_2), has the desirable properties of being insoluble and having the requisite, very high-melting point (2865°C). Once converted into UO_2 it is this that is manufactured into fuel pellets. These are inserted into the fuel clad (usually zirconium alloy for LWRs but for the AGRs used the United Kingdom, for example, stainless steel is used) to constitute fuel pins and these are combined with others into fuel assemblies.

To convert UF_6 to UO_2 again there are two options: either a *dry* or a *wet* process. The dry process yields less waste than the wet process (particularly since no waste liquid filtrate is produced) but can afford less flexibility in the behaviour of the product powder due to differences in particle size and morphology [1], such as a tendency towards agglomeration and large particle size. This can affect pellet manufacture or necessitate intermediate milling of the product.

In the dry process the UF_6 is vapourised in a kiln, combined with steam to yield uranyl fluoride (UO_2F_2) as per,

$$UF_6 + 2H_2O \rightarrow UO_2F_2 + 4HF \tag{12.4}$$

and then this is reduced by the introduction of hydrogen to remove the fluorine thus leaving UO_2. In the wet process(es), that is less common, UF_6 is mixed with water to form a slurry comprising uranyl fluoride (UO_2F_2). This is mixed with either ammonia (NH_3) or ammonium carbonate ($NH_4(2CO_3)$) to yield either ammonium diuranate (ADU—$(NH_4)_2U_2O_7$) or ammonium uranyl carbonate (AUC—$UO_2CO_3{\cdot}2(NH_4)_2CO_3$), respectively. The product is then filtered, dried and heated to get UO_2.

For those cases where UO_3 is the feedstock, water is added to form the hydrate (such as $UO_3{\cdot}6H_2O$) and this is then reduced in a kiln to yield the desired UO_2.

12.7.2 POWDER PROCESSING AND PELLET MANUFACTURE

Depending on which conversion route is adopted, and hence the qualities of the UO_2 produced from it, a variety of preparatory procedures are employed to refine the powder. These include: milling, homogenisation and the use of additives to ensure uniform distributions of particle size, density and specific surface area. Such properties can influence the retention of fission gas, for example. The addition of burnable poisons such as gadolinium may also take place at this stage.

After the conditioning described above, the UO_2 powder is pressed into cylindrical dies machined to the form of the pellet. To consolidate the pellets they are then sintered in a furnace at ~1750°C in a mixture of argon and 6% hydrogen that acts to reduce the chemical composition of the pellets. In particular, this reduces the oxygen-to-metal (O/M) ratio to that specified for the reactor the fuel is designed for, typically O/M ~ 2. After sintering, the pellets are machined to the more exacting dimensional tolerances required for their use as fuel, and they enter a process comprising several quality control steps to ensure their finish, size, shape and integrity comply with these requirements. Much of this activity is automated with extensive processing machines in order to achieve the required throughput and to limit the exposure of people to dusts, extreme temperatures and pressures.

12.7.3 FUEL PINS AND ASSEMBLIES

While the functions of the fuel are to produce heat and resist the effects of irradiation and temperature cycling, the framework maintains the spacing between fuel elements to high levels of precision and the clad prevents leakage of the fuel into the environment around it. To achieve this, the fuel pellets are loaded into tubes, usually of zirconium alloy, together each one constituting a fuel element. Each fuel element is backfilled with helium to inert the environment around the fuel pellets inside the cladding and then end caps are welded on to seal the fuel inside. A test of the element taken via x-ray and/or mass spectroscopy to confirm the integrity of the seal. A space is left to accommodate expansion of the fuel and the production of fission gas, and a spring is inserted to prevent movement of the pellets inside the element and to cater for the difference in the expansion coefficient between the fuel and the clad. Some of this assembly is done by machine and there are extensive quality control checks, some of which involve visual inspection. The fuel can be handled at this stage with appropriate personal protective equipment (glasses, foot protection, etc.) as it is low in radioactivity and not friable (Fig. 12.2A and B). Gloves are worn to keep the fuel clean.

A number of fuel elements are fitted into a precisely machined framework, usually made from zirconium alloy and steel. With the fuel, this framework is engineered to be compatible with the other reactor elements comprising moderator, control materials and burnable poisons. It is important that this framework provides for some flexibility but does not allow excessive movement of the fuel, since there is continual vibration when the reactor is operating from the coolant in turbulent circulation that can result in abrasion and wear, known as fretting. Specific details and characteristics of the fuel designs for particular reactors were given in Chapter 10.

FIG. 12.2

(A) Fuel pins for an advanced gas-cooled reactor being inspected during manufacture. (B) A near-complete fuel assembly for an advanced gas-cooled reactor under inspection prior to dispatch.

Courtesy of the Westinghouse Electric Company.

REVISION GUIDE

On completion of this chapter you should:

- understand the origin of uranium as a natural product that is recovered and refined like other heavy metal elements as a mineral ore
- appreciate that in its natural form the uranium is in too dispersed and impure a form to be used directly in energy generation
- understand why uranium recovery is a large-scale activity, and that there are several techniques for the extraction of uranium feedstocks from the ground
- be aware of the geographical regions where uranium is found in sufficiently high concentrations as an ore to warrant industrial exploitation
- be familiar with the need for the ore to be refined into a transportable form before being shipped off site for conversion into a gaseous feedstock for uranium enrichment
- know the names of the two principal approaches to enrichment and their advantages and disadvantages
- understand what is meant by *solvent extraction*
- appreciate the distinction between 'wet' and 'dry' processing routes
- understand the general processes and manufacturing principles associated with nuclear fuel production

PROBLEMS

1. Write down the performance requirements of nuclear fuel for use in fission reactors. Hence, describe why UF_6, UO_3 and metal fuels are not suitable for use in modern reactor systems. In this context, what advantage might accident-tolerant fuels bring to the performance of these reactor systems; why is thorium potentially of interest in this regard?
2. Explain the two salient incompatibilities of uranium ore with its direct use as a fuel. How do the ores used as feedstock for reactor systems now differ from the geological deposits associated with the natural reactor systems of 2 billion years ago?
3. Describe the benefits of the dry route over the wet route for the conversion of UF_6 to uranium dioxide.
4. Compare and contrast the enrichment process based on the centrifuge and gaseous diffusion in terms of the advantages and disadvantages of each.
5. Explain why yellowcake production tends to be done on the site of the uranium mine while conversion to UF_6, enrichment and conversion to UO_2 are done elsewhere.

REFERENCE

[1] A.P. Bromley et al., Changes in UO_2 powder properties during processing via BNFL's binderless route, http://www.iaea.org/inis/collection/NCLCollectionStore/_Public/28/047/28047123.pdf.

CHAPTER

NUCLEAR FUEL REPROCESSING 13

13.1 SUMMARY OF CHAPTER AND LEARNING OBJECTIVES

In this chapter, the scientific and technical basis of nuclear fuel recycling and reprocessing is considered. A perspective is adopted to provide the reader studying this from an engineering perspective with an understanding of the salient concepts and chemical principles by which uranium and plutonium are separated from spent fuel, by which they are separated from each other and then purified. An introduction to both hydrometallurgical and pyroprocessing is given.

The objectives of this chapter are to:

- introduce the concept of both *open* and *closed fuel cycles*, reviewing the merit of each
- consider a specific example of hydrometallurgical processing by which uranium and plutonium are partitioned, focussing on the plutonium uranium extraction (PUREX) process
- highlight each of the stages, indicating the principal engineering features and process chemistry that is exploited to separate and refine spent nuclear fuel (SNF)
- introduce an example of an alternative reprocessing technology that is rarely used for the reprocessing of commercial SNF but might be adopted in the future: pyroprocessing

13.2 HISTORICAL CONTEXT: SIR CHRISTOPHER HINTON 1901–83

Christopher Hinton is widely regarded as the person responsible, more than any other, for establishing Britain's nuclear energy industry: as Managing Director of what was then known as the Risley Industrial Group, he oversaw the design and construction of the nuclear fuel manufacturing plant at Springfields, the Windscale piles, the diffusion plants at Capenhurst, the first large-scale nuclear power station at Calder Hall and the fast reactor complex at Dounreay. Most significantly, with regard to the focus of this chapter, he led the development of the first generation chemical reprocessing facilities at Sellafield. He became chairman of the Central Electricity Generating Board when it was formed in 1957, was the first Chancellor of the University of Bath in 1966 and the first President of the Royal Academy of Engineering in 1976 (Fig. 13.1).

Nuclear Engineering. https://doi.org/10.1016/B978-0-08-100962-8.00013-5

FIG. 13.1

Sir Christopher Hinton.

13.3 INTRODUCTION

Perhaps one of the most significant achievements in nuclear energy development is the ability to separate uranium and plutonium from irradiated nuclear fuel. While the first recorded separation of relatively small amounts of such materials was reported a long time before the widespread expansion of commercial nuclear energy [1], this has been perfected to the extent that levels of purity are achievable that are compatible with these materials being used for power generation for a second time.

While reprocessing can attract political debate, as is discussed below, from an engineering perspective the ability to extract potentially useful materials from what is otherwise a useless and highly radioactive matrix leaves open options such as whether to recycle unused ^{235}U, extract ^{238}U for use in a fast breeder reactor or to utilise the inherent ^{239}Pu for mixed oxide (MOX) fuels; all of these options correspond to significant sources of energy made available by the ability to reprocess. There are a further range of applications associated with these materials, such as actinide burning, that are also been derived from our ability to separate radioactive streams on an industrial scale. The separation of long-lived radioactive components from wastes has the potential to reduce the volume of high-level waste that needs to be disposed of, and also to render it in a form suitable for further treatments.

The changes in terms of mass composition brought about by the irradiation of nuclear fuel are subtle: a low-enriched fuel based on uranium dioxide from, say, a PWR will typically exhibit significant reductions in ^{235}U content (typically from ~3% to ~1%) with a small amount (0.5% other uranium isotopes formed via neutron capture, etc.). The other prominent compositional effect is the formation of plutonium, accounting for typically ~1% of the spent fuel mass. From the perspective of radiotoxicity, the situation contrasts significantly: fresh fuel is relatively benign in this regard. In spent fuel, the majority of the radiotoxicity is associated with the fission product, plutonium and minor actinide components of the inventory. After long-term storage (say ~1000 years), this will be dominated by plutonium and the minor actinides. Interestingly, ~95% of the spent fuel has the potential to be recycled (essentially the combination of the uranium—94% and plutonium—1%). Radioactive waste corresponds to around 5% that comprises the fission products and the minor actinides. The percentage composition of a typical example of spent nuclear fuel (SNF) is given in Table 13.1.

In this chapter, we shall restrict coverage to a specific conceptual chemical process that is used most widely with a summary of the salient prospects for the future; the detail of *actual* processes tends to vary on a case-by-case basis and is often subject to commercial restrictions on information. Given here

Table 13.1 Percentage Composition by Mass of Spent Nuclear Fuel [2] Based on Uranium Dioxide of Primary Enrichment 3.5% wt. ^{235}U, Irradiated 33 GWd/t Initial Uranium[a] After 3 Years of Cooling[b]

Isotope or Chemical Family	% Composition by Mass	
^{234}U	0.02	Total U: 95.59
^{235}U	1.03	
^{236}U	0.44	
^{238}U	94.1	
^{238}Pu	0.018	Total Pu: 0.97
^{239}Pu	0.567	
^{240}Pu	0.221	
^{241}Pu	0.119	
^{242}Pu	0.049	
^{244}Cm	0.002	Total minor actinides: 0.08
^{243}Cm	3×10^{-5}	
^{242}Cm	1×10^{-5}	
^{243}Am	0.010	
^{242}Am	7×10^{-5}	
^{241}Am	0.022	
^{237}Np	0.043	
Rare gases	0.56	
Alkali metals	0.31	
Alkaline earths	0.25	
Lanthanides, Y	1.03	
Zirconium	0.37	Total fission products: 3.41
Chalcogens	0.05	
Molybdenum	0.34	
Halogens	0.02	
Technetium	0.08	
Ru, Rh, Pd	0.39	
Miscellaneous	0.01	

[a] *Burn-up expressed per tonne of* initial *uranium, that is when the fuel was fresh.*
[b] *Rare gases include: krypton and xenon, alkali metals: caesium and rubidium, alkaline earths: strontium and barium, chalcogens: selenium and tellurium, halogens: iodine and bromine, whilst the miscellaneous category includes the transition metal element composition of the fuel. Oxygen has been omitted.*

therefore is a conceptual overview of the science and engineering principles associated with reprocessing. It is worthy of note that implicit to this discussion is that individual processes are often repeated by cascading interconnected process plant with the product from one returned as the feed to others to achieve required levels of separation.

13.4 NUCLEAR FUEL REPROCESSING AND RECYCLING

13.4.1 CONTEXT

Although modern nuclear reactors used for commercial power generation are not refuelled very often, several factors require that the fuel is removed eventually, despite there being significant, residual quantities of unused uranium and plutonium left over. These include a variety of physical changes resulting in microscopic cracking, deformation and expansion as a result of irradiation that can degrade the thermal conductivity of the fuel and impede its performance in the reactor. As a result of the buildup of fission products and the products of neutron capture reactions, impurities are bred into the fuel that can perturb neutron transport and its reactivity response. Perhaps most importantly, the ^{235}U content eventually falls below a level that undermines the economic viability of power production. This is manifest, for example, in a sub-optimum core load profile in terms of its balance, uniformity and symmetry, thus compromising generation efficiency.

There are two technical options when nuclear fuel has been removed from a power reactor[1]:

- it can be stored indefinitely in shielded containers (often referred to as being *dry stored*), usually in steel casks in the form that it is extracted from the reactor as complete fuel assemblies *or*
- it can be reprocessed in order for the fissile content to be recycled for future use in another reactor

The former is associated with what is known as a *once-through* or *open* nuclear fuel cycle; this is the nuclear energy concept in which nuclear fuel is used once. The latter is associated with what is referred to as a *closed* fuel cycle in which the unused fissile isotopic content (uranium) and that which has been produced as a result of irradiation in reactor (plutonium) are fed back into the generation process. In some cases spent fuel has also been stored in ponds under water, but it cannot be kept in this state indefinitely. Clearly, with respect to reprocessing it is important that the efficiency of recovery is high, if it is to be worthwhile on a practical basis, and that the purity of the recovered uranium and plutonium products is sufficient to ensure that they are relatively free from residual radioactivity and impurities. Otherwise this might undermine the ease with which they might be used in fuel and their performance in a reactor.

The choice between open and closed fuel cycles has stimulated debate and resulted in contrasting national strategies. This is principally because, while reprocessing yields separated uranium and plutonium that constitute significant resources of energy, it can also raise concerns over the proliferation of nuclear material, amongst other factors. Conversely, the open cycle is usually part of a strategy to consign this resource beyond reach forever via permanent geological disposal. Further, such consignments might comprise what might be considered to be unnecessarily large volumes of unsegregated and long-lived radioactive waste, relative to that which might be achieved post-reprocessing. Via reprocessing, a resource

[1]Given the focus of this section on reprocessing, the option for the re-use of spent fuel in high neutron economy reactor designs such as CANDU has been set aside.

(in the form of both fissile and fertile residues) is recovered, and waste volumes and radioactivity levels can be reduced. Interestingly, both scenarios require long-term, national strategies for the treatment and disposal of SNF; such arrangements are not ratified easily. These issues are discussed further in Chapter 15.

By way of example, the United States stopped the reprocessing of commercial spent reactor fuel in 1977, generally due to concerns about its association with the threat of proliferation. France (La Hague), the United Kingdom (Sellafield) and Japan (Rokkasho) have all continued to reprocess SNF on a commercial basis. Understandably, the decision to recycle cannot be taken lightly because the infrastructure necessary is large, complex and constitutes a significant investment; it also must be operated under rigorous, regulatory control and is subject to international oversight. Further, the decision to reprocess also influences the long-term waste disposal options open to the spent fuel materials arising from nuclear reactors. While most reprocessing to date has led to the recovery of plutonium for recycle as MOX fuel, the original imperative was the recovery of ^{235}U when forecasts of its supply were a concern; now a variety of reprocessing avenues are feasible for several of the products that are recovered. For example, a priority in future might be to separate particular isotopes for specific opportunities such as for use in radioisotope thermionic generators (RTGs) for space missions. Alternatively, the strategy might be to extract combinations of isotopes together, such as the long-lived actinides for use in Gen IV fast spectrum reactor systems to facilitate actinide burning. Irrespective of the outcomes, spent fuel reprocessing needs to be economically viable, proliferation resistant and optimised to ensure that only minimal quantities of waste are generated.

13.4.2 RECYCLING OPTIONS

The separated products of reprocessing (assuming a mainstream flow sheet yielding separated uranium and plutonium) can be used again in nuclear reactors, although their uses differ dependent on the element concerned. In principle, plutonium can be utilised relatively quickly. Extracted in its final form as an oxide, it can be mixed with depleted uranium dioxide (UO_2) to yield what is termed *MOX* fuel. Such fuel accounts currently for ~5% of nuclear fuel in use today[2] associated with those reactors licensed to use it, particularly in Europe and Japan.

When used in an LWR reactor, MOX accounts typically for 30% of the fuel load. Small adaptations to the reactor are usually necessary, typically, for example, to modify the control rod arrangement. Where there is a requirement to up-rate existing reactors and increase burn-up (e.g. to meet targets for the reduction of CO_2 emissions and power demands), MOX use offers an alternative means by which to derive additional fissile content via the addition of plutonium in contrast to enriching further in ^{235}U. MOX use is also attractive in the context of reducing the volume of spent fuel and disposing of extant plutonium stockpiles for the generation of electricity. However, the military connotations, proliferation concerns and radio-toxicology of plutonium has led to controversy and public concern over the use of MOX in some cases. The disposal of SNF based on MOX is also more challenging because of a higher content of very long-lived actinides than arises for uranium-derived SNF.

In contrast, the re-use of uranium extracted via reprocessing in reactors is complicated by the presence of ^{232}U and ^{236}U. Further to the discussion in Chapter 11, and in contrast to the influence of ^{232}U on the thorium fuel cycle, ^{232}U from the irradiation of uranium fuels arises from the decay of ^{236}Pu, viz:

[2] www.world-nuclear.org.

$$^{235}\text{U}+n\xrightarrow{\gamma}{}^{236}\text{U}+n\xrightarrow{\gamma}{}^{237}\text{U}\xrightarrow[6.75\text{d}]{\beta^-}{}^{237}\text{Np}+n\xrightarrow{2n}{}^{236\text{m}}\text{Np}\xrightarrow[22.5\text{h}]{\beta^-}{}^{236}\text{Pu}\xrightarrow[2.86\text{y}]{\alpha}{}^{232}\text{U} \quad (13.1)$$

Subsequently, via a number of relatively short-lived decays, ^{232}U yields several emitters of high-energy γ rays (particularly ^{220}Rn, ^{212}Bi and most significantly the γ-ray emission from ^{208}Tl at 2.6 MeV). These make the recycled material very difficult to handle and to process if its re-use has been delayed for several years, allowing the contamination to become established. From a different perspective, ^{236}U has a relatively significant neutron absorption cross section that competes with ^{235}U in low-enriched fuels, and it is difficult to separate chemically from ^{235}U. Uranium-236 is also difficult to remove by traditional methods of enrichment because it tends to collect with the lighter, desirable ^{235}U isotope.

As a result, recycled uranium cannot be re-enriched with fresh uranium (because the ^{236}U would contaminate the fresh stream that is devoid of it) and consequently it can be necessary to enrich recycled fuel further, to yield recycled and re-enriched uranium, thus compensating for the effect of ^{236}U on neutron economy; ^{236}U can also prevent nuclear fuel from being recycled more than once. Only low burn-up fuels of natural enrichment tend to be recycled and enriched as a result because the ^{236}U content in these is sufficiently low to render this feasible.

13.4.3 HYDROMETALLURGICAL PROCESSES: THE PUREX PROCESS

The most established and widely practised hydrometallurgical method of reprocessing nuclear fuel is the PUREX process [3]. This was pioneered originally for military purposes for the recovery of pure plutonium, and it has been used to recover uranium and plutonium from both oxide and metal fuels. While several specific implementations (generally referred to as *flowsheets* as mentioned above) are feasible, we shall consider a generic example for the purposes of this text.

The first task[3] in nuclear fuel recycling usually comprises a combination of mechanical procedures to convert the irradiated fuel elements into a form suitable for the subsequent stages of reprocessing. Most importantly, these should prepare it to be rendered in the liquid phase (*dissolution*) prior to it being subject to several, subsequent levels of separation procedures. First, the fuel assembly structure and cladding are removed, and a variety of methods are used: this is often accomplished by the fuel being *sheared* by a large mechanical press into short cylindrical hulls (typically ~5 cm in length). Alternatively, the cladding can be stripped away by pushing fuel rods through a die but most often this step is a matter of disassembly and shearing. To separate the spent fuel into its constituent uranium, plutonium and waste products, it is necessary to form an aqueous solution of it compatible with the liquid–liquid extraction *solvent exchange* processes that follow; this is done by mixing the chopped fuel with hot nitric acid. This dissolution stage is usually performed with hot nitric acid in a special vessel often referred to as the *dissolver*. A schematic of the process is shown in Fig. 13.2.

At the start of the process (often referred to as *head-end*), the majority of the mass of the spent fuel is due to its uranium content whilst, conversely, the majority of the radioactivity at this point is due to the plutonium (particularly ^{238}Pu) and curium (^{244}Cm and ^{242}Cm) isotopes; over time, due to decay, the radioactivity profile migrates from the short-lived plutonium and curium isotopes to the longer-lived isotopes of plutonium, curium and americium. The spent fuel is stored for a period of time (months to years) to allow its resident short-lived radioactivity to decay away. This leaves a fission product

[3]After receipt, quality control, specification and burn-up estimates have been completed and assuming a period of cooling has already elapsed.

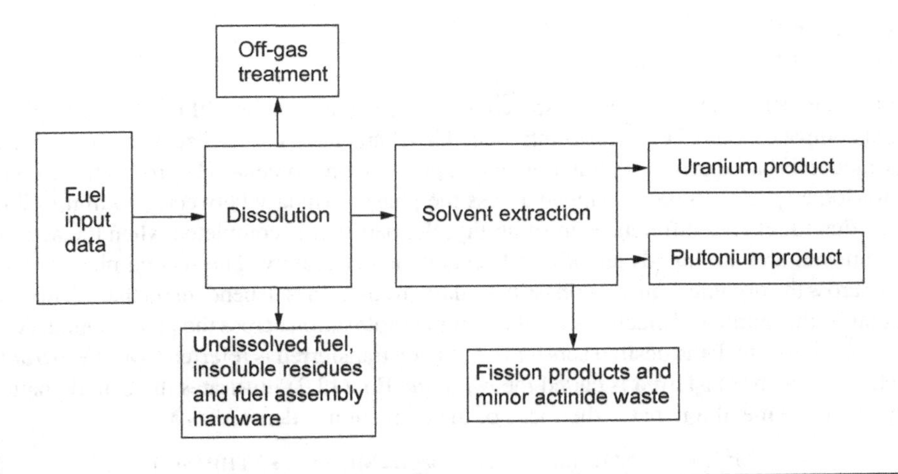

FIG. 13.2

A schematic representation of the various stages of the PUREX process.

Reproduced with permission of the OECD, Spent Nuclear Fuel Reprocessing Flowsheet, NEA/NSC/WPFC/DOC(2012)15, https://www.oecd-nea.org/science/docs/, 2012.

inventory dominated by ^{137}Cs (and the decay via ^{137m}Ba), ^{90}Sr (and subsequently ^{90}Y), ^{99}Tc and ^{129}I. The latter, volatile fission products along with ^{14}C (as CO_2), tritium and the noble gases (particularly krypton and xenon) are evolved during dissolution. These are routed for dedicated *off-gas* treatment and do not usually feature in the inventory of the aqueous phase that follows. Numerous technologies are employed for off-gas treatment where volatiles occur in the process, to ensure that regulatory constraints on gaseous, radioactive emissions are met.

Given the levels of radioactivity involved, and the requirement to control the processes of oxidation and reduction to provoke transfer between the organic and liquid phases at several stages of the solvent exchange, remote handling and fine process control are essential. Whilst the primary stages of the process deal with large amounts of material and significant flow rates, the partitioning of plutonium and its subsequent refinement presents a risk of a chain reaction occurring inadvertently (a criticality); this requires for example that process plant with restricted throughputs are used along with rigorous procedures by which the quantities of nuclear materials are accounted for. Further, the inherent radioactivity and acidity of the various media involved yield both oxidising and reducing radicals via radiolysis and hydrolysis, respectively, of media entrained with them that can influence process dynamics and efficiency. The output of the dissolution stage is an aqueous nitrate solution comprising the uranium, plutonium and the majority of non-volatile fission products. For the case of this example, it has been assumed that the partitioning of plutonium and uranium is the desired outcome since this is the most widespread objective of nuclear fuel reprocessing at present. However, there are other possibilities such as the separation of specific fission products or the production of a combined uranium/plutonium product stream that is free from minor actinides.

Following dissolution, there are six stages to the PUREX process; they are as follows:

1. the extraction of uranium and plutonium from the aqueous phase into the organic phase
2. the partitioning of uranium and plutonium
3. the recovery of uranium
4. the recovery of plutonium

5. the recovery of the solvent
6. the recovery of the nitric acid

Fundamental to PUREX is the process of solvent exchange and the ability of uranium and plutonium to form neutral complexes with TBP, as was introduced in Chapter 12. This is discussed in more detail here. Solvent extraction is based on the exploitation of two physical phenomena. The first of these concerns the tendency for ionic species to be exchanged across the phase boundary between two immiscible liquids when mixed due to relative differences in solubility; the partition is completed when the action causing mixing is removed and the liquids are allowed separate under gravity. The second phenomenon is that the transfer across the organic–aqueous-phase boundary in solution is dependent on the valence of the ions involved, that is, the number of electrons available for complexation across the phase boundary. Conventionally, the phase to which the desired substance has been transferred is referred to as the *extract* and that from which it has been transferred is called the *raffinate*. Eq. (13.2) illustrates the generic nature of this extraction where the metal ion that is the focus of the extraction is denoted as M,

$$M^{2+}(aq) + 2NO_3^-(aq) + 2\,TBP\,(org) \rightleftarrows M(NO_3)_2 \bullet 2\,TBP\,(org) \tag{13.2}$$

Following mixing, the extracted ions M are distributed across the aqueous and organic phases. The balance of this dynamic equilibrium is described by the corresponding distribution coefficients; these quantify the ratio of the concentration in the organic phase to that in the aqueous phase. Some examples are shown for relevant species in Table 13.2.

The data in Table 13.2 illustrate the range of exchange that is possible across the aqueous-solvent boundary. For example, uranium and plutonium-IV extract strongly to the organic phases relative to nitrate ions while the prominent fission products remain in the aqueous stream. A comprehensive summary of distribution coefficients is available in the corresponding index [6].

Table 13.2 Distribution Coefficients Corresponding to the Concentration Ratio of Organic-to-Aqueous for Various Ions Relevant to the PUREX Process [4,5]

	Distribution Coefficient	
Ion	25°C 3 M HNO_3, 30% TBP/OK	25°C 4 M HNO_3, 19% TBP/OK
U (IV)	8.1	–
Pu (IV)	1.66	11.5
Pu (VI)	0.62	2.5
Np (IV)	–	3.0
Np (V)	–	0.13
Np (VI)	–	11.0
Am (III)	–	0.08
HNO_3	0.07	–
Zr	0.02	–
Ru	0.01	–
Pu (III)	0.008	0.014
Nb	0.005	–
Rare Earths	0.002	–

The aqueous phase is maintained with the use of nitric acid. This has the advantage that it is recoverable via distillation thus reducing waste and the demand for reagent. Tri-*n*-butyl phosphate (TBP) is used as the organic extractant, due to its favourable distribution coefficients, low volatility and high flashpoint (and hence resistance to inadvertent combustion). To adjust the density and viscosity to ensure optimum flow characteristics through the aqueous phase under gravity, the TBP is diluted with odourless kerosene typically to levels between 20% v/v and 30% v/v. Mixing on a process-level scale can be performed by several types of machine: generally either *mixer settlers* or *pulsed columns*. In a mixer-settler, the fluids are mixed in one region and allowed to settle under gravity in another so that the separated phases can be drawn off independently. In a pulsed column the two liquids are mixed by the action of pulses actuated from outside the vessel. These have the advantages of limiting the quantity of fissile material that is processed at once to be safe by design, and they operate continuously without the need for individual extraction stages.

13.4.3.1 Extraction of uranium and plutonium

In the first stage of the PUREX process, any residual undissolved matter is removed by centrifugation (*clarification*) and the Pu(VI) produced as a result of dissolution is reduced to Pu(IV) (*conditioning*). Then the aqueous nitrate solution from the dissolution of the irradiated nuclear fuel is fed into a pulsed column, often referred to as a *contactor* in the sense that it facilitates contact between the two immiscible phases. This flows downward through TBP which flows in the opposite direction with mixing aided by the pulsing action of the column. The majority of the hexavalent uranium and tetravalent plutonium ions (given the significance of their distribution coefficients relative to the other ions present) are extracted to the organic phase. This leaves behind an aqueous raffinate comprising the majority of the fission products. The transfer of uranium follows Eq. (13.3),

$$UO_2^{2+}(aq) + 2NO_3^-(aq) + 2\,TBP\,(org) \rightleftarrows UO_2(NO_3)_2 \bullet 2\,TBP\,(org) \tag{13.3}$$

and that of plutonium as per Eq. (13.4),

$$Pu^{4+}(aq) + 4NO_3^-(aq) + 2\,TBP\,(org) \rightleftarrows Pu(NO_3)_4 \bullet 2\,TBP\,(org) \tag{13.4}$$

Aside from that of the uranium and plutonium, the solution chemistry of a proportion of the other elements present in the aqueous feed from dissolution also favours their transfer to the organic phase. This includes some fission products and also neptunium. These are extracted (or *scrubbed*) subsequently by further processing of the loaded organic phase with nitric acid. The concentrations of these washes are selected carefully so as not to cause significant transfer of the uranium and plutonium in the organic phase *back* to the aqueous phase, but nonetheless some re-extraction does occur. Hence, a further extraction stage with clean TBP is included to recover this. Overall, the yield from this stage in the PUREX process is an organic phase that contains the majority of the actinides from the spent fuel and a minority of fission products (particularly technetium and zirconium) that co-extract and escape scrubbing.

13.4.3.2 Partitioning of uranium and plutonium

The separation of plutonium is made possible by the exploitation of two further phenomena:

- Plutonium-IV extracts strongly to TBP whereas plutonium-III does not (as evidenced by the distribution coefficients in Table 13.2). This can be exploited by a further stage of solvent extraction to split the uranium and plutonium from one another without need to revert to an alternative separation process.

- The presence of uranium-IV causes plutonium-IV to be reduced to plutonium-III, enabling the extraction referred to above. This is favoured because extra metal-ion reductant additives are not needed that would require removal or destruction at a later stage.

To facilitate this, uranium-IV mixed with nitric acid is introduced into a contactor at the partitioning stage to mix with the loaded organic feed. The plutonium in the organic phase is reduced as a result and hence is much less soluble in TBP and thus transfers to the aqueous phase. Some of the uranium in the organic side also transfers; this is removed by a subsequent scrubbing of the aqueous phase with clean TBP. Also, since nitric acid can cause oxidation of the favoured plutonium-III, risking its re-extraction back to the TBP and the production of nitrous acid, hydrazine nitrate is introduced. This acts to remove nitrite ions from solution, arresting nitrous acid formation. The uranium fraction is recovered separately from the organic phase by contacting with hot, low-concentration nitric acid, causing it to transfer to the aqueous state. Hence two aqueous products arise from this stage of the process: one loaded with uranium and the other loaded with plutonium.

13.4.3.3 Recovery of uranium

In contrast to the discussion of the recovery of plutonium below, the recovery of uranium from the decontamination stage is essentially a task of purification based on further solvent extraction processes. Even small amounts of actinide- and especially fission-product impurities can complicate the ease with which the products of reprocessing might be handled, stored and used in future due to their associated radioactivity, necessitating further decontamination cycles. The primary contaminant of the uranium stream is neptunium. Most of this is of the form of ^{237}Np formed via neutron capture on ^{235}U and subsequently on ^{236}U, the β^- decay of ^{237}U arising from ^{238}U(n,2n)^{237}U, and also via decay of ^{241}Am in spent fuel; the half-life of ^{237}Np is long at 2.1×10^6 y.

Neptunium is undesirable in the uranium stream as it does not contribute significantly to fission; it is an α-emitter and is one of the more mobile actinides in the environment. In terms of distribution coefficient and hence extractability, neptunium-VI is greater than neptunium-IV with both regarded as being extractable. Neptunium-III and neptunium-V are not considered to be extractable. To exploit this, the uranium stream is concentrated by evaporation, the neptunium is oxidised to neptunium-VI (which is approximately as extractable to the organic phase as plutonium-IV) and neptunium-V (inextractable). Hydroxylamine nitrate is added to reduce the neptunium-VI proportion to neptunium-V so that when mixed with TBP, the recovered uranium is transferred to the organic phase leaving the neptunium behind. The uranium is then recovered from the organic phase as was done in the partitioning process. Nitrate-based extraction processes are particularly suitable for neptunium because other systems (fluoride, sulphate, etc.) result in the formation of complex ions that inhibit extraction.

13.4.3.4 Recovery of plutonium

The plutonium stream derived from the uranium/plutonium partitioning step in the aqueous phase is present in the plutonium-III state, in order to have minimised its extraction across to the organic side when partitioned from uranium. However, as with the uranium stream, further purification is necessary to remove fission product contamination (predominantly rhodium and ruthenium) and some neptunium (albeit usually less than is present in the uranium stream). To facilitate dispersion to the organic phase, the plutonium-III is oxidised to plutonium-IV by passing nitrous acid vapour through the aqueous phase to destroy the excess hydrazine.

13.4.3.5 Recovery of the solvent and nitric acid media

Both the organic and aqueous phases used in solvent extraction become contaminated and degraded in use. Rather than being used once through, which would result in greater consumption and the need for extensive requirements for waste storage and disposal of contaminated liquid media, the media used in reprocessing are refined and recycled. This is done for the most part continuously on-site and as an integrated part of the process.

The TBP and odourless kerosene that comprise the organic phase used in the PUREX process are subjected to degradation by the action of the acid and ionising radiation leading to hydrolysis and radiolysis, respectively. These processes can yield acidic compounds (including dibutyl phosphate, monobutyl phosphate acid and phosphoric acid) which, if left unchecked, can reduce the separation performance of the organic phase, particularly as a result of undesirable complexes formed with plutonium-IV and similarly with uranium by the organics. A range of light and heavy compounds are also formed, which constitute a further path by which the extraction potential is compromised. To recover the effectiveness of the solvent, it can be washed with a base to remove the acidic products; sodium carbonate is typically used since it has the advantage of forming soluble end products with any residual metallic ions that are present (such as plutonium, uranium, zirconium, niobium and ruthenium). The solvent is then evaporated and rectified[4] to remove the light and heavy products, respectively. Centrifugation has also been used to remove residual aqueous content and trace solids yielding purified solvent of a form compatible with re-use.

The approach taken for nitric acid is a little different to that of the solvent because the vast majority of the radioactive inventory is entrained with this phase. The long half-lives of several of the fission products and the minor actinides (in the absence of the possible use of actinide burning technologies heralded for the future) pre-dispose this material to very long-term storage. Thus, the priority in terms of the re-use of the aqueous phase is to produce a concentrate (usually by double distillation) comprising the waste radioactivity with ~90% of the nitric acid recovered by fractionation for re-use. The concentrate is neutralised with an alkali (sodium hydroxide) prior to long-term disposal.

13.4.4 PYROCHEMICAL REPROCESSING

Pyroprocessing is the term adopted for the reprocessing of SNF that is based on a variant of electro-refining and is currently at a developmental stage. Electro-refining is well established and used extensively in the metals refining industry. In this approach, a combination of electricity, complexation within a eutectic chloride salt and absorption is used. This enables the various ionic species comprising

[4] *Evaporation* refers to a single operation in which the more volatile components in a fluid are separated by heating as a vapour from the less volatile that remain in the liquid phase. *Distillation* refers to a single stage of evaporation and *condensation*, the latter usually driven by a dedicated condenser to return the separated vapour phase to liquid. *Rectification* in this context refers to the cascaded distillation process common in the refinement of organic liquids whereby escalating levels of evaporation and condensation are exploited to achieve highly purified vapour and liquid streams, respectively.

the spent fuel to be separated, transported through an electrolyte and then collected at a number of different locations in a vessel dedicated to the process.

Pyroprocessing is particularly relevant to fuel cycle options associated with future reactor concepts that might utilise fuel(s) from a reprocessing capability that is integrated with the reactor in question; a popular example is a fast reactor exploiting a ^{238}U-based blanket derived from an associated pyroprocessing facility. The start of the process requires a de-clad metallic fuel, i.e. that which follows mechanical removal of the cladding; oxide fuels can be used if they are reduced chemically to metallic form first. Unlike hydrometallurgical processes that can operate on a continuous basis, pyroprocessing favours a batch approach. Further, the high levels of product purity achieved with the PUREX process are not usually an essential requirement in applications considered to favour pyroprocessing.

Having separated the majority of the uranium, one of the residual streams comprises the majority of the available transuranic inventory. This includes the plutonium, placing it effectively beyond reach in terms of proliferation. Further, the level of fission product contamination of the uranium stream is high relative to, for example, the uranium stream derived from the PUREX process; this limits the ease with which the separated uranium from pyroprocessing might be handled for illicit purposes. While the usual benefits of reprocessing associated with it enabling fuel recycle and the reduction of waste volumes remain valid, pyroprocessing is particularly suited to the refinement of a wide range of spent fuel types. Consequently, it provides an option for the treatment of exotic fuel materials that remain from the legacy of early nuclear energy developments that are not otherwise easily disposed of.

For the purposes of this section, we shall draw in part on the case of the exemplar electro-refinery facility[2] at the Argonne National Laboratory, recognising that research continues on adaptations and developments of the principle worldwide. In pyroprocessing, separation is achieved by exploiting three physical phenomena that are exploited in a system that constitutes an electric circuit:

1. The more chemically active elements in the spent fuel (alkali, alkaline earths and rare-earth fission products) are oxidised by the addition of cadmium chloride to form very stable chlorides. These become fixed in the eutectic salt.
2. The application of a voltage to a steel basket in which the spent fuel is confined and immersed in the salt electrolyte (thus constituting an electrode—the anode—in the corresponding electric circuit) causes uranium and compounds of uranium and plutonium to be transported to one of two cathodes; uranium metal forms a dendritic deposit on a solid (steel) electrode while the compounds collect in an electrode made of liquid cadmium that is also suspended in the salt.
3. The noble metals present in the spent fuel are resistant to oxidation and are not transported or fixed in the salt. These remain in the basket.

A schematic of the Argonne embodiment of the process is reproduced in Fig. 13.3. The tank that constitutes the electro-refiner comprises the three electrodes, the lithium chloride/potassium chloride mixture and liquid cadmium. To maintain this liquefied state, it is heated to ~500°C. The chopped, metal spent fuel is inserted into the basket associated with the anode and lowered into the salt. The electrotransport process is then performed resulting in the partitioning described above. The elements that are not transported either achieve equilibrium with the surrounding salt or remain unaffected in the spent fuel basket. The products at the electrodes are recovered by melting (with any entrained electrolyte or cadmium being recovered and recycled).

The electrolyte and the basket residues can withstand the processing of several batches of fuel before they need to be refined; this is done with the addition of a reducing agent such as a mixture of

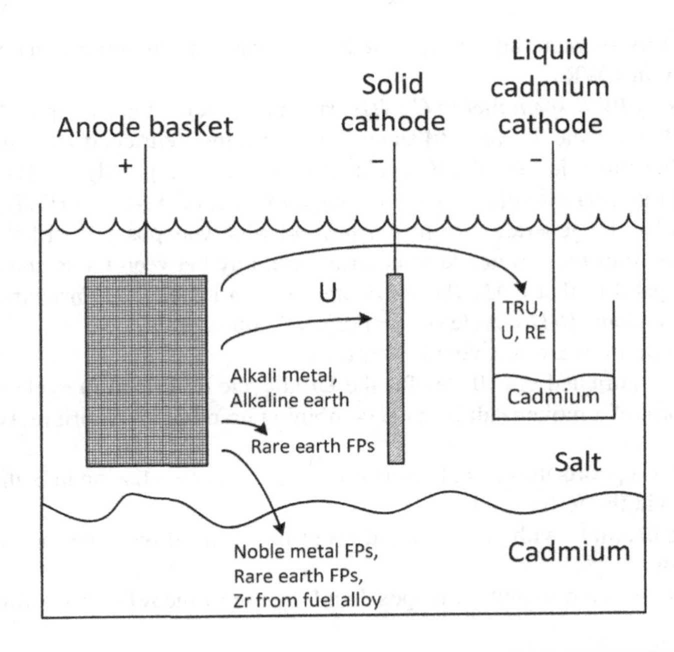

FIG. 13.3

A schematic diagram of the electro-refining process central to spent fuel pyroprocessing. Key: TRU—transuranic elements, U—uranium, RE—rare earth, FPs—fission products.

Reproduced with permission from J.J. Laidler, J.E. Battles, W.E. Miller, J.P. Ackerman, E.L. Carls, Development of pyroprocessing technology, Prog. Nucl. Energ. 31 (1/2) (1997) 131–140.

lithium and potassium. The entire process is accomplished remotely (due to the inherent radioactivity) with reference to criticality constraints and with the surrounding environment inerted due to the chemical reactivity of the substances involved.

13.5 A SUMMARY OF CLOSED FUEL CYCLES

At this stage it is instructive to summarise and compare the prominent closed nuclear fuel cycles under consideration in the world today. This summary is not exhaustive:

1. The most widespread fuel cycle referred to in this context, and also the most mature in a technical sense, is the *uranium/plutonium (U/Pu) cycle*. In this case, ^{235}U is the source of fission in the first instance with some ^{239}Pu yielded from fertile ^{238}U. The cycle is as follows:
   - **a.** Uranium is extracted from the ore or leachate, the specific compound that results being dependent on the process used.
   - **b.** This is converted to UF_6 to enable enrichment that follows.
   - **c.** The enriched UF_6 is converted into an oxide for fabrication into UOx fuel.
   - **d.** The UOx fuel is used in reactor systems.
   - **e.** The spent fuel is reprocessed by a hydrometallurgical (usually PUREX) process.

   f. The extracted fissile materials are recycled; in the case of plutonium, this would be via MOX fuel, typically in LWRs.
2. In the *Direct Use of PWR spent fuel in CANDU* (DUPIC) cycle, it is anticipated that the process is the same as the U/Pu cycle above until stage d, where lightly enriched UOx is used in reactor systems. After this stage, in the DUPIC cycle, the spent fuel (typically LWR) is dismantled and re-manufactured into fuel assemblies that are compatible with CANDU (PHWR) systems; it is used in these reactors for the generation of power beyond what was possible in LWR systems. This approach benefits from the difference in neutron economy between PWR and CANDU systems, exploiting the capability that CANDU can operate with a lower ^{235}U enrichment than PWR.
3. In the *uranium/thorium (U/Th)* cycle one anticipated approach is:
   a. Uranium *and* thorium are recovered from ores.
   b. The uranium is enriched via UF_6 as for the U/Pu cycle described in cycle #1 above.
   c. Fuel in the form of a molten salt is made combining uranium and thorium, typically as per UF_4 and ThF_4.
   d. The UF_4–ThF_4 supports the critical reaction to yield power via fission in both ^{235}U and ^{233}U, the latter arising via the fertile ^{232}Th.
   e. Reprocessing is online with an integrated spent fuel deposition route, potentially via pyroprocessing.
   f. Spent fuel products are routed for disposal and the use of the refined salt for power production continues.

REVISION GUIDE

On completion of this chapter, you should

- understand the context associated with the reprocessing of SNF
- be able to define what is meant by *open* and *closed* nuclear fuel cycles
- appreciate the advantages and disadvantages of nuclear reprocessing in terms of *nuclear proliferation*, the *efficient use of nuclear energy resources* and its relevance to the *environmental impact waste disposal*
- be able to describe the qualitative composition of low-enriched spent fuel in terms of its *mass distribution* and *radioactive inventory*
- appreciate that there are two main approaches by which SNF is reprocessed: *hydrometallurgical processes* and *electrometallurgical processes*
- be able to describe an example of each fuel cycle (i.e. U/Pu, DUPIC and Th/U) in detail and in terms of the *solvent extraction* (PUREX) and *molten salt* processes
- appreciate that, in addition to the long-established partitioning of plutonium and uranium, reprocessing and the separation of irradiated nuclear fuel offers the potential to extract combined streams of actinides
- understand that closed fuel cycles are of relevance to some Gen IV designs, the extraction of individual isotopes for specific purposes such as RTGs and some advanced waste management approaches

PROBLEMS

1. Describe the salient differences between *fresh* LWR oxide–based nuclear fuel and *SNF*, in terms of quantitative composition (mass) and radioactivity.
2. Summarise the organic–aqueous distribution behaviour of the relevant species of uranium, plutonium, neptunium and the nitrate ion.
3. Describe the design and function of mixer settlers and pulsed-column solvent extraction contactors. Why are the latter favoured over the former in most current industry-scale reprocessing systems?
4. Describe the effects of the interim period of storage and of off-gassing after dissolution on the isotope composition/radiotoxicity of SNF prior to reprocessing.
5. Summarise the advantages and disadvantages of hydrometallurgical methods of reprocessing and pyroprocessing methods.

REFERENCES

[1] B.B. Cunningham, L.B. Werner, The first isolation of plutonium, J. Am. Chem. Soc. 71 (5) (1949) 1521–1528.

[2] See for example:IAEA-TECDOC-1587, Spent fuel reprocessing options, 2008. www-pub.iaea.org/MTCD/publications/PDF/te_1587_web.pdf and http://www.radioactivity.eu.com/site/pages/Spent_Fuel_Composition.htm.

[3] See for example:E.R. Irish, W.H. Reas, The Purex Process—A Solvent Extraction Reprocessing Method for Irradiated Uranium, https://www.ipen.br/biblioteca/rel/R9904.pdf, 1957. HW-49483A. S.M. Stoller, R.B. Richards, Reactor Handbook: Fuel Reprocessing, vol. II, Interscience, 1961.

[4] E.R. Irish, W.H. Reas, The Purex Process—A Solvent Extraction Reprocessing Method for Irradiated Uranium, https://www.ipen.br/biblioteca/rel/R9904.pdf, 1957. HW-49483A.

[5] G.A. Burney, R.M. Harbour, Radiochemistry of Neptunium, United States Atomic Energy Commission, Oak Ridge, 1974.

[6] G. Petrich, Z. Kolarik, The 1981 PUREX Distribution Data Index, https://publikationen.bibliothek.kit.edu/270015615, 1981. KfK 3080 January.

FURTHER READING

[1] OECD, Spent Nuclear Fuel Reprocessing Flowsheet, NEA/NSC/WPFC/DOC(2012)15, https://www.oecd-nea.org/science/docs/, 2012.

[2] J.J. Laidler, J.E. Battles, W.E. Miller, J.P. Ackerman, E.L. Carls, Development of pyroprocessing technology, Prog. Nucl. Energ. 31 (1/2) (1997) 131–140.

CHAPTER

NUCLEAR SAFETY AND REGULATION

14

14.1 SUMMARY OF THE CHAPTER AND LEARNING OBJECTIVES

Safety is the most important aspect of nuclear engineering and nuclear power generation. It is testament to this philosophy that safety practices first developed and adopted for nuclear energy applications have influenced safety management and regulation in neighbouring sectors, such as oil and gas, chemicals and commercial airline operation, and vice versa. In this chapter, the focus is on the context of nuclear safety particularly with regard to the *hazard* associated with radiation emitted by radioactive material and the *risk*, that is, the likelihood of harm accruing from this hazard. The concepts of defence in-depth, the precautionary principle, radiation dose, radiotoxicity and a summary of the corresponding nuclear accident terminology is provided.

The objectives of this chapter are to:

- introduce the background on nuclear safety
- introduce the concept of the *no-linear threshold* and the *precautionary principle*
- provide a fundamental introduction to the concept of *radiation dose*
- introduce the concept of *radiotoxicity*
- review the *International Nuclear Event Scale*
- discuss the concept of *defence in-depth*
- review different event types such as *loss-of-coolant* and *loss-of-heat sink accidents*
- discuss a specific example of a nuclear accident: *the Windscale fire*
- provide a description of different regulatory approaches

14.2 HISTORICAL CONTEXT: LOUIS HAROLD GRAY 1905–65

Harold Gray's major contributions to the field of nuclear energy were to identify that different radiations deposit energy differently in biological tissue. He developed the concept of *radiobiological effectiveness* (RBE) and identified that the effect of radiation on a substance ought to be quantified in terms of the energy deposited per unit mass rather than the amount of ionisation produced in a given material in a given state. This enabled the effect of different radiations on living tissue to be quantified and hence their effects compared. This now forms the basis for the assessment of radiation exposure in terms of dose worldwide (Fig. 14.1).

Nuclear Engineering. https://doi.org/10.1016/B978-0-08-100962-8.00014-7

FIG. 14.1

Louis Harold Gray.

Gray studied physics and then specialised in what was, at the time, the emerging field of radiobiology. He explored the radiobiological basis of the radiotherapy of cancer, the oxygen effect and the relative biological effectiveness of neutrons. In doing so, he defined the unit of absorbed dose that is named after him: the Gray.

14.3 INTRODUCTION

In common with many landmark achievements of human civilisation such as fire, locomotion, modern medicine and electrification, the risks of nuclear technology were not fully appreciated when it was first discovered and exploited. A significant proportion of what has been learned in the 70 or so years since, has been derived from operating experience, research and from the small number of severe accidents that have occurred during that time (some of which are reviewed later).

Nuclear safety is distinct from what might be termed *conventional* safety, because it concerns the hazard associated with radiation emitted from radioactive materials. However, it is the conventional hazards that tend to be responsible for the majority of day-to-day accidents, loss of time at work by employees and, regrettably, in some extreme cases, fatalities. This state of affairs can be at odds with some preconceptions.

Records and the statistics associated with accidents resulting in an impact to the employee and/or the employer (e.g. through injury, loss of time at work, loss of production, litigation and cost etc.) demonstrate that the nuclear sector is among the safest of the energy-related industrial sectors by a significant margin. Consider, for example, the data provided in Table 14.1. This is derived from a related study by the Paul Scherrer Institute in Switzerland included in a comparison of energy-related accident risks by the

Table 14.1 Summary of Severe Accidents (≥5 Fatalities) That Occurred in Various Energy Sectors (the Chain of Activities Relating to a Given Energy Source), 1969–2000 [1]

	OECD			Non-OECD		
Sector	**Accidents**	**Fatalities**	**Fatalities/ GWey**	**Accidents**	**Fatalities**	**Fatalities/ GWey**
Coal	75	2259	0.157	1044	18,017	0.597
Coal (China 1994–99)				819	11,334	6.169
Coal (without China)				102	4831	0.597
Oil	165	3713	0.132	232	16,505	0.897
Natural gas	90	1043	0.085	45	1000	0.111
LPG	59	1905	1.957	46	2016	14.896
Hydroelectric	1	14	0.003	10	29,924	10.285
Nuclear	0	0	-	1	31[a]	0.048
Total	390	8934		1480	72,324	

[a]*Immediate fatalities associated with the Chernobyl accident only.*

Nuclear Energy Agency of the Organisation for Economic Co-operation and Development (OECD).[1] It demonstrates that for the period 1969–2000, the number of severe accidents, that is, those with at least 5 fatalities, was most significant for coal followed by oil, liquid petroleum gas and natural gas. While the coal chain of production had more than a thousand severe accidents, nuclear had one—the Chernobyl accident. That is not to say that Chernobyl was not a major catastrophe; it most certainly was. However, it is clear that a comparison of the safety performance across the energy sector has to be made if our understanding about the relative impact of one or the other of these activities is to be informed.

Nuclear *is* different predominantly because while the likelihood of a severe accident is very low, the impact can be very large, particularly in terms of social upheaval and economic cost. Furthermore, latent effects (those effects that follow some years after an accident as opposed to those that are prompt at the time of the accident) tend to dominate the distribution of effects forecast in nuclear incidents. This is particularly important, given the number of prompt fatalities is small relative to those of other energy activities, as per Table 14.1. Often, the provenance of latent effects (these are usually associated with the incidence of cancer) cannot be discerned from the prevalence of such effects that arise naturally in the population. Consequently, nuclear accident investigations are often reliant on extensive calculations with which to forecast the potential incidence of effects that might arise in the future. These are often subject to significant uncertainty and their results can be difficult to validate against real evidence; the populations involved are usually large, the timescales are long and the incidence is usually small.

Nuclear is not alone in its association with latent effects; fine particulate air pollution from fossil fuel burning for base load electricity production is suspected to yield significantly greater rates of chronic harm and fatality as a result of latent respiratory illnesses than those forecast from the latent effects of the most severe nuclear accidents. However, such evidence can be in contrast to widely held

[1]CDF is defined as an estimate of the frequency of a hypothetical accident resulting in a significant breach of fuel cladding derived on an analytical basis via probabilistic safety assessment.

views; only relatively recently, a number of countries have taken strategic decisions to scale back their nuclear electricity generation programs, despite the implications of climate change, largely as a result of safety fears raised by the Fukushima Daiichi accident. By contrast, continuous improvement in terms of systems design and operation, quantified on the basis of, for example, estimates of *theoretical core damage frequency*[1] (CDF) and theoretical *large release frequency*[2] (LRF), implies that the world's nuclear reactors are actually safer than they have ever been.

There are several contributing factors that cause an ostensibly 'safe' human activity to be regarded with a mixture of both founded but unfounded derision. First, nuclear energy was among the first human invention to combine the possibility of accidents that are of a very low likelihood, but nonetheless that have the prospect of very significant and far-reaching impact; it is perhaps most symbolic of the utter reliance of developed countries on advanced technology and, principally, electricity. For example, given a fleet of 450 power reactors worldwide and an operating experience exceeding 2000 person-years, it is difficult to list more than 10 accidents that might apportion consideration in terms of the International Nuclear Event Scale (INES): Chernobyl, Fukushima, Kyshtym, Three Mile Island (TMI), Tokai Mura and Windscale, as discussed later.

While Chernobyl demonstrated to the world that a serious nuclear accident can transgress international boundaries and its effects be felt the world over, lead to the displacement of large numbers of people and cause social and economic upheaval, it is also an example of a reactor design that would not have had licenced approval elsewhere and one that was driven recklessly to its limits without due regard for statutory regulation. The data referred to above are just one example that indicate such events are extremely unlikely occurrences, especially when compared with the production chains of all other fossil-fuel sources of baseload electricity.

Nuclear energy also deals almost exclusively with a hazard that is not as familiar to most of us as the more tangible hazards are that we encounter regularly—that of radiation. Although we might perceive day-to-day risks to be under our control (even though they may not be), radiation is invisible and anthropogenic 'nuclear' sources of radiation can be irrevocably intertwined with the arms race and military secrecy in the public conscious; nuclear has a legacy that does not engender trust easily.

Radiation in large quantities is certainly dangerous: the consequences of a significant exposure are well-known and its effects easily recognised. However, the instances of such hazards resulting in actual harm have a vanishingly small likelihood; the use of such sources (given that the use of radiation presents a hazard which extends beyond the nuclear industry) is regulated to ensure that this is the case, literally at all reasonable costs. By contrast, latent effects of low-level radiation exposure, in the very unlikely event that they arise, are long-term and, as mentioned earlier, often difficult to discern from much greater natural prevalence of cancer in the population. With this in mind, this chapter begins with a review of radiation in the context of nuclear safety.

14.4 RADIATION CONTEXT

14.4.1 THE LINEAR NO-THRESHOLD MODEL AND THE PRECAUTIONARY PRINCIPLE

Acute health effects that arise from high levels of exposure to radiation result from significant levels of cell death and exhibit a threshold below which they do not occur. Such effects include: nausea and vomiting, diarrhoea, headaches, fever, widespread damage to organs and, at very significant doses,

[2]LRF refers to an accident more severe than core damage alone that leads to a significant unmitigated release of radioactivity to the environment.

death. Wherever such effects arise within 24 hours of exposure, they are referred to collectively as *acute radiation syndrome* (ARS or radiation poisoning or radiation sickness). These effects are deemed *somatic* because they are manifest in the body of the individual who has been exposed in contrast to *genetic* effects that are defined as those that arise in the exposed individual's progeny. They are also deemed *early* because they arise in relatively close correlation with the time of exposure; late or *latent* effects arise *stochastically,* that is, without certainty, at a time after the exposure. The period of time between exposure and the manifestation of the effect can be of the order of decades.

Acute effects are clearly very serious and are avoided by a zero tolerance to the exposure of individuals to anything that might even approach such levels. This is ensured by systems that combine methods of elimination, isolation, control, protection and risk assessment: an approach known as the hierarchy of protection controls. Environments where a tangible risk of such exposure exists are usually heavily shielded in hot cells and glove boxes. Such facilities are protected by interlocks and essential procedures are carried out remotely with manipulators and robots. Such are the precautions taken, that accidents involving high doses of radiation are extremely rare.

However, it is conceivable (albeit extremely unlikely) that a latent effect such as the initiation of cancer *might* arise from the effect of a *single* quantum of radiation on a *single* cell. By definition, this is commensurate with radiation exposure at *very low dose levels.* At very low levels, radiation exposure from anthropogenic sources can be indistinguishable from the exposure to background radiation; and exposure to background is unavoidable. Furthermore, such considerations often do not include biological repair processes that are thought to be stimulated by low-level exposure, such as the production of enzymes that precipitate in the repair of damaged cells rather than necrosis or cellular lesions.

In contrast with acute effects, where an excess beyond a certain threshold is necessary to precipitate an effect, it is not known whether a threshold exists below which latent effects do not occur. Whilst it is known that the impact of radiation on health may vary from person to person, not all the factors that influence this sensitivity are known or understood. The study of the effects of low-level exposure is very difficult because latent effects arise, by definition, long after the exposure during which time the sources of other effects need to be taken into account. These might include: smoking, migration, socioeconomic influences and variation with geography (especially with regard to the effects of natural radon, for example). Interestingly, the question as to whether there is a threshold has been subject to a detailed study based on the inhalation of radon decay products [2], which provides a comprehensive example of the complexity associated with proving the incidence latent effects on the basis of a linear threshold, or otherwise.

In the absence of candid evidence of a threshold, because scientific understanding of this issue is currently incomplete, the *linear no-threshold theory* is adopted as the basis for low-level radiation protection. This theory is the logical extension of the one-quantum, one-cell hypothesis outlined above, in that as the number of potential cancer-initiating events increases, that is, dose, so does the probability of a latent effect occurring. To manage the risk associated with low levels of radiation exposure, the *precautionary principle* (or precautionary approach) is adopted. This principle implies that, in the absence of proof otherwise and where something (in this case low-level radiation) is suspected of causing harm, it is necessary to protect people from exposure to it.

As no lower limit below which a safe degree of radiation exposure has been proven, the precautionary principle is enacted by ensuring exposures to radiation are *as low as reasonably practicable* (ALARP) or *as low as reasonably achievable* (ALARA). Therefore, actions and policies are scrutinised at the development stage as to whether a practicable alternative exists by which the risk of exposure can be reduced further, provided the alternative is not grossly disproportionate in terms of the cost or effort required. The ALARP principle has also been adopted in risk management in nonradiation applications such as the petrochemical and rail safety sectors.

14.4.2 RADIATION DOSE

As with many scientific discoveries, our understanding of radiation and radioactivity developed gradually from the turn of the 20th Century through to the early 1930s, culminating in the discovery of the neutron. Consequently, a number of measurement units have been used to describe *radiation intensity*, *exposure* or *dose* and the reader may still encounter this variety especially in old texts and associated with vintage instrumentation. Some units of dose measurement are obsolete whereas others are used less widely than others, because they originated from measurement techniques that have been superseded by better approaches. The most prominent is perhaps the measurement of the amount of ionisation a given level of radiation exposure stimulates in a given medium; at the time this was a relatively easy property of radiation to measure via its effect on pieces of gold leaf or crystal fibres. We shall focus on the corresponding unit under the International System of Units (SI) for absorbed dose D, which is the *Gray* (Gy). This unit is defined as the amount of energy deposited by radiation per unit mass,

$$D = \frac{\text{Energy}}{\text{Mass}} \tag{14.1}$$

where $1\ \text{Gy} = 1\ \text{J kg}^{-1}$.

Given the discussion earlier in Chapter 1 concerning energy at the nuclear scale, it is clear that Gray is an enormous unit relative to the levels of energy likely to be imparted to matter by individual quanta of nuclear radiation. Consequently, occupational dose limits for example are often quoted in the realms of millionths (μGy) and thousandths (mGy) of a fraction of a Gy.

However, the use of Gray is limited because it does not include the specific interaction properties on matter of one radiation relative to another; this is referred to as the *relative biological effectiveness* (RBE) of different radiation types. Louis Gray observed shortly after the neutron was discovered that living cells exposed to the same number of neutrons and γ rays were killed at different rates. This highlighted that the effect of radiation on living cells was dependent on the type of radiation and that different radiations brought about different characteristic effects. He observed that neutrons were more effective at killing cells than γ rays, and we now know that similar contrasts are observed throughout the variety of radiation types, that is, photons, electrons, protons and so forth. Hence, the *absorbed dose* is defined as per Eq. (14.1) to be the radiation dose associated with the energy deposited per unit mass without reference to the type of radiation involved. This has units of Gray. By definition, this does not reflect the relative biological effectiveness that one type of radiation has over another.

As most environments in which radiation dose justifies measurement might comprise proportions of the range of possible radiation types, this variety is usually accounted for in its measurement; whereas the absorbed dose measured in each scenario might be the same, the effect on tissue may not be equivalent because the composition of the radiation environment in each case is generally not the same. Hence, the *radiation weighting factor* [3] w_R is introduced. This is an estimate of the biological effectiveness of a given type of radiation relative to that with a low degree of effectiveness, such as photons. A range of factors has been derived for each type of radiation as a result of radiobiological studies of their effects on living tissue. This is combined with the absorbed dose to bring an equivalence of dose estimation for all radiation types. The resulting quantity H is thus called the *equivalent dose*, represented as follows:

$$H = D w_R \tag{14.2}$$

for which the unit is the *Sievert*. Note that radiation weighting factors are dimensionless. An example for selection of radiation weighting factors is given in Table 14.2. It is noteworthy that, as the current state of knowledge of the effect of radiation on living tissue develops with research, these factors can be subject to change from time to time.

For neutrons, the radiation weighting factor is a more complicated, non-linear function of energy as shown in Fig. 14.2. This is because, for example, an interaction that results in the recoil of a nucleus with significant energy can result in significant ionisation in a small volume of tissue. In contrast, an

Table 14.2 Radiation Weighting Factors

Radiation Type	Radiation Weighting Factor, w_R
Photons	1
Electrons and muons	1
Protons	2
α particles, fission fragments, heavy ions	20
Neutrons	A continuous dependence with energy as included below.

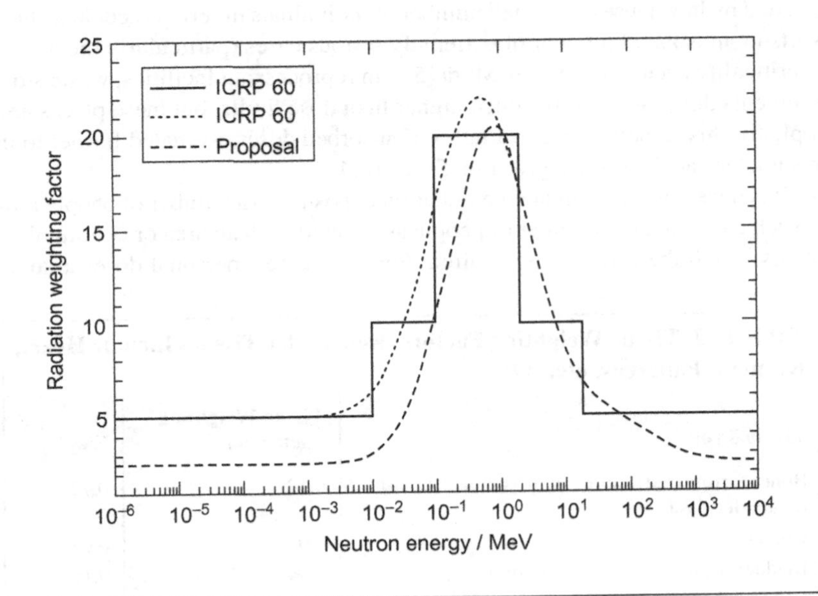

FIG. 14.2

Radiation weighting factor for neutrons as a function of energy.

Reproduced with permission from The 2007 recommendations of the International Commission on Radiological Protection, ICRP Publication 103, Ann. ICRP 37 (2–4) (2007).

interaction in which a neutron retains much of its energy might result in significantly less ionisation, over a greater volume. As the process is dependent on the interaction probabilities of specific isotopes, and because living tissue comprises a complex mixture of different elements and molecular structures, it can be important that the energy spectrum of the neutron field is also understood.

A further complication but nonetheless a critical aspect of the measurement of radiation dose—radiation dosimetry—is the contrasting sensitivity of different living tissues to radiation. While the majority of all living tissue is water, important distinctions arise from the variety in composition of living tissues that influences their response to radiation. For example, tissues (e.g., brain, bone, blood, liver, etc.) might respond differently due to the relative difference in the perfusion of blood, variation in Z in a given tissue (consider e.g. the abundance of calcium in bone) and the proportion of water (which might for example exacerbate the production of free radicals that are associated with damage to living cells).

To accommodate this, the *tissue weighting factor* w_T is used. A list of tissue weighting factors is given in Table 14.3. Combining the tissue weighting factors with the formulism for equivalent dose, we arrive at the *effective dose*. It is noted that, as the various weighting factors are dimensionless, the strict SI definition of a given dose equivalent remains the same; in this case, the unit for the dose is the same as for the absorbed dose; the Sievert, and defined thus,

$$H = D w_R w_T \tag{14.3}$$

As noted above, for the case of the Gray, in relative terms the Sievert is an enormous unit. Fortunately, exposures of the order of a Sievert are very rare and measurements made in the day-to-day management of the occupational rates of the rate of exposure to radiation would normally be of the order of μSv/h or less. Rates of mSv/h would normally be a cause for concern, particularly in terms of dose limits that cannot be exceeded by law. Doses to a small number of individuals in serious accidents have been of the order of Sieverts in an isolated number of extremely serious cases (particularly the accidents at Chernobyl and the criticality accident at Tokai-Mura [5]). In reprocessing facilities, waste storage silos and reactor environments dose rates can be much higher than dose limits, but these places are off limits to access by people for this reason. Some examples of absorbed doses estimated further to investigations of four serious nuclear accidents are given in Table 14.4.

Sometimes it is necessary to estimate the radiation exposure to a number of people rather than to an individual. In such cases, a large number of people associated with an area or community might be the population across which the estimate is required. For such cases, personal doses associated with the

Table 14.3 Tissue Weighting Factors (Remainder Tissues Include Heart, Kidneys, Pancreas, etc.) [4]

Tissue Type	Tissue Weighting Factor, w_T	Σw_T
Bone marrow (red), colon, lung, stomach, breast, remainder tissues.	0.12	0.72
Gonads	0.08	0.08
Bladder, oesophagus, liver, thyroid	0.04	0.16
Bone surface, brain, salivary gland, skin	0.01	0.04
Total	–	1.00

Table 14.4 Estimated Doses From Four Serious Nuclear Accidents

Incident	Dose Level	Notes
Chernobyl	0.8–16 Gy	Estimated to have been received by 134 workers.[a]
Fukushima	~10 mSv	Average effective dose to 99.3% workers.[b]
Tokai-mura	1.2–20 Gy	Range of exposure to three workers exposed.[c]
Windscale	0.01–0.1 Gy	Maximum estimated dose to the thyroids of local individuals[c]

[a]*United Nations Scientific Committee on the Effects of Atomic Radiation, www.unscear.org.*
[b]*UNSCEAR: The Fukushima accident, http://www.unscear.org/docs/revV1406112_Factsheet_E_ENG.pdf.*
[c]*Sources and effects of ionising radiation: Annex C Radiation exposures in accidents, UNSCEAR 2008, http://www.unscear.org/docs/reports/2008/11-80076_Report_2008_Annex_C.pdf.*

effective dose of a specific type of radiation to a particular part of the body are not very useful, especially when the details of localised exposures are poorly understood and the exposure might be across the whole population of a significant number of people. For these cases, the concept of *collective dose* is used. This is a summation estimated across the population in question, and has the units of man-Sv or person-Sv. Such estimates can be complex and related accident investigations by which they are derived often comprise several independent studies.

14.4.3 RADIOTOXICITY

An important concept in the context of nuclear safety is that of *radiotoxicity*. This is the term used to describe the hazard associated with a particular radioactive isotope and combines the physical (radiation), biological and chemical influences of its effects on living tissue. Radiotoxicity is an important concept because it enables the hazard associated with a specific isotope to be understood, and for these materials to be managed with this in mind, because some isotopes justify more concern than others in terms of the relative hazard potential that they present. While Gray's discovery highlighted that different radiation types affect living tissue differently, this is just one component of a number of properties that influence the radiotoxicity of a specific radioactive isotope.

By analogy with chemical toxins, we might draw a loose comparison between the concept of *concentration* and *radiotoxicity*. For example, a more concentrated toxin usually poses a greater risk than the dilute alternative. The distinction of radiotoxicity, however, is that several further dimensions need to be considered, including the effect of the radiation a radioisotope emits and also the chemical behaviour of the substance in the body. The latter influences how readily it is taken up and excreted. In addition, because radioactive substances decay, the hazard they pose is dynamic and dependent on the isotope involved. This is parameterised principally by the corresponding half-life.

14.4.3.1 Half-life and specific activity

To start with, let us begin with the half-life, $t_{1/2}$, as discussed in mathematical detail in Chapter 3; we recall half-life is *the time taken for a quantity of radioactive substance to decay by one-half*. An extremely large range of half-lives are present in the natural world, from the uranium isotopes that remain in the Earth's crust today many millennia after they were formed at the birth of the universe, through to those isotopes that last for a micro-second after they are formed in preparations made in the laboratory.

Very short-lived isotopes, with $t_{1/2} < 1$ min, are rarely a concern in terms of radiotoxicity as they are usually present in quantities too small to pose a significant hazard and they do not usually exist for long enough to be of concern.

However, an important caveat in this regard is that where such a short-lived isotope arises as the daughter product of a longer-lived isotope. The examples of ^{90}Y and ^{90}Sr, respectively, are explored in Case Study 2. In these examples, neither is sufficiently long-lived to arise naturally in any significant quantity. For the general case where the parent isotope is long-lived, there has been ample time for it to become dispersed widely throughout the Earth's crust. In this dilute form, isotopes related in this way do not usually present a significant risk; *long half-life* implies *low specific activity* and hence a *low rate of radiation emission*.

The exceptions to such cases are those where significant quantities have been made by man; anthropogenic sources such as those isotopes that comprise long-lived radioactive wastes or, to a lesser extent but still a concern, where man's activities have served to concentrate the products of such isotopes, such as the yield of *naturally occurring radioactive materials* (NORM) and technologically enhanced NORM from mineral recovery activities. Isotopes with intermediate half-lives such as ^{3}H (12.3 years), ^{137}Cs (30.2 years) and ^{60}Co (5.3 years) can be a concern, because they are almost entirely anthropogenic and thus can reside in sufficient quantities (having not had sufficient time to decay away since formed in the nuclear fuel cycle) to constitute a hazard that needs to be respected. Conversely, the half-lives of isotopes in this domain are not so long-lived to assume that their specific activity is sufficiently low not to warrant attention.

14.4.3.2 Radiation type

The type of radiation emitted by a given isotope (and the way it interacts with matter) is also reflected by its radiotoxicity. Gray observed more effective cell killing with neutrons than γ rays because neutrons generally[3] deposit their energy within a smaller volume of tissue than γ rays. Thus, the ionisation that arises from the interaction of a neutron is concentrated in a smaller volume of cells than for the corresponding γ ray. Consequently, the neutron can be more likely to compromise the ability of the cells to function, resulting in greater levels of biological damage. The interaction of radiation in matter in the context of dosimetry is often described by the average amount of energy that a specific type of radiation transfers per unit distance. This is known as the *linear energy transfer* (LET) and is closely related to the stopping power, dE/dx. Different types of radiation lose their energy differently and exhibit different LET; generally, the greater the LET, the greater the RBE of a given radiation and hence the more effective it is at killing cells, although radiobiological processes are complicated and by no means fully understood. A number of examples of LET for a variety of radiations is given in Table 14.5.

Clearly, an isotope that emits a type of radiation that has high LET is likely to have greater radiotoxicity than an isotope that emits a low-LET variant, notwithstanding the influence of processes and residence time in the body.

14.4.3.3 The body and the phase of the activity

Finally, there are two additional factors that influence our assessment of the radiotoxicity of a given isotope: the way in which the isotope behaves in the body and the phase in which the material is present. The first of these factors arises because some isotopes are sequestered by particular types of living

[3]Not forgetting that the deposition of energy by neutrons is dependent on the energy of the neutron.

Table 14.5 LET for a Variety of Contrasting Types of Radiation

Radiation Type	Type of Linear Energy Transfer	Typical Values [6]/keV μm^{-1}
Heavy ions, ^{12}C, ^{56}Fe, etc.	High	1000
α	High	166
Neutron	Intermediate	–
Proton	Intermediate	0.5→4.7
β^-	Low	–
β^+	Low	–
e^-	Low	–
X-ray photons	Low	2.0
γ-ray photons	Low	0.2

tissue and reside there longer than others that are not affected in the same way, due to differences in the way the associated element is processed by functions in the body. The element may not be identical to the one that the body needs but might, for example, be in the same group of the periodic table and therefore have similar chemical properties. Thus, by way of example, strontium is sequestered by bone as per calcium and radioactive iodine follows the iodine pathway to peptide-based hormones produced by the thyroid that contain iodine, such as tetraiodothyronine and triiodothyronine.

The second factor concerns the ease with which the body is able to acquire the isotope: gases and powders are usually acquired most easily and therefore can present the highest radiotoxicological hazard, followed by liquids with solids being the least significant. Powders and gases are generally more easily ingested via the stomach and respired via the lungs, particularly if they are dispersed or volatised, respectively, in the absence of appropriate control measures. These routes provide pathways by which vulnerable tissues in the body might be exposed to high-LET radiations such as α particles and β^- emitters that would not otherwise present a significant external exposure risk, especially if personal protective equipment (e.g. gloves, coveralls, etc.) are used. This is the principal reason why radioactive materials are classified as either 'open' (as in the case open quantities of gases, powders and liquids) or 'sealed'. Setting aside the radiotoxicity of the specific isotope in question, open sources of radioactive material constitute the more significant hazard of the two, as a greater number of pathways to expose living tissues in the body are open to them than is the case for sealed sources.

14.5 NUCLEAR ACCIDENT CLASSIFICATION AND TERMINOLOGY

14.5.1 THE INTERNATIONAL NUCLEAR EVENT SCALE (INES)

The INES provides a basis by which nuclear incidents and accidents can be compared with one another (see Table 14.6). It ranks event types in terms of off-site impact, on-site impact and the impact on defence in depth. It is referred to subsequently in this chapter.

14.5.2 ACCIDENT TYPES

The vulnerability of a reactor design (or indeed any nuclear plant for which a license is required to operate) is usually considered both *deterministically* and *probabilistically*. For the former, feasible accident scenarios are considered, assuming the reactor and associated emergency systems respond as

Table 14.6 The International Nuclear Event Scale (INES) [7]

Scale Level		Area of Impact		
		Off-site Impact	On-site Impact	Impact on Defence In-depth
Accident	7—Major incident	Major release—widespread health and environmental effects.		
	6—Serious accident	Significant release: likely to require full implementation of planned countermeasures.		
	5—Accident with off-site risk	Limited release: likely to require partial implementation of planned countermeasures.	Severe damage to reactor core/radiological barriers.	
	4—Accident without significant off-site risk	Minor release: Public exposure of the order of prescribed limits.	Significant damage to reactor core/radiological barriers/fatal exposure of a worker.	
Incident	3—Serious incident	Very small release: public exposure at a fraction of prescribed limits.	Severe spread of contamination/acute health effects to a worker.	Near accident—no safety layers remaining.
	2—Incident		Significant spread of contamination/ overexposure of a worker.	Incidents with significant failures in safety provisions.
	1—Anomaly			Anomaly beyond the authorised operating regime.
	0—Deviation	No safety significance		

designed, that is, they function as expected. The system is then scrutinised on the basis that the response of these systems is deemed sufficient to avert the possibility of core damage and radioactivity release, given the premise that the avoidance of such a release is the overarching priority. Such a process is referred to as *deterministic safety analysis* (DSA).

Often in parallel, the probability of the failure of the reactor system that results in a release to the environment or damage to the core is assessed (on the basis that the likelihood of the failure of each sub-system is estimated). The total probability is then quantified and considered against what is deemed tolerable. This is often quantified in terms of *failure-per-unit-time* and this approach is widely referred to as *probabilistic safety analysis* (PSA). As might be expected, the probability of catastrophic consequences such as core damage or a breach of containment are required to conform to extremely low values, cf. the core damage frequency for the AP1000 PWR design [8] of 2.13×10^{-7} per year.

It is useful to classify the range of incidents [9] that might occur, from the less significant through to the severe, because good practice requires that incident scenarios with a higher probability must correspond to those with the likelihood of causing the least harm, and vice versa. In order of significance, a hierarchy of incident scenarios is often considered as follows:

1. A nuclear plant must be designed to withstand conditions that are expected to arise frequently during *normal operation without shutting down.*
2. A nuclear plant design must mitigate the possibility of less frequent but nonetheless anticipated incidents to ensure that they *do not escalate to cause fuel damage.*

The two classes of events described above (1 and 2) correspond to anticipated changes in plant operation, and are thus defined as *transients.*

3. The class of events that are anticipated very infrequently, that is, once is several decades but do not result in *significant off-site effects.*
4. Those events that are not expected to occur at all and which could result in *off-site effects and injury.*

The latter two classifications (3 and 4) are clearly more significant in terms of potential impact, but much less probable. Hence, they are defined as *accidents.* Beneath these classifications lie the possibility of a wide range of specific, detailed accident scenarios beyond the scope of this text; the most widespread general examples are defined in the following.

14.5.2.1 Loss-of-coolant accidents

Loss-of-coolant accidents (LOCAs) are a type of reactor accident that has been studied extensively. They are the prominent scenario that protection systems and operational practice are designed to respond to. As the name suggests, a LOCA describes an event in which the coolant is lost from the reactor. The most dramatic scenario in which this might happen is known as a *large-break LOCA* (LBLOCA) in which a double-ended failure (often referred to as a *guillotine* rupture) of one of the main primary circuit coolant pipes serving the reactor pressure vessel occurs. LBLOCAs are usually specified for a flow aperture that is greater than $\sim 0.1\ m^2$.

In such an event, due to the pressure in the primary circuit, it is anticipated that the coolant would be lost—known as *coolant blow down*—very rapidly from the reactor vessel along with a very rapid rate of depressurisation. This would occur in a matter of seconds due to the pressure and the relatively large diameter of the hot-leg pipework in the primary circuit. In the absence of countermeasures, such an event would leave the core without a coolant and at risk of overheating with a likelihood of there being a great deal of vaporised coolant deposited in the containment. Most remedial actions designed to mitigate the effects of an LBLOCA involve rapid-response and sustained supplies of emergency sources of coolant to ensure the core remains covered in order to avert the possibility of core damage. These are usually designed to be sequenced according to the pressurisation state of the core during the blowdown.

Other LOCA scenarios are possible, and have also been studied. For example, where the rupture and thus the rate of coolant loss is much less significant than for the LBLOCA case, the associated class of accident scenarios is termed a 'small-break LOCA' (SBLOCA), corresponding on a technical basis to the flow area associated with the break being less than $\sim 0.1\ m^2$. Whereas small leaks with the potential to lead to SBLOCAs might be expected to occur more frequently than a major pipe break (the loss of coolant in the TMI accident discussed in Case Study 14.3b is a case in point), the implications of an SBLOCA can be serious, albeit the coolant loss occurs more gradually.

Due to the imperative of managing decay heat, some LOCA scenarios are also as applicable to a recently shutdown reactor as they are to an operating reactor. This can be an important feature of LOCA countermeasures because most LOCA scenarios involve an automatic shutdown of the reactor occurring in response to the core being over-pressure, over temperature or the detection of low coolant level. Hence, there is a significant likelihood that the reactor might be in the shutdown state by the time the LOCA is dealt with and it is the management of decay heat that is the priority. Other related definitions are used, particularly where the coolant loss does not arise from a pipe break *per se,* including small LOCA (SLOCA) and medium LOCA (MLOCA).

Most LOCA scenarios tend to have been studied and researched in terms of design for LWRs because light water is the coolant/moderator of choice for most new commercial power reactors and for the majority of existing, operable nuclear power plant. However, LOCA scenarios are feasible for all reactor types in which the fuel is vulnerable to overheating (all current power reactors in the world). Modern reactor designs often provide for additional means by which LOCA scenarios can be mitigated through, e.g., less pipework (affording a reduced probability of pipe break) and advanced countermeasures. The latter might include additional make-up tanks, high-pressure injection (HPI) systems, passive cooling systems based on natural circulation and isolation cooling systems that are designed to operate indefinitely; these were discussed in reference to specific reactor designs in Chapter 10.

14.5.2.2 Loss-of-heat sink accidents

A further categorisation of reactor accident is the *loss-of-heat sink accident* (LOHA) which differs from a LOCA because, although the coolant is not lost from the reactor core in a substantial sense, its role in heat removal is compromised. A LOHA might occur, e.g., as a result of loss of circulation as a result of coolant pump failure or loss of feedwater flow, the latter associated with the heat transfer function of the secondary circuit. In such a case, the coolant is still present in the system but the route by which heat is exhausted to the ambient is degraded or lost altogether. A subset of such scenarios are referred to as loss-of-flow accidents (LOFA) for the specific case where the loss of heat sink relates to flow being interrupted due to, e.g., repeated trips of the coolant pump system.

14.5.3 EMERGENCY CORE COOLING SYSTEMS

The term *emergency core cooling* is used to describe the collection of facilities or systems designed to ensure that the reactor core remains cooled in the event of an accident. The primary aim of these systems is to ensure that the reactor core does not become uncovered as a result of loss of coolant and that parts of the reactor system (particularly the containment) do not become pressurised to the extent that they might yield, allowing a pathway for radioactivity to reach the environment.

The function of emergency core cooling systems (ECCS) is often not as trivial as simply providing additional coolant to the reactor core. If the coolant loss is not immediate, the system pressure may remain high. In this scenario, it can be difficult to introduce emergency supplies of coolant unless it is introduced under high pressure by, for example, HPI systems. Conversely, if the coolant loss is large and rapid due a significant failure in the circuit, HPI systems may not provide the coolant in volumes sufficient to replace the lost coolant before core damage occurs; they usually draw the coolant from a tank of boronated water known as the *boronated water storage tank* (BWST). To cater to this scenario, a *low-pressure injection* (LPI) system is designed to supply water in large quantities to offset the loss.

In some cases, the LPI is the same system that serves as the decay heat removal system; under normal operational circumstances, this takes over the cooling operation when the temperature of the shut-down core passes below the temperature at which the steam generators are used to cool it, drawing coolant from a large BWST. To cater to the possibility that, in case of a large loss of coolant, there is a delay in the LPI or HPI making up the shortfall, greater resilience is often provided by having a further reservoir of coolant available stored in two large, *core flood* or *accumulator* tanks, usually located above the core in order to derive gravity assistance and passive operation. The water in these tanks is pressurised by a gas bubble, which acts to drive out the water into the core immediately that it is needed.

Reactor core isolation cooling systems (relevant in particular to the Fukushima Daiichi accident described in Case Study 14.3d) exploit a turbine driven by steam from the decay heat of the cooling core of a BWR to circulate the coolant through a condenser to sustain cooling, particularly in the event of a loss of power to coolant pumps or pump failure. A number of emergency coolant systems aim to operate for as long as possible by, e.g., recirculating coolant from a sump in the containment. In-containment sprays are configured not only to reduce pressure build-up in the event of coolant loss but also to retard the volatilisation and to enhance the condensation of the contaminated coolant.

The description of emergency core cooling systems given above tends to focus on those systems where core water state (i.e. gas or liquid) and loss are the issues to be mitigated associated with LWR systems. Other reactor variants draw on different systems as described in Chapter 10, where appropriate.

14.5.4 PROMINENT NUCLEAR INCIDENTS

The INES ranks nuclear accidents and incidents according to the relative order of magnitude across a range of criteria that are summarised in Table 14.6. Therefore, although a range of relatively minor anomalies and incidents are recorded on a continuous basis, there are five principal major accidents that have been ranked 5 on the INES above. Whilst not all of these examples are associated with commercially operated power plant, as per the focus of this text, they have all had an influence on nuclear safety practice to a greater or lesser degree. Chronologically, these are: Kyshtym (1957, in what was then the Union of Soviet Socialist Republics), Windscale (1957, United Kingdom), Three Mile Island (1979, United States), Chernobyl (1986, Ukraine) and Fukushima Daiichi (2011, Japan). Accidents resulting in casualties from exposures to radiation are listed[4] and widely available elsewhere, whereas the record of the *United Nations Scientific Committee on the Effects of Atomic Radiation* (UNSCEAR) provides a source of data and records associated with nuclear accidents.[5]

The five most-prominent nuclear accidents listed above have become, to a greater or lesser extent, infamous events in popular culture and have generated such extensive interest that there is now a bewildering amount of information associated with them, available via virtually all forms of media. This comprises that which has been refereed (e.g., there are more than 10,000 refereed scientific papers on the subject of Chernobyl alone) and also popular articles, TV documentaries, books, films, Internet coverage and journalism.

[4]http://www.johnstonsarchive.net/nuclear/radevents/radaccidents.html.
[5]See for example: http://www.unscear.org/docs/reports/2008/11-80076_Report_2008_Annex_C.pdf.

A comprehensive summary of this information is beyond the remit of this book especially as, by definition, there are texts dedicated to specific incidents and accounts of specific investigations. However, the legacy of these incidents that is relevant to this chapter is that they often illustrated shortfalls in how nuclear engineering systems have been designed and operated; it is important that this legacy is summarised. To accomplish this, a detailed discussion of the Windscale accident is provided by way of introduction. This has been selected as it is among the earliest recorded incidents, its legacy is well-known and it has been discussed widely both in the open literature. Then, each of the other major accidents is reviewed in terms of its causes, effects and legacy, as per case studies 14.3a through 14.3d.

14.5.4.1 An example of a nuclear accident: the Windscale fire

Context and technical background

A fire occurred at Pile No. 1 of the Windscale nuclear plant in the United Kingdom on 10–11 October 1957 following an attempt to anneal the graphite moderator. The fire spread to a number of fuel channels involving an estimated 10 t of uranium, and radioactivity was released primarily via the ventilation stack of the plant to the environment. The reactor was damaged beyond repair as a result and its twin, Pile No. 2, was closed shortly after the accident. The accident was rated 5 on the INES.

The Windscale piles were two nuclear reactors on the Sellafield nuclear site in Cumbria in the north west of England near its west coast. They were built shortly after the Second World War, primarily to produce plutonium for the UK atomic weapons programme. Power generation was not afforded by the plant design; the reactors comprised graphite piles in which metal uranium fuel, generally of natural enrichment, was placed horizontally in the form of short, aluminium-clad cartridges placed on graphite trays or *boats* as they were then known. The reactors were cooled by air forced through the piles at atmospheric pressure by motorised blowers. The waste heat was vented into the atmosphere via the pile stacks, each of 125 m in height with one for each of the reactors. Each of the reactor cores comprised a 2000-t graphite moderator comprising 3440 channels with each channel containing 21 fuel elements; thus, each core contained a total of more than 70,000 elements at full load. Control rods were also routed horizontally through the graphite with shutdown rods held vertically in electromagnetic clutches above the core with the facility to fall in under gravity in an emergency. The design power of each pile was 180 MWth and the maximum fuel temperature was 395°C.

During normal operations, each pile was brought to criticality and, after a short period of burn-up, fuel cartridges were pushed out of the back of the pile into skips in a water-filled duct leading to a storage pond. After a period of time necessary to allow the most prominent fission products to decay, the spent fuel cartridges were passed on to a neighbouring chemical plant for processing. New fuel was inserted into the front of the pile (the charge face) from a lift (the charge hoist). Filtered air from two sets of larger blower machines (one set on each side of the graphite core) entered the system via a duct at the front of the core (behind the charge face), was blown through the core and discharged at the back into the base of the chimney. The fan blowers had installed dampers to enable throughput to be adjusted with a separate set of shutdown blowers installed to remove decay heat. Each exhaust stack had a bank of filters fitted at the top; these were added to cater for the prospect of there being a radioactive release via the stacks.

Construction of the piles started in 1947 and they operated from 1951 through to the accident on Pile No. 1 in 1957. During this time, the UK nuclear programme developed from a focus on plutonium to the testing of new fuel designs and the production of specific isotopes. This reflected a rapidly expanding programme of activities driven by a significant increase in demand for plutonium, preparations for the subsequent generation power reactors (Magnox) and the development of the hydrogen bomb. Although the Windscale piles operated generally without any issue and at high levels of availability, a view expressed with hindsight is that the pressures of these demands exceeded the available staff resource, particularly among technical grades.

A specific technical issue prominent in understanding the sequence of developments that led to the fire at Windscale is that of *Wigner energy*. The Windscale piles were designed and built, like other reactors of the same era, based only the few years' experience in nuclear reactor technology that existed in the world at that time. This was complicated further by the limited flow of information across international borders due to military secrecy, particularly with regard to operational experience. Experience at Hanford in the United States had identified that graphite was susceptible to damage by the interaction of neutrons as a result of exposure in reactors, which had significant implications for their design and operation.

Known as the *Wigner effect*, after the Hungarian theoretical physicist Eugene Wigner who discovered it, this phenomenon arises due to the displacement of carbon atoms from their symmetrical positions in the graphite lattice by energetic neutrons (i.e. $E_n > 25$ eV). The displaced atoms occupy *interstitial* positions with energy absorbed from the associated neutron to provoke the displacement. The energy is stored as potential energy corresponding to the local instability of the position of the displacement relative to its original lattice position. The Wigner effect is manifest in two ways: (1) as a dimensional expansion of the graphite and (2) if left to go unchecked as an uncontrolled release of the stored energy in the form of heat.

The Windscale design team was alerted to the dimensional effect in time for the design of the graphite to be adapted. However, to accommodate a controlled release of the stored Wigner energy, operational procedures were improvised following a spontaneous release in 1952 to enable the temperature of the graphite to be raised to initiate a Wigner energy release; the associated increase in vibration of the atoms in the graphite lattice causes those atoms in interstitial positions to return to symmetrically located lattice sites. This is known as annealing.

Such anneals were done periodically at specified intervals on the basis of: burn-up (i.e. the product of the reactor power with the time at that power), research derived from samples of the graphite from the associated core and taking into account observations recorded from previous anneals. The temperature increase was initiated by *nuclear heating* from a shutdown state with the intent to establish a self-sustaining Wigner release that would then propagate throughout the core. Nuclear heating was initiated by withdrawing control rods slowly from the lower part of the pile to reach criticality at low power, with the main air blowers shut off and vents closed. A number of thermocouples were integrated into the graphite structure to assist the reactor operators in their control of the process on the basis of temperatures across the core.

The process described above had been accomplished 16 times previously during the operation of the Windscale piles and, although not routine in terms of formal written procedure (indeed a scant memorandum is all that remains as evidence of written instruction[1]), the team involved had extensive experience of the operation.

The chronology of the accident

On the occasion of the 9th anneal of Pile No. 1 at Windscale in Oct. 1957, a longer duration between Wigner releases had recently been adopted due to difficulties achieving a satisfactory anneal in previous attempts. It was also suspected that some residual Wigner build-up might remain from the previous anneal in some isolated parts of the core; thus, together with the longer interval potentially constituting a larger Wigner build-up than was the case on previous occasions.

The annealing operation began on Monday, the 7th of Oct. 1957 and at first progressed as planned with criticality achieved and thermocouple readings indicating that the release was progressing. When the maximum cartridge temperature of 250°C was observed; the core was shut down to allow the Wigner release to propagate through the graphite. However, by the following morning thermocouple readings caused the operators to suspect that the progression of the anneal might be subsiding, so the decision was taken to bring the core to criticality again. Nuclear heating continued until 5 p.m. of the next day when the core was again shut down.

The anneal progressed largely as expected, albeit with some temperature readings that were more erratic than usually encountered, until early on Thursday 10th Oct. when a very high reading of nearly 400°C was noticed on one thermocouple in particular. This specific reading remained high, despite attempts to reduce it by adjusting the airflow through the core. At this point, an increase in radioactivity was registered on the stack instrumentation of Pile 1, but this was confused at first with reports of excess readings from Pile 2. This was thought at the time to be due to a burst fuel cartridge in that reactor, but it transpired later that was actually due to an instrumentation fault.

Increasing airflow into the core by opening the fan dampers caused the excess temperature reading to fall slightly, but also coincided with significant radioactivity readings on site, for example, on the Pile 1 stack, on the roof of the site meteorological station and subsequently of air activity measurements made between 11 a.m. and 2 p.m. on the Thursday. This confirmed that something was indeed wrong with Pile 1, consistent with some cartridges having burst and so the shutdown fans were switched on to aid cooling of the core. Engineers put on protective equipment, dosimeters and respirators and went to the charge floor. When they removed the channel plug from the suspect channel in the charge face, at approximately 4.30 p.m., they could see fuel cartridges glowing red inside and the removal of the plugs surrounding it confirmed the same.

Looking down from the top of the pile, through an inspection port that provided a view of the space at the back of the reactor (of the void into which fuel cartridges were pushed when spent), at first a glow and then as the day progressed flames could be seen. Several attempts were made to arrest the progress of the fire. The first of these was to withdraw the affected cartridges. However, these were jammed due to being distorted by the fire. Subsequently, the cartridges around the fire were removed to effect a fire break and particularly to halt spread of the fire to unaffected channels above. This prevented the fire spreading, but did not lead to a reduction of temperature in the area of the fire.

At first, the use of water to extinguish the flames was rejected due to fears that this might produce an explosive atmosphere of hydrogen and oxygen. Instead, carbon dioxide was brought in tankers and injected into the core in an attempt to quench the fire (argon which would have been an alternative was not available in sufficient quantities). This had no effect, as the affected area was too hot at this stage. Subsequently, the decision to introduce water with the site fire engines was taken at approximately midnight due to fears of there being a second Wigner release as temperatures escalated further. Water was laid on at approximately 9 a.m. on Friday, the 11th of Oct. with the help of the fire brigade and fed into the discharged channels through scaffold poles. This had no effect until the shutdown blowers were switched off (these had been on to render the charge floor tolerable to the men who were trying to limit the spread of the fire by removing cartridges); when they were switched off, the fire subsided quickly. By midday on same day, the fire was out and 24 hours later, when the water was switched off, the core was cold.

Aftermath and legacy

There were two releases of radioactivity during the Windscale accident: The first occurred when the temperature escalation was first noted and the fan dampers were opened to attempt cooling of the core. The second occurred when water was introduced to extinguish the fire as steam was formed and carried radioactivity to the Pile 1 stack. Approximately 150 fuel channels were involved in the fire with a significant quantity of fuel being oxidised.

The most significant recorded release of radioactivity to the environment, that is, beyond the stack, on site and beyond this too, was at first associated with ^{131}I. Due to its volatility, iodine escaped the damaged fuel and the stack filters. Radiochemical analysis of milk from neighbouring farms to the site reflected the timescale of both the evolution of the fire and the half-life of ^{131}I.

It was appreciated rapidly that this posed a risk particularly to young children in the immediate vicinity of the Windscale piles due to their relatively high consumption of milk from local farms and the accumulation of radioactive iodine from the same in the thyroid gland. An assessment of the limit of exposure was made and the distribution of milk from farms immediately adjacent to the site was stopped on the evening of 12th October. Based on the analysis of the next 2 days, the ban was extended to a coastal zone extending 30 miles in length and 10 miles in breadth with the milk thrown away from the 15th of October. This constituted an area of 200 square miles with farmers compensated by the UK Atomic Energy Authority. The ban was lifted in stages and lasted approximately 6 weeks in total.

Over the months that followed, fuel unaffected by the fire was removed and sent for processing but some was stuck in its associated channels. The charge hoist was heavily contaminated, as a result of the efforts to push fuel out during the accident, with the radioactivity spread by the water used to extinguish the fire; contaminated water was collected and disposed of. Pile No. 1 was unserviceable and never restarted; Pile No. 2 was shut down a week later and never restarted because the adjustments deemed necessary in light of the accident were not practicable. Opinion remains divided as to the specific cause of the accident due to conflicting and inconclusive evidence: it could have been started by the oxidation of graphite, exacerbated by an increased, radiation-induced sensitivity to heating, or because a fuel cartridge failed allowing the contents to oxidise and form the seat of the fire. A debated perspective is that the second reheat followed the first too quickly and resulted in an oxidation of a fuel cartridge.

The Windscale fire was a serious accident. It resulted in the largest accidental release of radioactivity in the history of the UK nuclear industry. The prompt and decisive action of the people involved in controlling the fire and limiting the contamination averted what could have been a major catastrophe. It had several far-reaching effects:

- The significance of graphite behaviour under the extremes of radiation and temperature became of paramount interest, provoking an extensive research programme to improve the knowledge base in this area. This informed the design of the Calder Hall and Chapelcross power reactors as part of the Magnox generation that followed Windscale, particularly with regard to the provision of core instrumentation. It also stimulated correspondence about graphite issues with Belgium, France, Italy, Japan and the United States.
- It resulted in far-reaching changes and expansion of the organisation of the UKAEA and particularly with regard to safety. For example, it led to the formation of the *National Radiological Protection Board* (NRPB) to provide independent advice on radiological health and safety (latterly the *Radiation Protection Division* of *Public Health England*), albeit several years after the accident;

it consolidated a pre-existing view that nuclear health and safety should be independent from the industrial organisations. This culminated in the formation of the *Nuclear Installations Inspectorate* (NII) as the UK's independent nuclear regulator (latterly the *Office for Nuclear Regulation* (ONR)) and reinforced a culture in which safety is the overriding priority; it resulted in a significant contribution to the development of radiation protection criteria for nuclear accidents in the United Kingdom and wider, via the *International Commission on Radiological Protection* (ICRP).

- It removed two sources of radioisotope production. As the piles were near the end of their lives, their loss overall was less significant than might be assumed. However, the failed state and contamination in Pile No. 1 has complicated the ease with which it is decommissioned and has caused the cost of these operations to escalate significantly.
- Radioactive contamination released to the environment comprised ^{131}I (1800 TBq), but also ^{210}Po (42 TBq) and ^{137}Cs (180 TBq). An authoritative summary is provided by L. Arnold [10] with a more recent context available by R. Wakeford [11]. The earliest reference concerning the environmental monitoring programme following the incident is that of H. J. Dunster et al. [12] This considered the external, inhalation and ingestion risks (the latter culminating in the milk ban). Subsequent studies have explored pathways of ^{210}Po and the lung cancer risk; the health impact was revised when the threshold hypothesis was abandoned and the concept of collective dose was developed in the 1960s. In 1988, R. H. Clarke of the NRPB derived an upper bound for cancer deaths of 100 in the United Kingdom over the subsequent 40–50 years, but which is unlikely to be discernible from the overall cancer mortality rate. Several estimates of the atmospheric releases and movement following the accident have been made as summarised, for example, by J. A. Garland and R. Wakeford [13] in 2007 and N. Nelson et al. [14] in 2006. The mortality of those involved with the accident has been explored by D. Mc Geoghegan and K. Binks in 2000 [15] and in 2010 [16].

14.6 REGULATION AND NUCLEAR SAFETY PHILOSOPHIES

14.6.1 DEFENCE IN-DEPTH

The consensus in nuclear safety worldwide is that the strategic approach should be one of *defence in-depth* [17]. Conceptually, this pervades the safety culture and operations of many of the associated organisations. It describes the need, from a general perspective, for every point of perceived failure in a given system or process (which can be operational or administrative as well as relating to the physical or engineering integrity of the plant) to have a level of defence behind it; it is also referred to as the requirement for there to be *overlapping provisions* and for there to be *multiple levels of protection*. The philosophy is that, in the event that the first level of defence fails, the potential cascade of events that might lead to an incident is avoided by the preventative action of a second level of defence, and so on. The central objective of this approach is to ensure that radioactive materials remain isolated from people and the environment. Generically, with nuclear facilities the imperative is to ensure for example, there is *control, cooling* and *containment*. A related interpretation of the philosophy is in terms of five layers, defined as follows:

(i) Failures are *prevented* from occurring by ensuring design and operation are high quality.
(ii) Where failures do occur they are *detected* and *controlled* to prevent their consequences becoming significant.

(iii) If control were to be lost then the risk is *mitigated* through the provision of engineered safety features and accident response procedures to prevent core damage and radioactivity release. Hypothetical accident scenarios falling within the capabilities of such systems are referred to as *design basis accidents*.

(iv) The defence for accidents not catered for by engineered safety features (hence known as *beyond design basis accidents*) is the *containment*. This is designed to prevent a release of radioactivity to the environment.

(v) The final layer of defence in the event of a release is the *off-site emergency plan*, which acts to mitigate the radiological consequences via, for example, limiting the distribution of foodstuffs at risk of contamination, shelter, evacuation, etc.

Perhaps the most tangible example of defence-in-depth in practice is the engineering design of the reactor core with respect to the potential for a severe reactor accident and the release of radioactivity to the environment. Four barriers are usually[6] implemented from a generic perspective of a heterogeneous, thermal-spectrum reactor with low-enriched fuel, for example:

- The *fuel matrix* is designed to resist the dispersion of radioactivity, for example, to have a high melting point, be insoluble in water, etc.
- The *fuel cladding* prevents radioactivity from spent fuel pellets reaching the reactor coolant and other materials.
- The boundary of the *primary coolant system* that contains the reactor fuel (usually comprising the reactor vessel and associated primary circuit) prevents contaminated coolant from reaching the reactor containment (in the event that the cladding fails).
- The *reactor containment* that surrounds the vessel prevents contamination reaching the environment arising due to a failure in the vessel.

A further relevant aspect of the defence-in-depth philosophy is that it is adopted somewhat irrespective of how unlikely a perceived accident scenario might be. For example, a wholesale loss of primary coolant to the core is considered unlikely because, being the prominent design-basis accident that a reactor and its safety systems are designed to withstand, many counter measures are engineered such as HPI, LPI systems, etc. to prevent this occurring. However, in spite of the action of these systems, the four defences described above are put in place to ensure that contamination is isolated from the environment. A further example of the concept in action in this context is the provision of:

- scrutiny over the design of nuclear systems to *minimise the potential for system failures* (evident in the redundancy and integrity level required of such designs)
- additional systems that, in the event of a failure, *prevent the consequences of that failure escalating the situation to a major accident* (e.g., emergency injection sprays, containment cooling systems, etc.)
- design features to *mitigate the consequences of a major accident* (e.g., civil engineering features to limit the spread of radioactivity, core flood systems, etc.)

[6] Some first-generation reactor designs can be an exception to this norm.

14.6.2 REGULATORY APPROACHES

In common with all complex engineering systems that present a hazard with the potential to cause *danger*, *risk* or *injury*, nuclear reactor systems are subject to *regulation*. Regulation—the independent oversight of reactor design, construction and operation to ensure safety remains the priority—is necessary for a variety of reasons; put simply, its actions prevent accidents, administer licencing and ensure public confidence. Regulatory bodies often pursue independent research into vulnerabilities of designs, systems and new operational approaches, etc., which have not been appreciated previously. They also perform independent reviews of accidents and near misses to promote learning. Importantly, regulatory practice also provides quantitative performance assessments of various plants and operators, promoting self-improvement and public accountability. Recognising that nuclear power plants are operated to generate revenue for the commercial owner or utility operating them, *independent* regulation removes the potential (albeit unlikely) for there to be conflicts of interest between economic agenda and safety.

Nuclear reactor operation is different in a number of aspects with regard to other, highly regulated industries that rely on complex engineering assets. For example, commercial aviation and the automotive sector are both highly regulated, but have the benefit of knowledge derived from extensive performance data on the operation of their engineering systems. This enables issues to be detected more quickly and hence resolved more easily than might be the case of a specific nuclear power plant design; by contrast, less than 500 power plants are operable from a variety of generations and of a range of designs constituting relatively small statistical groups. Similarly, destructive testing of the whole reactor design or system is not generally possible; consequently, a focus on calculations to inform generous performance margins, expert judgement and careful consideration of the evidence and knowledge that arises from the few severe accidents that have occurred is more widespread.

Regulatory practice has also evolved from the era of the pioneering reactor designs that yielded significantly lower power outputs, were operated at lower levels of availability and were often used for research purposes. In the early period of the industry, a focus was targeted on the nuclear system itself, rather than the holistic power plant. In addition, the regulatory approach adopted in a particular country is influenced by the national structure of the corresponding industry; some countries have a single utility operating the nuclear plant whereas others, particularly the United States, have a significant number of independent commercial operators.

Regulation has also developed from early philosophies where the reactor core was deemed to be *the* safety-critical component of the plant, to a more holistic philosophy where everything is considered to have a relevance to safety, including the design, construction and operation. Overall, most regulatory bodies adopt this holistic approach supported by reference to defence in-depth and the overarching need for there to be a *questioning attitude* such that assumptions, claims and assessments are challenged. A useful summary of the ten IAEA fundamental safety principles is provided in Table 14.7, by way of related international context.

Relative to the early generations of nuclear reactors, today's nuclear power plants are large and complex, and hence for the regulation of them to be effective it needs to be far-reaching and can be expensive relative to other industries. However, the prominent reactor accidents of Windscale, TMI and Chernobyl demonstrated that the impact on people, clean-up and the loss of a plant far outweighs these costs. Thus, one way forward to achieve a more acceptable balance between economics and safety is to build smaller, less sophisticated reactor systems; it is this trend that is currently influencing discussions concerning the next generation of reactor designs, as was discussed in Chapter 11.

Table 14.7 The IAEA Fundamental Safety Principles [18]

Principle	Note
1: Responsibility for safety	The prime responsibility for safety must rest with the person or organisation responsible for facilities or activities that give rise to radiation risks.
2: Role of government	An effective legal and governmental framework for safety, including an independent regulatory body, should be established and sustained.
3: Leadership and management of safety	Effective leadership and management for safety must be established and sustained in organisations concerned with, and facilities and activities that give rise to, radiation risks.
4: Justification of facilities and activities	Facilities and activities that give rise to radiation risks must yield an overall benefit.
5: Optimisation of protection	Protection must be optimised to provide the highest level of safety that can be achieved reasonably.
6: Limitation of risks to individuals	Measures for controlling radiation risks must ensure that no individual bears an unacceptable risk of harm.
7: Protection of current and future generations	People and the environment, present and future, must be protected against radiation risks.
8: Prevention of accidents	All practical efforts must be made to prevent and mitigate nuclear or radiation accidents.
9: Emergency preparedness and response	Arrangements must be made for emergency preparedness and response for nuclear and radiation incidents
10: Protective actions to reduce existing and unregulated radiation risks	Protective actions to reduce existing or unregulated radiation risks must be justified and optimised.

EXAMPLE 1: THE NUCLEAR REGULATORY COMMISSION (US)

The NRC, as an independent body of the US government, is responsible for ensuring the safe use of nuclear materials for beneficial civilian purposes (including medicine) while protecting human health and the environment. The NRC has a large remit, given the number of nuclear plant and medical activities in the United States, across a significant number of individual utilities. It was established in 1974 along with what would become the US Department of Energy when the US Atomic Energy Commission (AEC) was abolished. The NRC regulatory process comprises five activities:

- The *development of regulations and guidance* for licensees and those applying to carry out new operations or procedures.
- The *authorisation to operate* (licensing) and termination of licenses where activities have ceased.
- *Inspection and assessment* to ensure compliance with NRC requirements, including responding to incidents and enforcement.
- *Review and analyse* incidents and generic issues (those affecting more than licensed facility).
- *Research, assessment and advisory activities* to inform decisions, the impact of waste disposal and to consider concerns of parties affected by decisions.

EXAMPLE 2: THE OFFICE FOR NUCLEAR REGULATION (UK)

In the United Kingdom, the *Office for nuclear Regulation*[7] (ONR) was formed by the merger of the *Nuclear Installations Inspectorate* (NII), the *Office for Civil Nuclear Security* and the *UK Safeguards Office and the Radioactive Materials Transport Team*; it was launched formally as an independent statutory corporation on 1st April 2014. Its regulatory predecessor, the NII, was formed in 1960 when it was appreciated that the UK nuclear industry justified its own dedicated inspectorate. The ONR is responsible for nuclear safety and security in the United Kingdom, holding those responsible (*licensees and duty-holders*) publicly accountable. It covers the activities of nuclear power generation, nuclear fuel cycle facilities, decommissioning, defence and new build.

The regulation of nuclear activities focusses on the definition of a *hazard*, which is anything that might cause harm, and *risk* which is the likelihood of the hazard occurring combined with the *effect* of the hazard. The primary hazard in the nuclear industry is the radiation emitted by radioactive materials; a strong focus is put upon on safety culture, and this spans both radiation-based and conventional hazards. The ONR regulates the risk from routine operations and those that might arise from accidents. Other statutory bodies (e.g., the Environment Agency[8] in an English context) regulate permitted discharges of radioactivity to the environment.

A central tenet of the ONR approach to regulation is that the responsibility for safety lies with the licensee or duty-holder, that constitutes the corresponding responsible person(s). Nuclear safety spans a number of requirements that can be required to be evidenced by the following: a *robust design* with appropriate limits of operation, a *rigorous operating regime*, an *experienced regulatory group* within the licensee's organisation, *external peer review* (such as by the World Association of Nuclear Operators[9] and the International Atomic Energy Agency[10]) and oversight by *a strong, independent regulator*. As per earlier discussion in this chapter, defence in-depth is central to nuclear regulation. The ONR ensures this via a set of *safety assessment principles*[11] (SAPs) that provide a framework within which inspectors are able to make consistent regulatory judgments.

Safety legislation of the nuclear industry in the United Kingdom draws on the *Health and Safety at Work Act* 1974 (HSWA), the *Energy Act* 2013 and the *Nuclear Installations Act* 1965 supported by the *Ionising Radiations Regulations* 1999 (IRR99) and *Radiation (Emergency Preparedness and Public Information) Regulations* 2001 (REPPIR). Safety is managed on the basis that risk is kept as low as reasonably practicable (ALARP) as per the philosophy described earlier in this chapter. Fundamental to the UK regulation is a *goal-setting regime* in which the onus is on licensees to demonstrate how best to meet the regulatory requirements set by the ONR, rather than prescribing how they should do it; this encourages innovation and continuous improvement. Regulatory requirements are set out in terms of 36 licence conditions spanning, for example, the consignment of nuclear matter through to organisational capability. Compliance is judged via inspection and assessment by the ONR and through consideration of a body of evidence that might include, for example, safety cases, reports of inspections, annual reviews and emergency exercises.

[7]www.onr.org.uk/.
[8]www.gov.uk/government/organisations/environment-agency.
[9]www.wano.info/.
[10]www.iaea.org/.
[11]www.onr.org.uk/saps/.

With reference to new nuclear power stations, operators seeking to build new reactors are required to obtain a site licence and consent from the ONR, environmental permission from the corresponding environmental statutory body and planning permission from the UK government. A process by which key issues such as health and safety and cost–benefit analyses can be considered ahead of planning consent is used by the ONR, called *Generic Design Assessment* (GDA). This establishes regulatory influence over designers early in a project before construction starts, so that issues are resolved at the design stage, it ensure that design and site-related issues can be considered separately and that the process is open and transparent. A generic design assessment of a given design is carried out via a number of steps, with the scrutiny at each step usually becoming more detailed as specific concerns or technical issues are explored.

CASE STUDIES

CASE STUDY 14.1: STRONTIUM-90

Strontium-90 (^{90}Sr) arises as a product of nuclear fission in reactors used for power and research. Its production is unavoidable. However, it is only released into the environment as a result of an accident that causes significant core damage. If dispersed, it can constitute a significant radiotoxicological concern for a number of important reasons. First, it is a fission product that is produced in quantities that are sufficient to pose a hazard if ingested with a half-life that is long enough (28.8 y) to render it a concern long after it is produced.

Second, strontium is a Group II element and thus the body sequesters it from the diet as per calcium for bones and teeth. These are some of the longest-lived substances in the body demonstrating that the body does not excrete such elements very effectively. If ingested, ^{90}Sr can reside in the said areas of the body for a long time and it can be very difficult to eliminate. In bone, radiation emitted by ^{90}Sr has the potential to damage bone marrow that is responsible for the production of red blood cells. The turnover of this type of tissue is rapid relative to other tissues, as these cells are replaced regularly by the bone marrow.

Strontium-90 is an emitter of β^- particles which, being high-energy electrons are of medium concern in terms of LET, given their maximum energy of 0.546 MeV. However, the distinct feature of ^{90}Sr is that it decays to ^{90}Y. Yttrium-90 is a rather special case because it is also radioactive, with a very short half-life (64.1 hours) and therefore has a high specific activity. Furthermore, it emits one of the highest known energies of β^- particles at 2.27 MeV. Thus, ^{90}Sr resident in living tissue poses a significant radiological hazard not just because of its decay, but also because of that of its ^{90}Y daughter product. In defuelling the stricken reactor at TMI in the 1980s following the reactor accident there in 1979, this became a significant issue. It necessitated the development of various technologies to ensure people were protected against the effects of ^{90}Sr and the penetrating ^{90}Y β^- emission.

CASE STUDY 14.2: POLONIUM-210

Polonium-210 (^{210}Po) is notorious for being the isotope responsible for the death of Alexander Litvinyenko in 2006. Conversely, whilst it is not relevant to nuclear power plant, *per se*, it provides a useful illustration of radiotoxicity. It has a relatively short half-life (138.4 d) and emits α particles in its decay of several MeV in energy. As a form of radiation, the α particle is relatively massive by the standards of other radiations we encounter, and charged. Thus, ^{210}Po has a very high specific activity and emits a high-LET radiation rendering it extremely radiotoxic. Although the benefit of its short half-life is that it quickly decays away to ^{206}Pb that is stable, this serves to complicate forensic studies following nuclear accidents because relatively soon after contamination by ^{210}Po occurs, there is little radioactivity left to betray its presence. Furthermore, α activity is difficult to measure in the field. For example, where its presence might be suspected, it has often long-since decayed away to nothing before satisfactory measurements have been possible. Also, note how this example illustrates how a subtle change between isotopic forms—in this case from ^{210}Po to ^{206}Pb—can radically change the radiotoxicity of a contaminated substance.

CASE STUDY 14.3A: NUCLEAR ACCIDENTS AND THEIR INFLUENCE ON NUCLEAR SAFETY—KYSHTYM

The *Kyshtym disaster* occurred on 29th September 1957 at the Mayak nuclear materials production facility on the south shore of Lake Kyzyltash in Chelyabinsk Oblast in the former Soviet Union. This military facility was located between the towns of Kyshtym and Kasli in the southern Urals and was a 'closed' institute. Therefore, it was not shown on maps prior to the dissolution of the Soviet Union and was simply known as Chelyabinsk-40. Knowledge of the disaster was not widespread until the Soviet exile, Zhores Medvedev, wrote about it in 1976 [19]. The area has a history of chronic radioactive contamination dating from 1949 from radioactive discharges to the Techa River emanating from an effluent holding pool to the Techa River valley[a]; several serious incidents occurred at the facility prior to the 1957 accident.

The accident occurred when a stainless steel, cylindrical tank containing radioactive residues of sodium nitrate and acetate overheated and exploded, releasing 740 PBq of radioactivity off-site and contaminating 20,000 km^2 of the region comprising a population of 272,000 people.[b] An area extending 300 km in length and 30–50 km wide was contaminated by ^{90}Sr to 3.7 GBq km^{-2}.

The tank was ~8 m in diameter and ~6-m height (volume 300 m^3). It was housed in a reinforced concrete liner of thickness 1–1.5 m with a 0.8-m thick reinforced concrete lid that was covered with 1–1.5-m soil. Although designed to be cooled by water, the cooling system had failed at some point prior to the accident. Evaporation driven by the heat evolved from the radioactivity in the absence of cooling caused the radioactive liquor to reduce to 70–80-t dry sediment. This, having continued to generate heat, underwent a chemical explosion [20]. Put concisely, the Kyshtym accident has been described as 'a record of the disastrous, long-lasting effects man can wreak on his environment if he fails to take adequate steps to protect it' [21]. The affected area has been the subject of medical and radioecological research that continues to this day. This benefited the long-term predictions of the radiation levels associated with the Chernobyl accident. Kyshtym has been rated at 6 on the INES.

CASE STUDY 14.3B: NUCLEAR ACCIDENTS AND THEIR INFLUENCE ON NUCLEAR SAFETY—THREE MILE ISLAND (TMI)

The accident at the TMI nuclear plant in the United States on 28th March 1979 caused extensive damage to the reactor preventing it from ever operating again. The primary containment vessel was not breached and the radioactivity released was limited largely to noble gas fission products. However, dismantling the plant was a lengthy and very expensive process. A great deal was learnt of the challenges involved. Significant lost revenues were incurred while other reactors were modified to ensure the accident was not replicated elsewhere and the accident had a significant and long-term impact on US energy policy. The TMI accident was rated 5 on the INES.

TMI is an island in the Susquehanna River near Harrisburg in Pennsylvania in the United States on which there are two PWRs designed by Babcock & Wilcox, each of approximate power rating 800–900 MWe. The events leading to the accident in Unit 2 started with a relatively minor operation to remove a coagulation of ion-exchange resin with compressed air from a pipe associated with the demineralisers in the feedwater system (these were used to purify the condensate before it was returned and used again). This operation caused the flow of feedwater to be interrupted. In turn, this caused the feedwater pumps, the main turbine steam stop valves and the turbine to trip, in this order. Without feedwater flow, the temperature and pressure in the primary circuit of the reactor started to increase. As per normal operation, this caused the pilot-operated relief valve (PORV) on the pressuriser to open which, following relief of the pressure as a result of the reactor having tripped, was actuated by a solenoid to close it. All these operations were completed automatically in a matter of seconds.

[a]http://www.unscear.org/docs/reports/2008/11-80076_Report_2008_Annex_B.pdf
[b]http://www.unscear.org/docs/reports/2008/11-80076_Report_2008_Annex_C.pdf

Unfortunately, although the action of closing the PORV was confirmed on the operators' control panel, it remained open. This went unnoticed for more than 2 hours while primary coolant leaked away, facilitating an SBLOCA. The coolant was lost at ~1000 litres per minute which reduced the system pressure. This was exacerbated by well-intentioned actions by operators based on the misinterpretation of diagnostic information. The core became uncovered and suffered major damage before the primary coolant circulation was regained some 16 hours into the accident timescale.

The open PORV was assumed closed because the control panel indicated that the solenoid had operated as opposed to the actual closure of the valve itself. Overall, the way in which the diagnostic information and the overwhelming number of alarms and indicators were presented, complicated the ease with which operators were able to make important decisions at critical points in the development of the incident. For example, there were two important pre-existing characteristics of the plant that influenced the decisions taken on the day of the accident. First, a pre-existing minor leak of coolant to a drain tank in the reactor basement was evident from temperature sensors on the associated drain line; this distracted operators from the more significant leak from the PORV when it occurred, because the coolant followed the same path but in much more significant quantities. Second, a pilot valve had been closed during maintenance that had not been re-opened. This was not noticed until 8 minutes into the accident, preventing emergency feedwater supplies reaching the core for this period.

Furthermore, operators had been trained on a simulator that did not account for the possibility that steam voids might form in the primary circuit and thus increase the primary circuit coolant volume. This single-phase model did not account for a rise in pressuriser level due to voiding, but rather one that indicates an increase in the single-phase volume, consistent with there being too much water in the primary. Thus, in response to the pressuriser being full or 'solid', operators switched off the high-pressure coolant injection system [22] when actually the pressuriser level was due to voiding in the core. The diagnostic information that the operators relied on, such as chart recorders to infer reactor temperature, complicated the interpretation on which they based their decisions. As a result of concerns over vibration in the coolant pumps due to mixed-phase flow and on discovering the open PORV, the former were switched off and the latter closed. This removed the two remaining sources of cooling to a reactor now low in primary coolant due to the SBLOCA, and led to widespread fuel damage.

The TMI accident stimulated decades of research, particularly on the design of human–machine interfaces and their influence on decision making. Novel methods for dismantling the stricken plant were developed as a result, to minimise the exposure of workers to radiation, particularly with regard to the energetic ^{90}Y β^- emission from ^{90}Sr, as referred to in Case Study 2. Despite the fuel melting, the containment remained intact and prevented the release of significant amounts of radioactivity. The retrieval of the damaged fuel from the core took many years and cost a great deal. Most significantly, the accident is a factor reputed to have quenched the growth of nuclear energy in the United States.

CASE STUDY 14.3C: NUCLEAR ACCIDENTS AND THEIR INFLUENCE ON NUCLEAR SAFETY—THE CHERNOBYL DISASTER

The accident at Unit 4 of the Chernobyl nuclear plant near Kiev in the Ukraine on 26th April 1986 is notorious in terms of its scale and long-lasting legacy. It has been the subject of more than 10,000 research papers and continues to be scrutinised from a wide variety of perspectives. It is widely regarded as the world's worst nuclear accident.

The accident is described in detail in a number of sources[a] [23]. It resulted in widespread contamination of large areas of the former Soviet Union and other countries in Europe (being responsible for the largest uncontrolled radioactive release for any civilian operation[b]: 1760 PBq ^{131}I and 86 PBq ^{137}Cs). This led to serious social and economic consequences for affected areas, including extensive and long-lasting evacuations. Two men were killed immediately by injuries associated

[a]http://www.world-nuclear.org/information-library/safety-and-security/safety-of-plants/chernobyl-accident.aspx
[b]http://www.unscear.org/docs/reports/2008/11-80076_Report_2008_Annex_C.pdf

with an explosion unrelated to radiation exposure. About 134 workers and emergency staff were exposed to high radiation doses resulting in ARS, which killed 29 of them with another 7 fatalities as a result of exposure or trauma in the immediate aftermath. The main effects on survivors of ARS were skin injuries and cataracts, with there being some evidence of increased incidences of leukaemia. The lack of adequate countermeasures for ^{131}I contamination in milk resulted in relatively large doses to the thyroid glands of the exposed population with a substantial proportion of the occurrence of thyroid cancers in this population considered to be due to this exposure. The health consequences have been summarised elsewhere[c]. These are reviewed and updated regularly.

The RBMK reactor of the type at Unit 4 had a number of design traits that were significant factors in the accident. The RBMK design has a propensity for a positive void coefficient of reactivity to dominate over the fuel temperature coefficient at low power levels. This rendered the reactor very sensitive to disruptions in the operation of coolant pumps and the temperature of the coolant at the inlet. At the time of the accident, the scram rod insertion mechanism (essential to shut down the reactor) was slow (18 s) relative to other designs at the time. The graphite followers (or *displacers*) on the lower end of the control rod design had the potential to cause a transient, positive insertion of reactivity due to displacing cooling water on insertion, dependent on spatial power distribution and density. Reactor control at low power levels (<10%) relied on detectors outside of the core which could not indicate the neutron flux distribution. The distribution is perturbed by poisoning, requiring significant operator involvement and experience to control the reactor at low power levels. This was exacerbated by the large size of the core (diameter 11.8 m × 7 m). At the time of the accident, some safety systems, automatic trip mechanisms and alarm signals could also be bypassed manually. The reactor did not have a civil engineering structure constituting a containment that could withstand a pressure release associated with the rupture of several fuel channels at once.

In anticipation of a routine shutdown, a test was planned to determine whether in the event of power loss the turbine could power the main circulation pumps as it slowed down until diesel-powered supplies took over. Power was reduced to approach shutdown, but halted at approximately half when demands for power were received from the grid. Some 9 hours later, the power reduction was resumed. Despite the intent to stabilise the reactor at approximately 30% power, the power level fell inadvertently to 30 MWth due to the significant influence of reduced boiling. Increasing the power from this level was complicated by xenon poisoning and many of the control rods were withdrawn in an attempt to do so, violating regulations, to reach 200 MWth. At this point, despite these difficulties and the known instability of the core at low power, it was decided to proceed with the test. At 01:23:04, the turbine stop valve was closed and the coolant pumps started to slow down. Power rose and exceeded 530 MWth in 3 s: It is disputed as to whether the reduced feedwater flow rate may have resulted in voiding in the core and whether this, combined with a reduction of xenon poisoning or insertion of the control rods when the reactor was scrammed at 01:23:40 caused the escalation that followed. The power is estimated to have then increased 100 times in 4 seconds. While the control rods were half-way in, the rapid power increase unsettled the core and caused them to become jammed; the pressurised channels containing the fuel began to rupture, causing depressurisation and a large yield of steam. Two explosions occurred: the first thought to be due to the yield of steam and a second suspected to be due to hydrogen from zirconium–steam reactions or a further power excursion. The 2000-t reactor lid was lifted by the explosions, with very significant quantities of radioactive material ejected into the atmosphere and scattered around the reactor site. The core itself burned for several days. The accident was rated 7 on the INES. The distribution of ^{137}Cs-based surface contamination from the accident in Europe is depicted in Fig. 14c.1.

The Chernobyl accident has generated a vast amount of scientific data, it confirmed the pasture–cow–milk pathway for ^{131}I and demonstrated the imperative for effective countermeasures in this regard (such as restricting the distribution of milk immediately following a nuclear accident and the use of potassium iodate tablets). It has also benefited the knowledge of the pathways of longer-lived radionuclides in the environment and led to dramatic improvements in the understanding of ARS and its treatment, particularly with regard to the lens of the eye. Chernobyl demonstrated to the world that, irrespective of political and national boundaries, the consequences of a severe nuclear accident are a global issue. It affected local communities and neighbouring countries significantly and affected the expansion of the nuclear power worldwide. The latent effects continue to be the subject of research and assessment.

[c]http://www.unscear.org/docs/reports/2008/11-80076_Report_2008_Annex_D.pdf

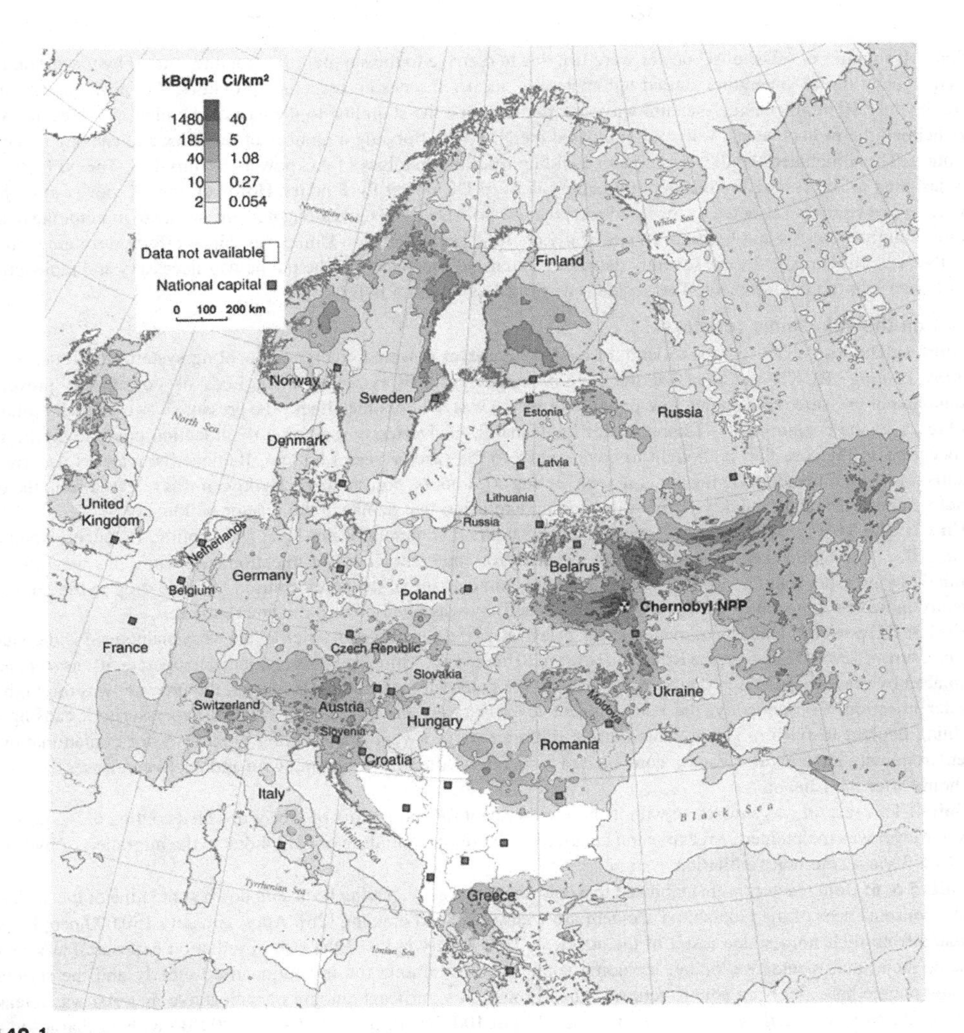

FIG. 14C.1

Surface contamination with ^{137}Cs in Europe after the Chernoby accident [24].

CASE STUDY 14.3D: NUCLEAR ACCIDENTS AND THEIR INFLUENCE ON NUCLEAR SAFETY—FUKUSHIMA DAIICHI

The accident at the Fukushima Daiichi nuclear power plant (NPP) on 11th March 2011 is unique among prominent nuclear incidents, as it occurred together with a major natural disaster and was what is known as a *station blackout event*. A station blackout event occurs when an NPP loses all electrical power due to a combined loss of off-site supplies and emergency sources. The Fukushima accident is described in detail in a number of sources and reports [25–28].

The Fukushima Daiichi NPP comprises six BWRs. Prior to the accident, three (Units 1–3) were at full power and three (4–6) had been shut down for refuel and maintenance. The Great East Japan earthquake (rated 9.1 on the Richter scale, the 4th largest earthquake recorded since records began in 1900) caused Units 1–3 to trip automatically in response to ground motion. All supplies of off-site AC power were lost due to damage to the supply infrastructure caused by the earthquake. The emergency diesel generators started automatically, as per the design response, providing AC power to all units. Approximately 40 minutes later, tsunami waves started to reach the shoreline to the east of the plant. The second wave overwhelmed the protective sea wall and inundated the NPP site, flooding a number of facilities and causing extensive disruption. These included the diesel generator building resulting in a loss of AC power to Units 1–5. The NPP systems were designed to survive station blackout on battery-derived DC power for 8 hours. However, the DC power systems to Units 1, 2 and 4 were also affected by the flood and power was lost to these units along with the ability to monitor the reactor systems, communications and lighting; subsequently cooling was also lost to Unit 3. Sustained efforts were made to stabilise the NPP but Units 1–3 were severely damaged, extensive evacuation of the public was necessary and radioactivity was released into the environment. The accident was rated as 7 on the INES.

Impact on the Fukushima reactors

Unit 1: The loss of DC power to Unit 1 isolated the steam-powered emergency cooling system (the *reactor core isolation cooling*, RCIC) and disabled the *high-pressure injection system* (HPIS). Loss of power also prevented in-containment pressure relief so that low-pressure injection was not possible either. The pressure was eventually relieved by a breach of the containment ~12 hours after the tsunami, inadvertently rendering the injection possible. Melting of fuel occurred in the core that is thought to have spread to the reactor vessel bottom. Radioactive releases occurred at 12 hours post-tsunami and following containment venting at 23 hours, post-tsunami. Hydrogen was released from the core but not vented sufficiently quickly. This resulted in an explosion that damaged the reactor building.

Unit 2: The RCIC system in Unit 2 operated for 68 hours maintaining water level and cooling. However, when this system failed, the internal pressure was at first too great for emergency coolant injection and, while this was achieved eventually by pressure relief as for Unit 1, rapid heating and damage of the core occurred. A rapid drop in pressure subsequently indicated a breach of containment and radioactivity was released into the environment.

Unit 3: DC power to Unit 3 was available for 2 days after the tsunami and water levels were maintained and containment pressure relieved. However, the RCIC failed after 20 hours and, while the HPIS continued to operate, it was stopped to be replaced by seawater injection due to concerns over the potential for it to fail. However, the pressure was too high for seawater injection. Consequently, the reactor water level dropped and the system became depressurised, causing the remaining coolant to flash to steam. This burst a rupture disc in a vent line yielding a pathway for contamination to the environment. Fuel in the reactor core melted ~43 hours after the tsunami, with radioactive releases occurring ~47 hours after the tsunami.

Unit 4: This reactor was shut down with its fuel in the spent fuel pool. Loss of power disabled cooling of the pool, but the water level was maintained. An explosion occurred in the reactor building of Unit 4 due to the migration of hydrogen from Unit 3 via a common ventilation system.

Units 5 & 6: Both reactors were stabilised to a cold shutdown state, having been shut down at the time of the accident.

The consequences of the Fukushima accident are widespread and ongoing [26]. Approximately 150,000 people were evacuated from their homes as a result of the accident. This greatly reduced the risk of radiation exposure but affected physical, social and mental wellbeing, including some fatalities among the institutionalised elderly and the seriously ill. Many people have not been able to return to their homes. A significant amount of radioactive material was released into the environment, including atmospheric releases: ^{131}I of 100–500 PBq and ^{137}Cs of 6–20 PBq with discharges to sea estimated to be 10%–50% of that of atmospheric releases.

There were no radiation-related deaths or acute effects among the public or workers. Doses to the public (measured and lifetime projections) are very low with no discernible radiation-related health effects anticipated, cf. lifetime effective dose (adults) <10 mSv (the number of infants receiving thyroid doses of 100 mSv is not known). An increased cancer risk is inferred for 12 workers with significant thyroid exposure via ^{131}I inhalation (2–12 Gy) and 160 workers with effective doses estimated >100 mSv. A survey (the Fukushima Health Management Survey) was started in 2011, which will run for 30 years, covering the 2.05 million people living in Fukushima Prefecture at the time of the accident.

Significant lessons have been learned from the Fukushima accident. Margins made for resilience in the design of the NPP for extreme seismic and flood events are considered to have been lacking. The possibility of simultaneous major hazards affecting multiple units was not anticipated in the design. The response to the nuclear emergency was complicated by the off-site consequences of the earthquake combined with the assumption, by both the operating company (TEPCO) and regulators, that the probability of an extended loss of power was considered to be very unlikely. Sympathetic effects due to the proximity of the reactor systems complicated the response to the accident. Misjudging the state of the isolation condensers and the delay in venting of the hydrogen in Unit 1 are considered to have been significant in exacerbating the development of the accident. Senior-level decision making was compromised by the breakdown of transport and communication links early in the accident [27]. Preparation and training measures in anticipation of a blackout incident are considered to have been deficient, as was the necessary degree of independence among the regulators, operators and government [28].

The scale of the earthquake and tsunami that afflicted Fukushima Daiichi was unprecedented in terms of prior NPP operational experience. However, the accident is regarded by some experts as 'manmade' and its causes as being 'foreseeable' [28]. This stance is taken because, although the NPP was built prior to the latest guidelines for seismic protection, the requirement for an enhancement of defences was understood but this had not been implemented by the time the accident occurred. Safety measures and procedures have been improved the world over as a result of the accident, including significant reform of the Japanese nuclear sector. Nuclear energy policies in Germany, Italy and Switzerland have also been reviewed and changed as a result.

REVISION GUIDE

On completion of this chapter, you should:

- understand the definition of *radiation dose*
- appreciate the parameters that might lead us to consider a particular radioactive substance to be a concern to health in terms of quantity, radiation energy and specific activity
- understand the definition of *absorbed dose*, *equivalent dose*, *effective dose* and *collective dose* and the units corresponding to these parameters
- have a general awareness of the range of *radiation weighting factors* and *tissue weighting factors* necessary in the calculation of the above dose quantities
- appreciate the range of exposures possible in terms of dose across background, occupational environments through to severe accidents
- appreciate the concept of *radiotoxicity* and how it is influenced significantly by a range of factors
- appreciate the general concept of *linear energy transfer* (LET)
- understand the international scale against which nuclear accidents are rated
- be aware of the prominent nuclear accidents and the detail of the Windscale fire as an example case study
- understand the concepts of *defence in-depth* and the *precautionary principle*
- understand the need for regulation and appreciate the difference in approaches based on two examples: the United States and the United Kingdom

PROBLEMS

1. Describe the differences between *absorbed dose*, *equivalent dose* and *effective dose*, stating the corresponding SI units for each in your answer.
2. Describe the difference between a *hazard* and a *risk*. Why is it difficult to estimate the risk associated with low levels of radiation? Hence, describe the philosophy that is adopted in the absence of better scientific data.

3. Why is it necessary to have a diversity of emergency core cooling approaches to cater for the event of coolant loss in a fission power reactor? Describe the systems you are aware of and name them.
4. Iodine-133 constitutes a significant radiotoxicological hazard where it occurs as discussed in Chapter 5. Research the properties of ^{133}I and hence describe why you might conclude this to be the case. Describe why we believe it to have been a significant cause for concern in the aftermath of the Chernobyl accident.
5. Krypton-85 was encountered as a result of the Three Mile Island accident: describe how it was dealt with in this case and the risk it posed to neighbouring populations in doing this.

REFERENCES

[1] Nuclear Energy Agency, Comparing Nuclear Accident Risks With Those From Other Energy Sources, ISBN 978-92-64-99122-4, 2010. Nuclear Energy Agency No. 6861.
[2] B.L. Cohen, Test of the linear-no threshold theory of radiation carcinogenesis for inhaled radon decay products, Health Phys. 68 (2) (1995) 157–174.
[3] J. Valentin, Relative biological effectiveness (RBE), quality factor (Q) and radiation weighting factor (W_R), Ann. ICRP 33 (4) (2003) 1–121.
[4] The 2007 recommendations of the ICRP publication 103, Ann. ICRP 37 (2–4) (2007) 1–332.
[5] IAEA, Report on the preliminary fact finding mission following the accident at the nuclear fuel processing facility at Tokaimura, Japan, www-pub.iaea.org/MTCD/Publications/PDF/TOAC_web.pdf, 1999.
[6] E.J. Hall, Cellular damage response, in: 4th International Conference on Health Effects of Low-level Radiation, British Nuclear Energy Society, 22–24 September, 2002.
[7] G.A.M. Webb, R.W. Anderson, M.J.S. Gaffney, Classification of events with an off-site radiological impact at the Sellafield site between 1950 and 2000, using the International Nuclear Event Scale, J. Radiol. Prot. 26 (2006) 33–49.
[8] ONR-GDA-AR-11-003, Generic design assessment—new civil reactor build, step 4 PSA assessment of the Westinghouse AP1000 reactor, http://www.onr.org.uk/new-reactors/reports/step-four/technical-assessment/ap1000-psa-onr-gda-ar-11-003-r-rev-0.pdf, 2011.
[9] L.S. Tong, J. Wiseman, Thermal Analysis of Pressurised Water Reactors, third, Am. Nuc. Soc., ISBN 0-89448-038-3, 1995.
[10] L. Arnold, Windscale 1957—Anatomy of a Nuclear Accident, Palgrave-Macmillan, London, ISBN 0-333-48252-2, 1992.
[11] R. Wakeford, The Windscale reactor accident—50 years on, J. Radiol. Prot. 27 (2007) 211.
[12] H.J. Dunster, H. Howells, W.L. Templeton, District surveys following the Windscale incident 1957, J. Radiol. Prot. 27 (2007) 217–230.
[13] J.A. Garland, R. Wakeford, Atmospheric emissions from the Windscale accident of October 1957, Atmos. Environ. 41 (2007) 3904–3920.
[14] N. Nelson, K.P. Kitchen, R.H. Maryon, A study of the movement of radioactive material discharged during the windscale fire in October 1957, Atmos. Environ. 40 (2006) 58–75.
[15] D. McGeoghegan, K. Binks, Mortality and cancer registration experience of the Sellafield employees known to have been involved in the 1957 Windscale accident, J. Radiol. Prot. 20 (2000) 261–274.
[16] D. McGeoghegan, S. Whaley, K. Binks, M. Gillies, K. Thompson, D.M. McElvenny, Mortality and cancer registration experience of the Sellafield employees known to have been involved in the 1957 Windscale accident: 50 year follow-up, J. Radiol. Prot. 30 (3) (2010) 407–431.
[17] International Atomic Energy Agency, Defence in depth in nuclear safety, INSAG-10, 1996, http://www-pub.iaea.org/MTCD/publications/PDF/Pub1013e_web.pdf.

[18] IAEA, Fundamental safety principles, Safety Standards Series No SF-1, IAEA, Vienna, 2006.
[19] Z. Medvedev, Two decades of dissidence, New Sci. 72 (1976) 264.
[20] M.I. Avramenko, et al., Radiation accident of 1957 and East-Urals radioactive trace: analysis of measurement data and laboratory experiments, Atmos. Environ. 34 (2000) 1215–1223.
[21] D.M. Soran, D.B. Stillman, An analysis of the alleged Kyshtym disaster, LA-9217-MS, http://www.iaea.org/inis/collection/NCLCollectionStore/_Public/14/724/14724059.pdf, 1982.
[22] S. Levy, Three Mile Island: a call for fundamental and real-time analysis, Heat Transfer Eng. 1 (4) (1980) 47–52.
[23] INSAG-7 The Chernobyl Accident: Updating of INSAG-1, IAEA Safety Series, 1992, http://www-pub.iaea.org/MTCD/publications/PDF/Pub913e_web.pdf.
[24] G. Steinhauser, A. Bran, T.E. Johnson, Comparison of the Chernobyl and Fukushima nuclear accidents: a review of the environmental impacts, Sci. Total Environ. (2014) 470–471. 800–817.
[25] Y. Amano, The Fukushima Daiichi Accident, IAEA, 2015. ISBN:978-92-0-107015.
[26] UNSCEAR, Levels and effects of radiation exposure due to the nuclear accident after the 2011 great east-Japan earthquake and tsunami, Annex A, http://www.unscear.org/docs/reports/2013/13-85418_Report_2013_Annex_A.pdf, 2013.
[27] Y. Funabashi, K. Kitazawa, Fukushia in review: a complex disaster, a disastrous response, Bull. At. Sci. 68 (2) (2012) 9–21.
[28] K. Kurukawa (chair), The official report of the Fukushima Nuclear Accident Independent Investigation Commission (NAIIC), The National Diet of Japan, 2012.

CHAPTER

RADIOACTIVE WASTE MANAGEMENT AND DISPOSAL

15

15.1 SUMMARY OF CHAPTER AND LEARNING OBJECTIVES

The focus of this chapter is the important topic of radioactive waste management and disposal. This area is covered from the perspective of the waste that currently exists from previous activities, principally nuclear energy generation but also military activities, and also from the point of view of ongoing generation and the forecast of radioactive waste from future operations. The composition and variety of materials that constitute radioactive waste are described, and the range in radiotoxicity and lifetime is considered. The means by which radioactive waste is managed is considered in terms of the common priorities including: minimising the production of waste, minimising the volume of the waste that is produced and rendering the waste in a form where the likelihood of contamination being dispersed is minimised (usually by immobilisation and containment). The principal means by which radioactive waste might be disposed of is also covered, from the perspective of the widely accepted approach of deep geological disposal, though to legacy activities that are now obsolete (principally marine disposal).

The objectives of this chapter are to:

- define the term *radioactive waste* as referring to a range of materials contaminated with radioactivity that no longer have a useful purpose and require management and disposal
- present the composition, timescales and origins of the principal forms of radioactive waste
- define and describe the various classifications of radioactive waste in terms of *specific activity*, *lifetime*, *heat-generating status* and the extent of contamination by *transuranic elements*, recognising that while general trends are common worldwide, specific classifications can vary from nation to nation
- introduce the main treatment options and concepts that are applied to radioactive waste, including packaging and immobilisation
- introduce the concept of *partitioning and transmutation* as an example of a technique under consideration that might reduce the hazard posed by radioactive waste to future generations
- introduce the final disposal options that exist and which have been considered with respect to radioactive waste, including deep geological disposal and marine disposal, recognising that the latter is now prohibited

Nuclear Engineering. https://doi.org/10.1016/B978-0-08-100962-8.00015-9

- introduce the concept of an *engineered emplacement* based on the idea of an engineered barrier system
- provide a summary of the various *natural analogues* that exist from which we can learn of the optimum way in which to consign radioactive waste for very long periods of time

15.2 HISTORICAL CONTEXT: JEAN-FRÉDÉRIC JOLIOT-CURIE 1900–58

Frédéric Joliot-Curie was an assistant to Marie Curie and completed a doctoral thesis on the electrochemistry of radioactive elements (subsequent to his thesis on engineering). He married Iréne Curie (Marie Curie's daughter) and together they investigated the effect of bombarding stable elements with α particles. They observed that this resulted in the creation of short-lived radioactive isotopes and received the Nobel Prize in Chemistry for this discovery, of *artificial radioactivity*, in 1935 (Fig. 15.1).

The significance of the Joliot-Curie's achievement was diverse. The amounts of the isotopes that they had created were too small to be quantified by the standard analysis methods, but since these isotopes emitted radiation, even the smallest quantities could be estimated. Frédéric Joliot-Curie summarised this eloquently: 'This special kind of chemistry in which one handles unweighable quantities' [1] referring to what was effectively the earliest form of synthetic radiochemistry.

Subsequently Joliot-Curie suggested that something similar might be possible with protons, deuterons and neutrons. This insight was to prove significant in the context of both the bombardment of ^{235}U that led to fission and the development of accelerator-based systems with which many of the transuranics were discovered. Frédéric Joliot-Curie also demonstrated the use of radioactive iodine as a tracer for the thyroid gland and suggested the potential for the use of artificial radioactivity in medical radiotherapy.

FIG. 15.1

Jean-Frédéric Joliot-Curie.

Joliot-Curie was focused on the production of energy from fission for much of his career. Perhaps most significantly, he measured the number of neutrons emitted by fission and showed that a nuclear chain reaction was possible, specifying the conditions necessary for such a reaction to be self-sustaining. He anticipated that this might lead to the liberation of enormous quantities of usable energy but also counselled the need for necessary precautions to be taken. His interest in fission was curtailed by the Second World War and the occupation of France, after which he became more involved with related diplomatic activities, such as serving as president of the World Peace Council.

15.3 INTRODUCTION

Waste is a substance that has no anticipated use in future. Most human activities produce waste of one form or another. Relatively recently, worldwide, interest across the fields of economics, scientific research and environmental management in relation to 'waste' and its consequences has increased due to concerns over the management of all kinds of waste and its disposal. Often the focus of these activities is on minimising the production of waste, ensuring that it is disposed of safely and ensuring that as much of it as possible is recycled into materials that *do* have future uses. Priorities in this regard can be *minimising the volume of waste*, *minimising the energy expended in its treatment and disposal* and *ensuring that it cannot contaminate the biosphere*.

In comparison with the quantities of waste produced from the spectrum of human activities, the quantities arising from the production of energy are significant. While nuclear energy production actually produces very small volumes of waste, relative to that generated as a result of most other forms of energy production (and especially that which is associated with fossil fuels), the priorities outlined above can be prominent concerns for the management and disposal of radioactive waste.

Radioactive waste (or nuclear waste) is defined as waste that contains radioactivity above levels specified by regulation. Material with less than the specified level of radioactivity (usually taking into consideration the route(s) by which exposure might occur) is, by definition, not radioactive waste and is thus deemed exempt or *exempt waste*. However, such material might still be considered hazardous due to its chemical properties and thus need to be considered within regulatory constraints applicable for non-radioactive waste products, as per general waste management legislation (beyond the scope of this chapter). Radioactive waste is produced across a range of activities such as energy production, defence, medicine and analytical processes. Although the focus of this chapter is on that associated with nuclear energy, many of the concepts that we will discuss are relevant and applicable to radioactive waste arising in other areas; most long-lived radioactive waste in the world today, and that which is projected to arise from future operations, will have arisen from nuclear energy production and defence.

Radioactive wastes are produced across the range of industrial activities that constitute the nuclear fuel cycle of one form or another and to a greater or lesser extent. An important distinction between the wastes produced *prior* to reactor use and those that arise *afterwards* is that the wastes associated with the former comprise only naturally occurring radioactive species, whereas those associated with the latter comprise anthropogenic species such as fission products, minor actinides and the products of neutron activation. Radioactive wastes can include material derived from the nuclear fuel irradiated in reactors but also a range of other materials. These include the tails and residues from uranium mining, manufacturing wastes from fuel fabrication, ion exchange agents from decontamination processes used for the treatment of moderators and coolants, and also activated components of the reactor infrastructure itself.

15.4 RADIOACTIVE WASTE COMPOSITION AND TIMESCALES OF DECAY

Further to the discussion in Chapter 14, concerning safety and regulation, materials without an anticipated future use that constitute a radioactive hazard need to be managed and disposed of to ensure that the hazard is either eliminated or isolated. This ensures that the risk of exposure to people and the environment is minimised and reduces the legacy to future generations from long-lived radioactive materials, as per the hierarchy of hazard control. From a practical point of view, the radioactivity associated with some wastes is sufficiently short lived that delaying their treatment and disposal results in its elimination by virtue of its natural disintegration; steel structural materials contaminated with ^{60}Co are sometimes a case in point in this regard. Alternatively, the radioactivity that is present can be very long lived such that elimination by decay is not a practical option; to do so would require active control and management over impractically long timescales.

For the latter case of long-lived wastes, the focus shifts from a case of *elimination* to the need for *isolation* and *immobilisation* instead; the challenge being to ensure the hazard is isolated over a period where the ultimate timescale is set by the time considered necessary for a *reference level* radiotoxicity to be reached. The reference state is usually taken to be the radiotoxicity of the natural uranium ore required to make the fuel that is the source of the waste, where it is assumed that secular equilibrium of the progeny in the ore has been established. This is usually quantified in terms of the *ingested dose per unit energy produced by the material*, that is, Sv/TWe-h, which varies for different enrichments and reactor types, particularly, for example, across the choice of thermal or fast spectrum reactor systems; the reference level falls in proportion with the amount of natural uranium required. By way of example, Fiorina [2] provides a derivation of the reference level associated with the EPR to give 5.9×10^6 Sv/GWe-yr. The possibility of shortening the time taken for radioactive wastes to reach reference levels, principally via transmuting long-lived nuclides to short-lived variants, is also being considered.

A comparison of the dependence of radiotoxicity versus time for spent light water reactor (LWR) fuel (the reference reactor being a French N4 pressurised water reactor (PWR)) with its corresponding separated high-level waste (HLW) after reprocessing[10] is given in Fig. 15.2, with reference levels for LWR and fast reactor (FR) scenarios. From this figure, several important features associated with the lifecycle of radioactive waste are apparent, as follows:

- The longest-lived waste scenario is that corresponding to the unprocessed spent fuel. This comprises the actinides and fission products and requires several hundred thousand years to decay to LWR reference levels.
- When uranium and plutonium are separated, the time taken for the residual HLW to reach reference levels is reduced in comparison with spent fuel, but it is still long at approximately 20,000 years.
- In the short term, the radiotoxicity is dominated by the fission product component of the spent fuel inventory, with the prominent heat-generating contributors to this being ^{90}Sr and ^{137}Cs that decay within the first 1000 years.
- In the long term, the actinides (principally ^{239}Pu and ^{237}Np) dominate.
- Long-lived fission products (LLFPs) also remain but at much lower levels of radiotoxicity than the actinides.

In a relatively small number of cases associated with some exotic forms of radioactive waste, a satisfactory route by which it is immobilised is not yet known. For example, an acceptable process

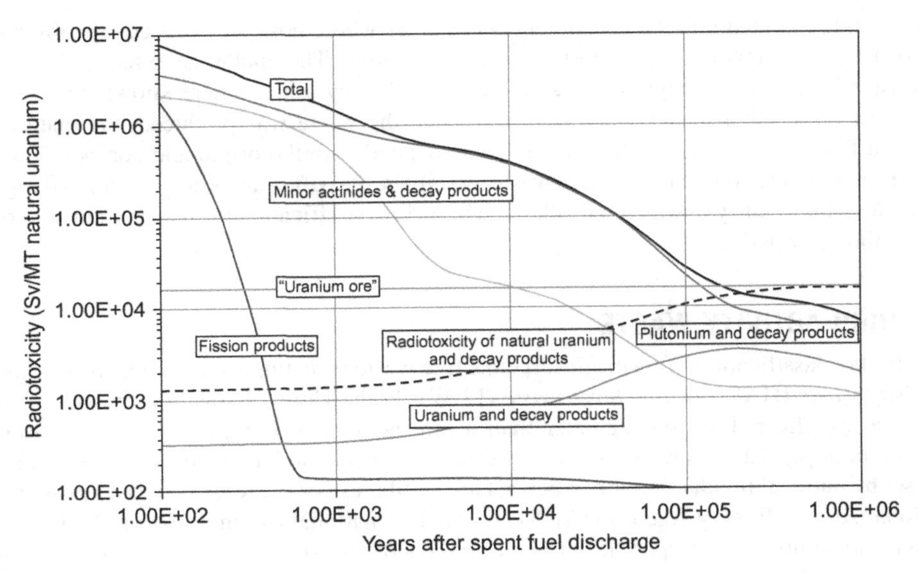

FIG. 15.2

Radiotoxicity as a function of time for uranium oxide fuel with an average burnup of 50 GWd/tHM for an LWR and its component nuclides. Spent fuel and HLW arising from 99.9% separation of uranium and plutonium are compared [3], reproduced with the permission of the OECD.

or immobilisation agent may not have been identified yet. These materials are often referred to as *orphan wastes*. While research continues, the strategy for these materials in the interim is on administrative and engineering controls to isolate the hazard that these wastes present while a permanent strategy is developed. Since radioactive waste can arise in solid, liquid and gaseous phases, and with a wide range of radiotoxicity, waste management operations often involve segregation, separation, sorting, precipitation, immobilisation and confinement. The focus of this chapter is on the engineering concepts associated with the requirements of these activities.

15.5 RADIOACTIVE WASTE CLASSIFICATIONS

Radioactive wastes arising from planned operations are termed as *controlled wastes*, whereas those arising from accidents and unplanned events (such as leaks, spillages and unplanned releases) are termed as *uncontrolled wastes*. In present day activities, controlled radioactive waste is often segregated near to the point at which it is produced; for example, articles contaminated with low levels of radioactivity (such as gloves, overshoes, swabs and laboratory consumables) are monitored at the point at which they are finished with to determine the extent of contamination and then disposed of as per the corresponding regulation.

For the case of uncontrolled wastes, the challenges can be different and are often more sophisticated because a wider combination of materials and radioactive species can be involved to an extent that is often not fully understood. Such materials need to be retrieved and extracted, often via decontamination techniques, and separated into the constituent waste types where possible. This can require the development of bespoke methods in the most significant of cases. The challenge associated with uncontrolled wastes is particularly significant with respect to legacy issues where knowledge was lacking as to the best route for disposal at the point at which the waste was produced, the materials were not sorted at that time or our current knowledge associated with the origin and composition of these materials is incomplete. While radioactive waste tends to be classified according to national regulations associated with the country in which it resides, the basis for classification from a qualitative perspective is often similar worldwide.

15.5.1 HIGH ACTIVITY WASTE

Aside from the classification of its origins, radioactive waste is distinguished in terms of whether it is *high-activity waste* (HAW) or *low-level waste* (LLW). In the United Kingdom, by way of example, HAW has a specific radioactivity greater than a threshold of 4 GBq per tonne α radioactivity or 12 GBq per tonne β/γ. LLW corresponds to wastes with a specific activity below this threshold. Thresholds are set by national regulatory authorities but are similar as per the reference cases provided by the International Atomic Energy Agency (IAEA) and the International Commission on Radiation Protection (ICRP). In addition to the specific activity of radioactive waste, it is also characterised in terms of the length of time that it will remain radioactive, with the half-life of ^{137}Cs (30.2 years) constituting the threshold between wastes that are deemed *long lived* or *short lived.*

An important distinction within the HAW classification is whether the inherent radioactive inventory of radioactive waste causes it to generate thermal heat. The presence of some prominent radioactive species (such as ^{137}Cs and ^{90}Sr) in sufficient quantity within the bulk substance comprising the waste can yield a noticeable elevation in the temperature of the material, due to the interaction of the radiation that is emitted with the host matrix. Such wastes tend to be associated with the highest levels of specific radioactivity and waste category, particularly the concentrated liquors from spent fuel reprocessing operations. This class of waste is often associated with the smallest volume of waste generated, typically 1% of the volume of all radioactive wastes, but most of the radioactivity. A conspicuous requirement of the management of heat-generating radioactive waste forms is that the thermal energy that is evolved has to be accommodated in the design of the corresponding waste management and disposal facilities, to ensure that it does not undermine the requirements of isolation and confinement. Radioactive waste in this category is usually termed as *high-level waste* (HLW).

Radioactive waste with a specific activity greater than the HAW threshold but that does not generate heat is termed as *intermediate-level waste* (ILW). By way of example, ILW in the United Kingdom constitutes 6% of the total waste volume and might comprise activated reactor components and materials arising from the treatment of radioactive liquid effluents.

15.5.2 LOW LEVEL WASTE

Contaminated materials with a specific activity of less than the threshold for ILW (i.e. <4 GBq t^{-1} α or <12 GBq t^{-1} β/γ) are defined as *low-level waste* (LLW). This comprises the largest volume of radioactive waste (typically ~94% of the total) and comprises materials from *operational* activities and *decommissioning* that have been contaminated in use, such as paper and plastics arising from, for example, contaminated personal protective equipment (gloves, protective suits) and building detritus (concrete, spoil), respectively. There is a sub-category of LLW termed *very low-level waste* (VLLW) that tends to comprise miscellaneous building materials arising from the decommissioning and demolition of power plant and related infrastructure; this can be disposed of with general industrial wastes at licensed (not meant to infer the requirement of a nuclear licence) landfill facilities.

15.5.3 SPENT NUCLEAR FUEL

In many countries, nuclear fuel retrieved from reactors at the end of its useful life is not reprocessed but is disposed of as *spent nuclear fuel* (SNF) instead. This material is highly radioactive but not to such a concentrated extent of the HLW liquor arising from reprocessing; conversely, spent fuel has a greater volume. The distinction in terms of the waste products from the reprocessing of spent nuclear fuel is that uranium and plutonium recovered from the dissolution from SNF (as described in Chapter 13) are deemed to have the potential for use in the future, principally as nuclear fuel, and are thus not waste, whereas the HLW liquid containing the fission products and minor actinides is the principal waste product in this regard in terms of specific activity. SNF in most cases is stored above ground at present in containers designed to shield people and the environment from the radiation it emits, to facilitate transport for its ultimate disposal at some point in the future and to limit degradation of the waste form with time. The barriers provided by the fuel pellet and the surrounding cladding provide the first layer of defence in addition to subsequent engineered barriers that are put in place as part of waste management and disposal solution.

15.5.4 TRANSURANIC WASTE (TRU)

Transuranic waste (TRU) is a classification of radioactive waste that is used predominantly in the United States for material that is contaminated with isotopes heavier than uranium, most often plutonium. These isotopes are typically longer lived than fission product constituents of waste and hence TRU will constitute the majority of the radioactive hazard posed by wastes that are produced now in 1000 years of time. Formally [4], TRU is waste containing more than 100 $nCi\ g^{-1}$ α-emitting transuranic isotopes, with half-lives greater than 20 years notwithstanding a number of specific exceptions. In the United Kingdom, material contaminated with plutonium is referred to as *plutonium-contaminated material* (PCM).

15.6 TREATMENT OPTIONS FOR RADIOACTIVE WASTES

Radioactive waste management is often necessary prior to its disposal because of the benefits that can be achieved in terms of reducing both the hazard and the cost associated with consigning the waste. Management practices are usually focussed primarily on characterising the waste in situ (where possible), retrieving it, sorting it, segregating it and conditioning it. Primarily, the aims of these activities are to reduce the ease with which the waste might be dispersed and to aid its transport to a site of storage and subsequent permanent disposal. There is often a focus on reducing the amount of waste (either via processing, volume reduction or optimising the processes that produce the waste to produce less) and on exploring options by which the material might be re-used or recycled; this approach is often termed as the *waste hierarchy*, as presented in Fig. 15.3. This depicts the overall strategic preference that waste

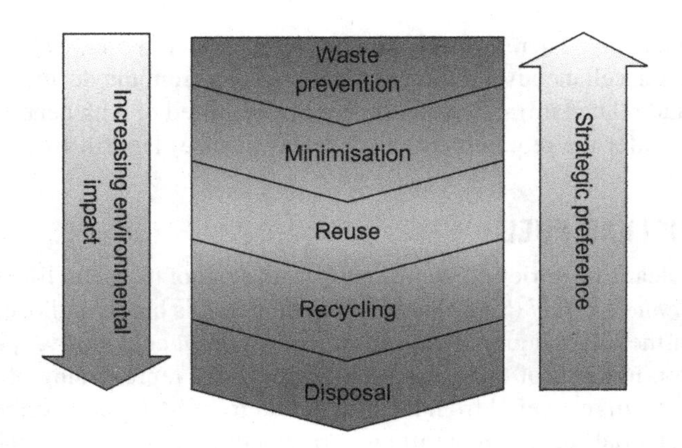

FIG. 15.3

The waste hierarchy [5].

production moves up the hierarchy, towards prevention rather than towards disposal, the latter which constitutes the last resort. Beyond these stages, the waste quantity and type are known, and the subsequent stages of *waste treatment*, *packaging*, *storage* and *disposal* can be pursued.

15.6.1 PACKAGING AND IMMOBILISATION

Historically, either due to the need for military expediency, a lack of insight or a lack of care, some radioactive wastes were not treated at source and were set aside instead. This has generated what are collectively referred to as a *legacy* of issues that are in some cases very challenging to deal with. By contrast, in the current era, radioactive waste treatment is often designed to commence very soon after the waste is produced and the methods employed have been developed in close correspondence with the type of waste that is involved. For example, the treatment of LLW can follow effectively immediately after it is produced. Conversely, for higher level wastes, a period of interim storage can be necessary to allow short-lived radioactive species to decline in abundance as this reduces the radiological hazard associated with the waste and renders it easier to handle.

Radioactive waste treatment can require a significant process engineering activity involving characterisation, decontamination, shredding, drying, compacting and immobilisation of the waste in inert matrices (e.g., bitumen, concrete, grout, glass, etc.). There is also often the need for transport both on and occasionally away from the site at which the waste was produced or treated. Understandably, the quality assurance associated with the design and compliance of the waste form and its package are necessarily rigorous and record keeping is a continual part of the regime.

15.6.1.1 HLW

HLW arising as liquor from the reprocessing of nuclear fuel has usually been concentrated by evaporation. It is very important that this liquid is converted to solid form to reduce the ease with which it might be dispersed at some point in the future because it is highly radioactive and long lived. The most widespread approach to doing this is to mix the HLW with glass under heating to constitute a molten product that can then be poured into stainless steel containers; specific methods vary, in terms of the glass compounds that are used and the melting process, but the general approach is termed as *vitrification* [6]. In this form, the glass provides the primary means of immobilisation and an engineered barrier to the migration of radioactivity. It is anticipated that this will be stable for a very long time to ensure that the waste is isolated from the biosphere when it is finally disposed of.

As indicated above, HLW generates heat in the relatively short period that follows immediately after its production and therefore, following vitrification, it is necessary to store the HLW containers in an above-ground facility to allow for the heat to be ducted away safely and passively for several years. Although no HLW has been consigned permanently yet, the accepted disposal route for HLW in this form is considered to be via *deep geological disposal* in a geological disposal facility (GDF). Depending on the specific approach that is adopted and the regulatory requirements for a given facility, this is usually anticipated to comprise the further containment of the HLW canisters in a form of second, double-hulled, corrosion-resistant container. These would then be consigned deep underground (typically >500 m depth) in a grouted chamber as part of a geologically stable rock formation.

15.6.1.2 ILW

ILW usually occupies larger volumes than HLW and thus there is often a greater emphasis in its treatment on optimising the space required for its disposal. Thus, the waste will often be subject to cutting, shredding, drying and compacting prior to disposal. In most applications, the waste material (often heterogeneous mixtures in the solid phase and immobilised resins, etc.) is placed in a container (typically 3 m^3 stainless steel boxes or 500 l; drums, for example) and immobilised with concrete [7]. These containers are then consigned with others in larger containers in a form that is then safe to transport and store. In some countries, strategies to consign ILW permanently are well advanced, whereas in others, the material is currently in interim storage above ground. The final destination in most cases is anticipated to be a GDF, as described for HLW above, albeit with different specifications consistent with the risk factors associated with the corresponding waste type.

For some ILW waste streams, especially those comprising water-soluble oils and resins, bitumen has been used as an immobilising agent because it provides a relatively cheap and resilient primary barrier that can then be consigned in secondary containers with a view to consignment schemes similar to those alluded to above. Bitumen immobilisation can be accomplished as an extrusion process and is a lower-temperature approach than vitrification, compatible with wastes that are volatile at higher temperatures. As bitumen is combustible, this needs to be considered in terms of the waste specification, whereas immobilisation in glass or concrete is generally not subject to this constraint.

15.6.1.3 LLW

Since LLW usually accounts for the largest volume of radioactive waste, volume reduction is often a key requirement of the corresponding treatment process. Historically, LLW has often been stored in bulk in above ground or in near-surface waste repositories overtopped with soil, historically in unlined trenches or silos (a practice now obsolete) and latterly grouted in metal containers in concrete-lined vaults; it is anticipated that such facilities will be controlled for 100 years or until the constituent radioactivity is no longer a concern. Approaches of this type—*shallow land burial*—have been adopted in Finland, France, Spain, Sweden and the United Kingdom.

However, in recent years different approaches have been adopted to reduce the quantity of waste that has to be consigned, particularly as existing LLW waste stores have approached full capacity and the justification for new facilities has met with resistance over public acceptability and cost. For example, techniques have been explored by which metals contaminated with low levels of radioactivity on their surfaces can have this removed by either mechanical or chemical decontamination methods. The material that is left is then scrutinised to ensure all the contamination is gone so that the residual metal is free to be recycled amongst the general metal scrap whilst the smaller volume of contaminated residue is consigned as LLW.

Incineration in a vessel designed to contain all of the combustion products is a possibility for organic substances (plastics and oils) leaving behind ash in greatly reduced volumes. LLW is also often subject to *supercompacting*, a process in which tremendous pressure ($\sim$2000 t m^{-2}) is applied to reduce its volume, after which it is grouted into large metal containers. Further areas of interest in future are to ensure appropriate sentencing, that is, that waste is characterised more effectively at source to ensure that its assignment and disposal route are appropriate and that, in particular, LLW is not inadvertently consigned as higher classifications due to shortcomings in assessment capabilities or in the absence of better data. This also supports possibilities where short-lived activity justifies *decay storage*, i.e. where the material is sorted to allow it to decay to reference levels, rather than consigned for permanent disposal.

A great deal has been learned about the waste hierarchy and SNF over the last 50 years, particularly with regard to the design and composition of nuclear fuel and waste management strategies, and the implications that these can have for disposal options in the future. For example, some early fuel designs comprised components that were very vulnerable to neutron activation. This resulted in thousands of them requiring special attention involving bespoke retrieval and consignment, often on a piecemeal basis. Also, some types of fuel cladding proved vulnerable to the effects of corrosion when stored under water to cool immediately after use in reactors (most notably Magnox cladding); this removed options as to how the fuel was dealt with and resulted in the formation of large quantities of corroded cladding/fuel material as uncontrolled wastes at the bottom of storage ponds, complicating the decommissioning of these facilities too. In this context, a strategic move towards fuel designs without reliance on components vulnerable to activation, with corrosion-resistant cladding and, in many cases, interim dry storage has greatly reduced the significance of these problems.

15.6.2 PARTITIONING AND TRANSMUTATION

As discussed in Chapter 5, and alluded to earlier in this chapter with reference to Fig. 15.2, relative to the full extent of the isotopic range of possibilities that arise as products of fission, only a few isotopes constitute a significant challenge in terms of their radiotoxicity; most of the others are short-lived and decay away quickly. This is highly relevant to radioactive waste management because it is often the long-lived group of isotopes that complicate the ease with which radioactive waste is isolated, consigned and disposed of, and that requires that waste is safeguarded for very long periods of time.

However, as discussed in Chapter 3, given a stimulus with sufficient energy to overcome the Q value for a specific nuclear reaction, it is possible to change one isotopic constituent into another. This is of great potential significance because, where a conversion of an isotope to more benign alternatives is possible, the nuclear properties of hazardous wastes might also be *transmuted* to yield a shorter-lived and less radiotoxic proposition. However, in order for this to be possible, the waste has to be processed to be suitable for transmutation to be possible (a stage termed *partitioning*) and then irradiated by some means to transmute it.

Partitioning and transmutation are techniques that are relevant to the requirements of an expansion of the world's nuclear energy supply infrastructure that are anticipated in order to appeal to future electricity demand. This is particularly relevant with regard to offsetting the potential for uranium supply constraints, as is discussed further in Chapter 16, and the possibility that energy conservation measures are not sufficient to offset the need for an expansion of nuclear energy provision needed to avert climate change. If some constituents of radioactive waste can be made more benign by this route, in principle its radiotoxicity can be reduced and its longevity as a hazard when disposed of can be shortened, thus reducing the burden on geological disposal. For example: the rate at which heat is generated by waste can impose capacity limits on repository designs because a degree of spacing is required between waste consignments to ensure the heat is dissipated safely; this might be ameliorated by reducing the inventory of those prominent heat-generating isotopes.

There is also a perceived risk of *intrusion*, that is the hazard to persons accessing waste stores at some point in the future, whether by accident or design. This might be reduced by reducing the longevity of waste radiotoxicity whilst greater assurance of waste store integrity is possible if the longevity over which a store must isolate materials from the biosphere is shortened. Last but not least, transmutation of the fissile constituents of wastes to non-fissile derivatives affords the potential for greater proliferation resistance. As a consequence, partitioning and transmutation could play a significant role in improving the public acceptability of radioactive waste management, particularly with regard to concerns over the long-term integrity of waste repositories and reducing the burden that long-lived wastes place on future generations.

However, to progress further with a conceptual discussion of partitioning and transmutation, it is instructive to revisit the composition of radioactive waste introduced in Section 15.4. As highlighted earlier, waste comprises two main forms: spent nuclear fuel and the products of waste management/fuel processing activities (i.e. LLW, ILW and HLW or their equivalents given national variances in radioactive waste classification). The principal challenges in terms of long-term waste disposal comprise the actinides and the LLFPs in high-level wastes. A summary of the specific isotopes of interest in these groupings is provided in Table 15.1.

It is worthy of note that a prominent mode of decay for LLFPs is β^- decay whereas α-decay is a prominent decay pathway for the actinides, notwithstanding a few short-lived exceptions for example where β^- decay occurs (^{238,239}Np, ^{241}Pu and ^{242}Am). The contribution to heat generation is a function of isotopic yield and specific thermal output (i.e. thermal power per unit activity W/Bq) with ^{90}Sr and ^{137}Cs dominating in the short term, plutonium in the long term. The longest-lived isotopes are ^{237}Np, ^{238,239,242}Pu and a number of the curium isotopes whilst the neutron yield is most significant for curium; this is an anticipated, complicating factor in the production of fuels that are then suitable for transmutation in a reactor because of the associated difficulty in handling them.

The concept behind partitioning and transmutation is that the most burdensome isotopes in radioactive waste (principally HLW) are isolated (usually via reprocessing schemes dedicated to sequester these specific groups of isotopes). These are then to be converted into more benign constituents, usually via their incorporation into fuels for reuse in reactors. This approach has spawned a number of advanced reactor concepts where, for example, the means for transmutation is either via the chain reaction in a reactor system or in a subcritical system in which the transmutation agent is introduced in the form of a particle beam from an accelerator. The principal isotopic components of interest are usually those that place the heaviest burden on the length of time over which repositories are required to isolate waste from the environment. These include: isotopes of americium, curium, neptunium and plutonium, and to an extent LLFPs such as ^{129}I and ^{99}Tc.

In general, the LLFPs are of less interest in terms of transmutation than the actinides because the yield is relatively small, they are generally less long-lived and of a lower radiotoxicity. However, they

Table 15.1 The Prominent Transuranic Isotopes and Long-Lived Fission Products (LLFPs) in Radioactive Waste From Nuclear Energy Production

TRU Nuclides	Half-Life	LLFP Nuclides	Half-Life	Typical LLFP Yield [8]/%
^{237}Np	2.14×10^6 y	^{14}C	5.70×10^3 y	<0.01
^{238}Np (β)	2.12 d	^{36}Cl	3.01×10^5 y	0.04
^{239}Np (β)	2.36 d	^{79}Se	2.95×10^5 y	0.11
^{238}Pu	87.7 y	^{90}Sr	28.9 y	11.78
^{239}Pu	2.41×10^4 y	^{90}Y	2.66 d	0.31
^{240}Pu	6.56×10^3 y	^{93}Zr	1.61×10^6 y	2.31
^{241}Pu (β)	14.3 y	^{99}Tc	2.11×10^5 y	19.32
^{242}Pu	3.75×10^5 y	^{107}Pd	6.50×10^6 y	4.71
^{241}Am	432.6 y	^{126}Sn	2.30×10^5 y	0.47
^{242}Am (β)	16.0 h	^{126}Sb	12.4 d	4.01
^{242m}Am	144 y	^{129}I	1.57×10^7 y	30.64
^{243}Am	7.39×10^3 y	^{135}Cs	2.31×10^5 y	25.92
^{242}Cm	163 d	^{137}Cs	30.1 y	<0.01
^{243}Cm	29.1 y	^{151}Sm	88.8 y	0.38
^{244}Cm	18.1 y			
^{245}Cm	8.56×10^3 y			
^{246}Cm	4.76×10^3 y			
^{247}Cm	1.56×10^7 y			
^{248}Cm	3.49×10^5 y			
^{249}Cm	1.07 h			
^{250}Cm	8.31×10^3 y			

Half-lives are stated to 3 significant figures and the approximate yield of LLFPs is given assuming a uranium-based fuel from a PWR. After M. Salvatores, G. Palmiotti, Radioactive waste partitioning and transmutation within advanced fuel cycles: achievements and challenges, Prog. Part. Nucl. Phys. 66 (2011) 144–166 (amended).

are generally very mobile in the environment which has implications for the assessed stability of the repository required to isolate them from the biosphere, especially where they arise in gaseous and liquid effluents. The main contributors to heat generation from the LLFP fraction (^{90}Sr and ^{137}Cs) are too short-lived to be transmuted effectively. Hence they are better stored in isolation to allow them to decay to safe levels.

With regard to the actinides, transmutation via irradiation with neutrons is the most widespread approach in order to stimulate fission of these constituents such that, in the form of shorter-lived fission products, the longevity of the waste is reduced. As a consequence, neutron economy once again becomes a relevant issue because poisoning by other isotopes (principally the lanthanides) has to be minimised (usually by removing the lanthanide component via chemical separation prior to transmutation). The possibility that neutrons are absorbed by the actinide component in preference to stimulating fission also has to be minimised. The latter scenario has the combined disadvantage of the net loss of the neutron that is consumed and the consequent increase of the actinide fraction; the latter acting to increase the radiotoxicity and longevity of the waste rather than to reduce it. This also precipitates neutron-emitting products (predominantly curium) that are undesirable in terms of fuel handling, as indicated earlier.

Several approaches have been explored to circumvent these challenges. Firstly, the partitioning component of the process is focused on isolating the isotopes that when transmuted benefit the reduction in radiotoxicity of the waste most effectively. This might comprise, for example, the separation of uranium, plutonium and neptunium from HLW waste liquors, and the extraction of the actinides and the lanthanides together followed by the separation of the lanthanides. If a thermal spectrum reactor is under consideration for the transmutation of the actinide component of long-lived wastes, then the poisoning effect of the absorption reactions of the actinides has to be offset by using greater levels of ^{235}U enrichment to achieve criticality; there can also be undesirable influences on reactivity feedback mechanisms that can imply control and reactor safety implications. To avoid these issues, a fast spectrum reactor can be used, such as a sodium fast reactor where the harder neutron field results in a higher rate of fission and consequently a reduction in the production of the higher-mass actinides than might occur otherwise via neutron absorption. This is especially relevant because a number of the more resilient actinides are susceptible to above-threshold fission that is enhanced with a fast spectrum, by definition.

Where there are insufficient neutrons to sustain transmutation *and* the chain reaction in a reactor, an alternative arrangement is to use an additional source of neutrons. This is usually derived from a particle accelerator in what are known as *accelerator-driven systems* (ADS). These are postulated to have the advantage that they avoid the reactivity penalties of transmutation described earlier, but this is at the expense of a degree of additional sophistication of the system infrastructure.

Partitioning and transmutation is an elegant concept that has the potential to reduce the time taken for radioactive waste to reach reference levels of radiotoxicity. However, it is not without its challenges. These are associated with, primarily, the separation of the constituents to be transmuted, the fabrication of fuels from these constituents that are suitable for transmutation and the physics of transmutation itself. For example, the neutron cross sections for capture and fission are critical to the viability of a partitioning/transmutation scheme but for several of the isotopes the nuclear data, and especially the neutron cross sections describing the trade-off between neutron absorption and fission, are currently subject to large uncertainties.

In terms of separations, high separation factors are necessary if wastes are to be sufficiently devoid of the very constituents that are the focus of the transmutation step for the advantage of transmuting these isotopes is to be realised to a satisfactory extent, and these processes are likely to require sophisticated civil engineering installations. Also, the repeated recycling of TRU is prevented by the build-up of ^{252}Cf due to its high level of neutron emission and the difficulties this poses on the fabrication of fuels that are suitable for transmutation.

15.7 FINAL DISPOSAL OPTIONS

Although the quantities of waste produced in the production of nuclear energy are small, some of the wastes are highly toxic and hazardous. These will have to be consigned and isolated for what is a very long time on the scale of human experience and historical record. As the production of these wastes has continued, pressure has started to accrue on storage facilities particularly in terms of capacity, whilst the storage of the most hazardous wastes (SNF, HLW and ILW) above ground in managed facilities is expensive and requires the overhead associated with the operation of secure facilities, at least in the near term. This approach is not sustainable in the long term given the length of time for which it is known that these wastes will be hazardous. Therefore, significant efforts have been made to determine safe and secure permanent methods of disposal for these materials.

15.7.1 DEEP GEOLOGICAL DISPOSAL

The concept of radioactive waste being disposed underground is almost as old as nuclear energy itself. The concept is relatively simple: the waste is placed in containers in the ground at depths of greater than several hundred metres. The containers are designed to be resistant to the influences of both the environment and the waste over very, very long periods of time and they are consigned in a stable geological location in the Earth's crust. The overarching objective of a deep, geological *engineered emplacement* of radioactive waste of this type is that the radioactivity remains *isolated* and *contained* from the biosphere until it has decayed to safe levels.

Whilst the concept described above is relatively straightforward, its realisation in practice has generally been more challenging. This is principally because of the long timescales over which it is desirable that the integrity of the emplacement is assured. On the one hand, relatively little evidence exists (although there is some that is discussed in Case Study 15.1) for systems engineered by man that have been resilient over such timescales to support the choice of materials and systems used for geological disposal. On the other, there is an unavoidable degree of uncertainty in our ability to forecast not only the likely integrity of the engineered systems but also geological stability and the structures and priorities of future societies. Therefore, because a degree of change is inevitable, the design of a GDF is required to accommodate this whilst ensuring that any dispersion of the waste that might occur at some point long in the future does not pose unacceptable health risks to future generations. In particular, the barrier system in a repository needs to function on a passive basis whilst the scenario of unintentional intrusion by future generations has to be considered so that their needs (particularly with regard to access to natural resources) are not compromised.

Two components of the disposal arrangement are subject to change in a physical sense in the context of deep geological disposal. Firstly, there is the radioactive waste itself which, as described earlier, will be subject to variation in terms of its volume, radiotoxicity, longevity, sympathy with hydrogeochemical influences and heat generating properties over long timescales. It is therefore very important that a deep geological waste repository is designed in sympathy with the type of waste that is to be consigned in it. Given that this is done, corresponding *waste acceptance criteria* have to be developed associated with a given store to ensure that only waste compatible with it is placed in it. Also, the radioactivity in the waste is decaying with time. Since this can be forecast very accurately, in comparison with the other factors referred to above, a permanent geological waste emplacement must be engineered to provide isolation and containment until the majority of the radioactivity has decayed away.

Second, the geological environment is changing as well, albeit usually extremely slowly; this can be due to tectonic drift, seismic disruption, volcanicity and also the effects of long-term changes in climate such as the effects of ice (both loading and permafrost) and sea level changes (especially for disposal sites near to coastlines). A repository therefore needs to be designed and located to accommodate these changes and thus ensure that the long-term influences of the geological environment on the waste deposit are sympathetic with the timescales necessary for the radioactivity of the waste to decay to safe levels, before dispersion is anticipated to have become significant.

Geological disposal facilities follow a design philosophy that comprises a number of barriers known as the *engineered barrier system* (EBS). In this approach, the waste is first immobilised in a solid matrix (via one of the treatments described earlier in Section 15.6.1); this constitutes the first barrier. This is then placed in a container which is often then encased in a second metal container or *overpack*, together providing two further layers of containment. The waste container is then deposited in the disposal facility comprising, for example, a specially-excavated, grout-lined vault. This is then backfilled with material selected to resist the ingress of groundwater and to form a long-lasting and intimate seal (usually bentonite clay [9] has been identified as the best choice for this purpose); finally further resilience and stability is ensured by the vault being located in a stable geological rock formation which constitutes the final barrier. Rather than a simple defence-in-depth philosophy as might be used in the operation of a reactor system that usually presupposes the various defences can be monitored and maintained over time, the role of each barrier in the EBS must be able to evolve with time. The aim is for the changes to each barrier to occur (albeit slowly) in sympathy with the others (particularly with regards to the effects of movement, groundwater flow, exacerbated corrosion, etc.).

Whilst each GDF is specific in its own right, dependent on the waste it will accommodate and the properties of the local geology, some general trends are being adopted. For example, for the metal container in which the waste in placed cast iron or carbon steel has been selected in the most mature designs thus far due to its resistance to corrosion and the potential for it to form a barrier to oxidation; in some cases this is placed in a thick copper vessel. It is anticipated that this will be resistant to corrosion for 100,000 years given the objective that the container is to remain intact until most of the radioactivity inside it has decayed away. In addition to the decay of the radioactivity, the disposal facility also needs to withstand the effects of the heat generated by some wastes (principally HLW and SNF); the thermal conductivity of the composition of the surrounding geological formation places constraints in this regard which influence the spacing required between waste packages and thus the capacity of a given waste store, as mentioned earlier.

The geological structure selected for the deposit needs to be deep, inaccessible and stable. It also needs to be of a type that is unlikely to be excavated in future for the recovery of mineral resources and it should be resistant to groundwater flow and rapid (albeit on geological timescales) movement. The protection and safety provided by the repository must be afforded on a passive basis because for most deep disposal designs operation and control are only anticipated for a maximum of 100 years, after which the assumption is that they will be closed and left. Thus, the lifetime of a permanent repository could well exceed widespread local knowledge of the existence and location of the waste deposit, say >1000 years hence.

Two design variants are envisaged for deep disposal facilities: the most widespread is the repository where waste is deposited in a vault that is accessed as part of an extensive civil engineered facility. When full, the facility would be closed with multiple engineering barriers and backfilled; the consignment is viewed as permanent but in some cases reversibility and the possibility of retrieval in extreme circumstances are also considered. Secondly but to a lesser extent, deep borehole disposal (DBD) is also being considered where waste packages are inserted into a deep, vertical, lined shaft in a stable and compatible geological formation and again backfilled and sealed. In general, the preferred geology is either one of hard crystalline rocks, clays mudstones/marls or dome/embedded halites. A comprehensive review of underground radioactive waste disposal is that by Chapman and Hooper [8].

15.7.2 MARINE DISPOSAL

There are two main routes by which radioactive waste has been disposed of to the sea: *dumping* and *effluent discharge*. Dumping is defined as any deliberate disposal of unwanted materials from man-made structures (e.g., vessels, platforms, etc.) to the ocean, and is now prohibited. Wastes disposed of in this way have mainly comprised low-level liquid and solid wastes (both of packaged and unpackaged objects), and also reactor vessels. Approximately two thirds of the radioactivity disposed of in this way is associated with nuclear-powered vessels dumped by the former Soviet Union in the Kara Sea in the Arctic and approximately one third is associated with packaged solid low-level wastes disposed of in the North Atlantic, mainly by the UK [10].

Effluent discharge comprises the controlled release of small amounts of liquids containing anthropogenic radioactivity from industrial facilities to adjacent rivers and coastal locations; in the context of the generation of nuclear energy discharges are mainly associated with nuclear plants and reprocessing facilities. Any disposal of such wastes to the sea must not result in unacceptable hazards to people and the environment. Marine discharges are thus regulated by both national and international standards with respect to radiological safety, and dedicated processing installations are often used to treat the effluent media prior to dispersal.

The conceptual basis for the marine discharge of radioactive effluents is that the reservoir that constitutes the ocean provides an environment in which radioactive material might be diluted and dispersed to below the levels of natural background concentrations. Hence, by this route noxious products from industrial activities might be rendered harmless trace residues. However, physical, geochemical and biological processes might also inadvertently constitute pathways by which harmless dilutions of non-degradable toxic agents are concentrated. This has the potential to increase the likelihood of exposure to living organisms including humans; the most prominent example being perhaps that of some heavy metals [11].

Some constituent long-lived isotopes of radioactive waste can be regarded as non-degradable on the timescales by which dilution and dispersion might be expected to occur. For effluents comprising shorter-lived radioactivity it can be preferable to store the material to allow for natural decay to remove any constraints (both technical and regulatory) with it being discharged. With regards to discharge operations, control at the point of discharge to the ocean *and* at the point of exposure when dispersed/diluted in the ocean is the most desirable option. However, control is often only practical at the point of discharge because of the extent of the range of exposure possibilities that are feasible once the contamination is dispersed.

Whilst it is relatively easy to monitor radioactivity levels prior to discharge, such an approach has to be applied in a way that ensures no-one receives doses that exceed statutory limits when the material is dispersed. This analysis results in a relatively simple set of discharge limits but, conversely, relatively large safety margins are necessary to account for uncertainties in marine behaviour. This, in turn, can precipitate significant waste treatment challenges (both technical and economic) if the materials is to be rendered below such margins prior to discharge which can favour alternative disposal methods.

As an alternative to at-source discharge assessment, the stable concentration limit of organs can be considered as a guide to the specific activity acquired by living organisms. Amongst a number drawbacks associated with this approach, perhaps the most significant is that it is not suitable for isotopic pollutants for which there are no stable alternatives (such as plutonium, technetium, etc.) which are often amongst to most relevant radiological pollutants. A further alternative is to perform a critical pathway analysis of the various environmental routes by which exposure might occur with discharge limits adjusted to suit. Examples of marine pathways include the consumption of contaminated fish or the accumulation of contaminated materials on beaches.

Whilst effects manifest in humans from dispersed radioactivity might be anticipated on the basis of exposure estimates in a few individuals, these can be impossible to discern from the natural incidence of such effects in the much greater host population. Consequently, the primary concern is usually on the preservation of marine populations that are deemed vulnerable via pathway analysis that demonstrates route(s) by which overall reproductive capacity might be affected [12]. It is worthy of note that most anthropogenic marine radioactivity derives from fallout from atmospheric nuclear weapons tests whereas the extent of controlled discharges have been reduced comprehensively as alternative approaches, especially on-site effluent treatment technologies, have been developed; a relevant summary of anthropogenic marine radioactivity in this regard is that by Livingston and Povinec [13].

Given the discussion above, it is perhaps no surprise that marine dumping of *packaged* radioactive waste is unsatisfactory due to concerns that wastes dumped at sea may at some point be mobilised due to corrosion or damage of the waste packages and thus pose a hazard to life and the environment. Further, radioactive waste that is dispersed, either intentionally or otherwise, has the potential to affect countries that are not responsible for dumping it, coupled with the constraint that retrieval is unlikely to be possible at some point in the future. Sea dumping of radioactive waste was first subject to regulations as per the London Convention 1972.[1] This resulted in several regional conventions which prohibited sea dumping outright in a number of oceans prior to 1989, with the disposal of radioactive waste at sea prohibited totally in 1994.

CASE STUDIES

CASE STUDY 15.1: NATURAL ANALOGUES OF RELEVANCE TO LONG-TERM RADIOACTIVE WASTE DISPOSAL

In contrast to forecasting the likelihood of radioactive materials being contained geologically over very long periods of time, by studying their half-lives and the resilience of engineered barriers, it has also been possible to study examples that exist today where it is evident that the mobility of substances has been constrained or preserved by geology or related influences. Such examples are known collectively in the context of waste disposal as *natural analogues* [14,15].

There are two prominent geological analogues that have been studied extensively [16]. These are the natural reactors at Oklo in Africa (as were discussed in Chapter 2) and the uranium ore deposit beneath Cigar Lake in Canada. Oklo comprises a particularly deep (300–400 m) uranium ore deposit that is believed to have sustained a critical chain reaction 2 billion years ago for a period of several hundred thousand years and which shares several features that might be imitated to confine a man-made deposit. Cigar Lake refers to a natural deposit of uranium ore formed over 1.3 billion years ago at a depth of approximately 400 m. Both deposits are embedded in a clay matrix known generally as a *clay halo*.

These analogues have informed our understanding of the disposal of radioactive waste because they combine the principal factors of a geological location, relevant mineralogy, a clay body that acts to limit groundwater penetration and also evidence of the chemical influence of the fluids on the composition of relevant elemental species. Whilst man-made UO_2 can be studied in the laboratory, geological deposits of uranium, such as uranitite, contain cation impurities that influence their stability and solution chemistry. Thus their study provides complementary insight into the long-term mobility of uranium in the ground. Further, accurate forecasting of the migration of radioactive species over ~100,000-year timeframes

[1] http://www.imo.org/en/OurWork/Environment/LCLP/Documents/LC1972.pdf.

consistent with a repository can be difficult whereas the analysis of uranium migration and, particularly for the Oklo case, the extent of the migration of fissiongenic lanthanide species can be estimated from the study of these mineral deposits. Whilst the geochemical processes at work are often complex, phenomena are observed that are common to both cases. For example, a reducing environment resulting in the long-term incorporation of radionuclides in some minerals in the clay (such as the uranium-based silicate *coffinite*, $U(SiO_4)_{1-x}(OH)_{4x}$). This is observed to have retarded mobility evidenced by, for example, the observation of very small migration distances of uranium and fission residues.

Several natural *archaeological* analogues also exist that provide evidence of the retarded degradation of objects over periods of several thousand years; these might be particularly relevant to transmutation-based disposal strategies because of the shorter timescales that are involved relative to those of the geological analogues. The first example is the discovery of over one million iron nails buried at the site of the Inchtuthil Roman fort in Scotland in the 1950s. These had been buried almost 2000 years earlier and were observed to have been rendered resistant to corrosion by a solid, iron oxide crust formed as a result of the corrosion of the outer proportion of nails surrounding the bulk. This is thought to have deoxygenated penetrating groundwater to constitute a reducing environment and thus impeded the degradation of the metal. The use of large quantities of iron is envisaged in a number of radioactive waste disposal scenarios to establish and maintain such a corrosion-resistant environment over the long periods that will be necessary.

A further example is known as the *Kronan cannon* which concerns a bronze cannon with a high copper content. This was discovered half-buried after being lost in the 1676 wreck of the Swedish warship, the Kronan. The extremely low corrosion rates evident from analysis of its surface (~0.15 μm per year) have been used to estimate long-term corrosion rates of copper waste disposal canisters. Similarly, evidence for the longevity and stability of cements can be drawn from the preservation of ancient constructions, such as Hadrian's wall, whilst the benefits of clay deposits in the preservation of 1.5 million-year-old trees at Dunarobba in Italy and of the body of Xin Zhui preserved in a Chinese tomb 2100 years ago have also been studied [17].

CASE STUDY 15.2: THE KBS-3 REPOSITORY

The majority of the nuclear waste [18] that is to be consigned in the KBS-3 repository in Sweden is spent LWR fuel from the Swedish fleet of power reactors, with there being similar plans in Finland. The Swedish fleet spans BWR and LWR designs. Most of the spent fuel material is in the form of assemblies of UOX pellets, in zirconium alloy clad and there is also a small amount of MOX fuel from one of the BWR plant. Some is stored underwater to cool at the Swedish interim storage facility, Clab, and the rest is forecast on the basis of current reactor operations in what is termed the *SKB reference scenario*. Spent fuel exists from the 1970s through to a projected end of life of existing operations to ~2045 of approximately 11,000 t in total. In addition there is also a smaller amount (~45 t) of what is referred to as *miscellaneous fuels* from the early activities of the Swedish nuclear industry and research reactor use. This comprises fuel residues, some MOX fuel, fuel from several small heavy water reactors that closed in 1974 and some spent metal fuel from a research reactor that closed in 1970. Only spent fuel deemed to be of a low-solubility form (i.e. oxide) will be accepted for disposal in the KBS-3 repository.

The KBS-3 system [19] comprises five main components: the waste *canister*, *buffer*, the surrounding rock (*host rock*), *backfill* and *closure*. It is important that the functions and properties of one barrier do not undermine those of another in the system. A schematic of the system and the deployment/plug arrangement is provided in Fig. 15C.1. The waste canisters will be placed in deep horizontal tunnels or vertical shafts (termed *openings*) that are then filled with buffer material, backfill and closed off. The openings have to be compatible with anticipated thermal, mechanical and hydro-geochemical conditions whilst ensuring the radioactivity in the waste is contained until it is no longer hazardous. They are then sealed with plugs to withstand the potential for increases in hydrostatic pressure and expansion of the backfill.

The canister is required to contain the waste, prevent its dispersion, resist the effect of external loads, resist corrosion, prevent criticality, shield the emission of radiation and be of a standard size to facilitate easy transport and deposition. In this case it is a thick-walled cylinder made of oxygen-free copper that is resistant to corrosion and is ductile; it is anticipated that it will remain intact for 100,000 years. To provide the necessary mechanical integrity, a load-bearing cast iron insert will be placed inside that will hold the spent fuel assemblies; two types of insert are anticipated to cater for the different geometries of PWR and BWR fuel assemblies, and to minimise the potential for a criticality to occur.

The buffer is the barrier between the canister and the surrounding host rock and is made of bentonite clay. It is required to prevent water flow and to limit any movement of the canister. It is also necessary in order to limit microbial activity that might cause corrosion and also to minimise the transport of radioactive substances.

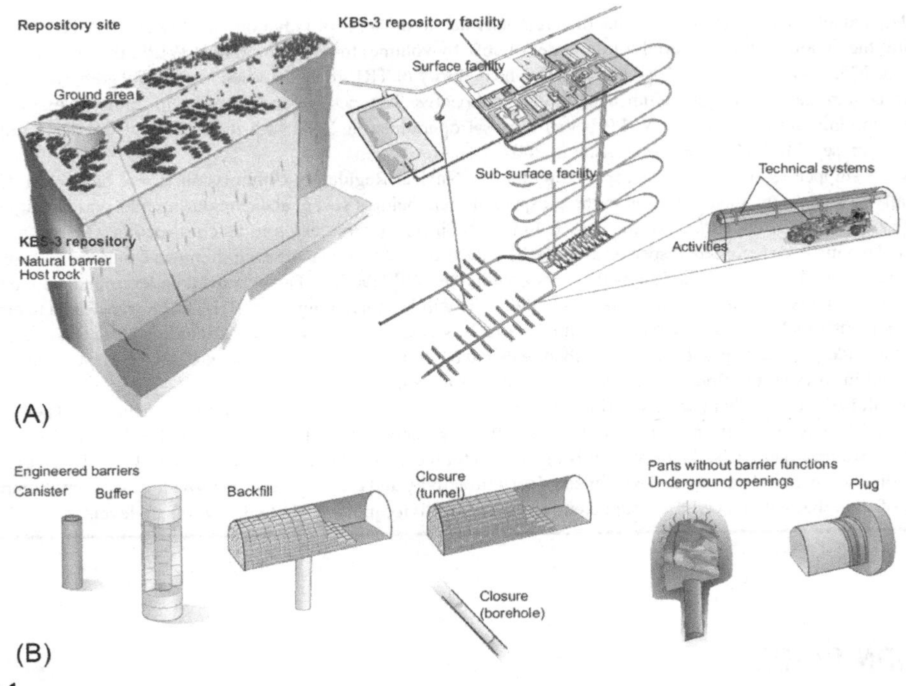

FIG. 15C.1

(A) A schematic of the KBS-3 repository after closure (left), its design and operation (centre and right) [19] and (B) the various barriers and components associated with the consignment of the waste.

Source: Posiva Oy.

Once a waste canister has been installed in a repository opening, and the buffer has been put in place around it, the access route will be backfilled with bentonite blocks and pellets to limit water flow and expansion of the buffer. The boreholes, shafts and access pathways will then be closed to restrict unintentional intrusion and to limit the flow of water through these facets; the closure will comprise pre-compacted clay, compacted rock material and blocks of crystalline rock. When full, a facility's production lines and transport systems would cease operation and be decommissioned. Finally the facility would be closed [20].

CASE STUDY 15.3: WASTE ISOLATION PILOT PLANT (WIPP)

Although there have been several repository designs, locations have been identified for them and public consultation exercises have been carried out throughout the world, the only operating deep repository at present is the Waste Isolation Pilot Plant (WIPP)[a]. This is located in the Chihuahuan desert outside Carlsbad, New Mexico in the United States.

The WIPP facility is located 700 m deep in a 250 million year old, ~600 m thick salt deposit. It was designed with waste acceptance criteria for TRU that is derived from US nuclear defence activities. WIPP accepts both TRU that can be handled by workers without any shielding, other than that provided by the containers in which it has been placed (termed

[a]http://www.wipp.energy.gov.

CH-TRU), and also that which has to be handled remotely due to the level of radiation that it emits (RH-TRU). CH-TRU constitutes the majority of the material (of the order of 96% by volume) to be disposed of at WIPP; the facility has a total capacity of 175,570 m^3 for CH-TRU and CH-TRU. The majority of TRU to be disposed of at WIPP comprises items and materials contaminated with plutonium and/or other radioactive isotopes beyond uranium in terms of mass, such as residues, clothing and other debris. WIPP began disposal operations in 1999 with the receipt of CH-TRU shipments and latterly those of RH-TRU were first made in 2007.

Waste is shipped to WIPP in casks certified by the US Nuclear Regulatory Commission. These have been designed specifically to contain the material, even in the prospect of there being a severe accident during transport whilst the dose emitted by the waste at the surface of the cask is limited by incorporating lead into its construction. Once at the WIPP facility and having been recorded, inspected and checked, the drums of waste inside each shipment container are transferred into a canister and then into a cask designed for disposal in the WIPP facility. This is moved underground via the facility elevator to a complex of underground rooms where the waste will be stored long-term. RH-TRU is transferred to emplacement equipment which pushes each canister out of the facility cask in order to deposit it into horizontal boreholes in the walls of the underground disposal spaces, and then seals the borehole with a concrete plug. CH-TRU waste containers are then stacked in rows on the floor of the underground disposal rooms.

Deep salt beds such as that used for WIPP were formed many millennia ago as a result of ancient seas. They have the advantages that they are easily mined, devoid of fresh flowing water and geologically stable. The National Academy of Sciences determined them to be the most promising method for the disposal of radioactive waste in 1957. This was done on the basis that they are rarely associated with areas of seismic activity and are anticipated to provide a stable environment for a period of time that will exceed the time taken for the radioactivity in the TRU to decay to safe levels.

REVISION GUIDE

On completion of this chapter, you should:

- understand what is meant by the term *radioactive waste* and appreciate that contaminated material without any intended purpose can arise from a variety of activities, in a range of quantities and in different physical forms
- appreciate that the *composition of radioactive waste* varies in terms of its chemical and physical composition and also in terms of isotopic content
- be familiar with the priorities that the *production* and *volume* of radioactive waste should be minimised, and that its vulnerability to inadvertent dispersion should be limited by *immobilisation* and *isolation*
- understand why our approach to managing radioactive waste varies from little more than is necessary for general, non-radioactive waste materials through to very significant and sophisticated processes and strategies dependent on the hazard associated with a given waste product
- be aware of the general approach taken worldwide to the *classification* of different types of radioactive waste in terms of *radiotoxicity*, propensity to generate *thermal heat*, its *transuranic composition* and *long-* and *short-lived isotopic constituents*
- appreciate the concept of *reference level* and its significance in estimating the time horizon by which radioactive waste might be deemed to have decay to levels consistent with ore used for the production of nuclear materials
- be able to describe the *isotopic content* you might associate with the different radioactive waste categories, particularly in terms of the *timescales of decay* and the major influencing constituents

- be able to summarise the variety of *treatment options* that exist for the *management of radioactive waste*
- similarly, be able to summarise the variety of *disposal options* that exist for radioactive waste including those that are defunct, such as the consignment in unlined near-surface trenches or marine disposal, and also those that are in plan or current use; the latter should include long-term solutions such as deep geological disposal as per the approaches of KBS-3 and WIPP
- appreciate the benefit that the study of *natural geological and archaeological analogues* can bring to our understanding of long-term disposal of radioactive waste

PROBLEMS

1. Explain the complementary purpose of the copper canister and the iron insert that comprise the design of the spent fuel canister that are to be used for disposal in the KBS-3 system.
2. Explain, with reference to the key challenges, why the disposal of packaged radioactive waste at sea is prohibited. Hence, explain how effluent discharges are safeguarded in terms of the dose to living things and describe some of the challenges in estimating this.
3. Identify which group of isotopes offer the greatest potential in terms of the likely benefits of processing radioactive waste by partitioning and transmutation and explain why this is so. What are these benefits?
4. Explain what is meant by the term *reference level* with respect to radioactive waste and radiotoxicity. What are the main dependencies in terms of a reference level?
5. One function of reprocessing spent nuclear fuel is that the actinides (for the purposes of this question uranium and plutonium) are removed. Describe the advantages and disadvantages of nuclear reprocessing in terms of long-term radioactive waste disposal.

REFERENCES

[1] F. Joliot, Chemical evidence of the transmutation of elements, Nobel Lecture, 1935. 12th December.

[2] C. Fiorina, The Molten Salt Fast Reactor as a Fast-Spectrum Candidate for Thorium Implementation (PhD dissertation), Politecnico di Milano, 2013. p. 32.

[3] OECD NEA, Accelerator-Driven Systems (ADS) and Fast Reactors (FR) in Advanced Nuclear Fuel Cycles, 2002, https://www.oecd-nea.org/ndd/reports/2002/nea3109-ads.pdf.

[4] Public law 102–579, The Waste Isolation Pilot Plant Land Withdrawal Act, 1992, amended 1996, http://www.wipp.energy.gov/library/CRA/BaselineTool/Documents/Regulatory%20Tools/10%20WIPPLWA1996.pdf.

[5] URN 15D/472, UK Strategy for the Management of Solid Low Level Waste From the Nuclear Industry, Department of Energy and Climate Change, 2016.

[6] M.I. Ojovan, W.E. Lee, An Introduction to Nuclear Waste Immobilisation, second ed., 2014, Elsevier, p. 362.

[7] W.E. Lee, M.I. Ojovan, C.M. Jantzen (Eds.), Radioactive Waste Management and Contaminated Site Clean-up: Processes, Technologies and International Experience, Woodhead, 2013, p. 879.

[8] N. Chapman, A. Hooper, The disposal of radioactive wastes underground, Proc. Geol. Assoc. 123 (2012) 46–63.
[9] R. Pusch, Use of bentonite for isolation of radioactive waste products, Clay Miner. 27 (1991) 353–361.
[10] K.-L. Sjöblom, G. Linsley, Sea disposal of radioactive wastes: the London Convention 1972, IAEA Bulletin, 2/1994.
[11] See for example: D.S. Woodhead, Marine disposal of radioactive wastes, Helgoländer Meeresun. 33 (1–4) (1980) 122–137.
[12] Effects of ionizing radiation on aquatic organisms and ecosystems, IAEA Technical Reports Series No. 172, 1976.
[13] H.D. Livingstone, P.P. Povinec, Anthropogenic marine radioactivity, Ocean Coast. Manag. 43 (2000) 689–712.
[14] Use of natural analogues to support radionuclide transport models for deep geological disposal repositories for long lived radioactive waste, IAEA-TECDOC-1109, 1999.
[15] W. Miller et al., (Ed.), Geological Disposal of Radioactive Wastes and Natural Analogues, vol. 2, Elsevier Science, Oxford, 2000.
[16] F. Gauthier-Lafaye, P. Stille, R. Bros, Special cases of natural analogues: the Gabon and Cigar Lake U ore deposits, in: Energy, Waste and the Environment: A Geochemical Perspective, vol. 236, Geol. Soc, London, 2004, pp. 123–134.
[17] Safety case for the disposal of spent nuclear fuel at Olkiluoto – complementary considerations 2012, Posiva 2012-11, ISBN 978-951-652-192-6, www.posiva.fi/files/2995/POSIVA_2012-11_web.pdf.
[18] Spent nuclear fuel for disposal in the KBS-3 repository, Technical report SKB TR-10-13, 2010, www.skb.se/upload/publications/pdf/tr-10-13.pdf.
[19] Design and production of the KBS-3 repository, Technical report SKB TR-10-12, 2010, www.skb.se/upload/publications/pdf/tr-10-12.pdf.
[20] R. Lidskog, G. Sundqvist, On the right track? Technology, geology and society in Swedish nuclear waste management, J. Risk Res. 7 (2) (2004) 251–268.

CHAPTER

PUBLIC ACCEPTABILITY, COST AND NUCLEAR ENERGY IN THE FUTURE 16

16.1 SUMMARY OF CHAPTER AND LEARNING OBJECTIVES

In this chapter, concepts associated with *public acceptability*, *economics* and the *future of nuclear energy* are introduced and discussed. These important topics span the fields of cost and the public perception of risk at an important time for the world in terms of sustainable development, globalisation and climate change.

The objectives of this chapter are to:

- introduce the historical context in terms of public consultation and cost associated with the nuclear power plant (NPP) in the world today
- introduce four examples in the context of public acceptability: *accidents*, *siting and local communities*, *intergenerational equity* and *radioactive waste*
- explain the prominent issues associated with how NPP are financed and the influence on cost of long timescales and fuel cycles
- summarise the conclusions of recent forecasts of nuclear energy production and what can be learnt from them
- explain the context of nuclear energy as a low-carbon component in the nuclear energy mix of the future
- provide a current summary of nuclear energy construction plans by geopolitical area

16.2 HISTORICAL CONTEXT: ALBERT EINSTEIN 1879–1955

Einstein is known universally as the stereotypical genius and physicist. In the scientific community, he is known most prominently for his four seminal works of 1905. These focussed on, in chronological order, the *photoelectric effect*, *Brownian motion*, *special relativity* and *the equivalence of matter and energy* (Fig. 16.1).

The work of most relevance to nuclear energy is the last example given in the list above. In this, Einstein produced what is often reputed to be the most famous equation: $E = mc^2$. This related the mass of a body to its energy, and vice versa. However, it also heralded the existence of nuclear energy in terms of the mass defect of nucleons bound together by the strong nuclear force. Einstein also lent his support to a group of physicists in 1939 by supporting the letter with Leó Szilárd to President

Nuclear Engineering. https://doi.org/10.1016/B978-0-08-100962-8.00016-0

FIG. 16.1

Albert Einstein.

Roosevelt described in Chapter 2. This highlighted the possibility of Nazi developing a nuclear weapon and was followed by several meetings with the President who instigated the Manhattan project; so began the long association between nuclear energy, public and political acceptability, and economics.

16.3 INTRODUCTION

Opinions can be divided about technologies that present hazards but also distinct benefits; nuclear energy being one such example. When brought together as a collective view, these perspectives can constitute the case as to whether the issue is deemed *publicly acceptable* or otherwise. Public consultations are often held to ascertain whether the public accept a specific technology option. Such activities can involve data gathering exercises including surveys, interviews, focus groups and, occasionally, referenda. Despite inevitable constraints as to the extent covered by such exercises, these activities are essential in order that preconceptions and assumptions that might be held about the view of the public are tested.

When the first, large-scale infrastructure for nuclear materials development was put in place in the World War II, the situation was very different and the public were not canvassed as to their view at all, in the interests of the military imperative and national security. In the most extreme cases, residents on land that was targeted for the developments received instruction, were compensated and given a date by when they should leave. Often, even the people actually involved in the work that followed did not know what the overall objective was; such were the imperatives of war time. While this may have contributed, eventually, to a perception of the industry as being autocratic and may have undermined public trust and confidence, nuclear energy was considered positively as a source of cheap and unlimited electricity by many, at least until the Three Mile Island (TMI) accident in 1979.

Public meetings with communities local to early civilian nuclear energy developments were often brief in comparison with the lengthy, consultative practices that are adopted now. However, this was not specific to nuclear energy; infrastructure development in the middle of the 20th century, particularly those developments associated with transport and energy, was less receptive to public concerns than it has to be today. The era immediately after war was different; relatively little was understood about the risks of large-scale developments or even how the view of the public might be best obtained. Often it was assumed that public attitude would be positive, due in part to there being perceived benefits such as employment and economic growth, but such assumptions were not always accurate.

However, public enquiries have for some time been much more extensive exercises, particularly following high-profile incidents such as TMI. For example, the Sizewell B public enquiry lasted more than 3 years in the early 1980s; it was the longest enquiry of its sort for some time afterwards. It has been evident for some time that society is changing to one in which we are 'becoming increasingly preoccupied with risks as part of everyday life' [1] and nuclear energy developments are no exception.

More recently, public acceptability and the methods by which it is assessed are becoming tested more frequently due to *climate change*; large-scale infrastructure developments associated with *renewable energy and transport*; *shale gas extraction* and, in this context, the *long-term disposal of radioactive waste*, *nuclear decommissioning* [2] and the *construction of new nuclear power stations*. While more is known about our impact on the Earth's geology and ecosystems than ever before, the choices open to us can appear more limited, and the implications of the decisions of greater potential significance both to us and our descendants; further, the choice to do nothing is no longer an option.

In this chapter, we consider the vocabulary of the processes by which public acceptability is considered and the scenarios by which nuclear power will be part of the world's future energy economies. This chapter follows later in this text so that it refers, where necessary, to the technical definitions provided earlier.

16.4 ISSUES OF PUBLIC ACCEPTABILITY AND RISK IN NUCLEAR ENERGY

The perception of *risk* (risk being the probability of a hazard resulting in harm) is often based on knowledge that is derived from a variety of sources. For example, from that which has been acquired formally (e.g., via education, training, etc.), from experience (either professional or personal), gleaned from sources (e.g., people, media, etc.) and so forth. Hazards that are perceived as involuntary, imposed or difficult to understand and particularly those that are perceived to pose a threat to human health tend to spawn negative attitudes. As Ahearne asserts [3], 'whether … these attitudes are technically valid is not … relevant for addressing public concerns in a democratic society'. This is a conceptually important point in our consideration of the public acceptability of nuclear energy and the examples that follow in this section.

16.4.1 ACCIDENTS

It is perhaps not surprising that events that impact the public directly (particularly with regard to infrastructure such as energy and transport) tend to reduce public confidence in the sectors associated with them, at least for a period of time that follows immediately afterwards. However, the manifestation of such effects can be more complicated than might first be assumed. For example, a train accident in one country perceived as being unlikely to occur in another due to a perception of there being a difference in operating systems, for example, might be unlikely to influence public attitudes elsewhere.

Nuclear energy is similar to other safety critical sectors, such as aerospace, oil, gas and so on, in some respects but contrasting in others. For example, there can be a perception of choice with regard to managing concern with respect to international transport, whereas the perception can be one of a risk imposed with respect to energy and its infrastructure, especially in contexts local to the associated infrastructure. Most significantly, as highlighted in Chapter 14, severe nuclear accidents can be characterised as extremely low-probability events that have the potential for significant disruption and upheaval. Reactor designs and regulator procedures can suggest that accident sequences are confined to specific, national contexts but, conversely, the dispersion of radioactive contamination does not respect national boundaries. As in the case of the airline industry, the nuclear industry shares experience and learning from incidents via, for example, the World Association of Nuclear Operators (WANO).[1] This ensures that best practice is maintained and distributed worldwide, and aims to avoid for example the possibility that generic scenarios that might constitute precursors to accidents are not replicated, albeit inadvertently.

By way of example of the significance of public acceptability, we shall return to the case of the Fukushima Daiichi accident in 2011, as discussed in case study 14.3d of Chapter 14. Following the accident, Poortinga et al. [4] reported in 2013 that, while an increase in scepticism over nuclear power as a solution to climate change was evident in Japan, this was not so in Britain. This was set in the context that nuclear power had, in the years prior to the accident in 2011, been considered widely as a means of low-carbon electricity production central to the climate change commitments of both the United Kingdom and Japan. Conversely, public trust in safety and particularly in the regulation of nuclear industry in Japan fell following the accident, whereas in the United Kingdom, it was relatively unaffected. There is evidence that Japanese public opinion had been affected by several accidents at NPP prior to the Fukushima Daiichi accident; these incidents were more significant in terms of direct fatalities but less so in terms of economic consequences and upheaval.

A key finding of Poortinga et al. is that the Fukushima Daiichi accident did not impact public opinion of nuclear power in Britain as significantly as had been expected, particularly in the context of it addressing what was perceived as the greater issue of climate change. This is not meant to infer that public acceptability is robust against nuclear accidents unless the public concerned are local to the accident, but rather that the influence of accidents on public opinion is complicated and can be contrary to expectations; thus, intuitive assumptions concerning public acceptability are no substitute for evidence-based research. Interestingly, as mentioned in Chapter 14, the response to the Fukushima accident in Germany (given a different political energy context to both the United Kingdom and Japan) was to schedule to closure of its nuclear stations; reviews of nuclear energy policies were also provoked in Italy and Switzerland.

Therefore, it can be concluded that severe nuclear accidents can have a significant, negative influence on the public acceptability of nuclear energy technology; however, predicting the long-term outcome of such influences is not easy to forecast and can be dependent on a number of factors of which that of regulation is often key. There is also evidence that NPP accidents can influence public views of related nuclear activities, such as waste disposal [5], as discussed later in this section.

[1]http://www.wano.info.

16.4.2 SITING AND LOCAL COMMUNITIES

NPP and related facilities are long-lived installations. Those being constructed now may operate for more than 60 years followed by lengthy periods of clean-up and decommissioning. Often, new power stations have been sited on the same sites as earlier ones. This is due to, for example: the relative ease of approval and environmental review of an already-licensed site, as opposed to a new location; physical geography that is understood to be sympathetic to such an installation; the availability of compatible infrastructure (particularly with regard to cooling and power take-off) and the availability of a nearby, experienced and nuclear-competent workforce. This has been appreciated for some time: for example, Burwell et al. stated in 1979 when they examined such a policy: '...a practical path to a nuclear system...is to place new reactors largely on existing sites' [6]. This can be of significant relevance to issues of public acceptability. Indeed, a further motivation towards existing sites can be the anticipation of an already sympathetic local community.

Places associated with industries stigmatised by an association with a perceived major hazard, that is, in this context, *nuclear energy* and *radioactivity*, respectively, can become infamous in popular culture. A proportion of this reputation in the current legacy of some nuclear facilities accrues from the era of military secrecy associated with the dawn of the industry and from the era that followed, particularly where it is perceived that decision making was more reckless than today. Further, the hazard(s) (at least those pertaining to ionising radiation being the main distinguishing feature of NPP) are often invisible and can be long lasting.

However, it would be wrong to conclude that the public local to such long-lived, infamous installations and brownfield developments reject them out-of-hand in their local communities. The *proximity effect* [3] describes the phenomenon associated with communities close to existing installations being more supportive of them than those further away. This is thought to arise because communities local to existing sites can be more accepting due to a number of factors such as familiarity and as a result of the recognition of the direct economic and social benefits that such developments may bring. By contrast, greenfield developments can generate greater levels of concern due to factors such as uncertainty and a perceived lack of control. In some reports, the proximity effect is referred to as the *halo effect* in that people 'reject negative associations with living close to such an environmental hazard' [7].

It is tempting to put supportive contexts down to the simple case that those benefiting directly, for example, via employment, increased revenues and so on, may have different views to those that do not. However, as for example Parkhill et al. [8] explain the situation with acceptability of host communities is much more complex than this, and it is often dynamic and transitory. This assertion is particularly relevant now, as many components of the worldwide nuclear industry seek to build new reactors on both existing sites and also on new sites; the latter associated in particular with emerging economies (particularly Brazil, Russia, India and China—the BRIC grouping[2]) that, for example, have been termed *embarking countries*.[3] It is also relevant to governments and utilities seeking sites for long-term waste disposal repositories.

The public acceptability of what Parkhill et al. for example term 'controversial technologies of modernity' has become a widespread focus for governments, technology vendors and planners. The

[2]Bric being the acronym used to refer to the grouping of emerging markets in the global economy defined by Jim O'Neil (Chief Economist, Goldman Sachs) in 2001.
[3]https://gnssn.iaea.org/regnet/embarking/Pages/default.aspx.

definition of these technologies goes further than NPP to encompass the siting of wind turbines, shale gas extraction facilities, waste incinerators and so on. The detail of the concerns of local communities can be different; concerns can be associated with the specific technology, that is, in this case, *nuclear energy*, or with specific issues or developments associated with what is often a pre-existing activity, that is, in a nuclear context the use of mixed-oxide fuel, waste disposal plans and other, specific perceived hazards.

The importance of public acceptance is much more accepted generally in the context of the 'location, adoption or rejection of technologies' [6], but the variety of views and perspectives are as significant if not more important than self-interest, as Parkhill et al. state, the 'importance and heterogeneity of the extraordinary in nuclear affairs' and what people consider important in where they choose to live is an important area of sensitivity when siting infrastructure near to people's homes. In common with all consultation and negotiation practices, openness, accuracy, trust and respect for the role that physical and historical experiences play in what concerns us about hazards in our neighbourhoods are of prime significance when making an assessment of the view of the public.

16.4.3 INTERGENERATIONAL EQUITY

The Organisation for Economic Cooperation and Development (OECD) defines *intergenerational equity* as referring '...within an environmental context, to fairness in the intertemporal distribution of the endowment with natural assets or of the rights to their exploitation'[4]; that is, it concerns the balance or otherwise of the impact of decisions made now by the current generation and their impact on future generations. It is an important ethical concept where there is the likelihood that the impact of the activities of the current generation might be significant and sufficiently long lived to bear an influence on future generations. Often, the reversibility or otherwise of the impact of a given decision on both current and future generations are important considerations.

Long-lived radioactive waste is often used to illustrate intergenerational equity. It was one of the earliest environmental issues to attract this association. Intergenerational equity is central to radioactive waste management, as per the definition by the International Atomic Energy Agency (IAEA) that waste must be managed 'without posing undue burdens on future generations' [9]. However, in today's *risk society* [10], there are a variety of impact contexts that require decisions *by man* (now) if the impact *on man* (in future) is not to constitute, an *undue burden*. These include, for example, *climate change*, *loss of biodiversity*, *air pollution* and other associated 'ecological megahazards'.[6] Use of resources, such as copper, oil and natural gas, is also attracting related attention in that, if used now, future generations may not be able to derive equitable benefit. Population growth has been regarded as 'perhaps the greatest intergenerational risk' [2] by some, particularly because of the implications of net growth on consumption and the widening of the divide between wealth and extreme poverty.

Intergenerational equity is fundamental to *sustainable development*, as per the definition of the latter as 'development that meets the needs of the present without compromising the ability of future generations to meet their own needs' [11]. As the extent of our reliance on technology continues to escalate, the significance of our influence on the lives on future generations appears at risk of escalation too. Perhaps, nowhere is this more apparent than in our search for energy and the infrastructure that this requires.

[4]http://www.oecd.org/.

At the focus of intergenerational equity is the ethical stance that the decisions taken by the current generation impact those that (might) need to be taken by generations in the future. This is a simple concept that can be complex in practice, especially in relation to rights and welfare, and costs and debts. For example, we may as a society *choose* to 'dispose' of long-lived radioactive waste in a geological disposal facility, but the interpretation of the term *disposal* requires careful consideration. For example, *deep disposal* methods usually imply permanence with no facility for retrieval of the waste. This might constitute the optimum technological solution by which future generations are isolated from the radiological hazard. However, in selecting such approach, it might be argued that current generations are denied the benefits they might have enjoyed had the cost of such a solution (rarely insignificant) been spent on their welfare instead, that is, intergenerational equity is being maintained at the expense of *intragenerational equity*. Further, were the waste to leak (albeit unlikely given this is what a consignment is designed to prevent at all costs) once consigned and risk contamination of the biosphere (e.g., groundwater, foodstuffs, etc.), then the impact on future generations could be significant. In this regard, the decision not to assure the waste was retrievable could be deemed to have imparted a burden on future generations unbalanced relative to that of current generations; a scenario the converse of welfare example. In terms of costs, perhaps the most immediate example is that of current estimates not covering what will be required in future, thus imposing further expenditure on those that have not benefited directly from the power that was generated that produced the waste.

The difficulty remains that it is impossible to anticipate the social structure of the world on the timescales necessary for long-lived radioactive waste to decay to benign levels; notwithstanding the fact that the levels of radioactivity associated with either low- or intermediate-level waste are drastically lower than those of high-level waste or spent nuclear fuel, and that the majority of the radioactivity associated with, for example, ^{90}Sr and ^{137}Cs decays relatively quickly. Such timescales exceed recorded human history, and also current estimates associated with other ecological megahazards such as the effects of climate change or shrinking biodiversity, although it is critical to state that our technological appreciation as to how to offset the impact of these nearer-term examples is arguably much less well evolved.

Conversely, not to secure waste from the risk of illicit use (terrorism, etc.) via permanent disposal in the relative short term might again deny current generations their rightful equity. As Shrader-Frechette states [12] '...if there is a risk that is unacceptable to present persons...then the risk would not be acceptable to future persons either', that is, espousing the view that intergenerational equity is a statement that current and future generations have equal rights. The IAEA summarise this in a radiological context that 'impacts on the health of future generations will not be greater than relevant levels of impact that are acceptable today'.[5] It is worthy of note that decisions taken with regard to radioactive waste management, early in the nuclear era, have resulted in very significant costs in only a few decades, because in some cases, they have exacerbated problems that have needed to be addressed now. However, this is arguably more a result of a lack of planning and understanding at that time rather than a case of disregard for intergenerational issues; much has been learned from this experience.

Also of relevance to intergenerational equity is the concept of *discounting*, in terms of the anticipated cost (price) of something required in the future being converted to an estimate of its worth now, usually in terms of its *net present value* (NPV). Money secured now for requirements in the future

[5]Principle 4, Protection of future generations, International Atomic Energy Agency (IAEA) Safety Fundamentals, Annex I, https://www.oecd-nea.org/rwm/reports/1995/geodisp/annex1.html.

will grow at a rate of inflation and constitute a greater sum. Discounting considers the value in the future (e.g., the cost of the long-term disposal of radioactive waste or decommissioning in this context) and converts it to its value today.

The debate concerning discounting with regard to long-term macroscopic issues, such as nuclear waste management and decommissioning, is that there are clearly several unknowns that influence present value. For example, an accurate forecast of inflation rates becomes more difficult to achieve as timescales extend further into the future, such timescales are long and it is not known very precisely *when* the present value that is subject to discounting will be required. This is particularly relevant where the discounted sum is intended to offset the costs of unforeseen, unscheduled and improbable events such as the clean-up of a severe accident; the element of risk complicates this scenario because it is likely that the sum would not be needed at all. Discounting *per se* over long timescales at estimated rates of inflation can mean that a relatively small sum invested now will be sufficient to cover very large financial costs incurred a long time in the future [13]. However, perspectives vary as to whether discounting is a valid basis for apportioning value or, conversely, is ethically suspect [14].

It might be argued, albeit naïvely, that a sum should be invested now at present day value to cover all eventualities. However, discounting over generations would imply that such a sum would be astronomical by the time it might be needed. As per discussions above, the generations between now and the eventual use of this investment would not have had the benefit at the time that they might have utilised it against 'real' welfare requirements as opposed to what are hypothetical needs (hypothetical in the sense that anything anticipated in the future is hypothetical until it is enacted).

Clearly, intergenerational equity is a complex issue: few people today wish to burden their descendants unreasonably with the consequences of their actions. However, examples exist where this is done somewhat implicitly: large capital projects benefiting people today (and sometimes large operational requirements) are often afforded by governments incurring a deficit, which is borne by future generations, discounted accordingly. Such scenarios are similar but different. For example, the timescales are usually shorter than those estimated to discount the cost of a catastrophe in 100 years' time, and hence the present value might be understood more accurately.

16.4.4 RADIOACTIVE WASTE

The management of radioactive waste arising from nuclear energy production is arguably the most prominent nuclear energy issue in terms of public acceptability, national variations notwithstanding. It is an issue that raises concerns, for example, over the potential for accidents, siting, intergenerational equity (introduced earlier) and the ethics of compensating host communities [15].

The awareness of the need for long-term, management strategies by which radioactive waste can be isolated from the biosphere is by no means a nascent realisation; the issue might not have been at the forefront of priorities during the Manhattan project, for example, but the advantages and disadvantages of temporary and permanent methods of disposal, and also the prospects for the use of neutralisation (transmutation), immobilisation in glass and even disposal in space were first discussed many decades ago [16].

For much of the waste that needs to be dealt with, robust technological solutions exist for both retrievable and permanent variants of disposal, as discussed in Chapter 15. In contrast, issues associated with the public acceptability and the cost of these waste disposal methods have endured despite the technical solutions themselves often being already several decades in development. Many countries have still not implemented disposal strategies because the public, especially those in prospective host communities, have not accepted proposals made to them.

Radioactive waste management can also have significant implications for the planning and operation of new NPP. For example, the proposal for a plant in Blythe, California (Sundesert), was deemed unable to proceed in 1978 until a permanent waste disposal method was demonstrated; the plant was never built. Similarly, in 1977, the Swedish government required plant owners to have developed a technical route for the disposal of radioactive waste before starting new reactors or refuelling existing ones.

Often, recognition of the benefits of nuclear energy (e.g., low-carbon production, security of supply, etc.) can be conflicted by a desire among both organisations and individuals for resolution of what is often regarded as the 'waste problem'. Public acceptability arguments concerning radioactive waste can span a range of concerns over whether the consignment is *permanent* (and thus not retrievable in the event that future technology might become available to neutralise it), and those over *temporary* solutions (perceived as being at risk of interference and terrorism). Not surprisingly, siting in local communities can be the subject of extensive consultation and debate, with voluntary communities often being sought for site investigations for waste repositories (referred to as 'voluntariness' [17]).

Views that are rarely debated are that waste from a given country ought to be dealt with within the borders of that country and that preparation ought to be made now rather than the problem be left for future generations to deal with; unsurprisingly public opinion can be accepting of the long-term, geological disposal concept so long as the preferred site is a long way away from the homes of those accepting it. Concerns that might be raised in public consultations include, for example, the possibilities of a *leakage of radioactivity* at the disposal site, an *accident at the store* or *an accident during transport* of the waste to a site, *terrorism*, a *decline in property values*[1] and so forth; contamination of the biosphere at some point in the future as a result of a leak is often the most significant concern. Simultaneously, and perhaps a little ironically, this is usually the principal design criterion of a given disposal technology.

Positive economic factors associated with the development of a repository might include the potential for diverse employment opportunities, and associated construction, management and monitoring activities. Negative economic factors might include, for example, the potential impact of waste storage on tourism and property values, and increased congestion of transport networks. Some opposition to nuclear waste siting has its roots in opposition to nuclear energy where, for example, agreeing a waste disposal solution is perceived as potentially condoning an indefinite commitment to nuclear power. Where *attitudes* towards nuclear energy are accepting (particularly in cases where alternative sources of energy are sparse), it has been shown that these can be a positive influence as to the siting of waste storage [18]. It is becoming widespread common practice to seek siting of repositories for the disposal of NPP waste (spent fuel) on an existing nuclear site, cf. Sweden and Finland.

While much waste management research has been done throughout this first century of nuclear energy (principally technological but also sociological), the implementation of long-term radioactive waste disposal solutions has stalled in many cases. This is particularly evident for the case of spent nuclear fuel and high-level waste. Progress on this matter is dependent on there being valid opinion studies, planned strategies by which risks are communicated, widespread participation, stakeholder involvement, transparency and the building of trust and confidence. It is a requirement that requires governments and utilities to understand public attitudes and concerns but also to evidence leadership.

From a positive perspective, extensive preparations for the permanent disposal of low- and intermediate-level wastes have been made in Finland and Sweden. However, less progress has been made particularly with regard to the latter in the United Kingdom, United States and Switzerland. The importance of public acceptability can limit site selection particularly if geology is also a discriminating factor. This can, at least in principle, precipitate the need for a better technical solution associated with the engineered barrier, demanding less reliance on the performance of the geological barrier. Then, however, the assurance of absolute safety can be called into question with regard to the geology at a given site, particularly if public confidence in the technical solution is lacking. In summary, long-term radioactive waste disposal requires an appropriate technological solution, a location meeting stringent geological requirements and the trust and confidence of the public who will host it. The technological requirements are generally regarded as well understood; the last-mentioned requirement is the focus of much effort at present.

16.5 THE ECONOMICS OF BUILDING NUCLEAR POWER PLANT

16.5.1 HISTORICAL CONTEXT

The early years of the development of nuclear energy are not separated easily from the associated development of nuclear weapons and what were state-sponsored programmes associated with the latter. As R. Rhodes comments [19], in the United States alone, the legacy of the nuclear arms race was 'a $4 trillion dollar national debt'. Thus the economic merit or otherwise of nuclear plant from this era do not help us judge the commercial viability of energy sources in today's liberalised electricity markets; major infrastructure projects are coordinated very differently now and the factors influencing the viability of these are very different. For example, international anxieties immediately following World War II were associated largely with the production of nuclear materials and, in the United Kingdom as case in point, an era in which British coal was considered an exportable commodity. Now imperatives include the need to offset climate change by decarbonising sources of electricity (and indeed energy), the provision of modern energy services to support a global population of an estimated ~10 billion by 2050[6] and security of supply across a variety of contrasting national contexts.

Immediately following World War II, nuclear plant were designed for the production of nuclear materials, or to produce power *and* to enable ready access to the fuel for the extraction of nuclear materials. They were usually financed by national government ministries and electricity was a fortuitous and appreciable by-product. The costs associated with decommissioning and waste management were rarely included in the lifetime costs of these early power plant and neither were these requirements factored into the designs of these early facilities; such was the haste to build these machines at that time. The emphasis on seeking a sound return for investors and cheap electricity was not as prominent as it is now and little was considered or anticipated about future liabilities. The economic risk was less prominent relative to the greater national security these developments promised in an era when security must have felt precious, transient and temporary.

There are other important distinctions, unrelated to the defence imperative, that contrast with the economics of power production today. For example, electricity markets were regulated. Most of the industries involved (the United States being the exception with an expanding fleet of commercially-operated LWRs) were nationalised with reactors operated and financed by national governments. As a result, the

[6]http://www.un.org/en/development/desa/news/population/2015-report.html.

cost of *capital*, that is, financial costs including interest, bank fees and so on, was more robust and often cheaper than in today's liberalised markets. The widespread use of electricity, by what would become known as 'consumers', was in its infancy. Electricity had only recently transitioned from being a luxury to a commodity essential to what is considered an accepted standard of living, as it is in the developed world today. Neither was the electricity supply market competed nor was it as crowded as it is now by both suppliers and technologies. Consequently, nuclear energy as a source of low-carbon baseload electricity offering 90% duty factor and yet with relatively limited *dispatchability*, that is, the ability of a source of power to follow changes in demand, did not need to compete on cost with more dispatchable alternatives offering lower load factors. Interestingly, greater dispatchability via the combination of small, modular reactors and cogeneration may now be more economically competitive [20].

In today's energy markets, nuclear power offers the only source of high-grade (i.e. heat and power), baseload, low-carbon energy with security of supply. It is able to follow demand more effectively than intermittent renewable alternatives but not as well as, for example, combined cycle gas turbines (CCGTs). In terms of costs, renewable sources have zero fuel costs, a relatively high build cost relative to power output (largely due to their low power density) and high systems costs (due to their intermittency but excepting hydropower in most instances). Fossil fuel power plants require significant amounts of fuel, supplies of some fossil-based sources can be vulnerable to political influence (especially gas) and their fuel costs are likely to be volatile, especially in the longer term.

By contrast, the economics of nuclear energy in the 21st century is characterised by a significant, upfront, capital investment that is necessary to build the plant followed by relatively low fuel and operating costs [21]. Construction periods are long (e.g., 5–10 years) but so are the periods over which the plant operate and thus generate revenues, and they do so at high duty cycles. While in engineering terms when costing a new nuclear plant, it is tempting to focus on the absolute cost of the necessary engineering infrastructure, equally important if not more so is the influence of the market and associated financial costs. The long lead times for investors (particularly private investors) to secure a return from nuclear energy production, given long construction periods before generation can begin, can engender significant financial uncertainty in comparison with competing, non-nuclear investment opportunities; the time between plant approval and a return via electricity revenues for the latter is generally shorter. Conversely, nuclear power plant offers reliable load factors over decades, as demonstrated by the data for load factor for era and corresponding nation in Fig. 16.2. This characteristic promises a sustained and consistent revenue stream.

16.5.2 DEFINING THE COSTS OF NUCLEAR POWER PLANT

In assessing the lifetime cost of nuclear energy, four components might be considered associated with a given NPP:

(i) the cost building the plant when new, which is influenced by whether it be first-of-a-kind (FOAK) or n^{th}-of-a-kind (NOAK)—see the detailed definitions of these terms given later in this section

(ii) the cost of *long-term operation* (LTO), should continued operation become desirable at some point beyond the anticipated life of the plant (also known as *life extension*)

(iii) the cost of *uprating* the performance of an existing plant to produce more power

(iv) the cost of decommissioning and waste management

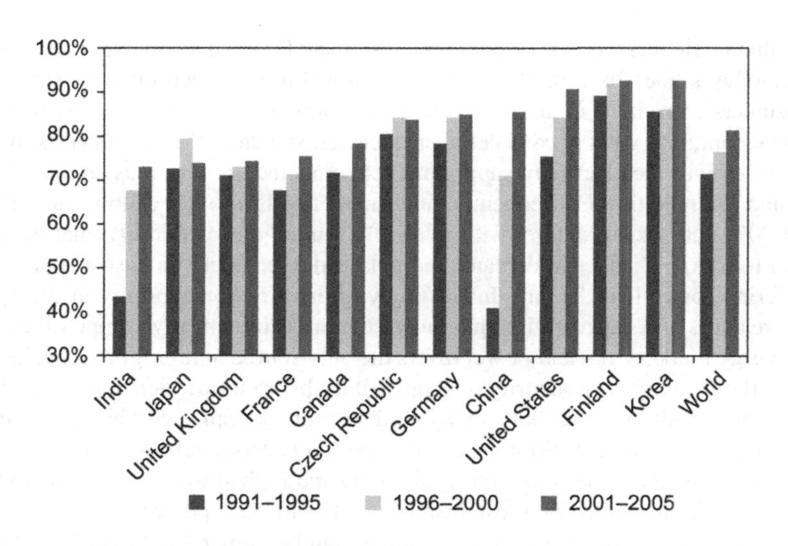

FIG. 16.2

Load factor as a function of era and nation [22].

Reproduced with permission and copyright of the OECD/IEA 2006 World Energy Outlook, IEA Publishing. Licence: www.iea.org/t&c.

All of these options have been or are being considered at present throughout the world. The focus in this section is on the cost of building a new plant only because LTO and uprating have rarely been considered at the point of build. In determining the lifetime cost of new plant (excluding LTO and uprating), it is necessary to extract a *levelized cost of electricity* (LCOE); this is the NPV[7] cost-per-unit-electricity generated for the lifetime of the plant, where costs usually comprise *construction, operation* and *decommissioning* over the lifetime of the plant [23], as per Eq. (16.1).

However, in the context of complete lifetime costs (i.e. what must be considered although this has not always been the case), the prospect of LTO or uprating are nonetheless likely to become relevant at some point in the future.

$$\text{LCOE} = \frac{\text{Net present value of costs over plant lifetime}}{\text{Total electrical energy generated over plant lifetime}} \tag{16.1}$$

Put simply, LCOE is the average price of electricity (£/MWh) necessary to cover all the lifecycle costs of the plant, that is, to build, operate and dismantle the plant.

Construction (capital expenditure or 'capex') costs usually comprise the outlay for preparation, construction and infrastructure of the NPP. Operational costs include the cost of insurance, fuel, services, manpower and so on. The information used to calculate the value of electricity generated usually comprises the generation capacity of the NPP, its availability, load factor and efficiency.

There is a diversity of views and approaches as to how to determine the cost of a NPP. This is because definitions of the components of the cost can vary, forecasts of interest rates on loaned capital can

[7]The term *net present value* implies that anticipated costs in the future have been discounted to their forecasted value today using estimates for inflation, as was discussed in Section 16.4.3.

be either pessimistic or optimistic, projected savings due to learning from previous projects can be at odds with reality (both positively and negatively) and figures provided on an *illustrative* basis are nonetheless used in the absence of more definitive estimates.

In this discussion, we shall adopt the approach defined in the work by W. D. D'haeseleer [24] in order to limit the interpretation of the various definitions. D'haeseleer defines the total 'cost' of building a NPP (for the purposes of this discussion, this refers to a conventional thermal spectrum fission plant) as the *social cost*, where,

$$social\ cost = private\ cost + external\ cost$$

In this definition, the *private cost* is the cost born directly by the firm(s) building the plant, that is, their costs as a business. The *external cost* comprises those costs brought about by the impact of an operation that is not fully accounted for in the private cost, often born by the owner of the plant (commercial operator, utility or government agency). This might include, for example, liability (such as being attributable due to the effects accidents and effluents) and proliferation control. Decommissioning and waste management are often internalised to the private cost and discounted as discussed in earlier. This has been assumed here.

The definition of private cost can be further subdivided, viz,

$$private\ cost = capital\ investment + operation\ \&\ maintenance + fuel\ cycle + decommissioning$$

There is some variety to be appreciated in this context. For example, as to whether *operation and maintenance* costs include the costs of major refurbishments to enable LTO or uprating (not usually the case at initial project cost estimation) or whether *fuel cycle* costs include the cost of the back end of the fuel cycle, that is, the interim storage of spent nuclear fuel, its encapsulation and disposal.

While these are important factors, the *capital investment* (i.e. the cost of the capital asset) is the most significant component. In defining this element of the private cost, the *overnight cost* is often considered. This is the cost of the nuclear plant, in terms of its constituent parts and services, if it were delivered immediately on receipt of the contract for it to be built, that is, overnight, as the title suggests. This is clearly a hypothetical basis used for the purposes of setting apart the cost of borrowing, because it is impossible to build a major, complex infrastructural asset such as a NPP overnight. The cost of the capital investment is spread through time (as with all major construction projects) and, as a result, becomes subject to inflation; funds borrowed to pay for the asset prior to it generating any revenue attract interest. Hence, the capital investment cost can be defined as,

$$capital\ investment\ cost = overnight\ cost + financial\ costs$$

The influence of inflation can of course be estimated with interest rate forecasts and estimates of the time necessary for construction. However, if the construction phase of the lifetime duration of the project is longer than is anticipated at the planning stage, because of overruns or interest rates that are higher than forecast, the investment cost will escalate accordingly. Further, revenue generation is of course postponed.

In addition to the overnight cost and the cost of capital, there are *owner's costs* (the owner, e.g. being a utility or a government-owned commercial organisation). These are costs not born by the vendor (or constructor) and might include preconstruction costs, such as the cost of upgrades to the transmission system (which can vary significantly depending on the inherent compatibility or otherwise), capital costs and financial charges. As with inflation, a variable rate of interest on loaned capital across a project that is taking longer to complete than anticipated can cause costs to escalate significantly relative to the overnight cost. Thus when projects are delayed and revised costs are reported relative to a prior estimate, that is, the overnight cost, the extent to which a project is claimed to be over budget can appear very significant. Clearly, the most naïve scenario is if the overnight cost is used as the original budgetary estimate.

Although there is international variation in the cost of nuclear plant and also of the time taken for them to be completed, the relative proportion of the individual components comprising external costs tend to be as follows [8]: capital investment: 60%–85%, operation and maintenance: 10%–25% and fuel cycle: 7%–15%. Given that the capital investment is the major cost, a key priority is thus to ensure that construction periods run to schedule to within the notional optimum of 5–6 years and do not escalate beyond what is forecast, if private costs are to be consistent with what is forecast.

External costs would normally include liability for accidents, proliferation and avoided CO_2 emissions; decommissioning and waste management are normally internalised through levies. For nuclear, the dividends associated with CO_2 could be positive rather than negative as is currently the case for fossil fuel plant. External costs also include *system effects*, which concerns the dispatchability of a given power plant (its ability to follow demand) relative to the performance of other sources of power. Often, in order to gain the confidence of creditors, shareholders (and the public given the importance of public acceptability), a LCOE is often identified as per Eq. (16.1). This is the long-term, break-even cost of electricity charged by the owner to ensure that the *social cost* of the plant (NPV sum of all costs) minus the *operating revenues* over the whole plant life is zero.

The definition of a LCOE suggests that there is a correlation between the *real* cost of generating electricity with the plant n years after plant completion. However, the LCOE is a largely illustrative cost because of the many variables and the relatively long timescales associated with NPP construction and operation. In particular, LCOE estimates are dependent on geography, the discount rate, load factor and project duration; all factors that can vary significantly. Load factor is now a particular area of variation given the large contribution from intermittent renewables and the potential for nuclear to exhibit greater dispatchability than was necessary in earlier build campaigns.

The overnight cost usually comprises a sum of the owners' costs, the construction cost (*engineering, procurement and construction*, often referred to as the 'EPC') and a contingency allowance. The composition of owner's costs can vary but might include, for example, land costs, site works, switch yards, licenses, permitting, environmental studies, electrical interconnection costs, plant-specific R&D, public relations and the salaries of the owner's engineering staff. Typical levels of owner's costs would be 15%–20% of the overnight cost; the financial cost that arises during construction is usually expected to be ~20% of the overnight cost.

16.5.3 NUCLEAR-SPECIFIC ISSUES

In addition to cost escalations due to time overruns and changes in interest rate, there are also nuclear-specific sources of escalation. Examples of these include the limited availability of nuclear-grade equipment, labour demands requiring nuclear experience and the demands of nuclear regulation and environmental certification, particularly for radiological matters. These factors are not encountered in non-nuclear power plant projects. Labour can account for a very significant proportion of the capital investment because, while equipment costs ought not to be too distinct from other power-generating installations (notwithstanding the higher value componentry associated with what is effectively a rather specialised source of heat on the nuclear island), the level of certification required across all assembly and construction requires that much higher tolerances are met. This results in a requirement for there to be a significant proportion of highly-skilled workers with a good knowledge of the needs of compliance.

Experience gained from recent new nuclear build projects and the associated costs (i.e. since the mid-1970s) suggests that benefit does not always accrue in terms of cost saving despite the prior experience of similar projects, notwithstanding the examples discussed below. This can be because projects are affected by revised regulation (such as the Flamanville EPR in France, for example), far-reaching reviews in light of accidents that occur during or just prior to the start of new-build projects (such as TMI, Fukushima and Chernobyl), the escalation of interest rates and poor management. These influences can lengthen construction periods and thus increase the cost of interest above that anticipated in overall completion cost estimates. Further, building plant capable of larger capacities has not always correlated to less cost-per-kW as might be expected.

However, while the latest, larger reactors are more expensive (in terms of overnight cost notwithstanding overruns and the impact of this in terms of interest), this can be because their design is more sophisticated. For example, the EPR has a greater diversity of installed safety facilities than the earlier designs from which it is derived; an important part of the acceptance criteria of a NPP can be to achieve higher ratings of probabilistic safety assessment than previous generations.

In terms of learning, an exception is apparent when associated with a similar suite of designs as significant benefits arise when moving from FOAK to NOAK. This highlights the benefit of factory-built modular designs and a strict adherence to standardisation across power reactor build. At the time of writing, benefits of this sort are becoming evident in South Korea and potentially also in China, in terms of cost and time to completion, arising from significant campaigns involving the replication of NPP designs, in this case the APR1400 and AP1000, respectively.

However, some recent experiences of costing nuclear build projects illustrate the distinction of FOAK designs. Two projects in Europe (Flamanville and Olkiluoto 3) are significantly behind schedule and over-budget relative to original estimates. Whilst they are arguably the most expensive nuclear power reactors ever built, they are also among the largest in terms of forecast power output. It is important to distinguish between the specific types of projects involved because the state of learning for a given design can exert a significant influence on the likelihood of unforeseen design and regulatory issues, set-backs, delays, the need for greater contingency and thus cost escalation. In this context, the following definitions are often made as extensions to those of FOAK and NOAK introduced above:

- *$FOAK_1$*. These tend to be more expensive to build than projects based on the same design that follows because only limited benefit can be had from learning from any similar design projects anywhere.
- *FOAK in particular country* (often termed $FOAK_2$). These are expected to be cheaper than $FOAK_1$ but more expensive than those that follow in the same country because of country-specific regulation, differences in the supply chain and skilled workforce availability.
- *NOAK*. These ought to represent the variant that conforms most accurately with initial forecasts of construction time and cost, assuming efforts are made to maintain standardisation across the fleet.

To reflect the nature of contingency in each of these scenarios, more contingency would be expected to be necessary for $FOAK_1$ compared with $FOAK_2$, and more for $FOAK_2$ than for NOAK, contingency allowances scale accordingly. For example, such allowances might of the order of 30%–50%, 15%–30% and 10%–15%, respectively [8].

16.5.4 THE INFLUENCE OF LONG TIMESCALES

An important distinction of the financing of nuclear energy projects from most other energy projects is that the timescales are long. By *medium term* a timescale of between 40 and 60 years is usually assumed; *long term* usually implies >80 years. The latter refers to the era after the plant is shut down and externalities such as waste management and decommissioning come into play. Returning to the discount rate, this is often expressed in terms of the *weighted average cost of capital* (WACC). This is defined as per,

$$\text{WACC} = r_{debt}\left(\frac{D}{V}\right)(1-t_c) + r_{equity}\left(\frac{E}{V}\right) \tag{16.2}$$

where it is assumed that the investment is shared between debt at an interest rate of r_{debt} and equity between shareholders at a rate r_{equity}. The total investment is V, the amount of debt is D, the amount of equity E such that $V = D + E$ and the corporate tax rate is t_c. Usually, in regulated markets, a lower discount rate can be assumed than in liberalised markets. In this case, the notation adopted by D'haeseleer [8] has been used for the purposes of consistency.

Nuclear power generation is a little distinct from other power generation sectors because *waste management*, *waste disposal* and *decommissioning* are more tangible components of the lifecycle of a plant; these issues are however starting to feature more prominently in the other parts of the energy sector. Beyond the already long timescales over which nuclear generation is possible (recalling that current reactors in build have projected operating lifetimes of 60 years), project lifetimes are longer compared to other generation methods because of these factors. Historically (and because of the complicated origins of the technology), neglect has caused costs associated with these requirements to be punitive and for which society was, with hindsight, ill prepared.

However, this situation ought not to be confused with the legacy of reactors built now. These will enter into these phases of their lifecycle some ~100 years hence. New commercial power reactors are designed and operated to produce relatively little waste compared to the experimental, exotic activities synonymous with the dawn of the nuclear era. Also, rigorous decommissioning plans are now required in order for the associated costs of these activities to be internalised and paid for by the electricity producer, for example, by the setting of LCOE. Thus, these only become additional costs if adequate funds are not available when they are needed, since the premiums for the funds are included in the market price for the electricity; the LCOE. There are conceptual externalities that are positive traits of nuclear electricity generation. These include: *security of supply*, the *ability to guarantee generation costs* and *baseload supply to guarantee electric system stability*. However, these are rarely costed directly in current projects.

Decommissioning and waste management were once economic liabilities regarded as specific to nuclear energy, although all capital infrastructure projects are subject to costs at the end of their lifecycle. The prospect of severe nuclear accidents presents significant liabilities too, although not different in scale to high-profile, non-nuclear environmental disasters such as the Deepwater Horizon explosion in 2010 or the Exxon Valdez oil spill in 1989. These demonstrated that non-nuclear industrial incidents can command similar post-operation requirements that can result in significant cost, comprising both private and external cost, albeit highly improbable.

The cost of decommissioning nuclear plant in the era when they were owned and operated by regulated, state-owned utilities or government was often paid for by the same. However, in liberalised electricity markets, government regulation (such as for example stipulated by the Energy Act

[2008] in the United Kingdom) requires that operators submit a funded decommissioning programme for approval *before* construction is allowed to commence. For example, a price is set for such a programme that includes a risk premium. This is allowed to escalate with inflation and operators are required to establish a secure and independent fund to cover this cost. The Nuclear Regulatory Commission in the United States have set up a similar arrangement in which money is required to be put aside while the plant is still operating and thus attracting revenues. The concept of *discounting* in this context was discussed earlier in this chapter.

Much of the nuclear decommissioning experience thus far has been associated with legacy plant that date from the industry's military era. These tend not to have been designed with cost-effective dismantling in mind. Most were built in haste and several ceased operation as a result of accidents or process malfunctions which left them in a state that exacerbated the challenges of decommissioning. Typical examples include holdups in throughput in processing facilities leading to blockages and corrosion; metals were used in the componentry of early reactor systems that were very susceptible to neutron activation compared to what is used now; waste was often stored without sorting or a record being kept of what had been discarded and land was contaminated resulting in difficult, high-volume characterisation and processing demands.

Issues like those described above have resulted in a diversity of decommissioning challenges that have caused costs to escalate significantly in some cases. This is because facilities affected in such ways can be difficult to clear out and bespoke waste management programmes can be required for the debris that is evolved. Consequently, nuclear decommissioning can attract the reputation of being a very significant, long-term economic *liability*. This contrasts significantly with a generating power plant which, due to electricity generating revenues, has the status of an *asset*.

The decommissioning requirements of modern reactor systems are now influential in their design. Once the spent fuel is removed (usually early in the decommissioning programme and taking with it the majority of the radioactive inventory of a NPP), accurate cost estimates for dismantling plant at the end of their generating phase can be made early in the project. With such estimates suitable reserves can be set aside by operators long before the capital is needed. The implications of labour costs and changes in regulation several decades hence however are sources of potential uncertainty.

16.5.5 THE ECONOMICS OF NUCLEAR FUEL CYCLES

Uranium enrichment provides an interesting example of the economic issues that can be inherent in the engineering options of the nuclear fuel cycle. Enrichment can account for around half of the cost of nuclear fuel and ~5% of the cost of electricity generated with it.[8] There are several factors that influence this cost. Firstly, the feedstock to enrichment plants can vary in natural enrichment dependent on the source, and this is reflected in its price: that with relatively low levels of natural enrichment being generally cheaper and less desirable than feedstock of relatively high natural enrichment. Correspondingly, there can be a variation in the cost necessary to achieve a customer's contractual requirement in terms of enrichment of the end product, relative to the cost of the ore itself.

As a simple illustration, it might be cost effective to enrich an ore of relatively low enrichment, that is, a low-cost feedstock, but by what is consequently a considerable amount requiring relatively significant costs in terms of electricity. Thus, a greater outlay is incurred on the cost of enrichment of the

[8]www.world-nuclear.org.

cheaper feedstock. Conversely, the preference might be to obtain an ore of greater enrichment at source (incurring higher feedstock costs) but consequently not need to enrich so far (lower enrichment costs).

The market [25] also depends on the relative demand at a given time and the supply capabilities of enrichment facilities. For example, the process selected by which the material is enriched also matters: as discussed in Chapter 12, gaseous diffusion requires a lot more energy (electricity) than do centrifuge systems, but it is flexible in terms of the amounts and the timescales over which it is operated. Conversely, gas centrifuges are expensive to start and stop and are best operated continuously and up to maximum capacity. However, they are more efficient in their use of electricity and hence cost less to run.

A further perturbation on global enrichment costs and operations is the level of demand. When demand is high, spare capacity can be limited and thus costs might escalate and vice versa. This can be affected by unforeseen developments; one example being the Fukushima incident. This resulted in a reduction in enrichment demand, a relative surplus of enrichment capacity worldwide and thus cheaper opportunities for the enrichment of relatively, low-cost feedstocks. Interestingly, enrichment constitutes the main pathway by which CO_2 is evolved from the nuclear fuel cycle due to its use of energy, although this remains a small fraction of that evolved by the combustion of fossil fuels. Further, this source of the emission of CO_2 for nuclear generation assumes that the power for enrichment is derived from fossil fuel sources relative to the amount of energy that is produced. In some cases, the power is derived in part from low-carbon alternatives, that is, hydroelectricity or nuclear energy itself, and strategies have been contemplated by which the latter can be achieved directly, such as via dedicated nuclear reactors on site to drive neighbouring enrichment facilities.

16.6 FUTURE OPTIONS AND NUCLEAR POWER

16.6.1 LONG-TERM FUTURE OPTIONS AND RECENT FORECASTS

A uniform expansion of nuclear energy production in the future, of the scale of there being, for example, 1000 NPPs in the United States as was mooted almost 40 years ago [6], appears unlikely. This is because the individual, domestic energy circumstances and requirements of each nation have a significant influence on the energy options open to that country. These include, for example, the extent of its natural resources, the ease with which it can import energy and its requirement for economic growth, and particularly whether it is an emergent or developed economy. These issues are discussed in more detail below, in reference to each country's current status with respect to NPP construction.

However, it is clear that some countries, especially the emergent economies, have a burgeoning requirement for clean energy over large areas of landmass that will have to be met in order for them to reach key development goals. In these cases, in particular, nuclear energy is likely to play a significant role. In several cases, especially China, this development is already in progress. Where economic development is more stable, nuclear energy has already been adopted and it is not in decline as a result of changes in energy policy, then a replacement of the existing, ageing nuclear infrastructure is likely in order maintain the contribution that nuclear energy makes to meet climate change emissions commitments and to ensure security of supply. However, cost and public acceptability will continue to be the primary factors in these developments.

Most countries with the benefit of high-grade energy generation continue to rely on fossil fuels for most of their energy requirements, even if they also use nuclear energy at present (France and Belgium are among the few much more heavily committed to nuclear). Fossil fuels are used for electricity generation (mostly coal, gas and some oil), heat (mostly gas) and transport (oil). Most of these resources remain in plentiful supply despite concerns of late about their medium-term viability; cheap oil and gas are generally regarded as being shorter lived than coal, with reserves of the latter anticipated to sustain for several hundred years yet. Shale gas extraction among other factors has relaxed pressures on gas supplies, at least in the short term, and has provided some countries with a renewed degree of security of supply (particularly the United States); in the 1970s, the OPEC oil crisis resulted in significant conservation in the use of oil whereas TMI and latterly Chernobyl tempered what were already wavering energy policies with regards to nuclear power, especially in the west.

However, nuclear energy is a particularly effective substitute for coal, especially where *climate change*, *air pollution* and/or *logistics* (i.e. the requirement to transport coal across continents to regions where it is not locally abundant) constrain the continued expansion of coal-fired electricity generation. Despite the concerns over *safety*, *waste management* and *safeguards* that are often associated with nuclear energy, its upfront *cost* and *public acceptability* are the main reasons that nuclear power is not taken up more widely.

By way of illustration of these influences, consider the use of natural gas. CCGT power stations can be constructed more quickly, cost less, can be operated more flexibly (offering greater dispatchability), attract less planning controversy and produce half of the CO_2 of coal. However, they require much greater quantities of fuel than their equivalent power derived from nuclear energy, which has implications in terms of storage and transport; there are limits to the amount that can be stored that correspond to typically a few weeks' operation where supplies are not indigenous and limits to the amount that can be transported. Perhaps most significantly, many countries are reliant for their supply of gas on regions of the world that are politically volatile which is a source of uncertainty over security of supply and stability of price. For some countries, forecasts of their economic growth and related electricity needs can be so significant that they will outstrip the capacity of indigenous reserves of gas. In these cases nuclear energy can present an attractive alternative too.

The availability of uranium has a degree of influence on nuclear energy forecasts but, in the short-to-medium term at least, this is not as significant an influence in comparison with the *cost* of building NPPs. Uranium shortages anticipated at the dawn of the nuclear industry have not been realised because the expansion of fission power did not follow early, optimistic forecasts of growth, partly because of the increase in the cost and time taken to build NPP in the 1970s and 1980s but particularly because of the widespread availability of cheap gas. Fusion power remains a long-term option of the order of 40–50 years away in terms of a commercial, fusion-based NPP.

Meanwhile, there exist significant quantities of nuclear materials associated with nuclear weapons (possessed by a relatively small number of countries) where political requirements to reduce the number of warheads can be an incentive for these materials to be used for power production. Nuclear power production constitutes a route by which these reserves of uranium and plutonium can be disposed of (a route often referred to as *swords to ploughshares*). However, for use in thermal-spectrum NPP, enriched uranium has to be blended down to low-enriched forms whereas plutonium has to be combined in mixed-oxide fuels, the latter route being generally less widespread.

Longer term, several further concepts are relevant with regards to fuel supplies of future NPP, drawing on the discussion Generation IV plans in Chapter 11:

1. Closed nuclear fuel cycles via which high-level radioactive waste volumes are reduced relative to source quantities of spent nuclear fuel and recovered ^{235}U and ^{239}Pu is recycled as fuel in subsequent generations of NPP.
2. The use of fast reactors to breed, for example, ^{239}Pu from ^{238}U.
3. Much higher burn-up performance, particularly in heavy-water moderated reactors, enabling lengthier performance cycles and more power to be accrued directly from plutonium derived from neutron capture in ^{238}U.
4. The use of small modular reactors as a flexible and adaptable means to exploit nuclear energy, potentially for power and heat.

Several forecasts of nuclear energy production in the 21st century have been made. Some of those made in the 1990s were reviewed by Hammond [26] in 1996; a study of these forecasts provides an interesting context to frame current projections. These reports forecasted growth in nuclear power production of between 30% (conservative) and 120% (optimistic) over the period 1990–2020, that is, a production range spanning 2500–4000 TWh by 2015, respectively.

In Fig. 16.3, nuclear energy production as a function of year and geopolitical grouping is given [9] from 1975 to the most recent survey of 2015. There are two salient features to these data: First, between 2002 and 2011, nuclear energy production stabilised at approximately 2600 TWh (notwithstanding fluctuations of the order ±100 TWh). Second, the Fukushima accident in 2011 caused worldwide nuclear energy production to fall to ~2400 TWh from 2011 to 2015. This occurred in Japan and in some parts of Europe because some NPP were subject to revised regulations and operation licensing, while others were closed due to changes in energy strategy, particularly in Germany. Overall these data indicate stabilised production at ~2500 TWh from 2001 to 2015, consistent with the most conservative, 30% growth estimate highlighted by Hammond in 1996. It is worthy of note that throughout 2011–15 a gradual increase is evident as production continued to increase, particularly in East Asia.

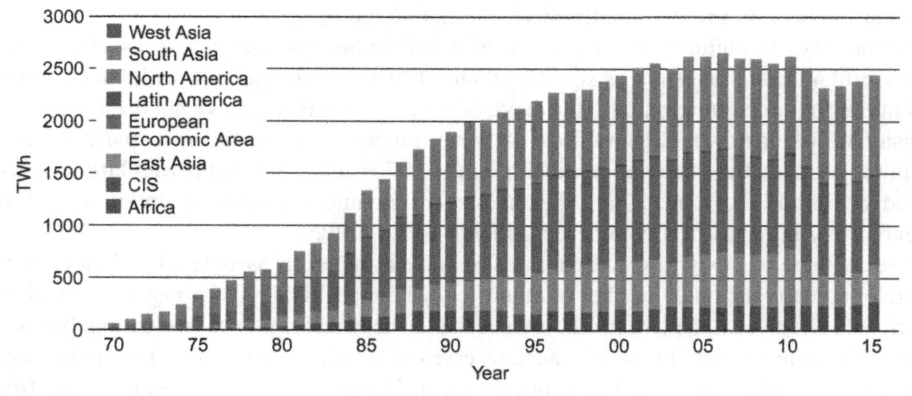

FIG. 16.3

Nuclear energy production as a function of year and geopolitical grouping, 1975–2015.

Reproduced with permission of the World Nuclear Association/IAEA (www.world-nuclear.org).

There are a number of reasons why growth has been towards the lower limit forecast since 1990. First, only a relatively small proportion of the ~150 fission reactors forecast for construction in the 1990s have been built and put into operation. Conversely, a smaller number have been shut down than was forecast with growth realised instead by LTO and uprating, while the growth that was forecast in Asia has taken longer to materialise although there is significant evidence of this, as highlighted earlier, especially in China. Growth was also stalled due to Fukushima, even where operation was continued after assessments were made, and it is feasible that the global economic crisis in 2008 has also quenched demand for electricity and the availability of funds for large capital projects.

Nonetheless, the qualitative aspects of these forecasts remain insightful. For example, the data provide candid evidence for the most significant factors affecting nuclear energy production being *cost* and *public acceptability*, as mentioned earlier. Whilst there has been growth in Asia, as forecast, this has taken longer than anticipated. This highlights another important, long-term qualitative feature of nuclear energy: the *duration* of nuclear construction projects remains an important issue; as Rickover stated when comparing 'paper' reactors with 'practical' ones, that is, those under construction: '(a practical reactor) takes a long time to build'. As anticipated, the majority of the world's nuclear power continues to be generated by a *small number of countries* (United States, France, Japan and China).

Interestingly, nuclear energy has not constituted a complete solution for climate change, as was also forecast albeit naïvely since it is impossible to build the number of reactors necessary in the time available, even if this was desirable. However, nuclear power remains an important *part* of a solution, especially where there is a lack of alternatives (whether due to an indigenous shortage of natural resources or as a result of the logistical constraints discussed earlier), where there is concern over security of supply and where *energy efficiency measures*—whilst an important component in combating climate change—might not realise the emissions targets of a particular place.

A relatively recent development, not anticipated in the 1990s, is the potential for the wholesale replacement of oil-fuelled transport systems by *electrically powered alternatives* (e.g., electric or hydrogen-powered vehicles). These technologies could draw significantly on nuclear energy production if current growth prospects in this sector are realised. Decarbonising transport in this way cannot be met by nuclear energy in isolation (once again cost and public acceptability will dictate the most suitable sources of electricity) but a significant contribution is feasible.

More recently, the challenges of sustainable development were reviewed in the context of sustainable consumption and production [27] with a focus on *sustainable cities*, *food security* and perhaps most significantly in this context, access to *modern energy services*. This highlights the potential for the control of emissions via investing 'decisively in energy efficiency'[4] and describes routes by which sustainable development might be achieved without the socio-political and technical challenges of *nuclear energy* and *carbon capture and storage* (CCS). However, it is cited that these would 'require special measures to improve energy efficiencies and reduce demand'.[4]

16.6.2 AS A SOURCE OF LOW-CARBON ELECTRICITY

In the 1990s and early 2000s, the evidence for climate change first began taking on an incontrovertible dimension. As a consequence, a change in the perception of nuclear power began to develop in a number of countries. The background to this development is summarised particularly well in the introduction to the paper by Bickerstaff et al. [28] as to the 'reframing' of the nuclear debate in the United Kingdom, as a case in point. However, the potential for nuclear energy to offset CO_2 production was actually appreciated a surprisingly long time ago. For example, in 1978, it was postulated [7] that a significant expansion in the United States alone might be necessary in the 21st century to offset the CO_2 problem if solar was to prove too expensive, fusion remained a long-term prospect and electricity continued to be used increasingly in place of other forms of energy (e.g., for heat, transport, etc.).

The context associated with climate change is summarised in a wide variety number of sources; the related NASA resource[9] is one such example. Further to the discussion of *intergenerational equity* earlier in this chapter, a perspective central to the low-carbon argument and nuclear power is that, whilst concerns over radioactive waste disposal tend to be associated with problems that might in several thousand years' time, the challenge posed by climate change is much closer at hand, of the order of decades and with the potential to be of much greater significance.

Under the Kyoto Protocol,[10] targets have been set for the reduction of carbon dioxide emissions (and those of other greenhouse gases [GHG][11]). These are to be met by countries through both national measures and market-based mechanisms. For example, for the first commitment period 2008–12, the 15 states who were members of the European Union in 1997 committed to a reduction of 8% of 1990 levels with the United Kingdom committing to a reduction of 12.5%. The most recent progress towards these targets [29] has been based on a number of initiatives not related to new nuclear build because none was physically possible within the timescales associated with this period. Existing NPP have, however, made an important albeit fortuitous contribution to keeping CO_2 emissions down, as has the expansion of generation with gas where the latter has served as a substitute for coal, lignite and other heavy CO_2-emitting fossil fuels.

Looking longer term in a climate management context, nuclear energy has attracted renewed government and public attention because of the prospect of the depletion of oil and gas, the closure of coal-fired power stations and because many existing nuclear power stations are reaching the end of their operational lives. The specific requirements of a country vary according to its current energy mix and its abundance of natural resources. Using the United Kingdom as an illustrative example, there is significant reliance on coal, gas and nuclear with a growing proportion of renewables and a much reduced margin of spare capacity than has been the case in recent years. The most likely scenario at the time of writing comprises a relatively rapid decline in coal, an interim replacement of this by gas (via CCGT) and the prospect of a relatively significant gap, long term, in baseload provision. It is this gap that it is anticipated that nuclear will fill because of the need to offset the loss of coal production, to ensure security of supply (which is less easy via gas alone) and to bring some baseload stability to a supply mix much more reliant on intermittent renewables.

As mentioned earlier, the principal consumption of energy associated with nuclear power that might yield GHG is the uranium enrichment process. This is influenced by whether the electricity used to power the centrifuges is derived from fossil-fuelled or low-carbon (nuclear or hydroelectric) sources. Also, the GHGs associated with waste management and decommissioning can be relevant to carbon accountancy, as is the *embedded carbon* associated with transport of fuel and wastes and so on; indeed, this is a point of consideration for all sources of electricity that is rarely factored in at present. For these

[9]http://climate.nasa.gov/scientific-consensus/.
[10]'an international agreement linked to the United Nations Framework Convention on Climate Change, which commits its Parties by setting internationally binding emission reduction targets', http://unfccc.int/kyoto_protocol/items/2830.php.
[11]The collective terms greenhouse gases (GHG) comprises: carbon dioxide, methane, nitrous oxide, halocarbons, perfluorocarbons and sulphur hexafluoride.

reasons, nuclear power (fission) is usually referred to as *low carbon*. However, all routes by which CO_2 is evolved in nuclear energy yield significantly smaller quantities than for the case of fossil fuels, principally because it is not evolved when nuclear energy is produced and because drastically smaller quantities of fuel have to be manufactured, transported and disposed of.

Nuclear energy offers an interesting component of what is often referred to as the 'low carbon energy mix'. It provides baseload provision that is less easily supplied by intermittent, low-carbon alternatives but with less dispatchability, certainly when compared to CCGTs. Nuclear power is suited to high-grade, baseload supply due to its high levels of availability, near-zero GHG emissions (particularly during operation), the availability of heat as well as power and because nuclear power stations are generally not suited to frequent changes in power demand.

Further, for countries already reliant on a fleet of nuclear power stations, the inevitable closure of these NPP at some point in the future will remove their contribution to carbon abatement. This will need to be transferred or replaced by low-carbon alternatives as otherwise the CO_2 benefits derived from current nuclear power operations will be lost. Just replacing like-for-like constitutes a significant nuclear build programme. The decarbonisation of transport and domestic heat goes beyond such a scenario. The closure of aged nuclear generation assets relinquishes existing nuclear-licensed sites and infrastructure that (as described earlier) can be compatible with replacement, new nuclear power stations. This association has led to nuclear fission power to be termed *sustainable*, although expert debate continues as to the detail implied by this classification.[12]

There are two further, important dimensions to carbon abatement debate. First, it is widely anticipated that societal changes resulting in greater efficiency and economy of energy use will play a significant role in averting climate change; as postulated earlier, the potential has been identified for these approaches to obviate the need for what are referred to as *high-technology solutions*, such as nuclear or carbon capture and storage (CCS). However, such changes could be very difficult to achieve in the short timescales necessary to meet Kyoto targets, particularly because of the prospect that they might compromise lifestyles (including wellbeing) and economic growth; factors that can be politically complex. On the other hand, the avoidance of these issues complicates the ease with which greater efficiency of use is achieved to the extent required and increases the likelihood of the need for contributions by nuclear and CCS. Such societal changes will not be without their own public acceptability concerns and costs, despite the possibility that a buoyant economy based on optimised utilisation is feasible.

Second, research is being pursued on intelligent electricity distribution. If adopted widely, this could reduce reliance on baseload provision. A significant amount of low-carbon, baseload power such as nuclear energy is still likely to be necessary as it increases the reliability to electricity networks now more reliant on a greater range of sources. High-grade sources such as nuclear can also be less vulnerable to fuel-price changes than for energy that is imported.

Some developments in nuclear power (perhaps the most prominent case being small modular reactors discussed earlier in Chapter 11) are being revisited to be more adaptable to distributed, sensitive and responsive electricity supply markets. A significant challenge is the link between *energy* (and thus GHG emissions) and *economic growth*. Currently, the latter drives greater economic prosperity and, hence, improved standards of living.

[12]Note that the term *renewable* would infer a recycle (reprocessing) stage which is another variant of the options for low-carbon operation.

Improved standards of living (the reduction of poverty in the case of emerging economies) currently provoke increased consumption and a corresponding escalation in GHG emissions. New sources of emissions might be felt particularly significantly from emerging economies that have made a relatively small contribution to the climate problem to date. Nuclear energy has attracted support in recognition of the benefit that existing nuclear stations make in terms of carbon abatement and also due to its potential for the future. This could be significant if energy efficiency schemes are found wanting in future because of the scale of the climate problem that needs to be addressed.

16.6.3 PLANS FOR NPP WORLDWIDE

The motivation to build new NPP varies from country to country. Some countries, such as the France, South Korea, the United States and the United Kingdom, already have extensive nuclear energy capacity. The construction of new plant in this case offsets the loss of the closure of old plant and can enable the associated industry to develop new certified and approved reactor designs for export (such as the EPR, APR1400 and AP1000). In other cases, the host nation may not build their own NPP but nonetheless seek to offset the prospect of an ageing reactor fleet and the need for low-carbon, high-grade, baseload power generation, as is the case in the United Kingdom. A prominent motivation can be *security of supply*, that is, ensuring greater independence from energy imports. Too great a reliance on electricity supplies via international interconnectors or on fuel imports can be a source of political tension, economic uncertainty and, in extreme cases, energy shortages.

The reduction of pollution is also a motivation; many countries seek to reduce their CO_2 emissions via nuclear energy, at least in part. China is heavily reliant on coal which, in addition to CO_2 emissions, yields toxins that affect air quality and also requires that large quantities of it are shipped across the country, presenting a limit on economic growth and constituting a further source of pollution and GHG emissions.

In the United Arab Emirates (UAE), whilst most of the electricity has been derived from gas to date, forecasts of significant escalation in demand cannot be met without the consideration of alternatives. In this case a critical requirement is for the production of drinking water via desalination. The motivation in several emerging economies is to advance economic growth and to alleviate poverty via the wider electrification of currently, under-developed territories. A high dependence on hydroelectricity (cf. Brazil and Finland) might be vulnerable to the effects of climate, especially during extended droughts, while the geographical scope for more capacity can be limited. This motivates policies towards a greater diversity of supplies.

Plans for NPP can range from a Government's voiced intent to *consider* nuclear energy for the first time through to *actual* construction projects nearing completion. With this range in mind, the following summary of current projects has been limited to those projects that are actively progressing towards completion, recognising that >160 power reactors are currently *planned* worldwide.[13] Further, the focus of this section is on the *build* of new plant; it is worthy of note that *uprating* of existing plant has also taken place in Finland, Spain, Sweden, Switzerland and the United States.

[13]http://www.world-nuclear.org/.

16.6.3.1 Asia

China currently has 35 operating NPP. Most of these are PWRs brought into operation since the mid-1990s, each with a capacity of ~1000 MWe. China is constructing a further 20 NPP with designs spanning a number of PWR variants including CPR1000s, AP1000s and EPRs. These are scheduled for completion by 2020–21. China also has firm plans for another 40 NPP with construction to start from 2017 onwards.

India has 22 operating NPP and has ambitious plans to expand its nuclear power generation capacity. The majority of its existing reactors are relatively small PHWRs with capacities ~200 MWe. Four more PHWRs (capacity ~700 MWe) are under construction. These are due to be completed in 2017 with the construction of a further 20 planned to start shortly afterwards.

Japan has two NPP under construction: Shimane 3 and Ohma 1. Both are of the ABWR type and of 1383 MWe capacity. Following the 2011 Great East Japan earthquake and the Fukushima Daiichi accident that followed, Japan has embarked on an extensive review programme of its 42 NPP with many of them shutdown whilst this takes place.

Pakistan has four operating NPP, comprising three PWRs and a PHWR. The latter is forecast to close shortly as it is the oldest plant operating in Pakistan. Three PWRs are under construction in Pakistan for completion 2017 through 2022, and there are further plans for additional NPP at 10 sites.

South Korea is currently completing three NPP (all APR1400s due to be finished 2017–19) and has plans for 8 more scheduled for completion 2021 through 2029. This will complement the existing South Korean fleet of 25 NPP that comprises PWR (Westinghouse) and PHWR (CANDU) designs from the 1980s and 1990s and the indigenous Generation II OPR design from the late 1990s.

Four APR1400 (of nominal capacity 1400 MWe) reactors are under construction in the UAE, to be completed by 2017 through 2020.

16.6.3.2 Mainland Europe

Finland has four operating reactors (two BWRs at Olkiluoto and two VVERs at Loviisa). A fifth reactor (a 1600 MWe Framatome-ANP EPR) is under construction (Olkiluoto 3). Construction on this NPP started in 2005 with an anticipated date of commercial operation then of 2009. However, the project has been delayed, partly due to the approval requirements of the reactor instrumentation and control system. Olkiluoto 3 is now scheduled for completion in 2018. Due to the similar EPR project ongoing at Flamanville in France which started a few months later (see below), Olkiluoto 3 is technically both a $FOAK_1$ and a $FOAK_2$.

Construction of Flamanville-3 (an AREVA 1650 MWe EPR in 2007 at Flamanville, Normandy, France) began in 2007. Completion and operation was expected by 2012 but this has been delayed, most recently due to further inspections required by regulators of the carbon content of the steel of the RPV. Further to earlier discussions regarding cost, the *overnight* construction cost of Flamanville-3 was estimated at €3.3 billion. With the overrun the estimate at the time of writing is €10.5 billion, including *financing*. Flamanville-3 is technically a $FOAK_2$, and completion is now expected in 2018.

Slovakia intends to operate its pair of VVER-440 V-213 reactors until 2024–25. Two further units of the same type were completed and started in the late 1990s at Mochovce, of a total of four planned at the same site. Construction continues on the latter two VVER-400s. These are expected to start in 2019.

Russia has a significant nuclear capability spanning the entire fuel cycle. It operates 34 power reactors and has ambitious plans to replace NPP as they are retired. These include four awaiting imminent start-up (2018–19) and many more planned and proposed. Most are of the VVER type.

16.6.3.3 South America

Electricity consumption in Brazil continues to grow sharply. Two PWRs are currently in operation: one of 600 MWe capacity from Westinghouse (USA), that has operated since 1982 (Angra 1), and another of capacity 1300 MWe (Angra 2) from Kraftwerk Union (KWU, Germany). The completion Angra 2 and subsequently another PWR project (Angra 3) have been delayed by economic problems. Plans exist for 8 more PWRs, each of a capacity of ~1500 MWe.

Similar growth in electricity consumption is apparent in Argentina where three NPP are in operation: Peron (335 MWe, formerly known as Atucha 1), Kirchner (700 MWe, formerly known as Atucha 2) and Embalse (600 MWe). All are PHWRs. However, the Atucha reactors are unusual in that they are a design that utilises a pressure vessel (Siemens); Embalse is a CANDU-6 (AECL). Argentina has plans to start construction of two more NPP in 2017 and 2019, probably of the PHWR type and Chinese PWR variants, respectively.

16.6.3.4 United Kingdom

The United Kingdom currently has 15 operating nuclear power stations: 7 pairs of AGRs and 1 PWR. The AGRs (being the older of the two variants) are scheduled to be subject to a phased programme of closure in the next decade with the last ones currently scheduled to close by 2030. These estimates already include life extensions (LTO) of between 5 and 10 years negotiated for a number of the older AGR plant. It is currently anticipated that the PWR at Sizewell in Suffolk will operate until 2035. Together, these plant constitute a total generation capacity of ~9 GWe. The UK Government currently anticipates 16 GWe of new nuclear capacity to be built by 2030 [30].

From a process beginning with the invitation of nominations for sites to be considered for new nuclear power stations in the United Kingdom, 8 sites were confirmed as suitable: Bradwell, Hartlepool, Heysham, Hinkley Point, Oldbury, Sellafield, Sizewell and Wylfa. Three technologies for new build are currently under consideration in the United Kingdom: EPR, AP1000 and the ABWR. In 2016, an agreement was signed between the UK Government and a joint venture NNB (New Nuclear Builds Generation Co. comprising the French utility EDF and the Chinese company CGN) to build two EPRs at Hinkley Point in Somerset. This is also the site of Hinkley A (a decommissioned Magnox station) and Hinkley B (one of the first AGRs). The capacity of Hinkley C will be 3.3 GWe. It is scheduled to be operational by 2026.

There are a number of other projects in plan in the United Kingdom:

- A further two EPRs are planned by EDF and CGN at Sizewell adjacent to Sizewell A & B and *NuGeneration* (NuGen—a venture comprising *Toshiba* and the French multinational *Engie*) have confirmed an intention to build three AP1000s (total capacity: 3.6 GWe) at Moorside near Sellafield in Cumbria.
- *Horizon Nuclear Power* (now a wholly-owned subsidiary of Hitachi Ltd.) won bids for land from the UK Nuclear Decommissioning Authority alongside the old Magnox power stations at Oldbury in South Gloucestershire, north of Bristol, and at Wylfa on the Isle of Anglesey, North Wales. The intention is to build a number (2–3) of Advanced BWR (ABWR) reactors each of capacity 1380 MWe on each of the sites; beginning with *Wylfa Newydd*, and with Oldbury following.

- *General Nuclear Systems* (a venture comprising China General Nuclear Power Group and EDF) has undertaken to bring forward a new reactor project 'Bradwell B' (adjacent to the partially decommissioned Bradwell Magnox power station in Essex) to the point of an investment decision. This initiative is based on the prospect of two *Hualong One*[14] reactor designs, each of capacity ~1150 MWe.

16.6.3.5 United States of America

The anticipated scale of the reactor build programme in the United States prior to the TMI accident in 1979 fell significantly afterwards. Coincidentally, natural gas became very economically attractive. More recently, domestic shale gas extraction has increased security of supply with respect to gas in the United States.

Nonetheless, nuclear energy production and uranium consumption have increased significantly of late due to uprating rather than the commissioning of new NPP; it is estimated that an increase of 6000 MWe capacity has been achieved via uprating in the United States alone since new build declined.[15] Uprating requires that the performance of the system be adapted to higher power outputs via: longer periods of availability, higher ratings and revised operational performance benefiting from, for example, improved understanding of the thermal hydraulic margins for operation.

Four NPP are currently under construction in the United States. These comprise four AP1000s (two at Vogtle, Georgia and two at the Virgil C. Summer site, South Carolina). Plans to build an EPR at Calvert Cliffs in Maryland by AREVA were suspended in 2015.

REVISION GUIDE

On completion of this chapter, you should:

- appreciate the significance of *public acceptability* and *cost* as selection criteria associated with nuclear energy
- understand the public acceptability contexts associated with *nuclear accidents, siting and local communities, intergenerational equity* and *radioactive waste*
- be familiar with the concepts associated with the cost of building NPP and, for example, the definitions of *social cost, private cost, external cost, overnight cost, owner's costs* and so forth
- appreciate the significance of *timescales* to the cost of nuclear build projects, including definitions of the WACC and the LCOE
- understand the context for nuclear energy to make a contribution to *carbon abatement* and *decarbonisation of baseload electricity supplies*
- be familiar with the current trend(s) in nuclear energy production and how they compare with recent forecasts
- appreciate the context worldwide in terms of the *construction of NPP construction, uprating* and *long-term operation*

[14]A Generation III, 3-loop PWR design originally related to the French PWR design of units 5 & 6 at Gravelines in France.
[15]'Cumulative Capacity Additions at U.S. Nuclear Facilities', Nuclear Energy Institute (NEI), Source: Nuclear Regulatory Commission. www.nei.org (updated: 10/2011).

PROBLEMS

1. Explain why, if the financial estimate of the cost of the construction of an NPP is based on the *overnight cost*, then the *capital investment cost* is always greater.
2. Explain what is meant by the LCOE. What relationship has to be met for an NPP to constitute a sound investment?
3. With reference to siting and local communities, explain why it is often assumed that the public local to existing nuclear facilities are more accepting of further developments than people near to new, proposed sites or further away from nuclear sites.
4. Based on current data, new NPP construction projects in the west appear to be following a strategy of replacement of existing NPP while in the east significant expansion is evident. Why is this?
5. Explain why the first generation of a nuclear reactor design tends to be more expensive than those generations that follow, and why NPP built in a country for the first time can be more expensive than those that follow. Specify the acronyms that describe each of these cases.

REFERENCES

[1] K.A. Parkhill, N.F. Pidgeon, K.L. Henwood, P. Simmons, D. Venables, Perceptions of risk when living near nuclear power in the UK, Trans. Inst. Br. Geogr. 35 (2010) 39–58.

[2] D.C. Invernizzi, G. Locatelli, N.J. Brookes, Managing social challenges in the nuclear decommissioning industry: a responsible approach towards better performance, Int. J. Proj. Manag. in press.

[3] J.F. Ahearne, Intergenerational issues regarding nuclear power, nuclear waste and nuclear weapons, Risk Anal. 20 (6) (2000) 763–770.

[4] W. Poortinga, M. Aoyagi, N. Pidgeon, Public perceptions of climate change and energy futures before and after the Fukushima accident: a comparison between Britain and Japan, Energy Policy 62 (2013) 1204–1211.

[5] A.H. Gallardo, T. Matsuzaki, H. Aoki, Geological storage of nuclear wastes: insights following the Fukushima crisis, Energy Policy 73 (2014) 391–400.

[6] C.C. Burwell, M.J. Ohanian, A.M. Weinberg, A siting policy for an acceptable nuclear future, Science 204 (4397) (1979) 1043–1051.

[7] D. Venables, N.F. Pidgeon, K.A. Parkhill, K.L. Henwood, P. Simmons, Living with nuclear power: sense of place, proximity, and risk perceptions in local host communities, J. Environ. Psychol. 32 (2012) 371–383.

[8] K.A. Parkhill, N.F. Pidgeon, K.L. Henwood, P. Simmons, D. Venables, From the familiar to the extraordinary: local residents' perceptions of risk when living with nuclear power in the UK, Trans. Inst. Br. Geogr. 35 (2010) 39–58.

[9] IAEA, The Principles of Radioactive Waste Management: Safety Fundamentals, Safety Series No. 111-F, IAEA, Vienna, 1995.

[10] U. Beck, World Risk Society, Polity Press, Cambridge, 1999. ISBN: 978-0-7456-2221-7.

[11] World Commission on Environment and Development, Our Common Future (The Brundtland Report), Oxford University Press, Oxford, 1987, p. 54.

[12] K. Shrader-Frechette, Duties to future generations, proxy consent, intra- and intergenerational equity: the case of nuclear waste, Risk Anal. 20 (6) (2000) 771–778.

[13] J. Lowe, Intergenerational Wealth Transfers and Social Discounting: Supplementary Green Book Guidance, ISBN: 978-1-84532-419-3, 2008. https://www.gov.uk/government/publications/green-book-supplementary-guidance-discounting.

[14] R.B. Belzer, Discounting across generations: necessary not suspect, Risk Anal. 20 (6) (2000) 779–792.
[15] M. Hannis, K. Rawles, Compensation or bribery? Ethical issues in relation to radwaste host communities, Radioact. Environ. 19 (2013) 347–374.
[16] See for example: F.K. Pittman, Management of commercial high-level radioactive waste, Adv. Chem. Ser. 153 (1976) 1–8.
[17] R. Lidskog, G. Sundqvist, On the right track? Technology, geology and society in Swedish nuclear waste management, J. Risk Res. 7 (2) (2004) 251–268.
[18] L. Sjöberg, Local acceptance of a high-level nuclear waste repository, Risk Anal. 24 (3) (2004) 737–749.
[19] R. Rhodes, Dark sun: The Making of the Hydrogen Bomb, Touchstone, ISBN: 0-684-82414-0, 1996.
[20] G. Locatelli, S. Boarin, F. Pellegrino, M.E. Ricotti, Load following with small modular reactors (SMR): a real options analysis, Energy 80 (2015) 41–54.
[21] See for example: Nuclear Energy Today, second ed., 2012, Organisation for Economic Cooperation and Development (OECD) and the Nuclear Energy Agency (NEA). ISBN: 978-92-64-99204-7.
[22] International Energy Agency, World Energy Outlook 2006, OECD, Paris, 2006.
[23] Electricity Generating Costs, https://www.gov.uk/government/publications/decc-electricity-generation-costs-2013, 2013.
[24] W.D. D'haeseleer, Synthesis on the Economics of Nuclear Energy. Study for the European Commission, DG Energy, 2013.
[25] OECD/IAEA, Uranium: Resources, Production and Demand (The Red Book), 2014 (Report OECD/IAEA NEA#7209). http://www.oecd-nea.org/ndd/pubs/2014/7209-uranium-2014.pdf.
[26] G.P. Hammond, Nuclear energy into the twenty-first century, Appl. Energy 54 (4) (1996) 327–344.
[27] Sustainable Development Challenges: World Economic and Social Survey 2013, Department of Economic and Social Affairs (DESA), United Nations, ISBN: 978-92-1-109167-0, 2013.
[28] K. Bickerstaff, I. Lorenzoni, N.F. Pidgeon, W. Poortinga, P. Simmons, Reframing nuclear power in the UK energy debate: nuclear power, climate change mitigation and radioactive waste, Public Underst. Sci. 17 (2008) 145–169.
[29] UK progress towards GHG emissions reductions targets, https://www.gov.uk/government/uploads/system/uploads/attachment_data/file/414241/20150319_Progress_to_emissions_reductions_targets_final.pdf, 19th March 2015.
[30] Nuclear industry strategy: the UK's nuclear future, BIS/13/627, https://www.gov.uk/government/uploads/system/uploads/attachment_data/file/168048/bis-13-627-nuclear-industrial-strategy-the-uks-nuclear-future.pdf, 2013.

Index

Note: Page numbers followed by *f* indicate figures, *t* indicate tables, *b* indicate boxes and *np* indicate footnotes.

C

D

J

K

L

M

O

S

T

U

V

W

X

Y

Printed in the United States
By Bookmasters